Molecular Biology in Cellular Pathology

Molecular Biology in Cellular Pathology

Edited by

John Crocker

Department of Cellular Pathology,
Heartlands Hospital, Birmingham, UK

Paul G. Murray

Department of Pathology,
The Medical School University of Birmingham, Birmingham, UK

WILEY

Telephone (+44) 1243 779777

Email (for orders and customer service enquiries): cs-books@wiley.co.uk
Visit our Home Page on www.wileyeurope.com or www.wiley.com

Other Wiley Editorial Offices

John Wiley & Sons Inc., 111 River Street, Hoboken, NJ 07030, USA

Jossey-Bass, 989 Market Street, San Francisco, CA 94103-1741, USA

Wiley-VCH Verlag GmbH, Boschstr. 12, D-69469 Weinheim, Germany

John Wiley & Sons Australia Ltd, 33 Park Road, Milton, Queensland 4064, Australia

John Wiley & Sons (Asia) Pte Ltd, 2 Clementi Loop #02-01, Jin Xing Distripark, Singapore 129809

John Wiley & Sons Canada Ltd, 22 Worcester Road, Etobicoke, Ontario, Canada M9W 1L1

Wiley also publishes its books in a variety of electronic formats. Some content that appears in print may not be available in electronic books.

Library of Congress Cataloging-in-Publication Data

Molecular biology in cellular pathology / edited by John Crocker, Paul
G. Murray. – 2nd ed.
p. ; cm.
Rev. ed. of: Molecular biology in histopathology, c1994.
Includes bibliographical references and index.
ISBN 0-470-84475-2 (paper : alk. paper)
1. Pathology, Molecular. 2. Pathology, Cellular.
[DNLM: 1. Genetic Techniques. 2. Cell Physiology. 3.
Cells – pathology. QZ 52 M718 20003] I. Crocker, J. II. Murray, Paul,
Ph. D. III. Molecular biology in cellular pathology.
RB43.7 .M6336 2003
611′.018 – dc21

2002154112

British Library Cataloguing in Publication Data

A catalogue record for this book is available from the British Library

ISBN 0-470-84475-2

Typeset in 10.5/13pt Times by Laserwords Private Limited, Chennai, India
Printed and bound in Great Britain by TJ International, Padstow, Cornwall
This book is printed on acid-free paper responsibly manufactured from sustainable forestry in which at least two trees are planted for each one used for paper production.

With love
to our families

Contents

Preface

Since the publication of the original edition of this book, there have been rapid advances in our understanding of disease, mainly as a result of the impetus provided by some of the newer technologies. In particular, the rapidly developing fields of genomics and proteomics are enabling an understanding of gene expression both at the mRNA and protein level on a global scale (i.e. the whole transcriptome or proteome) not previously imaginable. Whereas gene expression studies in pathology have frequently relied purely on immunohistochemistry and *in situ* hybridisation, in their own right still immensely invaluable procedures, they could only essentially give information on a single gene in a single experiment. Now information on expression from the whole of the genome can be assessed in a single experiment. Proteomics, in particular, is providing the tools not only to examine global protein expression, but also to dissect protein function, through the development of approaches to study protein activity. Likewise, in genetics, there is an impending revolution. Comparative genomic hybridisation, for a long time a difficult alternative to conventional cytogenetics, will blossom with the advent of array approaches to, allowing high resolution mapping of chromosomal changes across the whole genome without the need for difficult interpretation of chromosome morphology.

The key question is to what extent these developments will impact on diagnostic pathology in the future. The polymerase chain reaction was heralded years ago with the view that its introduction into routine diagnostic pathology practice was only a matter of time. Events have not proved this assumption correct; although the polymerase chain reaction does have applications in routine pathology it has not impacted directly in a significant way on routine histopathology. It is the authors' view that the same will not be true of the newer technologies. At the very least these newer approaches may identify a whole host of disease-specific markers for use in conventional assays for disease. At the other end of the spectrum, they may change forever the morphological assessment of disease to be substituted by an entirely objective set of array data providing detailed information on chromosome changes and global gene expression. We hope that this new edition will give some insights into some of the developments in molecular biology that provide us not only with immense opportunities for the future but also with considerable challenges.

We would particularly like to thank all those who have contributed to this text and also our families, to whom we are of course deeply indebted for allowing us to pursue this project, often at their expense. Mrs Ruth Fry supplied excellent secretarial support.

John Crocker
Paul Murray
21 October 2002

Preface to *Molecular Biology in Histopathology*

The past 20 years have witnessed numerous changes in the practice of histopathology, with many powerful techniques, such as immunohistochemistry and image analysis, aiding the accuracy and objectivity of diagnosis and research. However, perhaps the greater revolution, occurring in the past half decade, of the application of molecular methods, will be even more fruitful. Molecules related to, for example, hormones, immunoglobulins, infectious agents or chromosomes can be identified by means of gene probes. Furthermore, the molecular basis of cell replication has become more clearly understood, assisting in tumour prognosis. What, then, are these new methods? As the title of this book implies, the techniques included in it are those that can be performed on histological material, although not necessarily involving microscopic examination. The reader should not be led into the belief that 'molecular' must always imply 'DNA' and, as we can see in at least two chapters herein, 'molecular' should have a wider, more appropriate meaning.

The purpose of this series of volumes is to supply a guide to those just qualified and undertaking research or to those who have taken degrees some years in the past and who wish to glean new information rapidly. Thus, this is not a molecular 'recipe book'; such exist elsewhere.

In the first chapter, Mr Murray and Professor Ambinder have given an account of the methods available for the demonstration of infectious agents *in situ* in histological material. The applications of these techniques are also outlined. Chapters 2 and 4 by Drs Fleming, Morey and Yap, and by Drs Waters and Long, then describe these methodologies and others as applied to the examination of malignant tissues and chromosomes in histological material.

Chapters 3 and 5 give details of other methodologies, both of which are not histological (although one may become so). These techniques do, however, employ histological material, even of archival, paraffin wax-embedded type. Thus, Dr Young describes the value of the polymerase chain reaction in histopathology; indeed, this is already being adapted for use as an *in situ* method. Dr Camplejohn then describes the techniques of DNA flow cytometry and their applications. One of the latter is that of the assessment of cell proliferative status. Leading on naturally from this, in Chapter 6 I have given an account of the molecular basis of the cell cycle and of some of the antibody probes which can be applied to visualize some of the components of the cell cycle. Also highly related to cell proliferation is the activity of the interphase nucleolar organizer region. The full significance of this structure is not yet fully understood but in the subsequent chapter, Professors Derenzini and Ploton give an account of the morphological and molecular corollaries of the nucleolar organizer regions.

Just as we are realizing and understanding the importance of cell proliferation in disease, so we are appreciating that cell death is also central to many physiological and pathological conditions. Accordingly, in Chapter 8, Drs Arends and Harrison tell us of the molecular basis of 'programmed cell death' or apoptosis in health and disease.

Thus, this volume gives an introduction to the currently available molecular techniques in histology and an account of the molecular basis of certain phenomena of importance in everyday histopathology.

John Crocker
Birmingham
1994

List of Contributors

S. Jane Astley	Division of Biomedical Sciences, University of Wolverhampton, Wulfruna Street, Wolverhampton WV1 1SB, UK
Karl Baumforth	CRC Institute, The Medical School, University of Birmingham, Edgbaston, Birmingham B15 2TT, UK
Philip Bennett	Micropathology Ltd, University of Warwick Science Park, Barclays Venture Centre, Sir William Lyons Road, Coventry CV4 7EZ, UK
Werner Boecker	Gerhard Domagk Institute of Pathology, University Hospital Muenster, Germany
Horst Buerger	Gerhard Domagk Institute of Pathology, University Hospital Muenster, Germany
John Crocker	Department of Cellular Pathology, Birmingham Heartlands Hospital, Bordesley Green East, Birmingham B9 5SS, UK
Aoife Crowley	Department of Pathology, The Coombe Women's Hospital Dublin and The Department of Histopathology, Trinity College Dublin, Ireland
Michael J. Deery	Inpharmatica Ltd, 60 Charlotte Street, London W1T 2NU, UK
Massimo Derenzini	Università di Bologna, Dipartimento di Patologia Sperimentale, Via San Giacomo 14, 40126 Bologna. Italy
Paul I. van Diest	Department of Pathology, VU Medical Centre, Amsterdam, The Netherlands

Timothy Diss Histopathology Department, RF and UCL Medical School, University Street, London WC1E 6JJ, UK

Amanda Dutton Department of Pathology, The Medical School, University of Birmingham, Edgbaston, Birmingham B15 2TT, UK

Sara Dyer Regional Genetics Service, Birmingham Women's Hospital, Edgbaston, Birmingham B15 2TG, UK

David J. Harrison Department of Pathology, University of Edinburgh, Edinburgh, UK

Hermann Herbst Gerhard-Domagk-Institut für Pathologie, Westfälische Wilhems-Universität, Domagkstr. 17, 48149 Münster, Germany

Mario A.J.A. Hermsen Department of Pathology, VU Medical Centre, Amsterdam, The Netherlands

C. Simon Herrington Department of Pathology, Duncan Building, University of Liverpool, Daulby Street, Liverpool L69 3GA, UK

Lee B. Jordan Department of Pathology, University of Edinburgh, Edinburgh, UK

Ernst J. Kuipers Department of Gastroenterology and Hepatology, Erasmus University Medical Centre, Rotterdam, The Netherlands

Kathryn Lilley Cambridge Centre for Proteomics, University of Cambridge, Department of Biochemistry, Building O, Downing Site, Cambridge CB2 1QW, UK

Victor Lopez Department of Pathology, The Medical School, University of Birmingham, Edgbaston, Birmingham B15 2TT, UK

Fiona MacDonald West Midlands Regional Genetics Laboratory, Birmingham Women's Hospital NHS Trust, Edgbaston, Birmingham B15 2TG, UK

Cara Martin Department of Pathology, The Coombe Women's Hospital Dublin and The Department of Histopathology, Trinity College Dublin, Ireland

Gerrit A. Meijer	Department of Pathology, VU Medical Centre, Amsterdam, The Netherlands
Paul G. Murray	Department of Pathology, The Medical School, University of Birmingham, Edgbaston, Birmingham B15 2TT, UK
Paul N. Nelson	Division of Biomedical Sciences, University of Wolverhampton, Wulfruna Street, Wolverhampton WV1 1SB, UK
Gerald Niedobitek	Pathologisches Institut, Friedrich-Alexander-Universität, Krankenhausstr. 8–10, 91054 Erlangen, Germany
Marie-Françoise O'Donohue	CNRS UMR 6142, Faculté de Médicine, Reims Cedex, France
John J. O'Leary	Department of Pathology, The Coombe Women's Hospital Dublin and The Department of Histopathology, Trinity College Dublin, Ireland
Michael G. Ormerod	34 Wray Way, Reigate RH2 0DE, UK
Dominique Ploton	CNRS UMR 6142, Faculté de Médicine, Reims Cedex, France
Azam Razzaq	Cambridge Centre for Proteomics, University of Cambridge, Department of Biochemistry, Building O, Downing Site, Cambridge CB2 1QW, UK
Orla Sheils	Department of Pathology, The Coombe Women's Hospital Dublin and The Department of Histopathology, Trinity College Dublin, Ireland
Antoine Snijders	UCSF Cancer Centre, San Francisco, USA
Fiona J. Spence	GlaxoSmithKline Pharmaceuticals, The Frythe, Welwyn, Herts AL6 9AR
David Treré	Università di Bologna, Dipartimento di Patologia Sperimentale, Via San Giacomo 14, 40126 Bologna. Italy
Philip Warren	Division of Biomedical Sciences, University of Wolverhampton, Wulfruna Street, Wolverhampton WV1 1SB, UK

Jonathan J. Waters NE London Regional Cytogenetics Department, Great Ormond Street Hospital, London WC1N 3BG, UK

Fiona Watson Department of Pathology, Duncan Building, University of Liverpool, Daulby Street, Liverpool L69 3GA, UK

Marjan M. Weiss Department of Gastroenterology, VU Medical Centre, Amsterdam, The Netherlands

Sophie E. Wildsmith GlaxoSmithKline Pharmaceuticals, The Frythe, Welwyn, Herts AL6 9AR

1

Blotting Techniques: Methodology and Applications

Fiona Watson and **C. Simon Herrington**

1.1 Introduction

The study of many different types of biomolecules has been advanced by the ability to attach the molecule to a membrane support. The technique used to transfer the biomolecules to the membrane is known as blotting and there are many variations of it. The basic steps in the procedure include the following: isolation of a cell-free mixture containing the biomolecule of interest; resolving the mixture into its component parts (if necessary); transfer (blotting) of the component parts onto a suitable membrane; and detection of the biomolecule of interest.

In general the blots are named according to the type of molecule that is blotted onto the membrane and include the Southern, Northern and Western blot which are used for the detection of DNA, RNA and protein, respectively. Variations of these, such as the Southwestern, Northwestern and Farwestern techniques, have been developed and there is also a lesser known technique called the Eastern. In this chapter we discuss both the methodology used to perform these techniques and the applications of their use.

1.2 Blotting techniques

The Southern Blot

This technique, which is used to detect specific sequences within mixtures of DNA, was first described by E.M. Southern in 1975 (Southern, 1975). In a

Molecular Biology in Cellular Pathology. Edited by John Crocker and Paul G. Murray
 ISBN: 0-470-84475-2

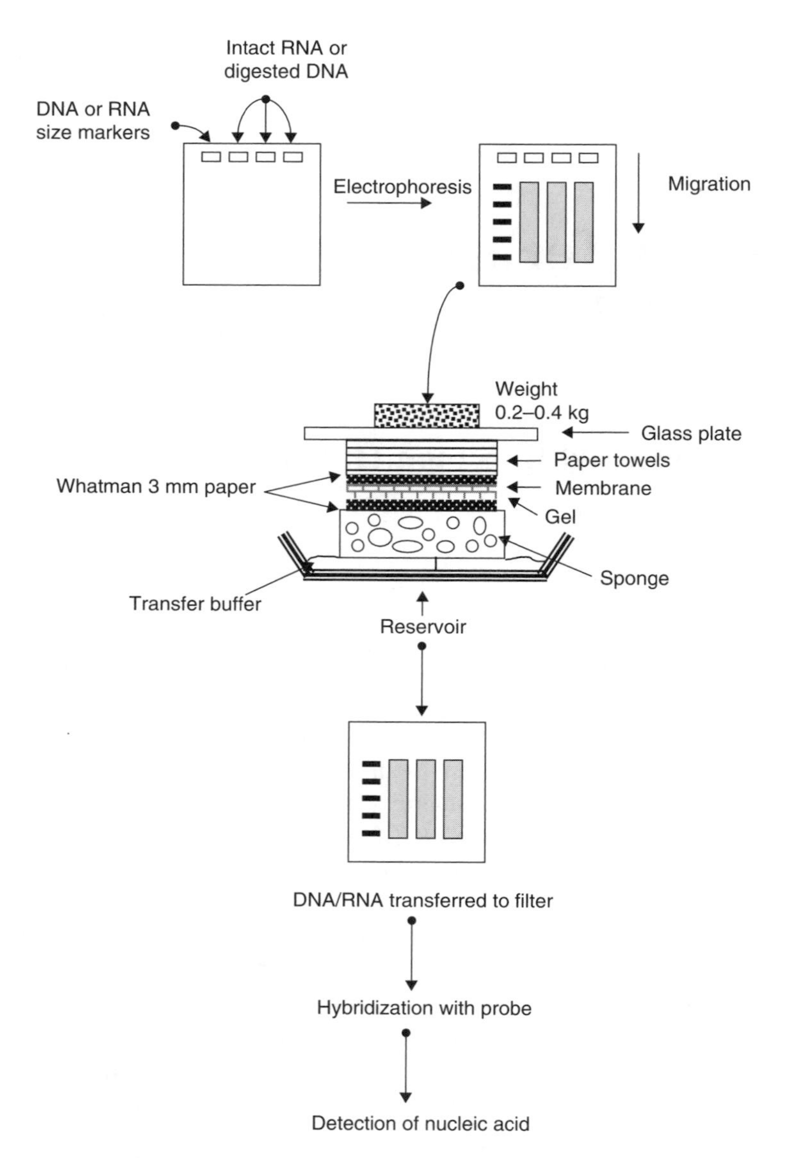

Figure 1.1 The steps involved in Southern and Northern blotting. Nucleic acids are resolved on a polyacrylamide gel prior to their upward capillary transfer onto a suitable membrane. The membrane is then hybridized with a probe which is suitable for the detection of the nucleic acid

Southern blot the DNA is size fractionated by gel-electrophoresis and then transferred by capillary action to a membrane (Figure 1.1). Membrane types and their uses are discussed in the section 'Membrane Types' below.

Non-specific binding sites on the membrane are then blocked and it is incubated with an appropriately labelled probe. Autoradiography or a phosphoimager is used to detect nucleic acid/probe hybrids when a radiolabelled probe is used but non-isotopically labelled probes require detection with a non-radioactive reporter system (described in the section 'Detection Methods' below). The size of the DNA recognized by the probe is determined by the co-electrophoresis of DNA fragments of known molecular weight.

The Southern blot technique has many applications. It has provided information about the physical organization of single and multicopy sequences in complex genomes, has expedited cloning experiments with eukaryotic genes, and was directly responsible for the discovery of introns (Doel *et al.*, 1977). Southern analysis is used to study the structure and location of genes by identifying restriction length polymorphisms (RFLPs) (Figure 1.2) and with the use of a pair of isoschizomers recognizing methylated and non-methylated nucleotides, gene methylation patterns can also be determined (Botstein *et al.*, 1980; Shaw *et al.*, 1993). Southern blotting has led to an increased understanding of the genomic rearrangements that are important in the formation of antibodies and T cell receptors, has identified numerous rearranged genes that are associated with disease, and has been used in the prenatal diagnosis of genetic disease

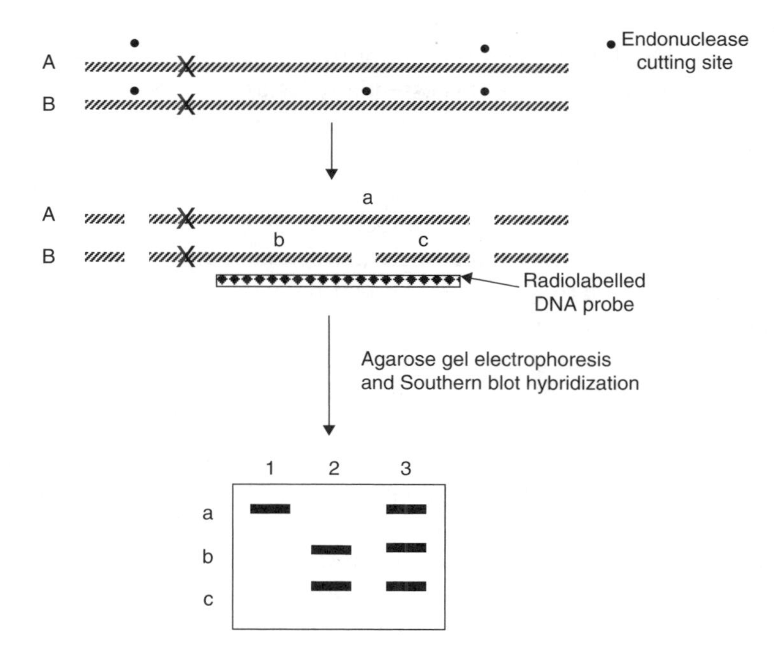

Figure 1.2 The principle of RFLP analysis. DNA restriction enzymes are used to create DNA fragments which are resolved by agarose gel electrophoresis. Differences in the size of the restriction fragments are detected by Southern blotting

(Chen *et al.*, 1999; Davies, 1986; Kramer *et al.*, 1998; Moreau *et al.*, 1999; Yu *et al.*, 2000). Activation of oncogenes by gene rearrangement, amplification or point mutation and inactivation of tumour suppressor genes by DNA rearrangements, point mutations, or allelic deletions can all be detected using Southern analyses (Munger, 2002). In addition, analysis of the allelic pattern of several polymorphic variable number of tandem repeats (VNTR) loci in an individual using the RFLP method can yield probabilities for investigating biological relationships or for matching forensic material found at a crime scene (Jeffreys *et al.*, 1985). The Southern blotting technique is also being utilized within the Human Genome Mapping Project to determine the order of the genes along the chromosome.

The Northern Blot

In a Northern blot, RNA is the target molecule blotted onto the membrane. The methodology is similar to that used in a Southern blot (Figure 1.1) but precautions should be taken to prevent RNA degradation that can be caused by the presence of ribonucleases, which are stable and active enzymes (Alwine *et al.*, 1977). To avoid RNase contamination, glassware should be baked and plasticware rinsed with chloroform prior to use. When possible, disposable plasticware should be used since it is essentially RNase free. Solutions that are made with ultrapure reagents that are reserved for RNA work in general are treated with diethylpyrocarbonate (DEPC) to 'inactivate' RNases and autoclaved prior to use. Tris-containing solutions are an exception. These solutions are made in DEPC-treated water and then autoclaved. It is important to remember that skin can be an important source of RNase contamination. For isolation of high-quality RNA, the starting material should be as fresh as possible and, if tissue or cells cannot be used immediately after harvesting, they should be flash frozen in liquid nitrogen and stored at -70°C until use. Commercial RNA isolation kits are available: these involve lysis of the cells with resultant inactivation of the ribonucleases as the first step.

A different type of gel from that used for Southern blotting is used for Northern blotting. In contrast to DNA, which usually is found as a double stranded molecule that migrates as a function of hybrid length, RNA has a significant secondary structure and must be electrophoresed under denaturing conditions if it is to migrate as a function of nucleotide length. The secondary structure associated with RNA would also reduce the efficiency of transfer to the membrane support. Thus RNA is denatured with either glyoxal and dimethylsulphoxide or formaldehyde and formamide (Lehrach *et al.*, 1977; Thomas, 1980). The formaldehyde gel is more common but both techniques are equally efficient. Estimation of the size of RNA can be achieved by comparing its migration with that of the 18S and 28S ribosomal components or with commercially available RNA markers of known size. Northern blots are used both to detect changes in gene expression levels and to detect possible alternative transcripts (Chen *et al.*, 2002; Sorensen *et al.*, 2002). Changes to either the transcript type expressed or to the level of expression of the mRNA may alter either the amount of protein product or its biological activity.

Northern blotting has been important in elucidating the physiological regulation of gene expression in both healthy and diseased tissues.

Technical Aspects of Southern and Northern Blotting

Choice and labelling of probe

The success of Northern and Southern blotting methods depends on the choice of probe type and consideration of how it should be labelled (Stickland, 1992). In both Northern and Southern analyses either DNA or RNA probes may be used. DNA and RNA probes were traditionally labelled using radioactively modified nucleotides, but most isotopes have a short half life and frequent probe preparation is necessary. In addition stringent safety procedures are required and the disposal of radioactive waste is expensive. A number of non-radioactive molecules such as biotin and digoxigenin have been used as alternatives in labelling reactions. They demonstrate increased stability compared with radioactively labelled probes and are relatively easy to handle. They can, therefore, be labelled in bulk and stored at −20°C. It has been suggested that the sensitivity, specificity and reproducibility of the non-isotopic alternatives are not equal to those obtained with radioactivity when used for filter hybridization. However, non-radioactive labelling appears to be the method of choice for techniques such as *in situ* hybridization.

Double-stranded DNA

Labelling of a double-stranded DNA probe can be performed using nick-translation (Rigby *et al.*, 1977), random primer labelling by primer extension (Feinberg and Vogelstein, 1983) or by the polymerase chain reaction (Mullis *et al.*, 1986). All of these procedures can be adapted to incorporate a radioactive or non-radioactive label. The nick-translation reaction is typically carried out on a DNA fragment that has been purified by gel electrophoresis and probably is the method of choice for biotinylating DNA. It involves the combined activities of DNase I and *Escherichia coli* DNA polymerase I. The nick-translation reaction is equally efficient with both linear and circular double-stranded molecules but is not appropriate for single-stranded DNA. The primer extension method of DNA labelling also utilizes the ability of DNA polymerase to synthesize a new DNA strand complementary to a template strand. In this method the DNA is denatured and annealed to random-sequence oligodeoxynucleotides which prime the DNA of interest at various positions along the template and are extended by activity of the Klenow fragment to generate double-stranded DNA that is uniformly labelled on both strands. Finally, the PCR may be used to generate a labelled probe from templates that have been subcloned into an appropriate vector using primers complementary to the regions just flanking the insertion sites of the vector, or directly from genomic DNA using specifically designed primers. Probe generation using PCR is useful for labelling subnanogram amounts of DNA of less than 500 bp. In the PCR reaction the labelling occurs through incorporation of an appropriately labelled nucleoside triphosphate. Double-stranded DNA probes require denaturation prior to hybridization.

RNA

RNA probes can be synthesized utilizing the ability of *E. coli* bacteriophage encoded RNA polymerases to synthesize specific single-stranded RNA molecules *in vitro* (Little and Jackson, 1987). The DNA template is cloned downstream of an appropriate bacteriophage promoter in a suitable vector. Many of these vectors are commercially available and choice is a matter of personal preference. If transcripts of both strands of the template are required, then it is beneficial to use a vector containing two different bacteriophage promoters. For Northern analysis the probe must be antisense, but in certain situations it is useful to generate sense RNA to be used as a negative control. Therefore, the point of cleavage used for linearization of the construct prior to probe synthesis depends on whether sense or antisense RNA is required.

Radiolabelled, biotinylated or digoxigenin labelled uridine triphosphate (UTP) can be used in the transcription reaction. The probes generated are interchangeable with DNA probes in all circumstances and have both increased sensitivity and lower background than DNA probes. Denaturation of RNA probes is advantageous because of the secondary structure associated with RNA molecules.

Oligonucleotide probes

Oligonucleotides are also used as probes in Southern blotting applications. Chemically synthesized oligonucleotides do not have a phosphate at their 5′ termini and can, therefore, be labelled with γ-^{32}P adenosine triphosphate (ATP) in a reaction catalysed by bacteriophage T4 polynucleotide kinase. ^{32}P is used frequently in this type of labelling as only a single labelled nucleotide is incorporated per oligonucleotide (Connor *et al.*, 1983). The enzyme terminal deoxytransferase can be used with both radiolabelled nucleotides and the non-radioactive labels, biotin and digoxigenin, to 3′ end label oligonucleotides for use as hybridization probes (Chu and Orgel, 1985). Depending on the reaction conditions, one or more labelled molecules may be added to the oligonucleotide. Probes should not be tailed with dTTP, as it may hybridize to poly (A+) sequences in mRNA, or dATP, as it may hybridize to poly (T) regions in genomic DNA. Oligonucleotide probes can also be labelled with alkaline phosphatase.

Hybridization

Hybridization is the reaction by which a target DNA or RNA nucleic acid sequence anneals to a complementary probe as a result of base pairing (Chan, 1992; Thomas, 1980). The hybridization conditions for both isotopically labelled and non-radioactive probes are similar but some experimentation with hybridization conditions is often necessary for optimal results. The hybridization procedure can be broken down into three steps.

1. Incubation of the membrane with prehybridization solution containing reagents that block non-specific DNA binding. Typically, Denhardt's solution which contains Ficoll, polyvinylpyrrolidone and bovine serum albumin (BSA) is used together with denatured salmon sperm DNA. Prehybridization for at least 3 h is recommended with nitrocellulose membranes and 15 min for nylon membranes. The blocking agents may need to be left out if using nylon membranes as they can interfere with the probe–target interaction.
2. The prehybridization solution is replaced with a high salt hybridization solution which contains the labelled probe and which promotes target–probe base pairing. Hybridization typically is carried out for 24 h at 68°C but, for RNA, lower incubation temperatures are favoured because high temperatures may result in RNA degradation. When genomic, cDNA or PCR synthesized probes are used, hybridization is performed at 42°C, while the hybridization temperature needs to be determined empirically for oligonucleotide probes as they are more easily destabilized by high temperatures and low ionic strength than conventional probes. However, oligonucleotide probes do not need denaturation and optimum hybridization time is often a matter of hours.
3. The membrane is washed until only highly matched hybrids remain. Hybridization reactions, which can be both formamide and phosphate based, are usually performed under conditions of relatively low stringency to allow rapid formation of the hybrids. Specificity is the function of the post-hybridization washes. The use of formamide probably confers no major advantage on DNA–DNA hybridization with nylon membranes and is not used for oligonucleotide probes. It does, however, allow the use of lower temperatures during RNA hybridizations as it destabilizes nucleic acid duplexes and reduces the heterologous background hybridization of RNA probes. For non-isotopic probes, Church and Gilbert (1984) hybridization mixture can be used.

Detection methods

Autoradiography is used for detection of radiolabelled probe–target hybrids and quantification can be performed using densitometry. Alternatively, use of a phosphoimager is recommended as it is more sensitive and can be quantified easily while the problem of overexposure, which can occur with X-ray film, is not an issue. In general, the detection of non-radioactive probes is more complicated than the detection of radioactive probes as the reporter molecules require immuno/affinity chemical techniques for their detection (Hughes *et al.*, 1995). Biotinylated probes are detected using streptavidin, which has a high affinity and specificity for biotin. The streptavidin can be attached to both the biotinylated probe and a reporter enzyme.

Alternatively, the multiple biotin binding sites on the streptavidin molecule can be used to sandwich it between a biotinylated probe and a biotinylated enzyme. Detection of digoxigenin is achieved by incubation with anti-digoxigenin antibodies directly coupled to a fluorochrome or enzyme. $F(ab)_2$ fragments are often used in an attempt to reduce non-specific binding and the use of uncoupled antibodies permits signal amplification methodologies to be utilized. Reporter enzymes are available that catalyse both colorimetric and chemiluminescent reactions. Alkaline phosphatase is probably the most widely used enzyme and can be used in both colorimetric and chemiluminescent reactions. The most common colour system involves the alkaline phosphatase-mediated reduction of Nitroblue tetrazolium salt (NBT) to diformazan resulting in the formation of a membrane-bound blue or brown precipitate. In chemiluminescent reactions, a light reaction is the end-point and several chemiluminescent compounds, such as the modified dioxetanes and luciferin derivatives, can act as the substrate. Horseradish peroxidase with hydrogen peroxidase are often used to oxidize luminol and form a chemically excited 3-aminophthalate dianion. As the excited molecule returns to the ground state, light is emitted at 428 nm for a short time. The use of enhancers increases and prolongs the light output to allow it to be captured on film.

Membrane types

Many different membrane types are available for use in blotting procedures (Kingston, 1987; Moore, 1987). The main advantage of a nylon membrane over nitrocellulose is its greater tensile strength and the ability to bind DNA covalently either by cross-linking or, in the case of positively charged nylon, by transfer with an alkaline buffer. DNA is attached to a nitrocellulose membrane by hydrophobic attachments induced by baking, which means that it can be reprobed only about three times. A nylon membrane can be reprobed up to 12 times but non-specific background can be a significant problem, especially when non-radioactive DNA probes are used. The method of detecting the target should, therefore, be considered when choosing which membrane to use. For colorimetric assays, nitrocellulose membranes probably are the best but are unsuitable for chemiluminescence unless used in conjunction with a specific blocking agent, e.g. Nitro-Block (Tropix). Nylon membranes are most suitable for use with digoxigenin-labelled probes in chemiluminescent reactions while it is better to use a membrane such as Immobilon-S with biotinylated probes. Many commercially available membrane types can be used to optimize the results of blotting experiments.

Dot and slot blotting

Dot and slot blotting is used to immobilize either bulk unfractionated DNA or RNA (Kafatos *et al.*, 1979). Hybridization analysis can then be carried out to determine the type and relative abundance of the target sequence. In this type of blotting the samples usually are applied to the membrane using a manifold and suction device which is quicker and more reproducible than manual blotting. However, manual blotting can be set up by spotting small aliquots of each sample on to the membrane and waiting for it to dry. Preparation of the dot blot is altered depending on the type of membrane used: uncharged nylon, positively charged nylon and nitrocellulose are all of use in this technique. However, meaningful comparisons of sequence abundance in different DNA samples can only be made if the DNA is fully denatured prior to blotting. In addition, dot/slot blotting with bulk DNA can result in the co-blotting of impurities which can have unpredictable effects on hybridization patterns.

Despite these limitations, the technique has numerous applications. For example, it is used in the sensitive and specific detection of human papillomavirus (HPV) DNA and has contributed greatly to the understanding of the aetiology and natural history of HPV infection (Bauer *et al.*, 1992). In this technique, degenerate or mixed consensus primers are used in PCR to amplify a broad spectrum of HPV types and the PCR products dotted onto nylon membranes. They are then hybridized with either long generic probes, such as with L1 products, or oligonucleotide probes (used for both E6 and L1 products). This method allows the rapid preparation and analysis of large sample numbers and the blots can be prepared in replicate and hybridized in parallel with different HPV-type specific probes. In addition, because the genome of every microorganism contains sequences that are unique to the species, it is possible to accurately identify any pathogen using well-designed DNA probes (McCreedy and Chimera, 1992). The DNA dot blotting technique is of particular use in the identification of pathogens that are difficult to detect using culture techniques and direct pathogen detection using DNA probes has reduced the time and expense associated with identifying many types of infectious agents. Pathogens are detected in clinical samples using polymerase chain amplification coupled with gene probe detection. Alternatively, the nylon or nitrocellulose membrane is placed on the surface of a nutrient agar plate and colonies transferred from growth plates of the unknown organism. The plates are then inverted and incubated for 12 h. The nucleic acids are denatured and then fixed to the membrane by baking or cross-linking prior to hybridization. DNA dot/slot blotting has also been used to identify point mutations in genes derived from clinical samples using an oligonucleotide hybridization method based on the principle that a duplex of oligonucleotides with even one base mismatch will display instability (Innis *et al.*, 1990). RNA slot blotting is often used to assess the expression profiles of tissue-specific genes and the RNA detection can be considered semi-quantitative, but it is important to analyse these types of experiments thoroughly and for the RNA that is used to be as free as possible from contaminants such as DNA and protein.

The Western Blot

Western blotting detects antigenic determinants on protein molecules using polyclonal or monoclonal antibodies and often is described as immunoblotting. The first step in Western analysis involves solubilization of the protein samples usually using sodium dodecyl sulphate (SDS) and reducing agents such as dithiothreitol (DTT) or 2-mercaptoethanol. Since it is important to avoid degradation of the proteins, lysis buffers usually contain a cocktail of protease inhibitors and cell lysis is performed at 4°C. Individual proteins are then resolved by SDS polyacrylamide gel electrophoresis prior to electrophoretic transfer. Most apparatus used to transfer the proteins employ the system illustrated in Figure 1.3. This uses vertical electrodes in contrast to 'horizontal blotting' apparatus which has two plate electrodes that can generate a uniform electric field over a short distance (Bjerrum and Schafer-Nielsen, 1986; Burnette, 1981; Harlow and Lane, 1988). In 'horizontal blotting' much less transfer buffer is required compared with the conventional blotting technique and multiple

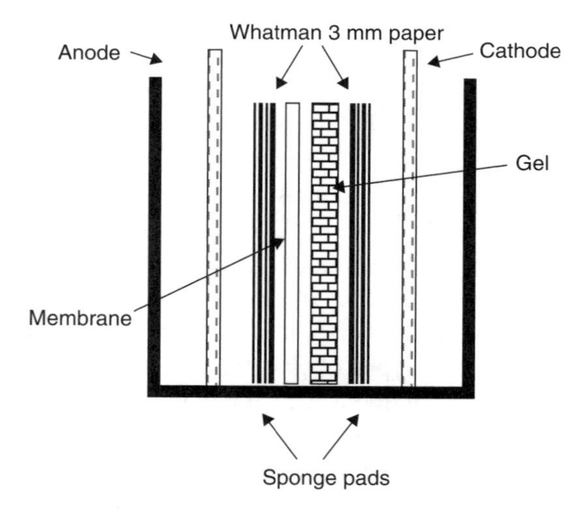

Figure 1.3 The apparatus typically used in a Western blotting experiment. The polyacrylamide gel is placed on filter paper and overlaid with a sheet of membrane followed by another sheet of filter paper. This is then sandwiched between two Scotch-Brite sponge pads and placed in a plastic support in transfer buffer. Transfer of a negatively charged protein is achieved by positioning the membrane on the anode side of the gel

gel/membrane and filter paper assemblies can be electrophoresed simultaneously. The electrodes can be made from cheap carbon blocks and less power is required for transfer, but prolonged transfers (>1 h) are not recommended. Following blotting, non-specific binding sites on the membrane are blocked with a concentrated protein solution (5% non-fat milk powder or gelatin) prior to incubation with a primary antibody, washing and incubation with a second conjugated probe antibody. These secondary antibodies might be conjugated anti-immunoglobulins, conjugated staphylococcal Protein A which binds IgG of various animal species, or probes to biotinylated/digoxigeninylated primary antibodies. Following the second incubation, the membrane is again washed prior to colorimetric/autoradiographic or chemiluminescent detection. The molecular weights of the proteins are determined by comparison with a set of molecular weight markers which are co-electrophoresed. These markers can be visualized in the gel, or if nitrocellulose or polyvinylidine fluoride (PVDF) membranes have been used the proteins can be reversibly stained with Ponceau S solution.

Alternatively, prestained SDS–PAGE markers, which can be transferred onto nitrocellulose, nylon or PVDF and are visible without any subsequent staining, are commercially available. This allows protein migration and electrophoretic transfer to be monitored in addition to the molecular weight of the protein of interest. The success of Western blotting is dependent on the availability of suitable antibodies. The generation of both polyclonal and monoclonal antibodies has the potential to be problematic and demands a significant investment in terms of time and money. However, the use of many commercially available antibodies that recognize the proteins involved in the regulation of the cell cycle, transcription, cell growth, oncogenesis and apoptosis has led to significant advances being made in these areas. The study of intracellular signalling has also benefited through the use of Western analysis. The expression of many different signalling molecules can be measured with antibodies and it is possible to detect post-translational modifications such as phosphorylation or glycosylation through the use of appropriately labelled antibodies. However, it is not always possible to raise good antibodies to all proteins and, in this scenario, the production of recombinant proteins tagged with an epitope to which antibodies are commercially available, is useful. Examples of possible tags are c-myc, FLAG (Asp–Tyr–Lys–Xaa–Xaa–Asp), haemagglutinin and histidine. Although this approach is open to the argument that the tag may interfere with the biological function of the native protein, it has advanced our knowledge of a number of biological systems (Zhang and Chen, 1987).

The Southwestern and Northwestern

Southwestern blotting is used to detect specific DNA binding proteins (Miskimins *et al.*, 1985). The initial step is to resolve the proteins on non-denaturing polyacrylamide gel. The separated proteins are then transferred to nitrocellulose and detected using a radiolabelled double stranded DNA that contains the putative protein binding site. Bound protein is then detected by autoradiography. A limitation of this technique is that it suffers from poor specificity so it is important to include a number of controls. If the object of the experiment is to determine the molecular weight of a DNA binding protein, then an alternative and perhaps more robust approach is to cross-link the proteins directly to a radiolabelled DNA probe using UV light and then to resolve the proteins by SDS–PAGE. The gel can be dried down and visualized using autoradiography. This method uses the same type of probes as the Southwestern blot and thus the same problem of poor specificity needs to be addressed. Both the Southwestern technique and protein/DNA cross-linking procedures have, however, underpinned major advances in the understanding of transcriptional regulation.

The Northwestern technique is similar to that of the Southwestern except that RNA binding proteins are detected using radiolabelled RNA probes (Schiff *et al.*, 1988). Following electrophoretic transfer and blocking of non-specific binding sites, the blots are probed with either ^{32}P-labelled RNA transcripts or double stranded probes. The blots are then washed and protein–RNA hybrids detected by autoradiography. This technique has revealed a multitude of RNA binding proteins that appear spatially and temporally in the cells of all organisms (Hall, 2002). The structures of these RNA–protein complexes are providing valuable insights into the binding modes and functions of these interactions. Some examples of RNA binding proteins and their function are: (a) the correct folding and packaging of pre-rRNA mediated through the direct binding of nucleolin to two mutually exclusive RNA sequences (Ginisty *et al.*, 2001); (b) Hu proteins, which are RNA binding proteins postulated to regulate gene transcription at the post-transcriptional level (Chung *et al.*, 1996); (c) proteins involved in pre-mRNA splicing such as the polypyrimidine tract binding protein and U2 auxillary factor (Patton *et al.*, 1991; Zamore *et al.*, 1992); and (d) the protein kinase PKR which is a serine threonine kinase activated by double stranded RNA (Tian and Mathews, 2001).

The Far Western

The Far Western blot detects protein–protein interaction (Edmondson and Roth, 1987). In this technique the protein of interest is immobilized on the membrane and then probed with a non-antibody protein. It is a useful technique for studying proteins that are difficult to solubilize or to express in cells. As a first step the protein mixture is resolved by SDS–PAGE prior to electrophoretic transfer to a membrane, although non-SDS gels, such as acid urea gels which separate on the basis of both size and charge, also have been used. Following transfer, the membrane may be stained with Ponceau S to help locate the proteins. The non-specific binding sites are then blocked with standard blocking reagents and usually incubated with a radiolabelled non-antibody protein probe. An *in vitro* translated probe of the protein of interest can be made using ^{35}S methionine, ^{14}C leucine and ^{3}H leucine or probes may be labelled enzymatically with ^{32}P. The proteins that bind the probe are detected by autoradiography. *In vitro* translated probes can be produced relatively quickly and are easily detected and quantitated. It is also possible to generate mutations in the protein using well-established cloning techniques. Peptide probes may also be of use. The Far Western technique is amenable to non-radioactive detection techniques. Biotin-labelled probes may be detected with streptavidin–biotin detection schemes and if an antibody to the interacting protein is available, then Western analysis can be used. This latter procedure is useful when a tagged recombinant protein and appropriate antibody are being used. The technique has been used to examine diverse protein interactions including: (a) the interaction of histones with regulatory proteins (Edmondson *et al.*, 1996); (b) keratin intermediate filaments with desmosomal proteins (Kouklis *et al.*, 1994); (c) proteins involved in transcriptional regulation (Chaudhary *et al.*, 1997; Kimball *et al.*, 1998); and (d) viral proteins and their host cell targets (Grasser *et al.*, 1993). Far Western analysis has also been used to study receptor–ligand interactions and to screen libraries for interacting proteins.

Additional uses of overlay blotting

Overlay blotting has been used in a number of diverse ways including for the identification of GTP-binding proteins (Celis, 1998). In this type of assay the proteins are resolved by SDS–PAGE prior to renaturation by incubation in an appropriate buffer and blotting. The blots are then incubated for 10 min with non-radioactive GTP at 25°C before incubation with γ-^{32}P GTP. The blots are then washed thoroughly before visualization by autoradiography. It is also possible to investigate calcium binding proteins by utilizing a similar approach. In this technique the blotted proteins are incubated with $^{45}Ca^{2+}$ before detection using autoradiography (Barroso *et al.*, 1996). This technique has been used to detect Ca^{2+}-induced conformational changes in proteins. Blotting assays can also be used to separate ^{32}P-labelled proteins or peptides from γ-^{32}P ATP following protein kinase assays. P81 phosphocellulose paper is used in this technique where the kinase reaction contents are blotted onto paper squares which are then washed at low pH to remove the excess γ-^{32}P ATP. However, if the phosphorylated peptide carries a net positive charge at low pH, then it binds to the paper. This can be achieved by using an appropriately modified peptide as the substrate. The kinase activity which is bound to the paper is quantified using scintillation counting (Carter, 1987).

DNA Arrays and Antibody Arrays

Nucleic acid arrays are used for the high throughput analysis of gene expression levels. Typically, thousands of cDNAs or oligonucleotides are bound to solid substrates such as glass microscope slides and hybridized to labelled cDNA probes which can be synthesized from RNA extracted from particular samples. These arrays can be difficult and costly to produce and require expensive equipment for their use. However, the set-up of dedicated microarray facilities is making them more available to the average researcher. The term 'DNA microarray' may refer to several different forms of the technology which differ in the type of nucleic acid applied and the method of attachment (Celis *et al.*, 2000). It is possible to buy cDNA expression arrays on which hundreds of cDNAs are spotted on to positively charged nylon membranes (Hester *et al.*, 2002; Levenson *et al.*, 2002). These commercially available membranes have specific cDNAs included as controls and can be hybridized to radiolabelled probes derived from RNA extracted from tissues or cell lines. Hybridization is performed in much the same way as with a Northern blot and the technology can be thought of as a Reverse Northern in which unlabelled probe is blotted to the membrane and hybridized to a radiolabelled target (similar to dot/slot blotting). The membranes from this type of array can be stripped and rehybridized for a limited number of times. Although the number of genes on each membrane is much less than that which can be spotted onto other array types, the membranes can be carefully designed to include genes that are associated with different research areas and therefore may have their use in specific applications. Antibody arrays also have recently become commercially available. This type of array contains hundreds of antibodies immobilized at predetermined positions on a membrane. These antibodies retain their ability to capture their target antigen and any proteins associated with it. An immunoblotting technique is then used to detect the immobilized proteins. Antibody arrays can be used to screen protein–protein interactions, study protein post-translational modifications, and examine protein expression patterns. In a similar manner to DNA arrays it is possible to attach the immobilized antibodies to glass slides. In this latter type of methodology the proteins that bind to the antibodies are labelled with Cy3 or Cy5 dyes (Walter *et al.*, 2002).

The Eastern Blot

Small molecular weight molecules can be visualized using Eastern blotting (Shan *et al.*, 2001). In this type of blotting the small molecules are resolved by Thin Layer Chromatography (TLC). The TLC plate is developed and dried prior to blotting at high temperature (120°C for 50 s) to a PVDF membrane. The membrane is then treated to reduce non-specific binding and the molecule of interest detected using an appropriate detection system. This method, as in Western blotting, is limited by the availability of specific antibodies to the molecules of interest.

1.3 References

Alwine JC, Kemp DJ and Stark GR (1977) Method for detection of specific RNAs in agarose gels by transfer to diazobenzyloxymethyl-paper and hybridization with DNA probes. *Proc. Natl. Acad. Sci. USA* **74**: 5350–4.

Barroso MR, Bernd KK, DeWitt ND, Chang A, Mills K and Sztul ES (1996) A Novel Ca^{2+} binding protein, p22, is required for constitutive membrane traffic. *J. Biol. Chem.* **271**: 10183–7.

Bauer HM, Greer C and Manos MM (1992) Determination of genital human papillomavirus infection by consensus PCR amplification, in: *Diagnostic Molecular Pathology: A Practical Approach* (eds CS Herrington and JO'D McGee), pp. 131–52. Oxford University Press, Oxford.

Bjerrum OJ and Schafer-Nielsen C (1986) Buffer systems and transfer parameters for semidry electroblotting with a horizontal apparatus, in: *Electrophoresis '86* (ed. MJ Dunn), pp. 315–27. Verlag Chemie, Weinheim.

Botstein D, White RL, Skolnick M and Davies RW (1980) Construction of a genetic linkage map in man using restriction fragment length polymorphisms. *Am. J. Hum. Genet.* **32**: 314–31.

Burnette WN (1981) Western blotting: electrophoretic transfer of proteins from sodium dodecyl sulphate-polyacrylamide gels to unmodified nitrocellulose and radiographic detection with antibody and radioiodinated protein. *Anal. Biochem.* **112**: 195–203.

Carter NA (1987) Assays of protein kinases using exogenous substrates, in: *Current Protocols in Molecular Biology, Vol. 4*, 18.8.1–18.8.19 (eds FM Ausubel, R Brent, RE Kingston, DD Moore, JG Seidman, JA Smith and K Struhl). John Wiley & Sons, Inc., New York.

Celis JE (1998) *Cell Biology. A Laboratory Handbook*, Vol. 4, pp. 454–6. Academic Press, San Diego.

Celis JE, Kruhøffer M, Gromova I, Frederiksen C, Østergaard M, Thykjaer T, Gromov P, Yu J, Pálsdóttir H, Magnusson N and Ørntoft TF (2000) Gene expression profiling: monitoring transcription and translation products using DNA microarrays and proteomics. *FEBS Lett.* **480**: 2–16.

Chan VT-W (1992) Molecular hybridization of nucleic acids, in: *Diagnostic Molecular Pathology: A Practical Approach* (eds CS Herrington and JO'D McGee), pp. 25–64. Oxford University Press, Oxford.

Chaudhary J, Cupp AS and Skinner MK (1997) Role of basic helix–loop–helix transcription factors in Sertoli cell differentiation: identification of an E-box response element in the transferrin promoter. *Endocrinology* **138**: 667–75.

Chen PM, Chiou TJ, Hsieh RK, Fan FS, Chu CJ, Lin CZ, Chiang H, Yen CC, Wang WS and Liu JH (1999) P53 gene mutations and rearrangements in non-Hodgkin's lymphoma. *Cancer* **85**: 718–24.

Chen W, Aoki C, Mahadomrongkul V, Gruber CE, Wang GJ, Blitzblau R, Irwin N and Rosenberg PA (2002) Expression of a variant form of the glutamate transporter GLT1 in neuronal cultures and in neurons and astrocytes in the rat brain. *J. Neurosci.* **15**: 2142–52.

Chu BC and Orgel LE (1985) Detection of specific DNA sequences with short biotin-labelled probes. *DNA* **4**: 327–31.

Chung S, Jiang L, Cheng S and Furneaux H (1996) Purification and properties of HuD, a neuronal RNA-binding protein. *J. Biol. Chem.* **271**: 11518–24.

Church GM and Gilbert W (1984) Genomic sequencing. *Proc. Natl. Acad. Sci. USA* **81**: 1991–5.

Connor BJ, Reyes AA, Morin C, Itakura K, Teplitz RL and Wallace RB (1983) Detection of sickle cell beta S-globin allele by hybridization with synthetic oligonucleotides. *Proc. Natl. Acad. Sci. USA* **1**: 278–82.

Davies KE (ed.) (1986) *Human Genetic Diseases: A Practical Approach*. IRL Press at Oxford University Press, Oxford.

Doel MT, Houghton M, Cook EA and Carey NH (1977) The presence of ovalbumin mRNA coding sequences in multiple restriction fragments of chicken DNA. *Nucleic Acids Res.* **4**: 3701–13.

Edmondson DG and Roth SY (1987) Identification of protein interactions by Far Western Analysis. *In Current protocols in molecular biology. Vol. 4*, 20.6.1–20.6.8 (eds FM Ausubel, R Brent, RE Kingston, DD Moore, JG Seidman, JA Smith and K Struhl). John Wiley & Sons, Inc., New York.

Edmondson DG, Smith MM and Roth SY (1996) Repression domain of the yeast global repressor Tup1 interacts directly with histones H3 and H4. *Genes Dev.* **10**: 1247–59.

Feinberg AP and Vogelstein B (1983) A technique for radiolabeling DNA restriction endonuclease fragments to high specific activity. *Anal. Biochem.* **132**: 6–13.

Ginisty H, Amalric F and Bouvet P (2001) Two different combinations of RNA – binding domains determine the RNA binding specificity of nucleolin. *J. Biol. Chem.* **276**: 14338–43.

Grasser FA, Sauder C, Haiss P, Hille A, Konig S, Gottel S, Kremmer E, Leinenbach HP, Zeppezauer M and Mueller-Lantzsch N (1993) Immunological detection of proteins associated with the Epstein-Barr virus nuclear antigen 2A. *Virology* **195**: 550–60.

Hall KB (2002) RNA – protein interactions. *Current Opinion in Structural Biology* **12**: 283–8.

Harlow E and Lane D (1988) Immunoblotting. In *Antibodies: A Laboratory Manual*. pp. 471–510. CSH Laboratory, Cold Spring Harbor, NY.

Hester SD, Benavides GB, Sartor M, Yoon L, Wolf DC and Morgan KT (2002) Normal gene expression in male F344 rat nasal transitional and respiratory epithelium. *Gene* **285**: 301–10.

Hughes JR, Evans MF and Levy ER (1995) Non-isotopic detection of nucleic acids on membranes. In *Non-isotopic Methods in Molecular Biology: A practical approach*. pp. 145–82. (eds ER Levy and CS Herrington), IRL Press at Oxford University Press, Oxford.

Innis MA, Gelfand GH, Sninsky JJ and White TJ (ed.) (1990). *PCR protocols, a guide to methods and applications*. Academic Press, San Diego, California.

Jeffreys AJ, Wilson V and Thein SL (1985) Hypervariable 'minisatellite' regions in human DNA. *Nature* **314**: 67–73.

Kafatos FC, Jones CW and Efstratiadis A (1979) Determination of nucleic acid sequence homologies and relative concentrations by a dot hybridization procedure. *Nucleic Acids Res.* **7**: 1541–52.

Kimball SR, Heinzinger NK, Horetsky RL and Jefferson LS (1998) Identification of interprotein interactions between the subunits of eukaryotic transcription factors eIF2 and eIF2B. *J. Biol. Chem.* **273**: 3039–44.

Kingston RE (1987) Preparation and analysis of RNA. In *Current protocols in molecular biology. Vol. 1*, 4.0.3–4.2.9 (eds FM Ausubel, R Brent, RE Kingston, DD Moore, JG Seidman, JA Smith and K Struhl). John Wiley & Sons, Inc., New York.

Kouklis PD, Hutton E and Fuchs E (1994) Making a connection: Direct binding between keratin intermediate filaments and desmosomal proteins. *J. Cell. Biol.* **127**: 1049–60.

Kramer MH, Hermans J, Wijburg E, Philippo K, Geelen E, van Krieken JH, de Jong D, Maartense E, Schuuring E and Kluin PM (1998) Clinical relevance of BCL2, BCL6 and MYC rearrangements in diffuse large B-cell lymphoma. *Blood* **92**: 3152–62.

Lehrach H, Diamond D, Wozney JM and Boedtker H (1977) RNA molecular weight determinations by gel electrophoresis under denaturing conditions: A critical reexamination. *Biochemistry* **16**: 4743–51.

Levenson AS, Kliakhandler IL, Svoboda KM, Pease KM, Kaiser SA, Ward JE 3rd and Jordan VC (2002) Molecular classification of selective oestrogen receptor modulators on the basis of gene expression profiles of breast cancer cells expressing oestrogen receptor alpha. *Br. J. Cancer* **87**: 449–56.

Little PFR and Jackson IJ (1987) Application of plasmids containing promoters specific for phage-encoded RNA polymerases. In *DNA cloning, Vol. III: A practical approach* (ed. DM glover), pp. 1–18, IRL Press, Oxford.

McCreedy BJ and Chimera JA (1992) Molecular detection and identification of pathogenic organisms. In '*Diagnostic Molecular Pathology: A Practical Approach*. pp. 173–82. (eds CS Herrington and JO'D McGee). Oxford University Press, Oxford.

Miskimins WK, Roberts MP, McClelland A and Ruddle FH (1985) Use of a protein-blotting procedure and a specific DNA probe to identify nuclear proteins that recognize the promoter region of the transferring receptor gene. *Proc. Natl. Acad. Sci.* **82**: 6471–744.

Moore DD (1987) Preparation and analysis of DNA. In *Current protocols in molecular biology. Vol. 1*, 2.0.1–2.1.10 (eds FM Ausubel, R Brent, RE Kingston, DD Moore, JG Seidman, JA Smith and K Struhl. John Wiley & Sons, Inc., New York.

Moreau EJ, Langerak AW, van Gastel-Mol EJ, Wolvers-Tettero IL, Zhan M, Zhou Q, Koop BF and van Dongen JJ (1999) Easy detection of all T cell receptor gamma (TCRG) gene rearrangements by Southern blot analysis: recommendations for optimal results. *Leukaemia* **13**: 1620–6.

Mullis K, Faloona F, Scharf S, Saiki R, Horn G and Erlich H (1986) *Cold Spring harbor Symp. Quant. Biol.* **51**: 163.

Munger K (2002) Disruption of oncogene/tumor suppressor networks during human carcinogenesis. *Cancer Invest.* **20**: 71–81.

Patton JG, Mayer SA, Tempst P and Nadalginard B (1991) Characterization and molecular cloning of polypyrimidine tract binding protein. A component of a complex necessary for pre-mRNA splicing. *Genes Dev.* **5**: 1237–51.

Rigby PW, Dieckmann M, Rhodes C and Berg P (1977) Labeling deoxyribonucleic acid to high specific activity *in vitro* by nick translation with DNA polymerase I. *J. Mol. Biol.* **113**: 237–52.

Schiff LA, Nibert ML, Co MS, Brown EG and Fields BN (1988) Distinct binding sites for zinc and double-stranded RNA in the reovirus outercapsid protein sigma 3. *Mol. Cell. Biol.* **8**: 273–83.

Shan S, Tanaka H and Shoyama Y (2001) Enzyme – linked immunosorbent assay for glycyrrhizin using anti-glycyrrhizin monoclonal antibody and an eastern blotting technique for glucuronides of glycyrrhetic acid. *Anal. Chem.* **73**: 5784–90.

Shaw DJ, Chaudhary S, Rundle SA, Crow S, Brook JD, Harper PS and Harley HG (1993) A study of DNA methylation in myotonic dystrophy. *J. Med. Genet.* **30**: 189–92.

Sorensen AB, Warming S, Fuchtbauer EM and Pedersen FS (2002) Alternative splicing, expression, and gene structure of the septin-like putative proto-oncogene Sint1. *Gene* **285**: 79–89.

Southern EM (1975) Detection of specific sequences among DNA fragments separated by gel electrophoresis. *J. Mol. Biol.* **98**: 503–17.

Stickland JE (1992) Probe preparation and labelling. In *Diagnostic Molecular Pathology: A Practical Approach*. pp. 25–64. (eds CS Herrington and JO'D McGee). Oxford University Press, Oxford.

Thomas PS (1980) Hybridization of denatured RNA and small DNA fragments transferred to nitrocellulose. *Proc. Natl Acad. Sci. USA* **77**: 5201–5.

Tian B and Mathews MB (2001) Functional characterization of and cooperation between the double-stranded RNA-binding motifs of the protein kinase PKR. *J. Biol. Chem.* **276**: 9936–44.

Walter G, Bussow K, Lueking A and Glokler J (2002) High-throughput protein array: prospects for molecular diagnostics. *Trends Mol. Med.* **8**: 250–3.

Yu J, Miehlke S, Ebert MP, Hoffmann J, Breidert M, Alpen B, Starzynska T, Stolte Prof. M, Malfertheiner P and Bayerdorffer E (2000) Frequency of TPR-MET rearrangements in patients with gastric carcinoma and in first-degree relatives. *Cancer* **88**: 1801–6.

Zamore PD, Patton JG and Green MR (1992) Cloning and domain structure of the mammalian splicing factor U2AF. *Nature* **355**: 609–14.

Zhang N and Chen J-L (1987) Purification of Recombinant Proteins and Study of Protein Interaction by Epitope Tagging. *In Current protocols in molecular biology. Vol. 2*, 10.15.1–10.15.9 (eds FM Ausubel, R Brent, Kingston RE, DD Moore, JG Seidman, JA Smith and K Struhl). John Wiley & Sons, Inc., New York.

2

In-situ Hybridisation in Histopathology

Gerald Niedobitek and **Hermann Herbst**

2.1 Introduction

In-situ hybridisation (ISH) allows the detection of nucleic acids, DNA or RNA, in tissue sections, cytological specimens, or chromosome preparations. Over 30 years after its first description (Buongiorno–Nardell and Amaldi, 1970; Gall and Pardue, 1969; John *et al.*, 1969), ISH has developed into an indispensable tool for morphology-based research, whereas it has found only a limited number of diagnostic applications. The spectrum of research applications ranges from gene expression studies and detection of viral genomes at the single cell level (e.g. in histo- and cytopathology), expression pattern analysis in whole-mount preparations (e.g. in embryology and neurosciences) to cytogenetic applications on interphase nuclei and metaphase chromosome spreads for the detection of numerical and structural chromosomal aberrations, including specialised applications such as comparative genomic hybridisation (CGH). In this chapter we focus on applications in histopathology and discuss some recent developments. For protocols and reviews of technical details the reader is referred to previous reports (Franco *et al.*, 2001; Herbst and Niedobitek, 2001; McNicol and Farquharson, 1996; Niedobitek and Herbst, 1991, 2001; Wilcox, 1993; Wilson *et al.*, 1997). Cytogenetic applications such as the direct visualisation of DNA on metaphase chromosomes as well as CGH are not covered in this review as they are considered elsewhere (see Chapters 4 and 14).

Molecular Biology in Cellular Pathology. Edited by John Crocker and Paul G. Murray
 ISBN: 0-470-84475-2

2.2 Experimental conditions

Tissues and Fixatives

ISH has been applied to almost any kind of cell or tissue preparation. Smears and cytocentrifuge preparations have been used as well as sections of snap frozen or formalin-fixed paraffin-embedded (FFPE) tissues. While DNA is quite stable, the biological half-life of most RNA species may be short. The half-life of some transcripts may be as short as 10 min (e.g. c-myc mRNA) or as long as many hours (e.g. immunoglobulin mRNA in plasma cells, Epstein–Barr viral EBER transcripts). Moreover, RNA is sensitive to digestion with ubiquitous RNases. Perfusion fixation is possible in animal experiments, but this is clearly not an option when using human tissues. Thus, the time interval between removal of the tissue and fixation should be kept short. Unfixed tissue samples should be snap frozen as quickly as possible and stored in liquid nitrogen, or at least at −80°C. Uninterrupted storage in liquid nitrogen is recommended if extraction of non-degraded cellular RNA for size determination by Northern blotting is intended. In most −80°C freezers, slight temperature fluctuations will result in mechanical disintegration of nucleic acid strands. However, this will not affect localisation of nucleic acids and detection of transcripts present in low copy numbers by ISH may still be possible after several years of storage. For routine processing, tissue samples should be transferred into formalin as soon as possible, although RNA ISH can be applied to post-mortem tissue collected up to 24 h after death (McNicol and Farquharson, 1996; Wilson *et al.*, 1997). All of this is related to the biological half-life of individual RNA species. Fixation is usually done overnight and should be complete before further processing. Poor fixation can result in a loss of signal, whereas over-fixation does not appear to affect RNA ISH when proteinase digestion of the tissues is extended appropriately (Franco *et al.*, 2001). If cytological preparations or frozen sections are used, then fixation with cross-linking fixatives, such as paraformaldehyde or neutral buffered formalin, is generally advocated (Franco *et al.*, 2001; McNicol and Farquharson, 1996; Niedobitek and Herbst, 1991). Precipitating fixatives, e.g. Carnoy's, Bouin's or Zenker's fixative, have been reported to result in poorer target retention (McNicol and Farquharson, 1996). However, in a recent study no advantage of formalin fixation over other fixatives was found (Tbakhi *et al.*, 1998). It has been estimated that the sensitivity of RNA ISH on FFPE material is reduced by 25% when compared with frozen sections (Frantz *et al.*, 2001). However, in our experience using suitably sensitive ISH methods detection of even low abundance transcripts is usually possible. In any case, it is advisable to confirm the preservation and accessibility of RNA with a suitable 'indicator' gene transcript by ISH (see below).

Prehybridisation and Hybridisation Conditions

In RNA ISH experiments, care must be taken to avoid RNase contamination. Baking of all glassware at 250°C can inactivate RNases. For the same purpose, all aqueous solutions (except Trishydroxy aminomethane [Tris]-containing solutions) should be treated with the cross-linking agent diethylpyrocarbonate (DEPC) and autoclaved.

To prevent loss of tissue sections or cytological specimens from slides during the lengthy ISH procedure, the use of adhesives is recommended. Several adhesives have been described in the literature but aminopropyltriethoxysilane (APES) appears to be the most efficient. This adhesive provides the glass with aminoalkyl groups that bind covalently to tissue sections (Rentrop *et al.*, 1986). Coating slides with APES is convenient and reproducible and can be used for DNA and RNA hybridisation as well as for immunohistological applications. Alternatively, suitably treated glass slides can be obtained commercially.

Before subjecting slides to *in situ* hybridisation a number of pretreatment steps are required, including treatment with hydrochloric acid and a protease (pronase or proteinase K) which are thought to remove proteins and make nucleic acids accessible to the probes. These are similar for DNA and RNA ISH with some modifications; DNA ISH protocols often include pretreatment with a detergent, e.g. Triton X-100, but this does not appear to be necessary for RNA ISH. Similarly, protease digestion may not always be strictly necessary for RNA ISH (Wilcox, 1993). In general, localisation of the nucleic acid to be detected will determine the pretreatment conditions: detection of target structures within the cytoplasm and karyoplasm will require less drastic conditions than detection of chromosomal DNA. On the one hand, protease digestion sufficient to permit hybridisation to cytoplasmic targets will leave chromosomal DNA coated by nucleoproteins and will thus prevent binding of the probe to the corresponding chromosomal locus. On the other hand, conditions suitable to allow hybridisation to chromosomal DNA will remove most of the cytoplasmic structures. Microwave irradiation as used for antigen retrieval in immunohistochemistry has been reported to improve the sensitivity of RNA ISH (Sibony *et al.*, 1995). In experiments using radiolabelled probes it is recommended to acetylate the slides with triethanolamine and acetic anhydride to reduce background signal due to non-specific interaction of probe with glass and tissue (Hayashi *et al.*, 1978). This does not appear to be necessary when using non-radioactive probes. A prehybridisation step, including all components of the hybridisation mixture except the labelled probe, has been recommended by some authors but has been eliminated from most published protocols. While the individual prehybridisation conditions have to be adapted by each individual laboratory, once established they can be applied in a fairly standardised manner to different tissues (Frantz *et al.*, 2001). Thus, the protocols published from our groups have been successfully employed for more than 10 years in different laboratories without significant modification (Herbst and Niedobitek, 2001; Niedobitek and Herbst, 2001).

In DNA ISH, probe and tissue DNA have to be denatured to form single strands in order to allow hybridisation to occur. This can be achieved by thermal, alkaline or acid treatment (Raap *et al.*, 1986). The most convenient method is to apply the probe to the tissue and than denature both simultaneously by heat treatment on heating blocks at temperatures between 90°C and 100°C for a few minutes (Brigati *et al.*, 1983; Niedobitek and Herbst, 1991). Heat treatment of probe and specimen is not required for RNA–RNA hybridisation. However, RNA may form secondary structures, and therefore moderate heat treatment of RNA probes is recommended before adding them to the hybridisation mixture. Denaturation of tissue DNA, however, should be avoided, as this may result in hybridisation of the probe to the gene rather than to its transcript.

Hybridisation and washing conditions vary depending on the stringency required. The melting temperature, T_m, is the temperature at which half of the double stranded molecules of a given DNA sequence are dissociated into single strands. This is a good indicator of the hybrid stability. High stringency hybridisation and washing conditions are close to T_m. The main factors affecting T_m are temperature, guanine/cytosine (GC) content, formamide and salt concentrations (Niedobitek and Herbst, 1991). Increased concentrations of sodium ions stabilise DNA–DNA hybrids by neutralising the negatively charged phosphate groups of DNA and by decreasing the solubility of the bases. The GC content contributes to hybrid stability because of the higher number of hydrogen bonds in GC base pairs as compared with AT base pairs. Formamide has the capacity to break up hydrogen bonds and therefore destabilises double stranded nucleic acids. Hybrid stability is also influenced by probe length. Thus, the impact of mismatches on hybrid stability is higher with short oligonucleotide probes than with longer cDNA or RNA probes. Temperatures close to the melting point prevent hybridisation of sequences with limited homology whereas lower temperatures will allow some cross-hybridisation between related but not identical sequences. The hybridisation mixture usually consists of formamide (up to 50% v/v), dextran sulphate (usually 10% w/v), 2 × SSC (0.3 M sodium chloride/0.03 M sodium citrate, pH 7.6), a carrier DNA (for DNA ISH) or tRNA (for RNA ISH), and the labelled probe. Dextran sulphate binds water and thus reduces the effective volume of the hybridisation mixture leading to enhanced hybridisation (Lawrence and Singer, 1985).

Liquid hybridisation kinetics are not easily transferable to ISH and therefore optimum hybridisation time and probe concentration have to be determined empirically. In general, the hybridisation is driven by the probe concentration (Lawrence and Singer, 1985). Therefore, increasing the probe concentration allows a reduction of hybridisation time although this may be at the expense of an increased background. It is generally held that hybridisation is complete after approximately 4 h (Wilcox, 1993). Nevertheless, it has been suggested that replacing the usual overnight hybridisation by hybridisation for approximately 40 h may improve sensitivity (Ky and Shughrue, 2002). After hybridisation, slides are washed in solutions containing formamide and standard saline citrate (SSC). Dithiotreithol (DTT) should be added to the hybridisation mixture and to all washing solutions to a final concentration of 10 mM when ^{35}S-labelled probes are employed. Usually, washing conditions are chosen closer to T_m than hybridisation conditions. In general, washing has to be more extensive for ISH with radioactive probes than for experiments with non-isotopic probes.

2.3 Probes and labels

Probes for ISH are labelled nucleic acid molecules with a complementary sequence to the target nucleic acid. Labelling usually is achieved by enzymatic incorporation of labelled nucleotides (see below). Alternatively, chemical labelling procedures have been described (Niedobitek and Herbst, 1991). The most recent development in this respect has been the introduction of the 'universal linkage system' (van Gijlswijk *et al.*, 2001). This method employs binding of a platinum complex to nucleic acids and allows labelling of nucleic acids with a variety of non-radioactive tags (van Gijlswijk *et al.*, 2001).

DNA targets usually are detected using double stranded DNA probes. These are commonly available as whole plasmid DNA which is labelled using nick translation or by random priming with commercially available kits. The use of inserts alone may reduce background staining since plasmid DNA may hybridise to bacterial contaminants (Ambinder *et al.*, 1986). However, this is more of a problem when studying extracted DNA by Southern blot hybridisation. The use of total plasmid DNA, on the other hand, may increase the signal intensity by formation of networks at the site of hybridisation. This is mediated by probe fragments consisting of both insert and plasmid sequences (junction pieces) (Lawrence and Singer, 1985). Since the plasmid sequences do not hybridise to the target, they remain available for hybridisation with other plasmid sequences, thus increasing the number of reporter molecules at the site of hybridisation. Because the effect produced by the network formation cannot be calculated, total plasmid DNA is not useful for quantitative ISH. In order to achieve hybridisation, probe DNA as well as target DNA have to be denatured, usually by heat treatment.

For the detection of RNA targets, the use of single stranded probes has become common practice. The most widely applied type is single stranded RNA probes (so-called riboprobes) which are generated by *in vitro* transcription from appropriate plasmids using DNA-dependent RNA polymerases. Alternatively, polymerase chain reaction (PCR) products can be obtained with primers incorporating appropriate promoter sequences at their 5′ ends thus obviating the need for laborious cloning procedures (Logell *et al.*, 1992; Wilson *et al.*, 1997). In one study using this approach, labelled probes were efficiently generated with

T3 and T7 but not SP6 RNA polymerases, while in another paper both T7 and SP6 RNA polymerases were reported to work well (Divjak *et al.*, 2002; Logel *et al.*, 1992). RNA probes usually are generated by including labelled uridine triphosphate (UTP) in the transcription reaction (Herbst and Niedobitek, 2001; Niedobitek and Herbst, 2001; Poulsom *et al.*, 1998). Protocols for labelling radioactive probes usually recommend using ^{35}S-UTP without addition of the unlabelled nucleotide (Herbst and Niedobitek, 2001; Niedobitek and Herbst, 2001; Wilcox, 1993; Wilson *et al.*, 1993). By contrast, commercial kits for the generation of digoxigenin-labelled probes contain an excess of unlabelled UTP in addition to the labelled compound (Pohle *et al.*, 1996). Riboprobes should be between 400 and 1000 bp in length although shorter as well as longer probes have been used (Niedobitek and Herbst, 1991; Poulsom *et al.*, 1998). In labelling reactions containing an unbalanced ratio of nucleotide triphosphates, such as in reactions with isotopically labelled nucleotides, the transcription reaction may be prematurely terminated, thus creating a bias of transcripts to sequences adjacent to the promoter region. This problem is exaggerated with longer inserts. When designing a probe for RNA ISH, the use of adenine/thymine (AT)-rich sequences is preferable since these will incorporate more labelled UTP and show a lower melting temperature than GC-rich sequences, thus improving the signal-to-noise ratio (Poulsom *et al.*, 1998). ISH with single stranded RNA probes has several theoretical advantages. In addition to producing 'antisense' RNA probes with a sequence complementary to the target RNA such constructs can be used to generate 'sense' RNA probes with a sequence identical to the RNA target. The latter should not hybridise to the target and therefore provides a good negative control (see below). RNA–RNA ISH results in highly thermostable RNA–RNA hybrids which are resistant to most ribonucleases. This allows post-hybridisation RNase digestion which may reduce background signal by removing non-specifically bound single stranded RNA sequences. However, it has been reported that RNase A digestion after hybridisation may also lead to a loss of signal (Wilcox, 1993; Yang *et al.*, 1999). To increase penetration into tissues and cells, it has been recommended that RNA probes should have an average length of between 50 and 150 bp (Cox *et al.*, 1984). This is achieved by controlled alkaline hydrolysis of labelled probes. However, the use of non-hydrolysed, longer probes has been reported to improve the signal-to-noise ratio (Poulsom *et al.*, 1998; Wilcox, 1993; Yang *et al.*, 1999).

One problem with RNA probes is their lack of stability due to their sensitivity to RNase digestion. Single stranded DNA probes have been used in an attempt to circumvent this problem. Such probes may be generated with M13 phage production systems. This, however, is tedious. Use of DNA oligonucleotide probes has, therefore, been proposed as an alternative. These are usually short (20–30 bp) and thus cover only a small proportion of the available target sequence resulting in a lower sensitivity. Use of oligonucleotide mixtures has therefore been advocated (Pringle *et al.*, 1990); however, this is relatively expensive. Oligonucleotide probes often have been used for whole-mount specimens to resolve relative levels of gene expression in embryonic tissues.

Single stranded DNA probes also have been generated by unilateral PCR incorporating labelled nucleotides. Analogous to the post-hybridisation removal of non-specifically bound single stranded RNA probes with RNase A, such probes may be digested with S1 nuclease (Kitazawa *et al.*, 1999). Recently, circularisable oligonucleotide probes, termed 'padlock probes', have been applied to ISH (Baner *et al.*, 1998; Lizardi *et al.*, 1998; Nilsson *et al.*, 1994). These probes consist of two target-specific sequences at the 5′ and 3′ ends separated by an irrelevant sequence. The two target-specific sequences either hybridise next to each other on the target or are separated by a small gap which is filled by hybridisation with an appropriate short oligonucleotide (Lizardi *et al.*, 1998; Nilsson *et al.*, 1994). Following proper hybridisation, the hybridised sequences are joined by enzymatic ligation leaving a circular molecule catenated to the target (Baner *et al.*, 1998; Lizardi *et al.*, 1998; Nilsson *et al.*, 1994). The applicability of this probe type for ISH has not been fully investigated.

A crucial problem faced by anyone using DNA or RNA ISH is the choice of label. Fundamentally, this is a question of whether to use radioactively or non-radioactively labelled probes. This has been subject to a debate conducted with an almost religious intensity. Nevertheless, there are rational arguments for and against the use of either probe type. Radioactively labelled probes have been used for more than 30 years. ^{32}P, ^{35}S, ^{3}H and, more recently, also ^{33}P are the most frequently used labels. Of these, ^{32}P has the highest energy and the shortest half-life, thus resulting in the shortest exposure time. However, owing to the high energy of ^{32}P, resolution is also relatively poor, which may be a distinct disadvantage in ISH where localisation of the signal is important. In whole-mount specimens, where resolution at the single cell level is not required, use of ^{32}P may be adequate. Moreover, labelled probes can only be stored for a short period of time. At the other end, ^{3}H-labelled probes are characterised by low energy and a long half-life, resulting in relatively good resolution and probe stability. On the other hand, ^{3}H-labelled probes require lengthy exposure times, measured in months rather than days or weeks, and thus these probes may unbearably test the patience of the investigator. ^{35}S provides a compromise with reasonable tissue resolution and exposure times and therefore is now the most widely used radionuclide in ISH (Franco *et al.*, 2001; Niedobitek and Herbst, 1991; Poulsom *et al.*, 1998; Wilcox, 1993; Wilson *et al.*, 1997). ^{35}S has a half-life of about 87 days and thus probes labelled with this nuclide can be used for up to 4 weeks, although it is recommended to check if degradation has occurred after longer storage (Poulsom *et al.*, 1998). Probes should be stored at $-20°$C in aliquots to prevent repeated freezing and thawing and DTT should be added to the probe to inhibit oxidation. ^{33}P may represent another useful nuclide with respect to resolution and exposure time; however, this has not been widely used in ISH and, in our limited experience, has no significant advantage over ^{35}S (Poulsom *et al.*, 1998). The detection of radioactively labelled probes is straightforward and is achieved by dipping slides in a photographic emulsion. Use of radiolabelled probes offers the possibility of a quantitative assessment of the signal. This can be achieved by counting grains over labelled cells. Alternatively, imaging techniques using a charge-coupled device camera have been used for this purpose (Laniece *et al.*, 1998). This approach has been used to examine the expression of chemokines in thyroid glands from patients with Grave's disease (Romagnani *et al.*, 2002). Moreover, sections hybridised to ^{33}P can rapidly be evaluated using a phosphoimager (Frantz *et al.*, 2001). Finally, use of radiolabelled probes may be advantageous for double-labelling techniques (see below).

Handling of radioactive nuclides requires special facilities which may not be easily available. Together with the often lengthy exposure time and the relatively poor resolution, this has stimulated research into the application of non-radioactive probes for ISH. These efforts have led to the description of various non-radioactive tags. The first of these was biotin, which offers easy detection using well-established immunoenzymatic techniques (Brigati *et al.*, 1983). However, biotin is endogenously present in many tissues, e.g. liver, and this may lead to false positive results (Niedobitek and Herbst, 1991; Niedobitek *et al.*, 1989a; Pringle *et al.*, 1990). Nevertheless, biotin is still widely used. To overcome the disadvantages of biotin, a number of other non-radioactive reporter molecules have been employed, e.g. digoxigenin, bromodeoxyuridine, and fluorochromes. Of these, digoxigenin has proved the most useful and versatile. Detection of these tags is usually achieved using specific antibodies and immunoenzymatic or immunofluorescent techniques.

The comparative sensitivity of ISH with radioactive vs. non-radioactive probes is still a matter of controversy. Detection of single copy viral genomes in well-characterised cell lines has been reported using both radioactive and non-radioactive probes (Niedobitek *et al.*, 1991; Zehbe *et al.*, 1997). In tissue sections, sensitivity will inevitably be lower because only parts of any one cell will be present. With regards to RNA ISH, an assessment of the relative sensitivities of different ISH methods is more difficult since the exact copy number of a gene transcript in an individual cell usually is not known. Nevertheless, detection of abundant RNA transcripts can be achieved by non-radioactive methods, while for low copy transcripts, in our hands, use of radioactively labelled probes is required. In spite of technical advances in non-radioactive ISH in recent years, we and others believe that radiolabelled probes are still superior to non-radioactive probes with respect to sensitivity and reproducibility (Franco *et al.*, 2001; Niedobitek and Herbst, 1991; Wilcox, 1993).

2.4 Controls and pitfalls

It is important to bear in mind that not all labelling obtained by ISH with radioactive or non-radioactive probes is a specific signal. In fact, the scientific literature is riddled with false positive ISH results. As has been pointed out by Wilcox (1993), the most important control is to employ common sense when interpreting ISH slides. For example, DNA ISH generally should result in a nuclear labelling, while RNA ISH targeting an mRNA should produce a predominantly cytoplasmic reaction product. On the other hand, certain RNA transcripts are known to fulfil their function in the nuclei and thus ISH, e.g. for the detection of the telomerase RNA component, should result in nuclear staining. All unexpected labelling patterns must be interpreted with extreme caution and ideally should be confirmed by an independent technique. In addition to applying common sense, it is important to carry out control experiments to ascertain the specificity of ISH results. In DNA ISH, this can be achieved by hybridising control sections to an unrelated probe. Use of total plasmid DNA may give unspecific results due to hybridisation to bacterial DNA (Ambinder *et al.*, 1986). Although mainly this is a problem in filter hybridisation experiments, it is advisable to use labelled plasmid DNA without the specific insert as a control. In virus research, hybridisation to unrelated viral DNA provides an additional negative control. Similarly, when doing RNA ISH, control sections should be hybridised to appropriate negative control probes. The most widely used are probes transcribed in a sense direction from the same plasmid used for generating the specific antisense-strand probes. However, there is a possibility that sense RNA probes may hybridise to regulatory antisense transcripts (Lipman, 1997). Moreover, in virus research it has been reported that sense probes may generate ISH signals by hybridisation to replicating viral DNA (Niedobitek *et al.*, 1991). Therefore, it may be advisable to hybridise control sections to a probe for a gene known not to be expressed in a particular tissue or cell type. Similar considerations also apply to the use of oligonucleotide probes. Additional specificity controls may be carried out such as to test the sensitivity of the signal to RNase digestion. Cellular RNA will be easily digested by this enzyme. When using RNA probes, however, this enzyme should be substituted by micrococcal nuclease (Williamson, 1998), since residual RNase may attack the probe RNA thus leading to false negative results. Furthermore, it may be helpful to correlate ISH with immunohistological results if suitable antibodies are available. Size determination by Northern blot analysis of the transcript to be detected by ISH should always be carried out if heterologous nucleic acid probes are used. When used as a negative control probe, the sense RNA must not produce a signal in Northern blots. Probing for different regions of a particular transcript in different reactions with non-overlapping probes should result in identical expression patterns before accepting unexpected ISH results.

To avoid false negative results, a number of positive control experiments should be performed. Hybridisation of labelled probes to tissues known to contain the nucleic acid of interest ensures that the probe is labelled properly and that the hybridisation conditions are appropriate. Particularly when doing RNA ISH, it is important to rule out RNA degradation as a cause of a negative result. Hybridisation to an appropriately labelled oligo-d(T) probe may indicate the presence of polyadenylated mRNA. The detection of highly expressed RNA species, e.g. immunoglobulin light chain mRNAs in plasma cells, may allow a rapid assessment of RNA conservation. However, this may not be informative with regard to the detection of low abundance transcripts. Hybridising control tissue sections to probes specific for 'house keeping' genes has been recommended. Probes detecting β-actin are used frequently (Poulsom *et al.*, 1998). Transcripts for elongation factor (EF)-1α, a universal co-factor for protein synthesis in all living cells, has also been recommended as a suitable target to control RNA accessibility (Grubar and Levine, 1997).

An argument of authors favouring non-radioactive ISH techniques is the high background attributed to radiolabelled probes (Seyda *et al.*, 1988; Syrjanen *et al.*, 1988). While acetylation of slides has been recommended to reduce unspecific binding of probes (Hayashi *et al.*, 1978), probe concentration and exposure time are the main factors affecting background in radioactive ISH. If high background occurs with radioactive probes, we recommend reducing the amount of radioactive probe added per slide while increasing the exposure time. In our hands, the detection of low copy RNA transcripts in paraffin sections may require up to 40 days of exposure time when using ^{35}S-labelled probes (Meru *et al.*, 2002). Overexposure of slides may lead to an increased radioactive background. However, this problem is also encountered in immunoenzymatic detection of non-radioactive probes as overdevelopment of enzymes may also increase background staining. For non-radioactive techniques microscopic control of the enzymatic reaction is suggested to prevent overstaining. Since this is not suitable for radioactive ISH, it is recommended to expose at least three sets of slides from every experiment and develop them at different exposure times.

Grain formation in nuclear track emulsions may be induced by factors other than radioactive decay. Light, heat, electrostatic discharges, inappropriately fast drying of the emulsion, and mechanical factors such as scratches, pressure, or an uneven surface of the section may lead to a diffuse or focal aggregation of silver grains. Also chemical factors in the tissues or solutions may have a negative or positive influence on the formation of silver grains (positive or negative chemography). Occasionally, macrophages may contain foreign material that can induce positive chemography. Fading of latent images before development can be induced by heat or humidity (Rogers, 1979).

Many non-radioactive ISH protocols employ biotin either as a probe tag or in the methodology used for probe detection. It is well known that many tissues, e.g. liver and kidney, contain endogenous biotin, mainly as a prosthetic group of various enzymes (Niedobitek and Herbst, 1991). This has been recognised as a cause of background staining in immunohistology and methods for blocking endogenous biotin employing incubation of sections with avidin and subsequent saturation of free biotin-binding sites have been described (Wood and Warnke, 1981). Endogenous biotin may also give rise to non-specific staining in ISH experiments and this cannot be reliably blocked (McNicol and Farquharson, 1996; Niedobitek *et al.*, 1989; Pringle *et al.*, 1990). Thus, ISH procedures employing biotinylated probes or detection reagents have to be particularly carefully controlled.

The presence of eosinophilic granulocytes may also give rise to non-specific labelling in ISH (Niedobitek *et al.*, 1989; Patterson *et al.*, 1989). This probably is caused by the major basic protein of eosinophilic granulocytes which can precipitate nucleic acids (Gleich *et al.*, 1974). In haematoxylin and eosin (H&E) stained tissue sections this artefact is easily identified by the intense cytoplasmic staining of eosinophilic granulocytes and rare eosinophils usually do not interfere with the evaluation of ISH. However, in tissues infiltrated with numerous eosinophilic granulocytes, e.g. Hodgkin lymphoma or bone marrow sections, it can result in massive background labelling making evaluation of a specific signal impossible. This background usually is stronger with DNA probes than with RNA probes, possibly because RNase digestion may remove some of the labelled probe bound to eosinophils. On the other hand, the presence of a few eosinophils can serve as an internal control for some parameters of the ISH protocol, e.g. the detection system. This can be useful, particularly for radioactive ISH, since an accumulation of grains over eosinophils excludes that factors such as negative chemography or latent image fading have influenced ISH results. Another cell type prone to non-specific probe binding is the Paneth cell (Garrett *et al.*, 1992). Finally, lipofuscin has been reported to be a cause of misinterpretation of ISH results, particularly in neuronal tissues (Steiner *et al.*, 1989).

2.5 Double-labelling

The combination of ISH with immunohistochemistry allows the simultaneous detection of nucleic acids and proteins. This approach has been successfully used in a variety of settings. ISH for the detection of DNA requires heat denaturation of tissue and probe DNA which may destroy antigens. Similarly, the prehybridisation treatment and the formamide used in many ISH protocols (DNA or RNA) may interfere with subsequent antigen detection. Also, dextran sulphate, which is a constituent of hybridisation mixtures in most published protocols, can bind to proteins and impair their antigenic properties. Therefore, it is usually recommended to perform immunohistology before ISH (Herbst *et al.*, 1992; Niedobitek *et al.*, 1997a; Roberts *et al.*, 1989; Van der Loos *et al.*, 1989). However, methods performing ISH prior to immunohistology have been described for some stable antigens (Wolber and Lloyd, 1988). When immunohistochemistry is conducted prior to ISH, RNase inhibitors must be added to antibodies and other solutions to prevent RNA degradation. Heparin, placental RNase inhibitor and yeast tRNA are useful to block RNase (Höfler *et al.*, 1987). Also, low concentrations of diethylpyrocarbonate (DEPC) may be used, but high DEPC concentrations can also degrade immunoglobulins and antigens. However, in spite of all precautions, RNA degradation will occur to some extent during immunohistology. Therefore, control slides hybridised to the probe without previous immunohistology should be included to assess RNA loss and avoid erroneous evaluation.

Immunohistochemistry can be readily combined with ISH employing either radioactive or non-radioactive probes and various immunohistochemical detection systems. Use of radioactive probes is advantageous in some settings. The detection of radiolabelled probe requires only dipping of slides into a nuclear track emulsion. By eliminating the immunohistochemical detection reagents necessary for immunoenzymatic probe detection, there is no risk of cross-reactivity of any reagents. However, in some situations it may be more appropriate to use non-radioactive instead of radioactive ISH. The tissue resolution usually is better when enzymatic procedures are used for the demonstration of bound probe. Also, overexposure of radiolabelled probes may result in an intense signal leading to saturation of the emulsion which may hide the underlying immunohistochemical staining product.

Immunohistochemistry on frozen or paraffin sections as well as on cytological preparations is performed as required for the respective antibody. After completion of immunohistology, including enzymatic development of the reaction, a proteolytic treatment may be required to unmask the target nucleic acid. Treatment of sections with a 3M KCl solution in addition to the proteolytic digestion has proved helpful in our hands (Niedobitek and Herbst, 1991). Some attention has to be given to the choice of enzymes and chromogens. If both immunohistology and ISH are developed by enzymatic procedures, then chromogens have to be chosen to give good contrast. Diaminobenzidine (DAB) for peroxidase and new fuchsin or Fast Red for alkaline phosphatase result in stable precipitates still clearly visible even after photographic development of radioactive ISH. Nitro blue tetrazolium (NBT) results in a dark blue to brown precipitate which contrasts well with the red colour of new fuchsin or Fast Red. It is important to bear in mind that some of these chromogens are not alcohol resistant. This may make modifications of the ISH protocol necessary.

When setting up double staining experiments using two immunoenzymatic detection systems, it is important to chose reagents so as to avoid cross reactivities. Using DAB as the chromogen for the first step is advantageous since the reaction product will mask all underlying antibody reagents thus reducing the risk of cross reactivity. Double staining using two alkaline phosphatase-based detection systems is possible but must be carefully controlled to exclude that the alkaline phosphatase used to detect the first reaction is still reactive in the second step. In our experience this is never entirely possible. Of course, detection systems employing two different fluorochromes may also be employed (Zaidi *et al.*,

2000). Double-labelling techniques can also be applied to the simultaneous detection of two different nucleic acids. In this setting, usually one radioactively labelled probe is combined with one non-radioactive probe. Both probes are hybridised simultaneously. The non-radioactive probe is detected by immunoenzymatic techniques including colour development followed by dipping into nuclear track emulsion. Finally, it is possible to perform double-labelling ISH using probes labelled with ^{3}H and ^{35}S or ^{33}P (Salin *et al.*, 2000). However, discrimination of the signals generated by these probes requires sophisticated technical equipment which is not widely available (Salin *et al.*, 2000).

2.6 Increasing the sensitivity of ISH

Attempts have been made in recent years to increase the sensitivity of nucleic acid detection *in situ*. Basically two different strategies have been employed: amplification of target and amplification of signal.

Amplification of target, RNA or DNA, by *in situ* PCR, either with direct incorporation of labelled nucleotides or followed by detection using labelled oligonucleotide probes, has been reviewed elsewhere (Hawkins and Dodd, 2000; Long, 1998; Nuovo, 2001). Alternatively, Höfler *et al.* (1995) described a method for the *in situ* detection of measles virus RNA employing reverse transcriptase with primers including a T7 RNA polymerase site to generate a cDNA intermediate followed by T7 RNA polymerase-directed generation of RNA transcripts *in situ* and detection using ^{35}S-labelled oligonucleotide probes. A related technique, primed *in situ* labelling (PRINS), is based on the hybridisation of oligonucleotide probes followed by extension using Taq polymerase in the presence of labelled nucleotides (Coullin *et al.*, 2002; Wilkens *et al.*, 1997).

Signal amplification has been another focus of research and these approaches will be discussed in more detail. When using radioactive probes, using more than one labelled nucleotide, e.g. ^{35}S-ATP in addition to ^{35}S-UTP, in the transcription reaction may improve sensitivity (Franco *et al.*, 2001; Ky and Shughrue, 2002; Sedlaczek *et al.*, 2001), although others have not found this to be beneficial (Poulsom *et al.*, 1998). Increasing the photographic development time may also lead to a more intense signal. However, this also increases the background and a better signal-to-noise ratio is achieved by increasing the exposure time (Franco *et al.*, 2001).

With regards to non-radioactive probes, application of multiple layers of detection systems (e.g. mouse anti-digoxigenin antibody followed by biotinylated horse anti-mouse reagent and an enzyme-labelled avidin–biotin complex) may improve the signal intensity (Niedobitek *et al.*, 1989b; Speel *et al.*, 1998). However, such systems are also prone to an increased background (Speel *et al.*, 1998). The most significant and practically useful advance in this field has been the description of tyramide signal amplification (TSA) systems. Initially developed for membrane immunoassays (Bobrow *et al.*, 1989, 1991), this method has been successfully applied to immunohistochemistry and ISH (Adams, 1992; Aigner *et al.*, 1999; Herbst *et al.*, 1998; Niedobitek *et al.*, 1997; Speel *et al.*, 1998; Zaidi *et al.*, 2000; Zehbe *et al.*, 1997). TSA is based on the horseradish peroxidase-catalysed deposition of hapten-labelled tyramide molecules at the site of antibody or probe binding. In the case of ISH, peroxidase usually is brought to the site of hybridisation by employing a hapten-labelled probe, followed by binding of a peroxidase-labelled hapten-specific antibody (Speel *et al.*, 1998). Since any of the reagents employed for probe detection may contribute to background labelling, use of probes directly labelled with peroxidase followed by TSA has been advocated (van de Corput *et al.*, 1998). However, such probes are not widely available yet. Use of biotin-labelled probes and biotinylated tyramide can result in excessive background staining due to the presence of endogenous biotin in many tissues (Niedobitek and Herbst, 1991; Niedobitek *et al.*, 1989a; Speel *et al.*, 1998). This can be avoided by using tyramides linked to other markers e.g. digoxigenin, fluorochromes, or dinitrophenol (DNP) (Kolquist *et al.*, 1998; Lewis *et al.*, 2001; Schmidt *et al.*, 1997; Speel *et al.*, 1998; Zaidi *et al.*, 2000). Application of TSA systems in ISH results in an increased sensitivity with an estimated amplification factor of 5–25 (Schmidt *et al.*, 1997; Speel *et al.*, 1998). Accordingly, probe concentration may be reduced significantly (Yang *et al.*, 1999). Using TSA, detection of rare RNA transcripts by ISH has been reported, for example including those encoding for the reverse transcriptase component of human telomerase (Kolquist *et al.*, 1998), insulin (Speel *et al.*, 1998), and collagens (Aigner *et al.*, 1999). In the latter study, two rounds of TSA using biotinylated tyramide have been employed (Aigner *et al.*, 1999). At the DNA level, detection of single copies of human papillomavirus (HPV) genomes by ISH has been reported with biotin-labelled DNA probes using TSA followed by streptavidin nanogold and autometallography with silver acetate (Zehbe *et al.*, 1997). A commercially available detection system using a secondary antibody linked to a dextran polymer coupled to numerous enzyme molecules has been reported to achieve almost the same sensitivity as TSA detection reagents in HPV-specific ISH (Wiedorn *et al.*, 2001). Another method of potential use for ISH has been designated rolling circle amplification (RCA). This method is based on the observation that circularised oligonucleotides can support a replication reaction analogous to the replication of certain viral genomes. Application of this technique to RNA ISH has been reported by Zhou *et al.* (2001), who have used digoxigenin-labelled probes specific for β-actin. Bound probe was detected using an anti-digoxigenin antibody covalently linked to an RCA primer. Alternatively, RCA can be supported by primers which, in addition to a target-specific sequence, contain an RCA primer sequence (Lizardi *et al.*, 1998; Zhong *et al.*, 2001). A circular oligonucleotide is then hybridised to the RCA primer, and RCA is initiated by DNA polymerase generating a large DNA molecule which remains attached to the site of hybridisation (Baner *et al.*, 1998; Zhong *et al.*, 2001; Zhou *et al.*, 2001). This is detected either by direct incorporation of labelled nucleotides or by hybridisation to a labelled probe (Lizardi *et al.*, 1998; Zhong *et al.*, 2001; Zhou *et al.*, 2001). The potential usefulness of this approach for increasing the sensitivity of ISH requires further investigation.

2.7 What we do in our laboratories

The following is a guide to ISH procedures as used in our laboratories. Our detailed protocols have been published elsewhere (Herbst and Niedobitek, 2001; Niedobitek and Herbst, 2001). Our protocols are fairly standardised and are equally applicable to sections from formalin-fixed, paraffin-embedded tissue blocks, frozen sections and cytological preparations. The only variable is the concentration of protease required (see below).

The detection of RNA transcripts in frozen sections may be more sensitive due to better RNA preservation (Frantz *et al.*, 2001). However, this advantage is offset by the poorer morphology making identification of labelled cells more difficult. Thus, whenever possible, we prefer using paraffin sections even for the detection of low copy RNA species (Meru *et al.*, 2002). Tissue sections are mounted on aminopropyltriethoxysilane (APES)-coated slides to prevent loss of sections during the lengthy hybridisation procedure. Preparation of sections necessary to make nucleic acids, DNA or RNA, accessible for ISH includes treatment with hydrochloric acid and a protease, usually pronase. The protease concentration required is lower for frozen sections and cytological specimens than for paraffin sections, but optimum conditions have to be established in each individual laboratory.

Occasionally, ISH is employed to detect viral DNA genomes in tissue sections [Figure 2.1(a)] (Herrmann *et al.*, 2002; Niedobitek *et al.*, 1989c; Reiss *et al.*, 2002). For this purpose, we use plasmids carrying appropriate virus-specific inserts. Total plasmid DNA is labelled by nick translation using either ^{35}S- or digoxigenin-dCTP and commercially available kits. In situations where detection of lytic virus replication is sufficient, we use digoxigenin-labelled probes. Hybridised probes are detected using standard immunohistochemical procedures. In experiments aimed at the detection of latent viral genomes, e.g. persistent Epstein–Barr virus (EBV) infection, the use of ^{35}S-labelled probes is preferred because in our hands this methods offers higher sensitivity [Figure 2.1(a)].

RNA ISH generally is done using ^{35}S-labelled riboprobes in our laboratories. This approach has proved useful in a variety of settings, e.g. for the detection of various cytokines [Figure 2.1(b)] (Beck *et al.*, 2001; Herbst *et al.*, 1996, 1997), recombination activating genes (Meru *et al.*, 2001b, 2002), or collagens (Pohle *et al.*, 1996). For transcripts expressed at high levels, non-radioactive probes may suffice (Pohle *et al.*, 1996) but we recommend using radiolabelled probes when commencing ISH experiments for the detection of a new gene. Only when a distinct labelling pattern has been established and a signal is detected after only a few days of exposure, are attempts to use non-radioactive probes likely to be fruitful.

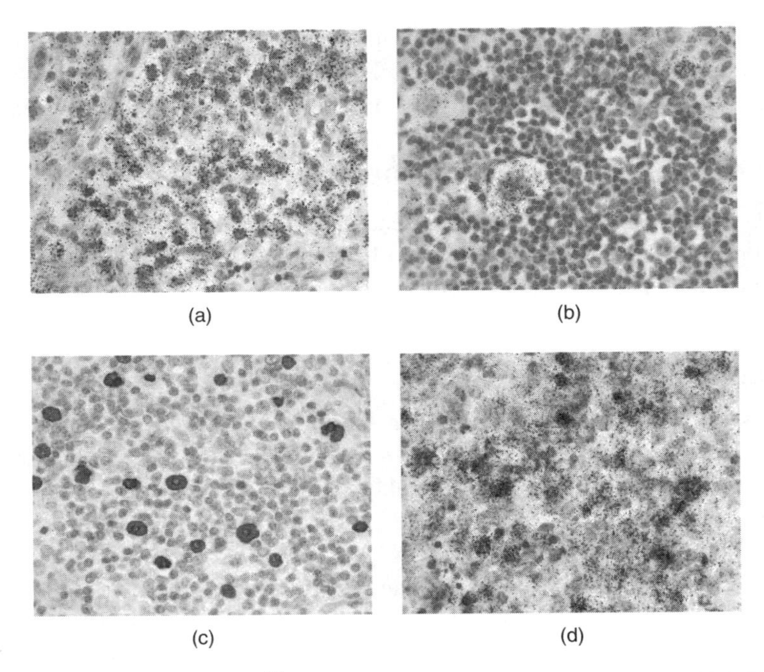

Figure 2.1 (a) DNA ISH using ^{35}S-labelled probes reveals the presence of EBV DNA in tumour cells of a nasopharyngeal carcinoma (black grains). (b) RNA ISH with ^{35}S-labelled riboprobes demonstrates expression of interleukin-10 in Hodgkin and Reed–Sternberg cells of a Hodgkin lymphoma (black grains). (c) Expression of the EBERs in Hodgkin and Reed–Sternberg cells of a Hodgkin lymphoma is visualised using digoxigenin-labelled RNA probes (red nuclear staining). (d) Double-labelling immunohistochemistry and ISH reveals expression of CD30 (red staining) in a proportion of EBV-positive cells (black grains) in infectious mononucleosis. A colour version of this figure appears in the colour plate section

Exceptions to this rule are certain viral transcripts which are expressed at high copy numbers. A major focus of our laboratories is the study of EBV infection. Initially, we and others used DNA ISH for the detection of the virus (Anagnostopoulos *et al.*, 1989; Niedobitek *et al.*, 1989c). This method has been replaced by ISH for the detection of the small EBV-encoded RNAs (EBERs). The EBERs are transcribed at high copy number and therefore can be detected using non-radioactive riboprobes or commercially available oligonucleotide probes. In our laboratories, detection of the EBERs is achieved using digoxigenin-labelled RNA probes in the setting of malignant tumours and reactive conditions [Figure 2.1(c)] (Beck *et al.*, 2001; Niedobitek *et al.*, 1992; Spieker *et al.*, 2000). However, we feel that for the detection of latently infected B-cells, which may be very rare, radioactive ISH with ^{35}S-labelled riboprobes is more robust [Figure 2.1(d)] (Meru *et al.*, 2001a; Niedobitek *et al.*, 1992).

As discussed, ISH can be used in double-labelling experiments combined with immunohistochemistry or applying more than one probe. We have employed this technique to determine the phenotype of EBV-infected cells by combining

EBER-specific ISH with ^{35}S-labelled or digoxigenin probes with the immunohistochemical detection of lineage-specific antigens [Figure 2.1(d)] (Niedobitek *et al.*, 1992, 1997). In this scenario, immunohistochemistry is carried out first using RNase-free conditions followed by ISH. ISH can also be performed simultaneously applying different probes. This approach has been used for the detection of cytokine or light chain transcripts in EBV-infected cells (Beck *et al.*, 2001; Herbst *et al.*, 1992, 1997; Spieker *et al.*, 2000). For such experiments, digoxigenin-labelled EBER-specific probes and ^{35}S-labelled cytokine- or light chain-specific probes are simultaneously applied to sections. Following hybridisation and washing, the non-radioactive probe is detected first using standard immunohistochemical procedures followed by dipping of slides into a nuclear track emulsion. Finally, triple-labelling experiments have been carried out using immunohistochemistry for the detection of lineage-specific antigens followed by double ISH with digoxigenin-labelled EBER-specific probes and ^{35}S-labelled light chain-specific probes (Spieker *et al.*, 2000).

2.8 Applications of ISH: examples

In contrast to molecular biological techniques based on the analysis of nucleic acid extracts from tissues or cells, e.g. PCR, Northern or Southern blot hybridisation, ISH allows the detection of nucleic acids in their morphological context. In the following paragraphs; selected examples of the application of ISH to the detection of DNA and RNA will be used to illustrate the power of this technique. No attempt is made to provide a comprehensive overview of the available literature, which would be beyond the scope of this chapter.

RNA ISH permits the analysis of numerous cells present in one section for expression of the same gene while single cell RT-PCR is suited for the analysis of multiple genes in one (or a limited number) of cells (Todd and Margolin, 2002). For phenotypic analysis in a diagnostic setting, immunohistochemistry clearly is the method of choice. Antibodies suitable for staining of paraffin sections are available against a vast array of relevant antigens, and the techniques required to obtain reproducible immunohistochemical staining results are widely available. Nevertheless, RNA ISH has maintained its role as a powerful tool in various research applications.

First, the role of RNA ISH in the validation of immunohistochemical reagents must not be underestimated. We have recently used ISH to localise transcripts of the w, which are involved in antigen receptor gene rearrangements during lymphocyte development. We found expression of these genes in a small subset of tonsillar lymphocytes localised at the interface between lymphoreticular tissue and stroma and adjacent to, but not within, germinal centres (Meru *et al.*, 2002). These results were unexpected in view of several other papers reporting conflicting immunohistochemical staining patterns using RAG-'specific' antibodies but agreed well with the expression pattern of terminal transferase, another protein expressed in developing lymphocytes (Meru *et al.*, 2002). Thus, if unexpected or novel results are obtained with immunohistochemistry, it appears a good idea to use RNA ISH for validation (and vice versa!).

In spite of the development of a large and still growing number of highly specific reagents, immunohistology may fail to provide information about the cellular source of a protein. This is a particular problem with secreted proteins. Thus, RNA ISH has been successfully employed to study the cellular origin of different extracellular matrix components and spatial and temporal expression patterns of collagen RNA transcripts, e.g. in liver fibrosis (Milani *et al.*, 1995). Similarly, the identification of cytokine- or chemokine-producing cells may be difficult because these polypeptides may be secreted by one cell and bound by receptors on another cell, leading to false negative or false positive results, respectively. RNA ISH has proved useful in this context. Thus, Hodgkin and Reed–Sternberg (HRS) cells were demonstrated to produce a number of cytokines, e.g. IL-6 and IL-10, using this technique (Beck *et al.*, 2001; Herbst *et al.*, 1996, 1997). Similarly, expression of chemokines has been analysed by RNA ISH in tumours as well as in autoimmune disease (Romagnani *et al.*, 2002; Teruya-Feldstein *et al.*, 1999).

The intracellular demonstration of secreted proteins may not necessarily reflect protein synthesis at this location, but may be the result of active uptake due to membrane receptors or phagocytosis, or of passive influx as a supravital or fixation artefact. Such issues can be resolved by RNA ISH because RNA is unlikely to be subject to diffusion processes from one cell to another. For example, immunoglobulins of polyclonal origin may occasionally be detected by immunohistology in normal squamous epithelial cells or in giant cells of various histogenetic origins, such as anaplastic carcinomas, glioblastomas, sarcomas, or large cell lymphomas. Such reactivities can be observed in paraffin sections stained for immunoglobulins, but not in corresponding frozen sections, strongly suggesting an influx of serum proteins, probably as the result of poor fixation. The problem is more complex in the case of cells expressing certain types of receptors for constant domains of Ig (Fc-receptors), such as macrophages or B-lymphocytes. ISH for the detection of immunoglobulin gene transcripts, however, may not only be valuable in such situations, but within limits also allows analysis of the clonal composition, isotype usage and class switching at the single cell level.

Recent progress in the cloning of the human genome has opened a new field of application of RNA ISH (Frantz *et al.*, 2001). Available databases contain sequences of roughly 30 000 mRNAs. Most of these are not characterised, their functions are unknown, and there are no antibodies available for immunohistological detection of the corresponding proteins. Using RNA ISH, signals detected, for example in Northern blots, can be attributed to specific cell types. Moreover, the identification of spatial and temporal expression patterns of newly identified genes may provide clues as to their function. In conjunction with tissue microarrays, RNA ISH allows the rapid generation of expression data for new and potentially interesting genes (Bubendorf *et al.*, 2001; Frantz *et al.*, 2001). RNA ISH has also been used for rapid confirmation of data on differential gene expression, e.g. in tumour-derived vs. normal endothelial cells (St Croix *et al.*, 2000). In these situations, preparation of gene-specific probes suitable for RNA ISH is quickly achieved while generation of specific antibodies suitable for immunohistology would require months of work with uncertain outcome.

Finally, RNA ISH can be useful for the detection of RNA transcripts not translated into proteins, e.g. U6 RNAs. This has proved a useful approach to the detection of certain infectious agents. Thus, ISH for the detection of the small EBERs has become the method of choice for the detection of latent EBV infection (see below). Moreover, detection of ribosomal RNA species by ISH can be used for the identification of other microorganisms, such as mycobacteria and fungi (Boye *et al.*, 2001; Hayden *et al.*, 2002).

ISH has been used for the detection of specific DNA sequences in many different settings (for a review, see Niedobitek and Herbst, 1991). In recent years, two main areas of application of DNA ISH have emerged.

Detection of Infectious Agents

One major field is the detection of infectious agents. DNA ISH has been successfully employed for the demonstration of several bacteria, e.g. *Helicobacter pylori*, *Chlamydia trachomatis*, *Haemophilus influenzae* and *Mycoplasma pneumoniae* (Horn *et al.*, 1988; Saglie *et al.*, 1988; Terpstra *et al.*, 1987; van den Berg *et al.*, 1989). More recent developments in this field include the use of DNA probes for the detection of ribosomal RNA sequences specific for bacterial or fungal organisms (Boye *et al.*, 2001; Hayden *et al.*, 2002). Furthermore, DNA ISH has been used to study acute viral infections, e.g. in immunocompromised individuals, and to investigate possible associations of DNA tumour viruses with human malignancies. A major application remains the detection of HPV in anogenital neoplasia. In addition to identifying virus types associated with low or high risk of progression, it has been demonstrated that the pattern produced by HPV DNA ISH using non-radioactive probes may be of relevance. Thus, a diffuse labelling of nuclei has been reported to indicate the presence of episomal viral DNA, while a punctate pattern is suggestive of virus integration into the host genome (Cooper *et al.*, 1991). This may be of relevance, since it has been reported that a punctate pattern may indicate a poor prognosis (Gomez Aguado *et al.*, 1996).

DNA ISH has also been used for the detection of EBV DNA in latent and lytic infection (Herrmann *et al.*, in press; Niedobitek *et al.*, 1989c). For the demonstration of latent EBV infection, this method has largely been replaced by ISH for the detection of the EBERs. These are expressed at very high copy numbers in all established forms of latent EBV infection (Niedobitek *et al.*, 2001).

Therefore, they represent an ideal target for ISH using radiolabelled or non-radioactive probes (Niedobitek *et al.*, 2001; Wu *et al.*, 1990). Recently, the possibility of the existence of an EBER-negative form of EBV latency in hepatocellular and breast carcinomas has been raised (Bonnet *et al.*, 1999; Sugawara *et al.*, 1999, 2000). This, however, has not been confirmed by others (Chu *et al.*, 2001a, 2001b; Herrmann and Niedobitek, 2003; Junying *et al.*, in press). Thus, at present it would appear that EBER-specific ISH is a reliable method for the detection of latent EBV infection.

Interphase Cytogenetics

The second major field of application of DNA ISH with implications for diagnostic histopathology is interphase cytogenetics (Wolfe and Herrington, 1997). This term refers to the detection of chromosomal abnormalities in interphase nuclei and, in contrast to classical cytogenetics, does not require viable cells for the generation of metaphase spreads. Thus, interphase cytogenetics is applicable to conventional cytological or histological preparations, including paraffin sections (Wolfe and Herrington, 1997). Several groups have used interphase cytogenetics for the detection of chromosomal abnormalities in haematological tumours. Mantle cell lymphomas are characterised by t(11;14) juxtaposing the *CCND1* gene to the immunoglobulin heavy chain locus. This results in overexpression of the cyclin D1 protein. Immunohistochemical detection of this protein, however, is temperamental. By contrast, interphase cytogenetics by ISH using fluorochrome-labelled probes (FISH) has demonstrated the t(11;14) in over 95% of cases (Belaud-Rotureau *et al.*, 2002; Remstein *et al.*, 2000). Amplification of the *HER2* gene occurs in a significant proportion of breast carcinomas and is associated with poor prognosis. Moreover, information on HER2 amplification and overexpression has become relevant in view of immunotherapeutic approaches to breast cancer treatment using a humanised monoclonal antibody directed against the HER2 protein. Conventional assays for the immunohistochemical detection of HER2 protein expression are convenient but subject to technical and interpretational variability. Reliable assessment of *HER2* gene amplification by ISH using fluorochrome (FISH) or colorimetric detection (CISH) is currently employed to identify breast cancer patients that may benefit from a therapy with HER2-specific antibodies (Figure 2.2) (Dandachi *et al.*, 2002; Zhao *et al.*, 2002). Similarly, amplification or deletion of the TOP2A gene coding for topoisomerase IIα may be visualised by FISH (Järvinen *et al.*, 2000).

Both DNA and RNA ISH can be combined with the immunohistological detection of proteins. Moreover, in RNA ISH, radioactive and non-radioactive probes can be combined in the same assay. Such techniques have been used in various settings. In virus research, double-labelling experiments have been used for the simultaneous demonstration of viral nucleic acids and virus-encoded proteins (Niedobitek *et al.*, 1997b). Concurrent detection of virus DNA or RNA

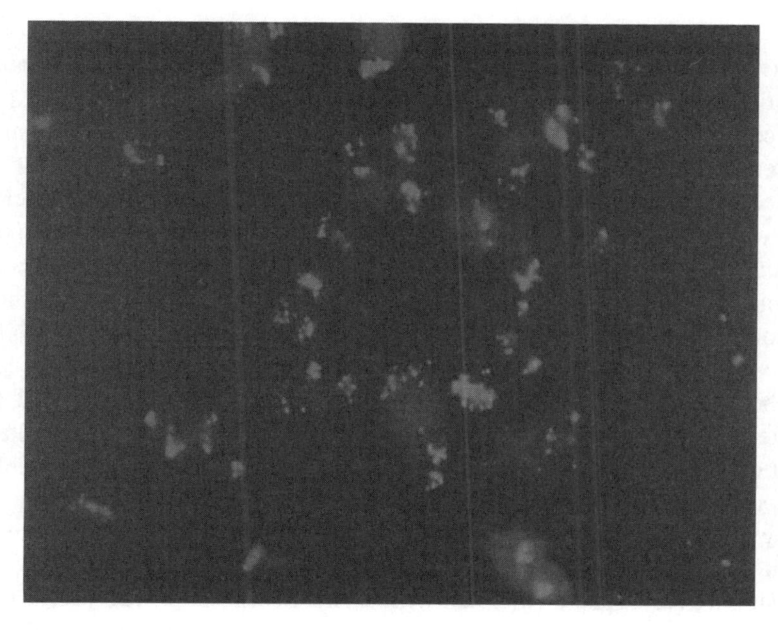

Figure 2.2 Fluorescence ISH demonstrates amplification of the *HER2* gene in a breast carcinoma (red signal). A colour version of this figure appears in the colour plate section

transcripts and proteins associated with viral latency or replication may allow an assessment of the state of the virus in an infected cell. The simultaneous demonstration of viral DNA and transforming viral proteins may allow an assessment of the potential significance of a virus for the pathogenesis of malignant tumour. Combining ISH with immunohistochemical detection of lineage specific antigens, e.g. intermediate filaments or leukocyte differentiation antigens, may be used for determining the phenotype of virus-infected cells. This approach has proved useful for identifying B-cells as a major site of EBV persistence (Anagnostopoulos *et al.*, 1995; Karajannis *et al.*, 1997; Niedobitek *et al.*, 1992, 1997a).

2.9 Perspective

Classical histological techniques have relied on the identification of cell and tissue components by their specific binding of certain dyes or by their function in enzyme histochemistry. The introduction of immunohistology has revolutionised histology and histopathology because it has made possible the demonstration of cellular and extracellular tissue constituents by their antigenic properties. This has greatly expanded the range of detectable molecules and has added a new level of specificity. Thus, immunohistology has developed into an indispensable tool particularly for the histopathologist. Nevertheless, there are some limitations to immunohistochemistry which have stimulated further methodological developments.

The specificity of some monoclonal and polyclonal immunohistological reagents may be ill-defined and not all specific antibodies are suitable for immunohistological staining. Monoclonal antibodies detect only short peptide sequences or carbohydrate residues of an antigen which may be shared by unrelated antigens.

Molecular biological techniques, on the other hand, employ nucleic acid probes of defined specificity and sequence. Using public databases, it is possible to obtain defined probe sequences that can be rapidly generated, either synthetically in the case of oligonucleotides or using PCR for longer probes. Choosing stringent hybridisation conditions and appropriate controls, the possibility of cross-hybridisation to unrelated sequences can largely be excluded. In contrast to methods based on the extraction of nucleic acids from tissues, ISH allows the detection of signals in their morphological context and the simultaneous evaluation of any underlying pathology. Thus, ISH has become a valuable tool in various areas of research. This technique is a useful complement to immunohistology; it may be used to verify immunohistological results and vice versa. ISH can be combined with immunohistology to identify the phenotype of a cell expressing a given gene or infected with a virus. ISH has proved useful in identifying the cellular sources of secreted proteins, e.g. cytokines or collagens, a question that is difficult to address by immunohistology. Moreover, ISH extends immunohistochemistry, allowing the generation of expression data of genes for which immunohistochemical reagents are not available. This aspect has become particularly important with the cloning of the human genome which has led to the identification of numerous genes of unknown function. ISH can provide expression data on such genes and thus contribute to defining their function.

At the same time, after more than 15 years of application, ISH arguably has made little impact on diagnostic histopathology. Although the introduction of robust non-radioactive methods has promoted the use of ISH, there are very few diagnostic applications. The detection of virus infections and interphase cytogenetics is likely to remain the major diagnostically relevant area.

Nevertheless, ISH has proved to be an important tool in morphology-based research and is certainly here to stay. It is increasingly being recognised that morphological analysis is an important tool in medical research, particularly in cancer research. ISH will continue to make important scientific contributions in these areas.

2.10 References

Adams JC (1992) Biotin amplification of biotin and horseradish peroxidase signals in histochemical stains. *J. Histochem. Cytochem.* **40**: 1457–63.

Aigner T, Zhu Y, Chansky HH, Matsen FA, Maloney WJ and Sandell LJ (1999) Reexpression of type IIA procollagen by adult articular chondrocytes in osteoarthritic cartilage. *Arthritis and Rheumatism* **42**: 1443–50.

Ambinder RF, Charache P, Staal S, Wright P, Forman M, Hayward SD and Hayward GS (1986) The vector homology problem in diagnostic nucleic acid hybridization of clinical specimens. *J. Clin. Microbiol.* **24**: 16–20.

Anagnostopoulos I, Herbst H, Niedobitek G and Stein H (1989) Demonstration of monoclonal EBV genomes in Hodgkin's disease and Ki-1 positive anaplastic large cell lymphoma by combined Southern blot and *in situ* hybridization. *Blood* **74**: 810–6.

Anagnostopoulos I, Hummel M, Kreschel C and Stein H (1995) Morphology, immunophenotype, and distribution of latently and/or productively Epstein–Barr virus-infected cells in acute infectious mononucleosis: implications for the interindividual infection route of Epstein–Barr virus. *Blood* **85**(3): 744–50.

Baner J, Nilsson M, Mendel-Hartvig M and Landegren U (1998) Signal amplification of padlock probes by rolling circle replication. *Nucleic Acids Res.* **26**: 5073–8.

Beck A, Päzolt D, Grabenbauer GG, Nicholls JM, Herbst H, Young LS and Niedobitek G (2001) Expression of cytokine and chemokine genes in Epstein–Barr virus-associated nasopharyngeal carcinoma and Hodgkin's disease. *J. Pathol.* **194**: 145–51.

Belaud-Rotureau M-A, Parrens M, Dubus P, Garroste C-D, de Mascarel A and Merlio J-P (2002) A comparative analysis of FISH, RT-PCR, and immunohistochemistry for the diagnosis of mantle cell lymphomas. *Mod. Pathol.* **15**: 517–25.

Bobrow MN, Harris TD, Shaughnessy KJ and Litt GJ (1989) Catalyzed reporter deposition, a novel method of signal amplification. *J. Immunol. Methods* **125**: 279–85.

Bobrow MN, Shaughnessy KJ and Litt GJ (1991) Catalyzed reporter deposition, a novel method of signal amplification. II. Application to membrane immunoassays. *J. Immunol. Methods* **137**: 103–12.

Bonnet M, J-MG, Kremmer E, Grunewald V, Benhamou E, Contesso G and Joab I (1999) Detection of Epstein–Barr virus in invasive breast cancers. *J. Natl Cancer Inst.* **91**: 1376–81.

Boye M, Jensen TK, Ahrens P, Hagedorn-Olsen T and Friis NF (2001) *In situ* hybridisation for identification and differentiation of Mycoplasma hyopneumoniae, *Mycoplasma hyosynoviae* and *Mycoplasma hyorhinis* in formalin-fixed porcine tissue sections. *APMIS* **109**: 656–64.

Brigati DJ, Myerson D, Leary JJ, Spalholz B, Travis SZ, Fong CK, Hsiung GD and Ward DC (1983) Detection of viral genomes in cultured cells and paraffin-embedded tissue sections using biotin-labeled hybridization probes. *Virology* **126**: 32–50.

Bubendorf L, Nocito A, Moch H and Sauter G (2001) Tissue microarray (TMA) technology: miniaturized pathology archives for high throughput *in situ* studies. *J. Pathol.* **195**: 72–9.

Buongiorno-Nardelli M and Amaldi F (1970) Autoradiographic detection of molecular hybrids between RNA and DNA in tissue sections. *Nature* **225**: 946–8.

Chu PG, Chang KL, Chen Y-Y and Weiss LM (2001a) No significant association of Epstein–Barr virus infection with invasive breast carcinoma. *Am. J. Pathol.* **159**: 571–8.

Chu PG, Chen YY, Chen W and Weiss LM (2001b) No direct role for Epstein–Barr virus in American hepatocellular carcinoma. *Am. J. Pathol.* **159**: 1287–92.

Cooper K, Herrington CS, Strickland JE, Evans MF and O'D McGee J (1991) Episomal and integrated human papillomavirus in cervical neoplasia shown by non-isotopic *in situ* hybridisation. *J. Clin. Pathol.* **44**: 990–6.

Coullin P, Roy L, Pellestor F, Candelier J-J, Bed-Hom B, Guillier-Gencik Z and Bernheim A (2002) PRINS, the other *in situ* DNA labeling method useful in cellular biology. *Am. J. Med. Genet.* **107**: 127–35.

Cox KH, DeLeon DV, Angerer LM and Angerer RC (1984) Detection of mRNAs in sea urchin embryos by *in situ* hybridization using asymmetric RNA probes. *Develop. Biol.* **101**: 485–502.

Dandachi N, Dietze O and Hauser-Kronberger C (2002) Chromogenic *in situ* hybridization: a novel approach to a practical and sensitive method for the detection of HER2 oncogene in archival human breast carcinoma. *Laboratory Investigation* **82**: 1007–14.

Divjak M, Glare EM and Walters EH (2002) Improvement of non-radioactive *in situ* hybridization in human airway tissues: use of PCR-generated templates for synthesis of probes and an antibody sandwich technique for detection of hybridization. *J. Histochem. Cytochem.* **50**: 541–8.

Franco D, de Boer PAJ, de Gier-de Vries C, Lamers WH and Moorman AFM (2001) Methods on *in situ* hybridization, immunohistochemistry and β-galactosidase reporter gene detection. *European Journal of Morphology* **39**: 3–25.

Frantz GD, Pham TQ, Peale FV and Hillan KJ (2001) Detection of novel gene expression in paraffin-embedded tissues by isotopic *in situ* hybridization in tissue microarrays. *J. Pathol.* **195**: 87–96.

Gall JG and Pardue ML (1969) Formation and detection of RNA–DNA hybrid molecules in cytological preparations. *Proc. Natl Acad. Sci. USA* **63**: 378–83.

Garrett KL, Grounds MD and Beilharz MW (1992) Nonspecific binding of nucleic acid probes to Paneth cells in the gastrointestinal tract with *in situ* hybridization. *J. Histochem. Cytochem.* **40**: 1613–8.

Gleich GJ, Loegering DA, Kueppers F, Bajaj SP and Mann KG (1974) Physiochemical and biological properties of the major basic protein from guinea pig eosinophil granules. *J. Exp. Med.* **140**: 313–32.

Gomez Aguado F, Picazo A, Roldan M, Corcuera MT, Curiel I, Munoz E, Martinez R and Alonso MJ (1996) Labelling pattern obtained by non-isotopic *in situ* hybridization as aprognostic factor in HPV-associated lesions. *J. Pathol.* **179**: 272–5.

Gruber AD and Levine RA (1997) *In situ* assessment of mRNA accessibility in heterogeneous tissue samples using elongation factor-1a (EF-1a). *Histochemistry and Cell Biology* **107**: 411–6.

Hawkins AJ and Dodd PR (2000) Localisation of GABAA receptor subunits in the CNS using RT-PCR. *Brain Research Protocols* **6**: 47–52.

Hayashi S, Gillam IC, Delaney AD and Tener GM (1978) Acetylation of chromosome squashes of *Drosophila melanogaster* decreases the background in autoradiographs from hybridization with [125I]-labeled RNA. *J. Histochem. Cytochem.* **26**: 677–9.

Hayden RT, Qian X, Procop GW, Roberts GD and Lloyd RV (2002) *In situ* hybridization for the identification of filamentous fungi in tissue sections. *Diagnostic Molecular Pathology* **11**: 119–26.

Herbst H, Foss H-D, Samol J, Araujo I, Klotzbach H, Krause H, Agathanggelou A, Niedobitek G and Stein H (1996) Frequent expression of interleukin-10 by Epstein–Barr virus-harboring tumor cells of Hodgkin's disease. *Blood* **87**: 2918–29.

Herbst H and Niedobitek G. Phenotype determination of Epstein–Barr virus infected cells in tissue sections. In: JB Wilson, GHW May, editors. *Epstein–Barr Virus Protocols*. Totowa: Humana Press; 2001. pp. 93–102.

Herbst H, Samol J, Foss HD, Raff T and Niedobitek G (1997) Modulation of interleukin-6 expression in Hodgkin and Reed–Sternberg cells by Epstein–Barr virus. *J. Pathol.* **182**: 299–306.

Herbst H, Sauter M, Kühler-Obbarius C, Löning T and Mueller-Lantzsch N (1998) Human endogenous retrovirus (HERV)-K transcripts in germ cell and trophoblast tumors. *APMIS* **106**: 216–20.

Herbst H, Steinbrecher E, Niedobitek G, Young LS, Brooks L, Muller-Lantzsch N and Stein H (1992) Distribution and phenotype of Epstein–Barr virus-harboring cells in Hodgkin's disease. *Blood* **80**: 484–91.

Herrmann K, Frangou P, Middeldorp J and Niedobitek G (2002) Epstein–Barr virus replication in tongue epithelial cells. *Journal of General Virology* **83**: 2995–8.

Herrmann K and Niedobitek G Epstein–Barr virus-associated carcinomas: facts and fiction. *J. Pathol.* 2003, **199**: 140–145.

Höfler H, Pütz B, Mueller JD, Neubert W, Sutter G and Gais P (1995) *In situ* amplification of measles virus RNA by the self-sustained sequence replication reaction. *Laboratory Investigation* **73**: 577–85.

Höfler H, Pütz B, Ruhri C, Wirnsberger G, Klimpfinger M and Smolle J (1987) Simultaneous localization of calcitonin mRNA and peptide in a medullary thyroid carcinoma. *Virchows Archiv B Cellular Pathology* **54**: 144–51.

Horn JE, Kappus EW, Falkow S and Quinn TC (1988) Diagnosis of Chlamydia trachomatis in biopsied tissue specimens by using *in situ* DNA hybridization. *Journal of Infectious Diseases* **157**: 1249–53.

Järvinen TAH, Tanner M, Rantanen V, Bärlund M, Borg A, Grenman S and Isola J (2000) Amplification and deletion of topoisomerase IIa associated with ErbB-2 amplification and affect sensitivity to topoisomerase II inhibitor doxorubicin in breast cancer. *Am. J. Pathol.* **156**: 839–47.

John HA, Birnstiel ML and Jones KW (1969) RNA-DNA hybrids at the cytological level. *Nature* **223**: 582–7.

Junying J, Herrmann K, Davies G, Lissauer D, Bell A, Timms J, Reynolds GM, Hubscher SG, Young LS, Niedobitek G and Murray PG. Absence of Epstein–Barr virus DNA in the tumor cells of European hepatocellular carcinoma. *Virology*, in press.

Karajannis MA, Hummel M, Anagnostopoulos I and Stein H (1997) Strict lymphotropism of Epstein–Barr virus during acute infectious mononucleosis in nonimmunocompromised individuals. *Blood* **89**(8): 2856–62.

Kitazawa S, Kitazawa R and Maeda S (1999) *In situ* hybridization with polymerase chain reaction-derived single-stranded DNA probe and S1 nuclease. *Histochemistry and Cell Biology* **111**: 7–12.

Kolquist KA, Ellisen LW, Counter CM, Meyerson M, Tan LK, Weinberg RA, Haber DA and Gerald WL (1998) Expression of TERT in early premalignant lesions and a subset of cells in normal tissues. *Nature Genetics* **19**: 182–6.

Ky B and Shughrue PJ (2002) Methods to enhance signal using isotopic *in situ* hybridization. *J. Histochem. Cytochem.* **50**: 1031–7.

Laniece P, Caron Y, Cardona A, Pinot L, Maitrejean S, Mastrippolito R, Sandkamp B and Valentin L (1998) A new high resolution radioimager for the quantitative analysis of radiolabelled molecules in tissue sections. *Journal of Neuroscience Methods* **86**: 1–5.

Lawrence JB and Singer RH (1985) Quantitative analysis of *in situ* hybridization methods for the detection of actin gene expression. *Nucleic Acids Res.* **13**: 1777–99.

Lewis F, Maughan NJ, Smith V, Hillan K and Quirke P (2001) Unlocking the archive – gene expression in paraffin-embedded tissues. *J. Pathol.* **195**: 66–71.

Lipman DJ (1997) Making (anti)sense of non-coding sequence conservation. *Nucleic Acids Res.* **25**: 3580–3.

Lizardi PM, Huang X, Zhu Z, Bray-Ward P, Thomas DC and Ward DC (1998) Mutation detection and single-molecule counting using isothermal rolling-circle amplification. *Nature Genetics* **19**: 225–32.

Logel J, Dill D and Leonard S (1992) Synthesis of cRNA probes from PCR-generated DNA. *BioTechniques* **13**: 604–10.

Long AA (1998) *In-situ* polymerase chain reaction: foundation of the technology and today's options. *European Journal of Histochemistry* **42**: 101–9.

McNicol AM and Farquharson MA (1996) *In situ* hybridization and its diagnostic applications in pathology. *J. Pathol.* **182**: 250–61.

Meru N, Davison S, Whitehead L, Jung A, Mutimer D, Rooney N, Kelly D and Niedobitek G (2001a) Epstein–Barr virus infection in paediatric liver transplant recipients: detection of the virus in posttransplant tonsillectomy specimens. *J. Clin. Pathol.: Mol. Pathol.* **54**: 264–9.

Meru N, Jung A, Baumann I and Niedobitek G (2002) Expression of the recombination activating genes in extrafollicular lymphocytes but no apparent re-induction in germinal centre reactions in human tonsils. *Blood* **99**: 531–7.

Meru N, Jung A, Lisner R and Niedobitek G (2001b) Expression of the recombination activating genes (Rag1 and Rag2) is not detectable in Epstein–Barr virus-associated human lymphomas. *Int. J. Cancer* **92**: 75–8.

Milani S, Herbst H, Schuppan D, Grappone C and Heinrichs OE (1995) Cellular sources of extracellular matrix proteins in normal and fibrotic liver. Studies of gene expression by *in situ* hybridization. *Journal of Hepatology* **22**: 71–6.

Niedobitek G, Agathanggelou A, Herbst H, Whitehead L, Wright DH and Young LS (1997a) Epstein–Barr virus (EBV) infection in infectious mononucleosis: virus latency, replication and phenotype of EBV-infected cells. *J. Pathol.* **182**: 151–9.

Niedobitek G, Finn T, Herbst H and Stein H (1989a) Detection of viral genomes in the liver by *in situ* hybridisation using 35S-, bromodeoxyuridine-, and biotin-labeled probes. *Am. J. Pathol.* **134**: 633–9.

Niedobitek G, Finn T, Herbst H and Stein H (1989b) *In situ* hybridization using biotinylated probes. An evaluation of different detection systems. *Pathol. Res. Practice* **184**: 343–8.

Niedobitek G, Hamilton-Dutoit S, Herbst H, Finn T, Vetner M, Pallesen G and Stein H (1989c) Identification of Epstein–Barr virus-infected cells in tonsils of acute infectious mononucleosis by *in situ* hybridization. *Hum. Pathol.* **20**: 796–9.

Niedobitek G, Herbst H, Young LS, Brooks L, Masucci MG, Crocker J, Rickinson AB and Stein H (1992) Patterns of Epstein–Barr virus infection in non-neoplastic lymphoid tissue. *Blood* **79**: 2520–6.

Niedobitek G and Herbst H (1991) Applications of *in situ* hybridization. *International Review of Experimental Pathology* **32**: 1–56.

Niedobitek G and Herbst H. *In situ* detection of Epstein–Barr virus DNA and of viral gene products. In: JB Wilson, GHW May, editors. Epstein–Barr virus protocols. Totowa, NJ: Humana Press; 2001. pp. 79–91.

Niedobitek G, Kremmer E, Herbst H, Whitehead L, Dawson CW, Niedobitek E, von Ostau C, Rooney N, Grasser FA and Young LS (1997b) Immunohistochemical detection of the Epstein–Barr virus-encoded latent membrane protein 2A (LMP2A) in Hodgkin's disease and infectious mononucleosis. *Blood* **90**: 1664–72.

Niedobitek G, Meru N and Delecluse H-J (2001) Epstein–Barr virus infection and human malignancies. *Int. J. Exp. Pathol.* **82**: 149–70.

Niedobitek G, Young LS, Lau R, Brooks L, Greenspan D, Greenspan JS and Rickinson AB (1991) Epstein–Barr virus infection in oral hairy leukoplakia: virus replication in the absence of a detectable latent phase. *Journal of General Virology* **72**: 3035–46.

Nilsson M, Malmgren H, Samiotaki M, Kwiatkowski M, Chowdhary BP and Landegren U (1994) Padlock probes: circularizing oligonucleotides for localized DNA detection. *Science* **265**: 2085–8.

Nuovo G (2001) Co-labeling using *in situ* PCR: a review. *J. Histochem. Cytochem.* **49**: 1329–39.

Patterson S, Gross J and Webster AD (1989) DNA probes bind non-specifically to eosinophils during *in situ* hybridization: carbol chromotrope blocks binding to eosinophils but does not inhibit hybridization to specific nucleotide sequences. *Journal of Virological Methods* **23**: 105–9.

Pohle T, Shahin M, Gillessen A, Schuppan D, Herbst H and Domschke W (1996) Expression of type I and type IV collagen mRNAs in healing gastric ulcers – a comparative analysis using isotopic and non-radioactive *in situ* hybridization. *Histochemistry and Cell Biology* **106**: 413–8.

Poulsom R, Longcroft JM, Jeffery RE, Rogers LA and Steel JH (1998) A robust method for isotopic riboprobe *in situ* hybridisation to localise mRNAs in routine pathology specimens. *Euro. J. Histochem.* **42**: 121–32.

Pringle JH, Ruprain AK, Primrose L, Keyte J, Potter L, Close P and Lauder I (1990) *In situ* hybridization of immunoglobulin light chain mRNA in paraffin sections using biotinylated or hapten-labelled oligonucleotide probes. *J. Pathol.* **162**: 197–207.

Raap AK, Marijnen JG, Vrolijk J and van der Ploeg M (1986) Denaturation, renaturation, and loss of DNA during *in situ* hybridization procedures. *Cytometry* **7**: 235–42.

Reiss C, Niedobitek G, Hör S, Lisner R, Friedrich U, Bodemer W and Biesinger B (2002) Peripheral T-cell lymphoma in herpesvirus saimiri infected tamarins: tumor cell lines reveal subgroup-specific differences. *Virology* **294**: 31–46.

Remstein ED, Kurtin PJ, Buno I, Proffitt J, Wyatt WA and Hanson CA (2000) Diagnostic utility of fluorescence *in situ* hybridization in mantle-cell lymphoma. *British Journal of Haematology* **110**: 856–62.

Rentrop M, Knapp B, Winter H and Schweizer J (1986) Aminoalkylsilane-treated glass slides as support for *in situ* hybridization of keratin cDNAs to frozen tissue sections under varying fixation and pretreatment conditions. *Histochemical Journal* **18**: 271–6.

Roberts WH, Sneddon JM, Waldman J and Stephens RE (1989) Cytomegalovirus infection of gastrointestinal endothelium demonstrated by simultaneous nucleic acid hybridization and immunohistochemistry. *Archives of Pathology and Laboratory Medicine* **113**: 461–4.

Rogers AW. *Techniques in autoradiography*. Amsterdam: Elsevier; 1979.

Romagnani P, Rotondi M, Lazzeri E, Lasagni L, Francalanci M, Buonamano A, Milani S, Vitti P, Chiovato L, Tonacchera M, Bellastella A and Serio M (2002) Expression of IP-10/CXCL10 and MIG/CXCL9 in the thyroid and increased levels of IP-10/CXCL10 in the serum of patients with recent-onset Grave's disease. *Am. J. Pathol.* **161**: 195–206.

Rowlands DC, Ito M, Mangham DC, Reynolds G, Herbst H, Hallissey MT, Fielding JWL, Newbold KM, Jones EL, Young LS and Niedobitek G (1993) Epstein–Barr virus and carcinomas: rare association of the virus with gastric adenocarcinomas. *Brit. J. Cancer* **68**: 1014–9.

Saglie R, Cheng L and Sadighi R (1988) Detection of *Mycoplasma pneumoniae* – DNA within diseased gingiva by *in situ* hybridization using a biotin-labeled probe. *Journal of Periodontology* **59**: 121–3.

Salin H, Maitrejean S, Mallet J and Dumas S (2000) Sensitive and quantitative co-detection of two mRNA species by double radioactive *in situ* hybridization. *J. Histochem. Cytochem.* **48**: 1587–91.

Schmidt BF, Chao J, Zhu Z, DeBiasio R and Fisher G (1997) Signal amplification in the detection of single-copy DNA and RNA by enzyme-catalyzed deposition (CARD) of the novel fluorescent reporter substrate Cy3.29-tyramide. *J. Histochem. Cytochem.* **45**: 365–73.

Sedlaczek N, Jia JD, Bauer M, Herbst H, Ruehl M, Hahn EG and Schuppan D (2001) Proliferating bile duct epithelial cells are a major source of connective tissue growth factor in rat biliary fibrosis. *Am. J. Pathol.* **158**: 1239–44.

Seyda M, Scheele T, Neumann R and Krueger GR (1988) Comparative evaluation of non-radioactive *in situ* hybridization techniques for pathologic diagnosis of viral infection. *Pathol. Res. Practice* **184**: 18–26.

Sibony M, Commo F, Callard P and Gasc J-M (1995) Enhancement of mRNA *in situ* hybridization signal by microwave heating. *Laboratory Investigation* **73**: 586–91.

Speel EJM, Saremaslani P, Roth J, Hopman AHN and Komminoth P (1998) Improved mRNA *in situ* hybridization on formalin-fixed and paraffin-embedded tissue using signal amplification with different haptenized tyramides. *Histochemistry and Cell Biology* **110**: 571–7.

Spieker T and Herbst H (2000) Distribution and phenotype of Epstein–Barr virus-infected cells in inflammatory bowel disease. *Am. J. Pathol.* **157**: 51–7.

St Croix B, Rago C, Velculescu V, Traverso G, Romans KE, Montgomery E, Lal A, Riggins GJ, Lengauer C, Vogelstein B and Kinzler KW (2000) Genes expressed in human tumor endothelium. *Science* **289**: 1197–202.

Steiner I, Spivack JG, Jackson A, Lavi E and Fraser NW (1989) Effects of lipofuscin on *in situ* hybridization in human neuronal tissue. *Journal of Virological Methods* **24**: 1–9.

Sugawara Y, Makuuchi M and Takada K (2000) Detection of Epstein–Barr virus in hepatocellular carcinoma from hepatitis C-positive patients. *Scandinavian Journal of Gastroenterology* **35**: 981–4.

Sugawara Y, Mizugaki Y, Uchida T, Torii T, Imai S, Makuuchi M and Takada K (1999) Detection of Epstein Barr virus (EBV) in hepatocellular carcinoma tissue: a novel EBV latency characterized by the absence of EBV-encoded small RNA expression. *Virology* **256**: 196–202.

Syrjanen S, Partanen P, Mantyjarvi R and Syrjanen K (1988) Sensitivity of *in situ* hybridization techniques using biotin- and 35S-labeled human papillomavirus (HPV) DNA probes. *Journal of Virological Methods* **19**: 225–38.

Tbakhi A, Totos G, Hauser-Kronberger C, Pettay J, Baunoch D, Hacker GW and Tubbs RR (1998) Fixation conditions for DNA and RNA *in situ* hybridization. A reassessment of molecular morphology dogma. *Am. J. Pathol.* **152**: 35–41.

Terpstra WJ, Schoone GJ, ter Schegget J, van Nierop JC and Griffioen RW (1987) *In situ* hybridization for the detection of *Haemophilus* in sputum of patients with cystic fibrosis. *Scandinavian Journal of Infectious Diseases* **19**: 641–6.

Teruya-Feldstein J, Jaffe ES, Burd PR, Kingma DW, Setsuda JE and Tosato G (1999) Differential chemokine expression in tissues involved by Hodgkin's disease: direct correlation of eotoxin expression and tissue eosinophilia. *Blood* **93**: 2463–70.

Todd R and Margolin DH (2002) Challenges of single-cell diagnostics: analysis of gene expression. *Trends in Molecular Medicine* **8**: 254–7.

van de Corput MPC, Dirks RW, Gijlswijk RPM, van Binnendijk E, Hattinger CM, de Paus RA, Landegent JE and Raap AK (1998) Sensitive mRNA detection by fluorescence *in situ* hybridisation using horseradish peroxidase-labelled oligonucleotides and tyramide signal amplification. *J. Histochem. Cytochem.* **46**: 1249–59.

van den Berg FM, Zijlmans H, Langenberg W, Rauws E and Schipper M (1989) Detection of *Campylobacter pylori* in stomach tissue by DNA *in situ* hybridisation. *J. Clin. Pathol.* **42**: 995–1000.

Van der Loos CM, Volkers HH, Rook R, Van den Berg FM and Houthoff HJ (1989) Simultaneous application of *in situ* DNA hybridization and immunohistochemistry on one tissue section. *Histochemical Journal* **21**: 279–84.

van Gijlswijk RP, Talman EG, Janssen PJ, Snoeijers SS, Killian J, Tanke HJ and Heetebrij RJ (2001) Universal linkage system: versatile nucleic acid labeling technique. *Expert Reviews in Molecular Diagnostics* **1**: 81–91.

Wiedorn KH, Goldmann T, Henne C, Kühl H and Vollmer E (2001) EnVision+, a new dextran polymer-based signal enhancement technique for *in situ* hybridization (ISH). *J. Histochem. Cytochem.* **49**: 1067–71.

Wilcox JN (1993) Fundamental principles of *in situ* hybridization. *J. Histochem. and Cytochem.* **41**: 1725–33.

Wilkens L, Komminoth P, Nasarek A, von Wasielewski R and Werner M (1997) Rapid detection of karyotype changes in interphase bone marrow cells by oligonucleotide primed *in situ* hybridization (PRINS). *J. Pathol.* **181**: 368–73.

Williamson DJ (1988) Specificity of riboprobes for intracellular RNA in hybridization histochemistry. *J. Histochem. Cytochem.* **36**: 811–3.

Wilson KH, Schambra UB, Smith MS, Page SO, Richardson CD, Fremeau RT and Schwinn DA (1997) *In situ* hybridization: identification of rare mRNAs in human tissues. *Brain Research Protocols* **1**: 175–85.

Wolber RA and Lloyd RV (1988) Cytomegalovirus detection by nonisotopic *in situ* DNA hybridization and viral antigen immunostaining using a two-color technique. *Hum. Pathol.* **19**: 736–41.

Wolfe KQ and Herrington CS (1997) Interphase cytogenetics and pathology: a tool for diagnosis and research. *J. Pathol.* **181**: 359–61.

Wood GS and Warnke R (1981) Suppression of endogenous avidin-binding activity in tissues and its relevance to biotin–avidin detection systems. *J. Histochem. Cytochem.* **29**: 1196–204.

Wu T-C, Mann RB, Charache P, Hayward SD, Staal S, Lambe BC and Ambinder RF (1990) Detection of EBV gene expression in Reed–Sternberg cells of Hodgkin's disease. *Int. J. Cancer* **46**: 801–4.

Yang H, Wanner IB, Roper SD and Chaudhari N (1999) An optimized method for *in situ* hybridization with signal amplification that allows the detection of rare mRNAs. *J. Histochem. Cytochem.* **47**: 431–45.

Zaidi AU, Enomoto H, Milbrandt J and Roth KA (2000) Dual fluorescent *in situ* hybridization and immunohistochemical detection with tyramide signal amplification. *J. Histochem. Cytochem.* 1369–75.

Zehbe I, Hacker GW, Su H, Hauser-Kronberger C, Hainfield JF and Tubbs R (1997) Sensitive *in situ* hybridization with catalyzed reporter deposition, streptavidin-nanogold, and silver acetate autometallography. Detection of single-copy human papillomavirus. *Am. J. Pathol.* **150**: 15553–1561.

Zhao J, Wu R, Au A, Marquez A, Yu Y and Shi Z (2002) Determination of HER2 gene amplification by chromogenic *in situ* hybridization (CISH) in archival breast carcinoma. *Mod. Pathol.* **15**: 657–65.

Zhong X-b, Lizardi PM, Huang X-h, Bray-Ward PL and Ward DC (2001) Visualization of oligonucleotide probes and point mutations in interphase nuclei and DNA fibers using rolling circle DNA amplification. *Proc. Natl. Acad. Sci. USA* **98**: 3940–5.

Zhou Y, Calciano M, Hamann S, Leamon JH, Strugnell T, Christian MW and Lizardi PM (2001) *in situ* detection of messenger RNA using digoxigenin-labeled oligonucleotides and rolling circle amplification. *Experimental and Molecular Pathology* **70**: 281–8.

3

DNA Flow Cytometry

M.G. Ormerod

3.1 Introduction

Measurement of the DNA content of a cell gives information about the cell cycle and detects aneuploidy; the latter normally is only observed in tumours. There are several dyes whose fluorescence is enhanced on binding to nucleic acids and accurately reflects the DNA content. Flow cytometry is routinely used to measure the DNA content of cells using such dyes.

Flow cytometry is the method of choice for these measurements because:

- it measures single cells;
- it is a multi-parametric measurement; that is, the measurement of DNA can be combined with other parameters, such as protein expression; and
- large numbers of cells can be measured, giving good statistics.

The disadvantage of flow cytometry is the requirement for single particles (cells or nuclei). With solid tissues, this can create a problem.

In this chapter I discuss the different aspects that should be considered when measuring a DNA histogram. For a simple introduction to flow cytometry see Ormerod *et al.* (1999). Protocols for preparing single cells or nuclei and for recording a DNA histogram have been given by Ormerod (2000a, 2000b), Darzynkiewicz *et al.* (1994) and Radbruch (1992). Shankey *et al.* (1993) and Ormerod *et al.* (1999) have published consensus guidelines for the standardisation of the measurement of a DNA histogram in clinical samples.

3.2 Definitions and terms

Ploidy. The term ploidy refers to the numbers of chromosomes in a cell. In particular, aneuploid refers to an abnormal number of chromosomes. When the

Molecular Biology in Cellular Pathology. Edited by John Crocker and Paul G. Murray
 ISBN: 0-470-84475-2

DNA content of a cell is used as a measure of ploidy, it should be described as DNA ploidy. For example, a tumour cell with a normal quantity of DNA should be referred to as DNA diploid. The tumour may not be strictly diploid since it is possible that there is a change in the chromosomes that is too small to be detected by flow cytometry.

DNA Index. DNA ploidy in tumours is described in terms of the DNA Index (DI), defined as the ratio between the mean DNA content of the test cells and the mean DNA content of normal diploid cells, in the G0/G1phase of the cell cycle. Hence, a cell that is DNA tetraploid has a DI of 2.

Coefficient of variation. An estimate of the quality of a DNA histogram is given by the width of the peak in the DNA histogram of cells in G1/G0 of the cell cycle. The width is expressed as the coefficient of variation (CV) across the peak, defined as:

$$\frac{\text{standard deviation across the peak}}{\text{mean channel number across the peak}}$$

It normally is expressed as a percentage.

The CV is used, rather than the standard deviation, because it is dimensionless and independent of the position of the peak in the DNA histogram.

3.3 Dye used for DNA analysis

The properties of dyes used for measuring a DNA histogram are shown in Table 3.1. With the exception of the *bis*-benzimidazole, Hoechst 33342, and the anthraquinone, DRAQ5, these dyes do not cross an intact plasma

Table 3.1 The properties of some fluorochromes used to label nucleic acids

Fluorophore	Excitation maxima (nm)	Emission maximum (nm)
Propidium iodide	535, 342	617
Ethidium bromide	518, 320	605
Acridine orange	503	530 (DNA)
		640 (RNA)
DRAQ5	646	681
TO-PRO-3	642	661
Hoechst 33342	395	450
DAPI	372	456

The fluorescent properties of most of these dyes change on binding to nucleic acid. The wavelengths given are those of the dye–nucleic acid complex.

membrane. The cells have to be fixed or permeabilised or nuclei prepared before staining.

The most commonly used dye is propidium iodide (PI). It can be excited by the blue line (488 nm) of an argon-ion laser, which is fitted to a large majority of flow cytometers. When bound to double-stranded nucleic acids, PI fluoresces red. It can conveniently be used in combination with an antibody labelled with fluorescein (see below). PI binds to double-stranded regions of RNA and samples are usually treated with RNase prior to analysis.

Ethidium bromide has similar properties to PI.

7-Aminoactinomycin D (7-AAD) is ideally excited by the green line of an argon-ion laser (513 nm) or by a green diode laser (539 nm). However, it has sufficient absorption at 488 nm to allow excitation at that wavelength. It has been used in combination with two immunofluorescent stains, one antibody being labelled with either fluorescein or Alexa488, the other with phycoerythrin (PE) (Schmid *et al.*, 2000). It is difficult to use PE in combination with PI as the fluorescence emission spectrum of PI has considerable overlap with that of PE. 7-AAD fluoresces a deeper red than PI and shows less spectral overlap with PE. The disadvantage is that it is often difficult to obtain as good a DNA histogram using 7-AAD as compared with PI.

Another alternative is DRAQ5 (Smith *et al.*, 2000), which is ideally excited at 630 nm using a He–Ne laser or a red diode laser. However, it has a high quantum efficiency that gives sufficient fluorescence emission when excited at 488 nm. It fluoresces in the deep red (maximum emission at 681 nm).

Acridine orange has two modes of binding to nucleic acids. It fluoresces green when intercalated in double-stranded nucleic acids and red when stacked on the charged phosphates in single-stranded nucleic acids. These properties permit the simultaneous measurement of DNA (double stranded) and RNA (single stranded) (Darzynkiewicz and Kapuscinski, 1990).

4′,6-Diamidino-2-phenylindole (DAPI) and the *bis*-benzimidazole, Hoechst 3342, are both DNA specific and both need to be excited by UV. DAPI has been used extensively in instruments fitted with a mercury arc lamp but few of these instruments are still in routine use. Hoechst 33342 is taken up by viable cells and can be used to sort them according to the cell cycle phase. The uptake is dependent on time and concentration of the dye. Because Hoechst 33342 is removed from the cell by the p-glycoprotein pump (pgp), the conditions needed to give a good DNA histogram should be determined for each type of cell used. If the cells over-express the pgp, then it may be necessary to inhibit the action of the pump with verapamil or cyclosporin A.

TO-PRO-3 is a cyanine dye, produced by Molecular Probes (`www.probes.com`), which fluoresces deep red. It is excited by the red line of either a He–Ne laser or a red diode laser. It has been used in conjunction with PI to shift the fluorescence emission wavelength of PI to that of TO-PRO-3 and thereby enable PE immunofluorescence to be easily observed (Corver *et al.*, 1997).

3.4 Sample preparation for DNA analysis

The object of sample preparation is to obtain a suspension of cells or nuclei with as few clumps and as little debris as possible.

Fresh Tissue

Tissues biopsies can be used fresh or stored frozen at −80°C. Single cells can be prepared by mechanical disaggregation, which can be achieved by several methods. In one, cells are removed from the biopsy by taking a fine needle aspirate. In another, the tissue is minced in tissue culture medium, aspirated through a plastic pipette tip and finally filtered through a 30-μm nylon mesh. This procedure can be combined with treatment with collagenase. A similar effect can be obtained by a commercially available machine (Medimachine from Dako, Ltd).

The preparation of single cells can be fixed, if the DNA stain is to be combined with an antibody stain, or cells permeabilised with a detergent, or nuclei released, also with a detergent. Vindeløv *et al.* (1983) have published a method for producing stabilised nuclei stained with PI, which gives DNA histograms of high quality.

If cells are fixed, then either ethanol or methanol are the preferred fixatives. Aldehyde-based fixatives generally give poor quality DNA histograms.

Mechanical disaggregation and release of the nuclei can be combined by mincing the tissue and then agitating it in a buffer containing detergent and PI (Petersen, 1985).

Nuclei can be prepared directly from fine needle aspirates, blood samples, peritoneal aspirates and similar samples, in which there are already mainly single cells, without any mechanical treatment. Figure 3.1 shows a typical DNA histogram obtained from a fine needle aspirate of a human breast carcinoma.

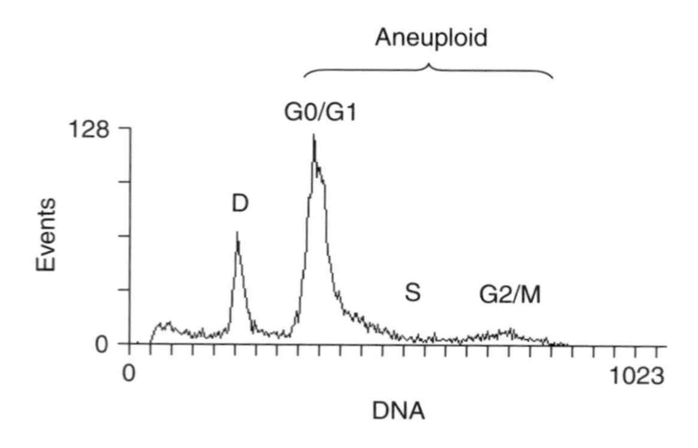

Figure 3.1 DNA histogram from a fine needle aspirate (FNA) of a breast carcinoma. The aspirate was centrifuged and the cells resuspended in a buffer containing PI and RNase. The diploid peak is marked 'D'; the cell cycle phases of the aneuploid (A) cells are marked. Data recorded by Mrs Jenny Titley, Institute of Cancer Research, Sutton, UK

Nuclei from Paraffin Wax Embedded Blocks

Nuclei can be extracted from routinely processed pathological material (formalin-fixed tissue embedded in paraffin wax) (Hedley, 1994). Thick (~50 μm) sections are cut, de-waxed, taken to water and incubated with pepsin at pH 1.5 to release the nuclei. The quality of the histograms obtained, which can be surprisingly good, depends on the way in which the tissue was initially handled in the histopathology laboratory. Figure 3.6 below gives an example of a DNA histogram of nuclei prepared from a paraffin block.

3.5 Analysis of the DNA histogram

The DNA fluorescence should be recorded using linear amplification in instruments using analogue electronics. In those instruments using digital electronics, the data should be displayed using a linear display. The reasons for this are that the cell cycle is linear, not logarithmic. The relevant information will occupy more of the data space in the histogram and algorithms for analysis of the cell cycle phases assume a linear scale.

The electronics will only recognise a signal as being of interest if it rises above a preset threshold. It is advisable to use the DNA signal as the threshold (or discriminator) parameter so that only those particles that contain DNA will be analysed and any debris ignored. The value of the threshold should be set between 5% and 10% of the channel number of a diploid cell in G1.

The sample should be run at a low flow rate. High flow rates can give rise to small perturbations in the sample stream, which may broaden the width of the peaks in the DNA histogram.

3.6 Quality control

The quality of a DNA histogram is measured from the width of the G0/G1 peak as measured by its CV. Small CVs give better resolution of small differences in ploidy in tumour samples and give a more reliable estimate of the different cell cycle components. Using normal lymphoid cells from blood or bone marrow, CVs close to 1% often can be obtained. Tumour samples or cultured cells sometimes have a higher CV and 2.5% might be typical (see Figure 3.2).

The quality of the DNA histogram is affected by:

- instrument alignment,
- sample preparation, and
- correct analysis of the data.

Sample Preparation

Sample preparation can be checked microscopically. If there is an excess of debris or an excessive number of cell clumps, then the method for initially

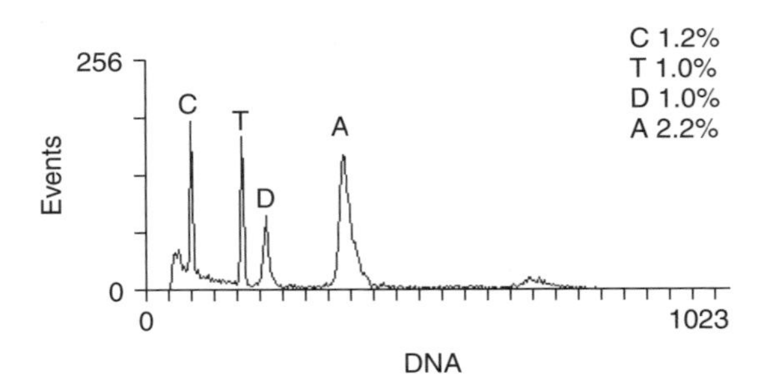

Figure 3.2 DNA histogram from a FNA of a human breast carcinoma. Nuclei were prepared using the Vindelov method; for further details, see Ottesen *et al.* (1995). C and T mark the position of the chicken and trout erythrocytes and D the diploid DNA peak. The cell cycle phases of the aneuploid (A) cells are marked. The percentages show the CVs for the different peaks. Note that the aneuploid tumour cells have a higher CV. Data supplied by Ib Jarle Christensen, The Finsen Laboratory, Copenhagen

preparing the sample should be checked. Cell clumping can also be caused during fixation (particularly if the cells are fixed in 70% ethanol) and it is important that there is a good suspension of single cells before fixation. Cell clumps can sometimes be reduced by passing the sample through a 26 gauge needle. Some workers filter cells through nylon mesh (50–80 μm mesh size) to remove large clumps; if a large proportion of cells is removed by this procedure, the final result could be affected by preferential removal of a particular type of cell.

If a DNA histogram on its own is required, that is, there is no antibody stain, then problems associated with cell clumping can be avoided by using a preparation of nuclei.

Sufficient time must be allowed for the dye to equilibrate with the sample. If PI is used as the DNA stain, then time is also needed for the RNase to remove all double-stranded RNA. With fixed cells, leaving samples overnight in the cold often will improve the quality of the DNA histogram.

The concentration of cells or nuclei should be about 10^6 cells/ml. If it is too high, there may be insufficient dye to maintain stoichiometry. Generally this will manifest itself in the DNA histogram by a poor resolution of the G2 phase of the cell cycle.

Instrument Alignment

The instrument alignment should be checked daily using fluorescent beads sold for that purpose. Using appropriate beads, a ‘half-height’ CV of 1.5% or better should be obtained for the fluorescence parameter being used for the DNA analysis. The value of the CV together with the peak fluorescence channel number for the beads (using standard settings) should be recorded and plotted. If there is an abrupt increase in the CV from one day to the next, there may be some dirt in the flow cell and the cleaning procedure should be repeated. A gradual increase in the CV and/or a decrease in the peak channel number over several days or weeks suggests that the instrument needs to be re-aligned.

Data Analysis

Clumping of cells can cause a problem in analysis, particularly when working with fixed cells. Two cells in G1 will have the same DNA content as a cell in G2. These events can be distinguished if the analysis is performed correctly. This is achieved by focusing the laser beam down to a narrow ellipse and analysing the pulse of light falling on the detector.

A nucleus in G2 will have a diameter about 30% larger than a nucleus in G1. Its time of flight through the laser beam (determined by the beam diameter plus the cell diameter) will be only slightly greater than that of a G1 nucleus. The increased DNA fluorescence will manifest itself in a larger peak signal falling on the detector. By contrast, two nuclei in G1 (aligned one behind the other in the flow system) will have twice the diameter of a nucleus in G1 but will have a signal with the same peak height.

Displaying either the peak height of the DNA signal or its width against the total DNA fluorescence (signal area) will distinguish between single cells and clumps (Figures 3.3 and 3.4).

3.7 Computer analysis of the DNA histogram

The DNA histogram gives information about the G0/G1, S and G2/M phases of the cell cycle. Because of the width of the distributions in G0/G1 and G2/M

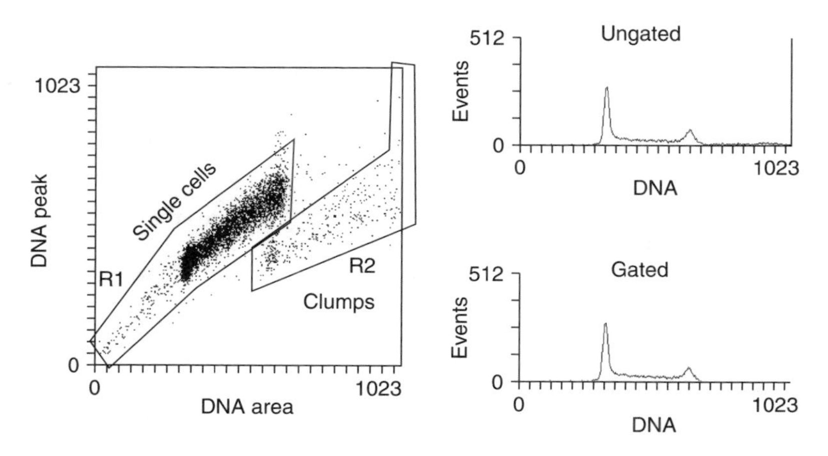

Figure 3.3 Murine leukaemic cells (L1210) fixed in 70% ethanol, rehydrated and stained with PI. A plot of the peak of the red fluorescence signal vs. the integrated area discriminates between clumps (R2) and single cells (R1). The DNA histograms show the effect of setting a gate on region, R1. Data recorded on a Coulter Elite ESP

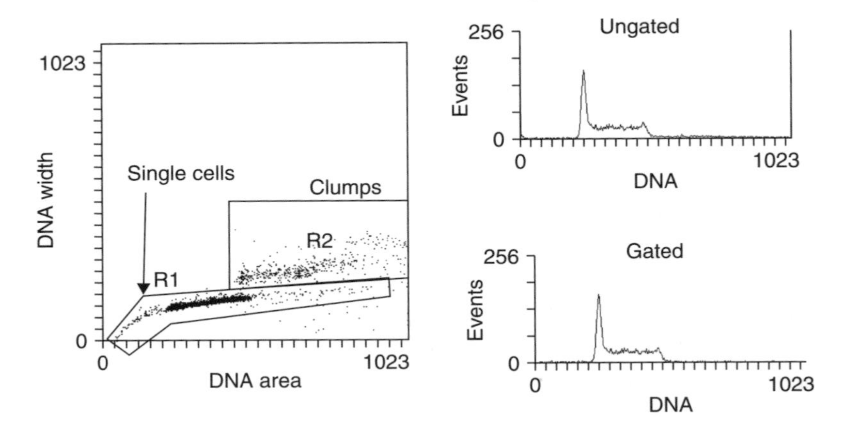

Figure 3.4 Murine leukaemic cells (L1210) fixed in 70% ethanol, rehydrated and stained with PI. A plot of the width of the red fluorescence signal vs. the integrated area discriminates between clumps and single cells. The DNA histograms show the effect of setting a gate on region, R1. Data recorded on a BD FACSort during the Royal Microscopical Society's Course on Flow Cytometry, Cambridge, 1994

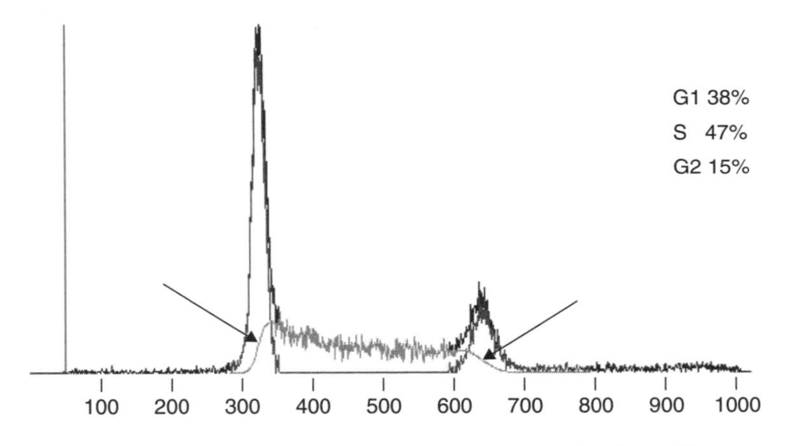

Figure 3.5 Results of cell cycle analysis using the program, Cylchred. The percentage of cells in the G1, S and G2/M phases of the cell cycle are shown. The arrows mark the overlap of early S phase with G1 and late S phase with G2

nuclei (as measured experimentally and given by the CV), there is considerable overlap between cells in the G0/G1 and early S phases and cells in the G2/M and late S phases. Various computer programs have been written to resolve these overlaps and to deconvolute the DNA histogram into the three separate phases. See, for example, Ormerod *et al.* (1987) and Rabinovitch (1994). Programs are available commercially and at least one program is available free on the internet (Cylchred: `www.uwcm.ac.uk/uwcm/hg/hoy/software.html`). An example of an analysis using this program is shown in Figure 3.5.

A typical DNA histogram obtained from a clinical sample with a low, unperturbed S phase can be analysed without recourse to a computer program by assuming a rectangular-shaped S phase. The method for carrying out this analysis has been described by Ormerod (2000b).

There are a variety of computer programs available for the analysis of DNA histograms. Often they offer a choice of algorithms. The different algorithms in the various computer programs make different assumptions and may give slightly different results. When analysing a set of histograms it is important to use the same program and the same algorithm for the whole data set.

Some programs have additional features. Most will calculate the number of cells in a 'sub-G 1' peak – a measure of apoptosis. They may attempt to calculate the number of aggregated cells present. They may also attempt to subtract debris from the DNA histogram and, in material from paraffin blocks, correct for the presence of sliced nuclei. In my view, it is always preferable to concentrate on cell preparation and the correct analytical procedures rather than rely on a computer program. For example, aggregates should be eliminated during analysis using pulse shape analysis (see above). If one of the more advanced features of a computer program is used, it is important that the correct gating strategy is employed. If debris subtraction is used in the computer analysis, then care should be taken to ensure that the debris is not excluded from the DNA histogram when gating on a peak/area or width/area plot; light scatter gating should not be used.

Controversially, some programs will analyse overlapping cell cycles. Estimates of S phases in polyploid samples containing overlapping DNA histograms may not be reliable and should be treated with caution.

3.8 Multiparametric measurement

The strength of flow cytometry is that several parameters can be measured on each of the cells or nuclei being analysed.

Light Scatter

Forward and right-angle (or side) scatter should be measured routinely. Although these parameters often yield no extra information, occasionally they can improve the analysis. Figure 3.6 shows an example in which right-angle light scatter was used to distinguish between normal and neoplastic nuclei in a breast tumour.

Antigen Expression

Surface antigens

For single surface antigen, the cells should be labelled with the antibody conjugated to fluorescein and fixed in 70% ethanol before adding PI and RNase. If

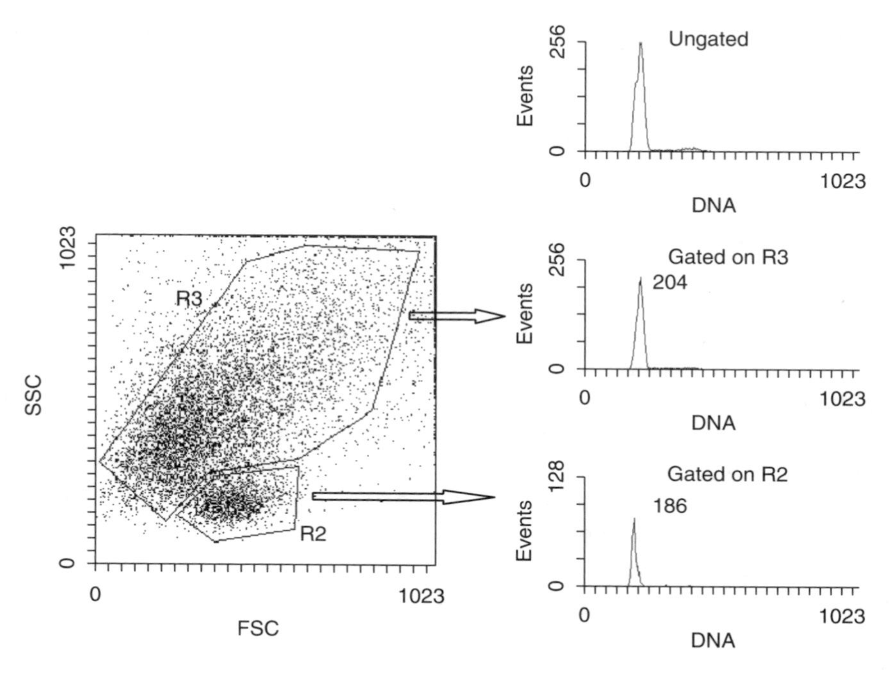

Figure 3.6 Nuclei extracted from a paraffin block of a breast carcinoma and stained with PI. Region R2 defines the normal, diploid, cells and Region, R3, the aneuploid, tumour, cells. The DNA histograms show the effect of gating on the two regions, R2 and R3. The numbers are the mean channel numbers for the G1 peak. It can be seen that the tumour was DNA aneuploid with a DI of 1.1 (204/186). Material supplied by Dr R. Camplejohn and data recorded during the Royal Microscopical Society's Course on Flow Cytometry, Cambridge, 1994

two surface antigens are needed, then fluorescein and phycoerythrin can be used as the antibody labels; alcoholic fixatives should be avoided as denaturation of PE will destroy its fluorescence. The unfixed cells can be labelled with DRAQ5 or, alternatively, they can be permeabilised with 0.02% saponin before adding 7-AAD (Schmid *et al.*, 2000).

Intracellular antigens

The cells need to be fixed and/or permeabilised before labelling with the antibody. The method of fixation has to be chosen so that the epitope, with which the antibody reacts, is preserved. The best method has to be determined by trial and error. Different fixatives have been discussed by Larsen (2000) and Schimenti and Jacobberger (1992).

A combined measurement of DNA with an intracellular antigen often is used to determine the variation of protein expression through the cell cycle. In clinical studies, it is frequently used to distinguish between different cell types, in

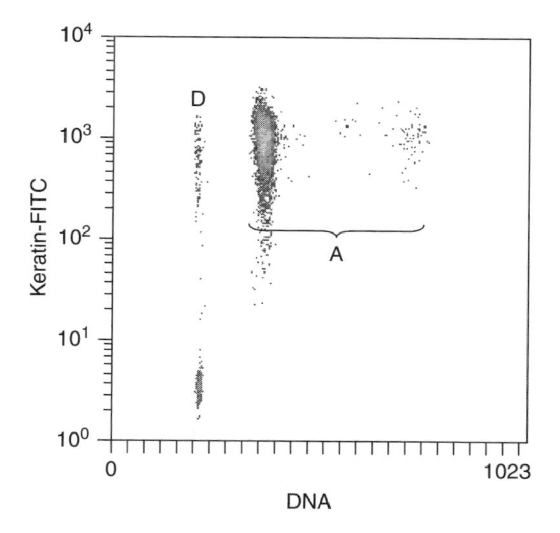

Figure 3.7 Ovarian carcinoma cells fixed and stained with FITC-anti-keratin 8/18 and PI. The diploid cells are labelled 'D' and the aneuploid, 'A'. It can be seen that there are a small number of diploid epithelial cells (keratin positive). Data supplied by Willem Corver, Leiden

particular between epithelial and non-epithelial (presumptively normal) cells in carcinomas (see Figure 3.7).

3.9 Acknowledgements

I thank the colleagues who gave me the data files from which I constructed Figures 3.1, 3.2, 3.6 and 3.7.

3.10 References

Corver WE, Fleuren GJ and Cornelisse CJ (1997) Improved single laser measurement of two cellular antigens and DNA-ploidy by the combined use of propidium iodide and TO-PRO-3 iodide. *Cytometry* **28**: 329–36.

Darzynkiewicz Z and Kapuscinski J (1990) Acridine orange: a versatile probe of nucleic acids and other cell constituents, in: *Flow Cytometry and Sorting*, 2nd edn (eds Melamed MR, Lindmo T and Mendelsohn MI), pp. 291–314. Wiley–Liss, Inc., New York.

Darzynkiewicz Z, Robinson JP and Crissman HA (eds) (1994) *Flow Cytometry. Methods in Cell Biology*, Vols 41 & 42. Academic Press, Inc., San Diego.

Hedley DW (1994) DNA analysis from paraffin-embedded blocks, in: *Flow Cytometry. Methods in Cell Biology*, Vol. 41 (eds Darzynkiewicz Z, Robinson JP and Crissman HA), pp. 231–40. Academic Press, San Diego.

Larsen JK (2000) Measurement of cytoplasmic and nuclear antigens, in: *Flow Cytometry. A Practical Approach*, 3rd edn (ed. Ormerod MG), pp. 133–58. IRL Press at Oxford University Press, Oxford.

Ormerod MG (2000a) Preparing suspensions of single cells, in: *Flow Cytometry. A Practical Approach*, 3rd edn (ed. Ormerod MG), pp. 35–45. IRL Press at Oxford University Press, Oxford.

Ormerod MG (2000b) Analysis of DNA. General methods, in: *Flow Cytometry. A Practical Approach*, 3rd edn (ed. Ormerod MG), pp. 83–97. IRL Press at Oxford University Press, Oxford.

Ormerod MG, Payne AWR and Watson JV (1987) Improved program for the analysis of DNA histograms. *Cytometry* **8**: 637–41.

Ormerod MG, Tribukait B and Giaretti W (1999) Consensus report of the task force on standardization of DNA flow cytometry in clinical pathology. *Anal. Cell Path.* **17**: 103–10.

Ottesen GL, Christensen IJ, Larsen JK, Hansen B and Andersen JA (1995) Flow cytometric DNA analysis of breast cancers with predominance of carcinoma *in situ*: a comparison of the premalignant and malignant components. *Clin. Cancer Res.* **1**: 881–8.

Petersen SE (1985) Flow cytometry of human colorectal tumors: nuclear isolation by detergent technique. *Cytometry* **6**: 452–60.

Rabinovitch PR (1994) DNA content histogram and cell cycle analysis, in: *Flow Cytometry. Methods in Cell Biology*, Vol. 41 (ed. Darnzynkiewicz Z, Robinson JP and Crissman HA), pp. 263. Academic Press, San Diego.

Radbruch A (ed.) (1992) *Flow Cytometry and Cell Sorting*. Springer-Verlag, Berlin.

Schimenti KJ and Jacobberger JW (1992) Fixation of mammalian cells for flow cytometric evaluation of DNA content and nuclear immunofluorescence. *Cytometry* **13**: 48–59.

Schmid I, Cole SW, Zack JA and Giorgi JV (2000) Measurement of lymphocyte subset proliferation by three-color immunofluorescence and DNA flow cytometry. *J. Immunol. Methods* **235**: 121–31.

Shankey TV, Rabinovitch PS, Bagwell B, Bauer KD, Duque RE, Hedley DW, Mayall BH and Wheeless LL (1993) Guidelines for implementation of clinical DNA cytometry. *Cytometry* **14**: 472–7.

Smith PJ, Blunt N, Wiltshire M, Hoy T, Teesdale-Spittle P, Craven MR, Watson JV, Amos WB, Errington RJ and Patterson LH (2000) Characteristics of a novel deep red/infrared fluorescent cell-permeant DNA probe, DEAQ5, in intact human cells analyzed by flow cytometry, confocal and multiphoton microscopy. *Cytometry* **40**: 280–91.

Vindeløv LL, Christensen IJ and Nissen NI (1983) A detergent-trypsin method for the preparation of nuclei for flow cytometric DNA analysis. *Cytometry* **3**: 323–7.

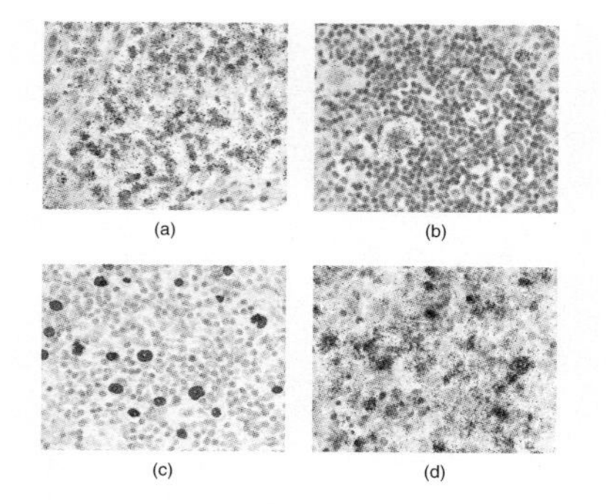

Figure 2.1

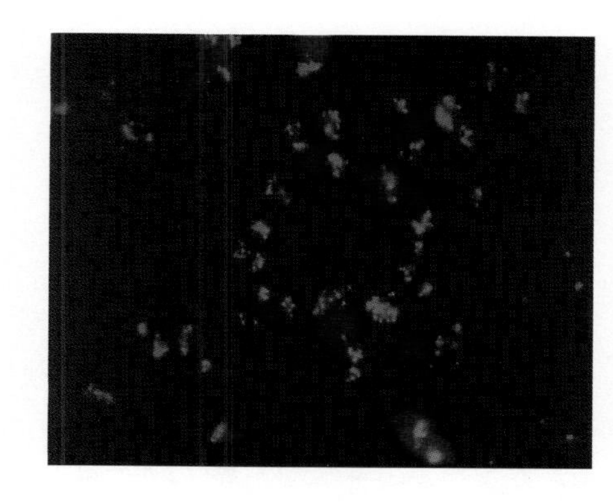

Figure 2.2

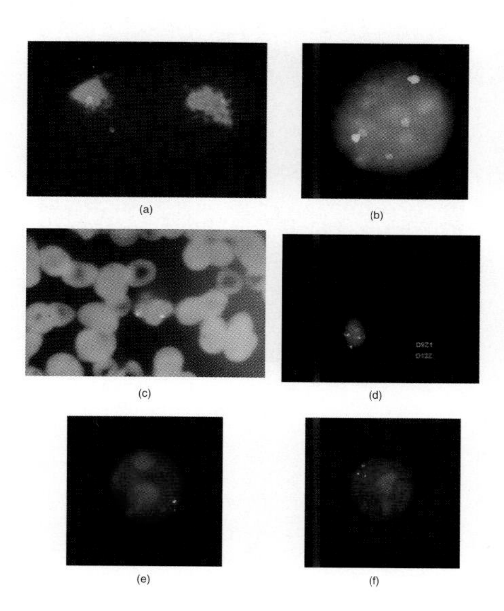

Figure 4.2

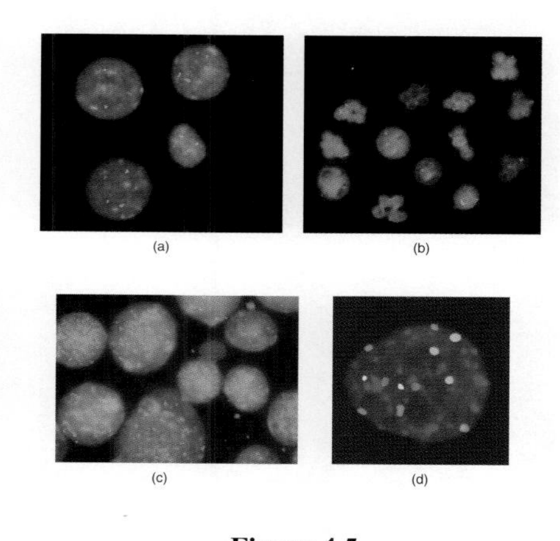

Figure 4.5

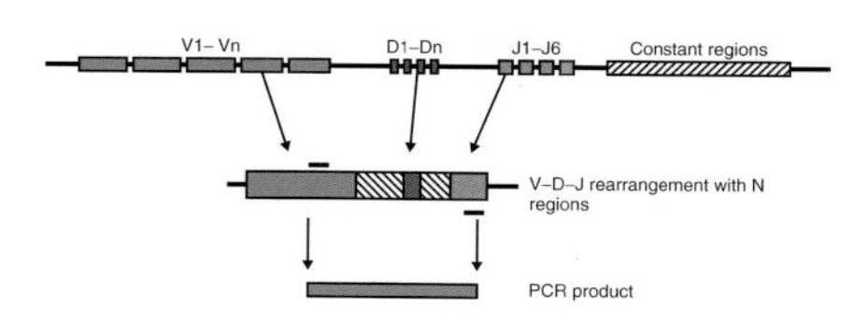

Figure 9.3

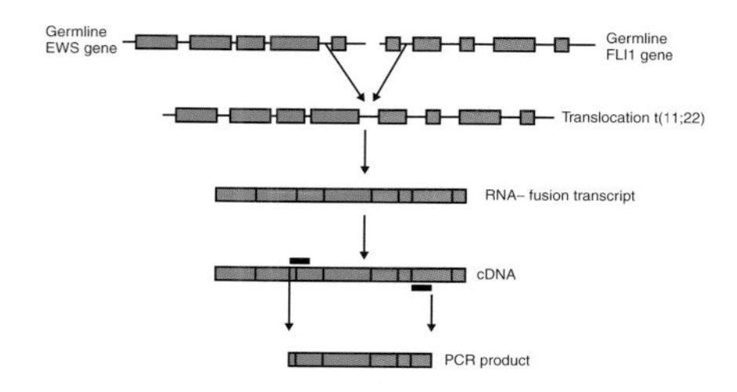

Figure 9.6

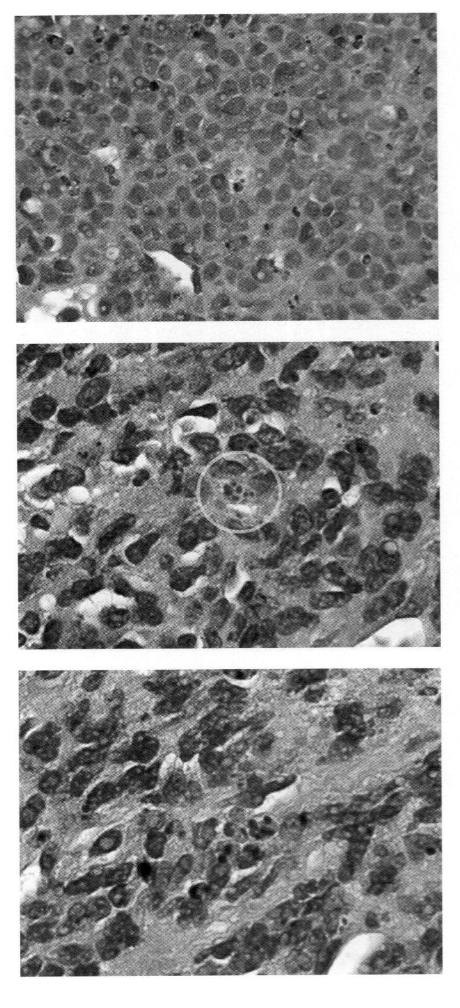

Figure 8.1

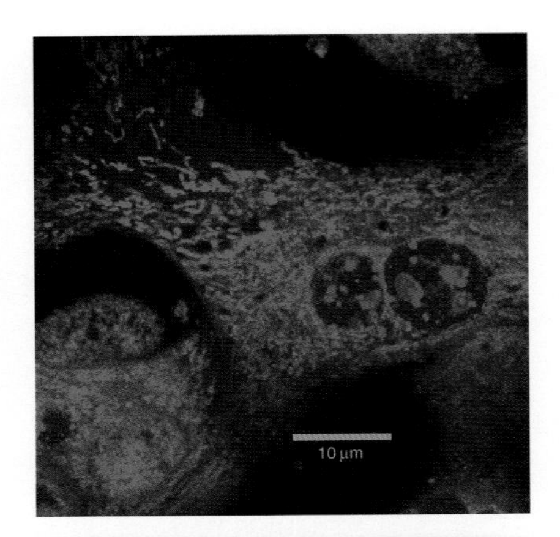

Figure 8.3

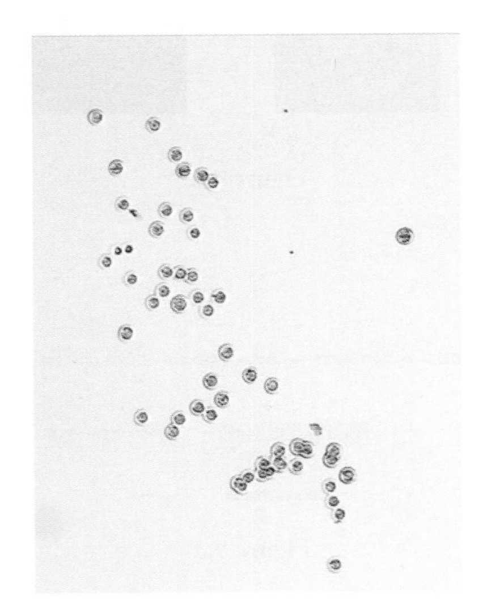

Figure 10.4

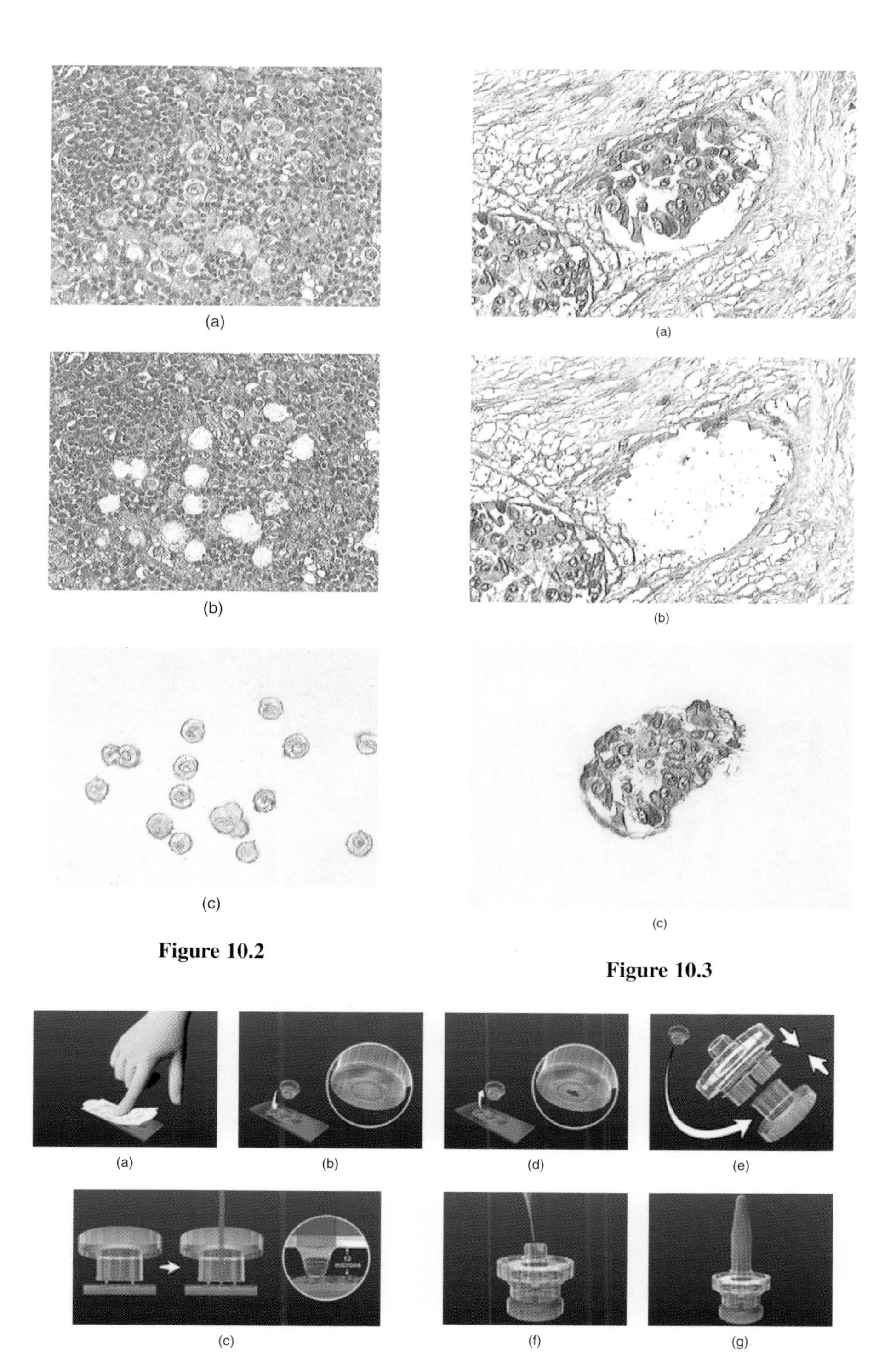

Figure 10.2

Figure 10.3

Figure 10.1

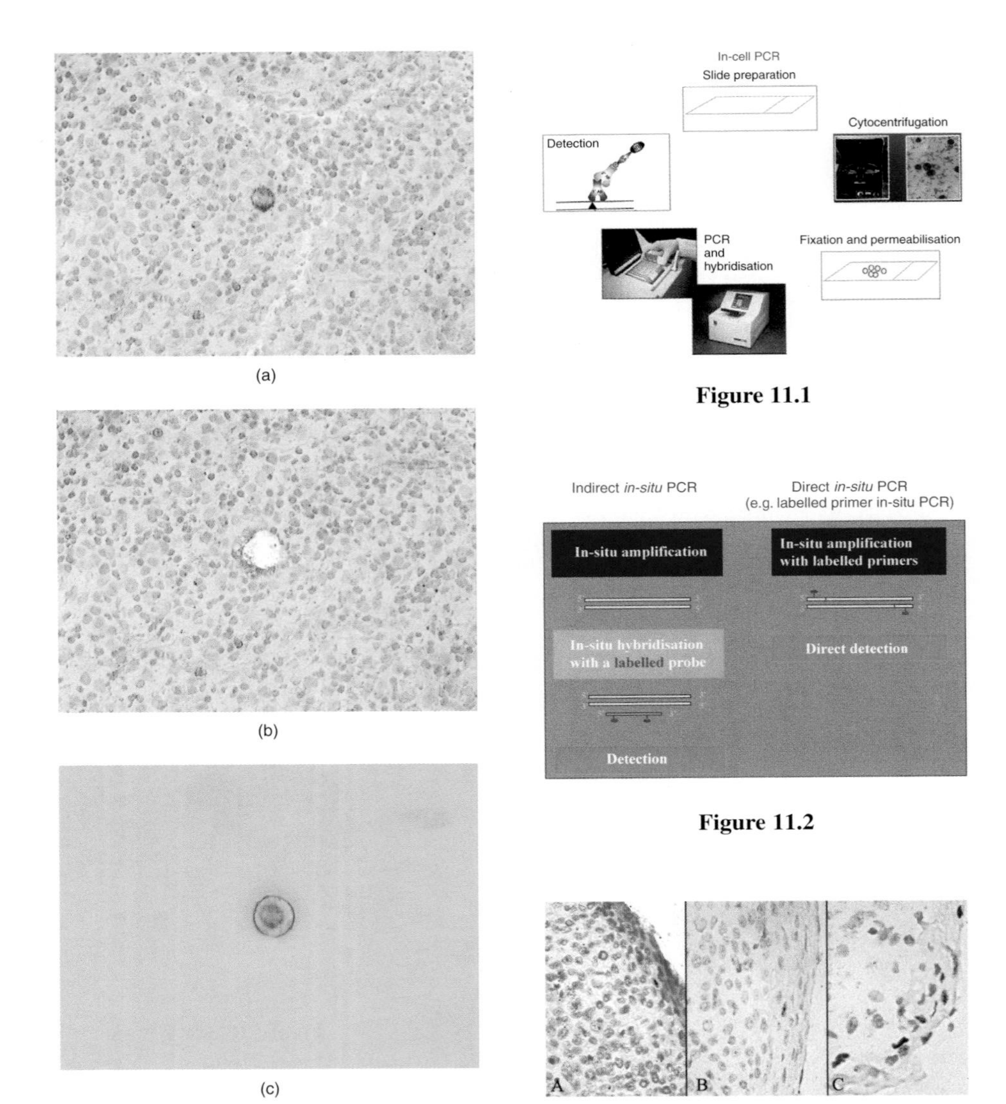

Figure 10.5

Figure 11.1

Figure 11.2

Figure 11.6

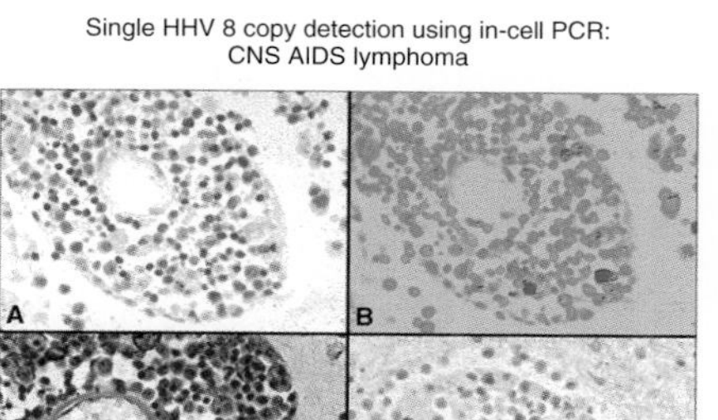

Figure 11.7

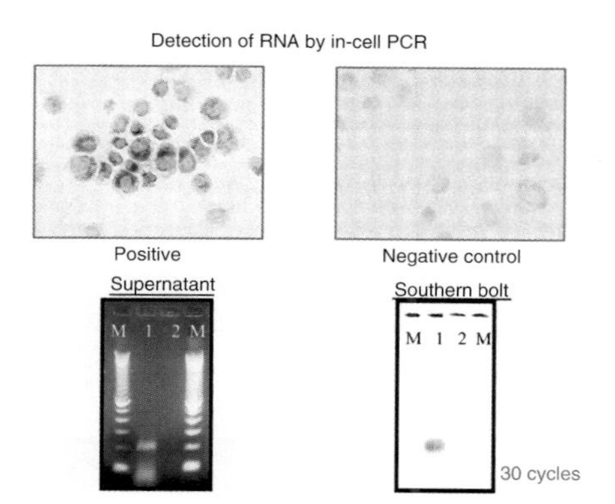

Figure 11.8

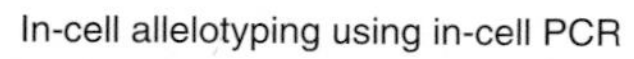

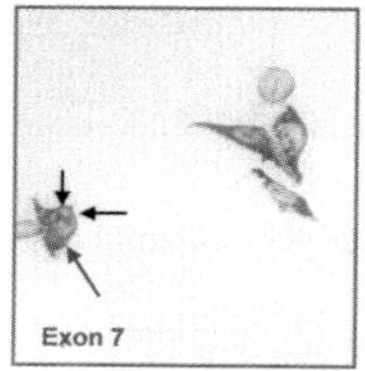

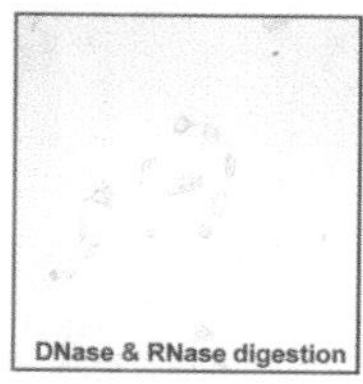

RT 112 carcinoma cells

Figure 11.9

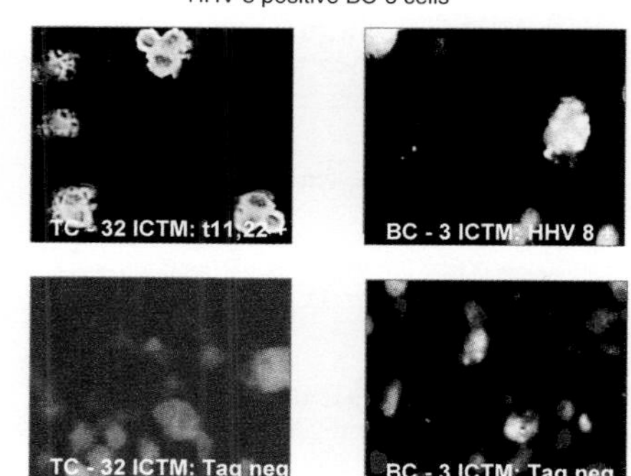

Figure 11.10

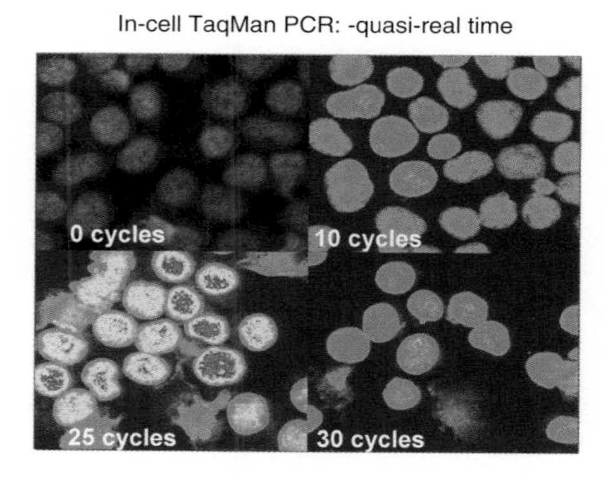

Figure 11.11

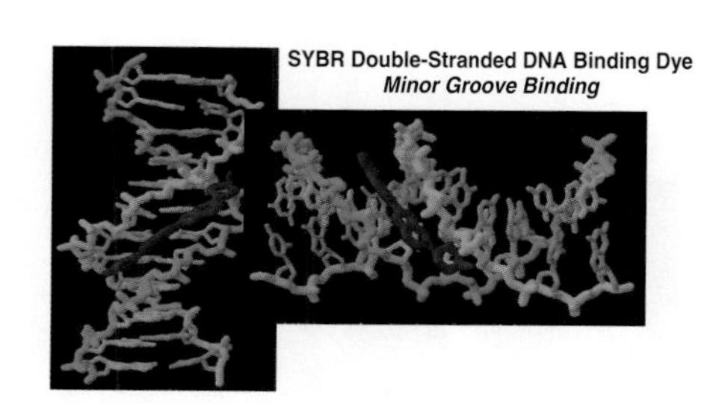

Figure 12.2

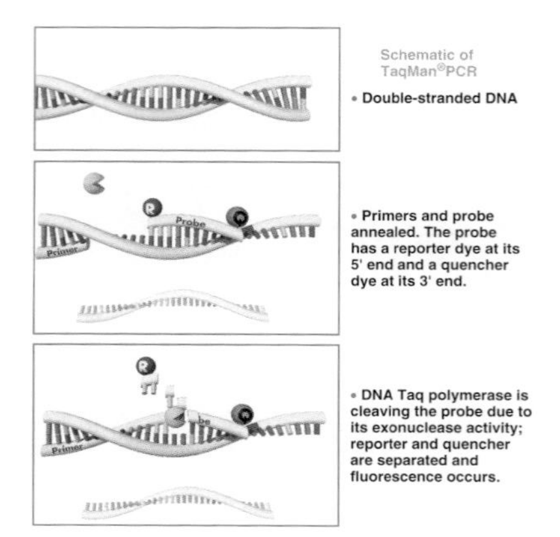

Figure 12.3

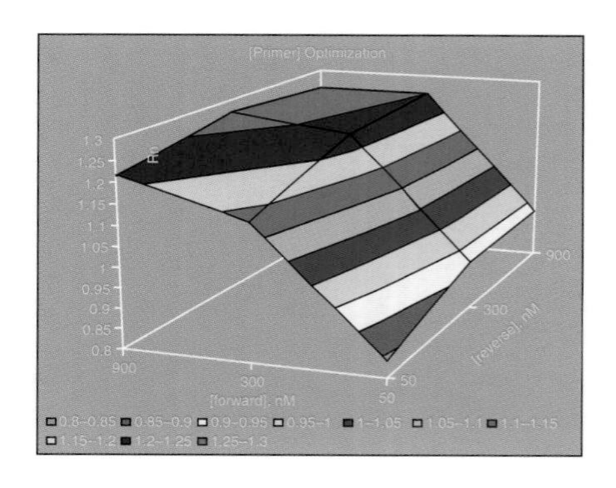

Figure 12.4

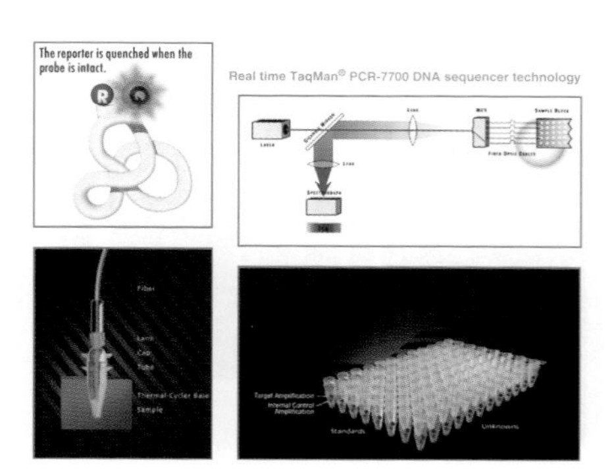

Figure 12.5

Cytokine gene expression plate: Relative quantification results

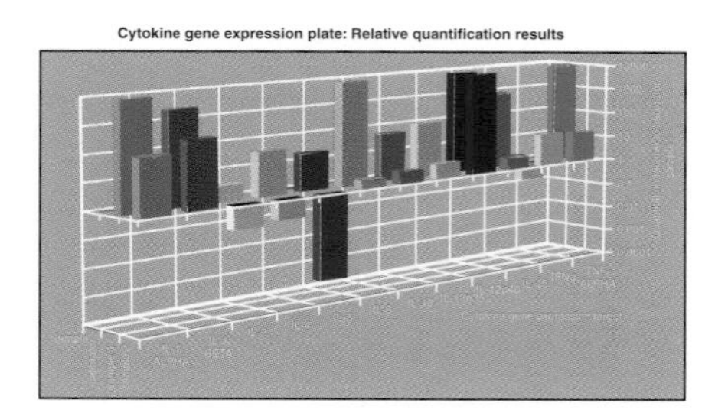

Figure 12.7

Allelic discrimination using TaqMan PCR

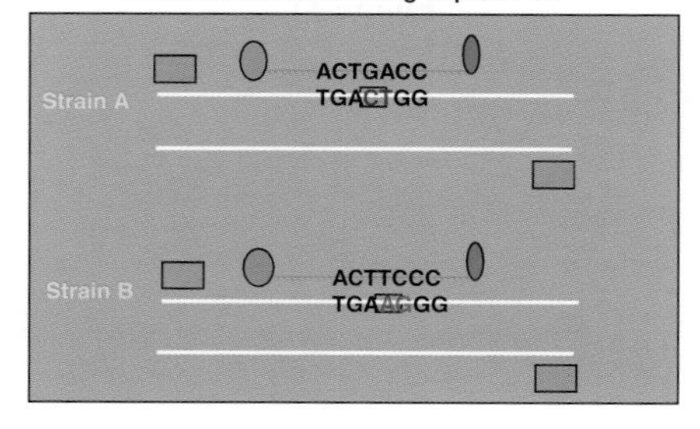

Figure 12.8

Allele Discrimination Data View

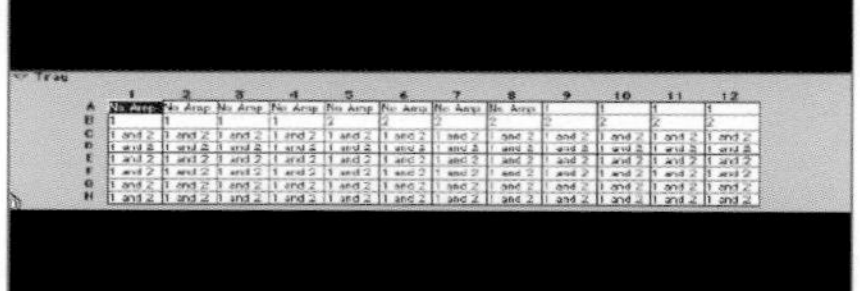

Figure 12.9

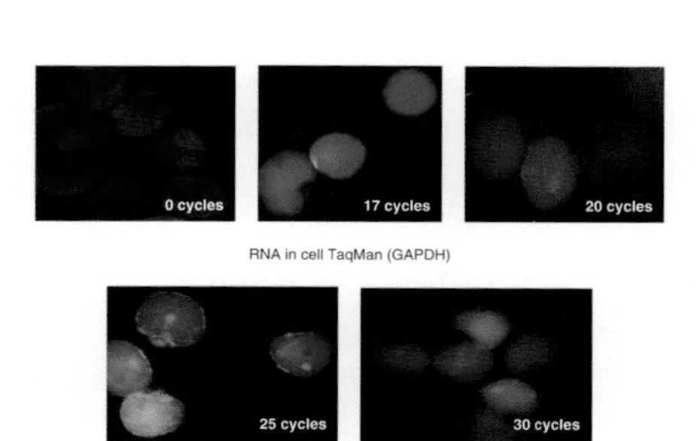

Figure 12.10

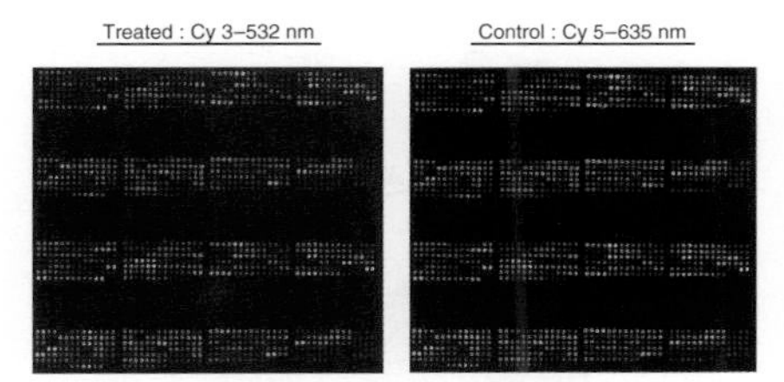

Figure 13.2

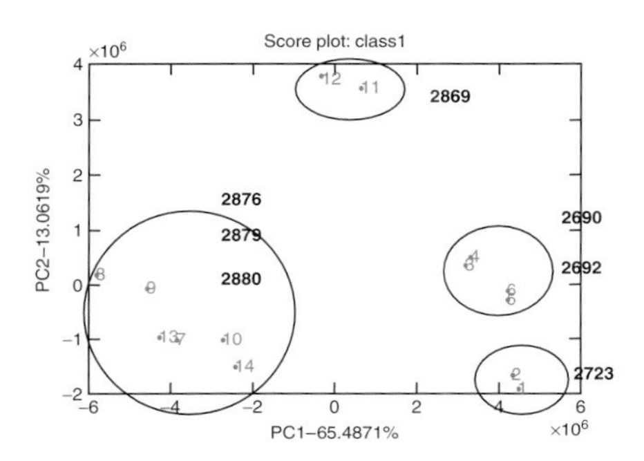

Figure 13.8

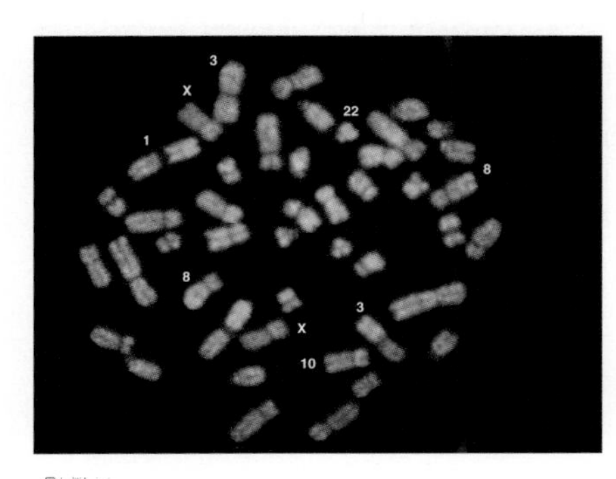

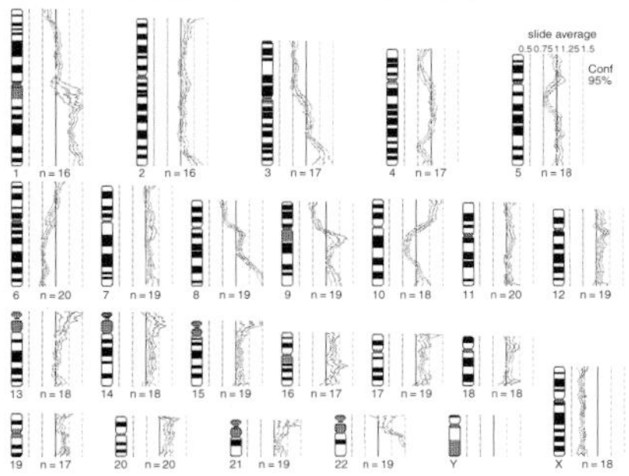

Figure 14.2

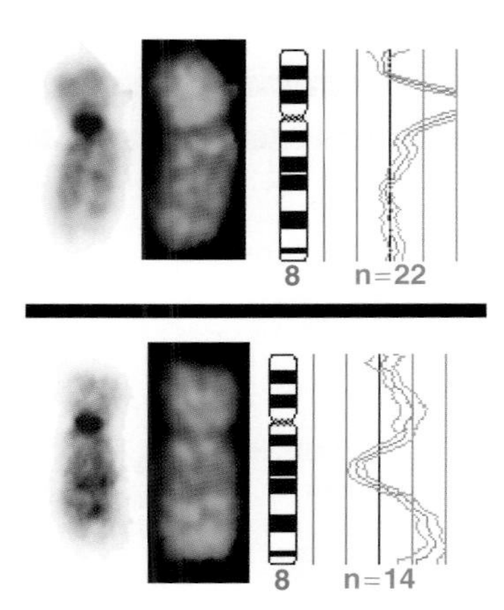

Figure 14.3

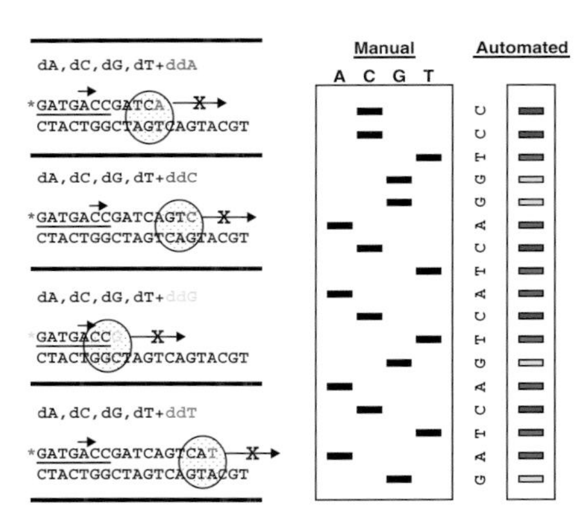

Figure 15.2

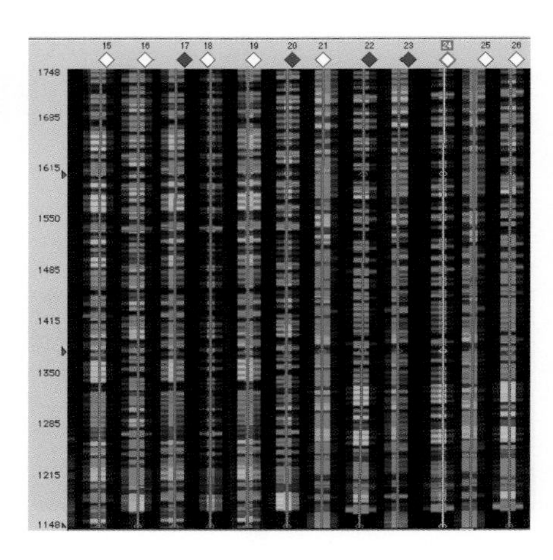

Figure 15.3

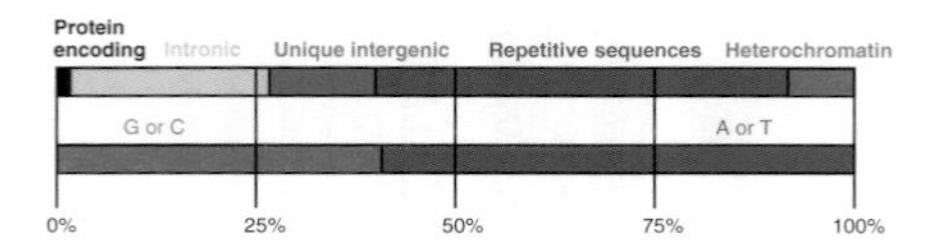

Figure 15.6

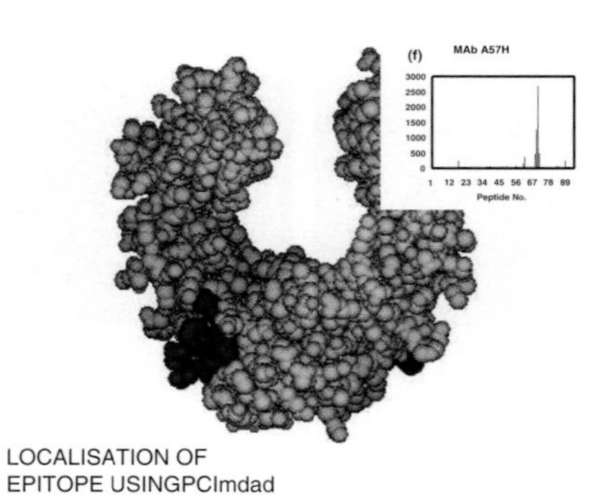

Figure 16.4

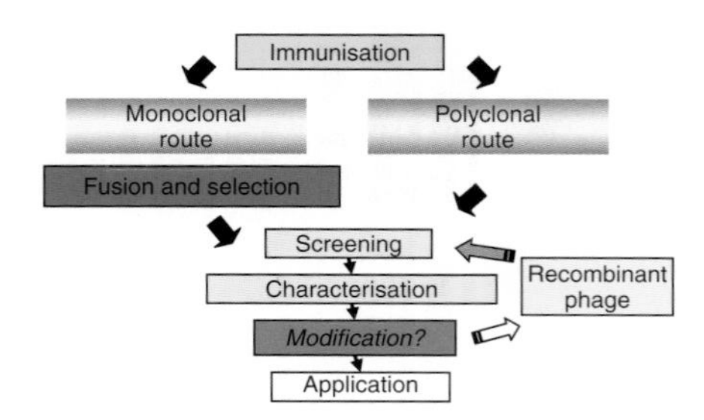

Figure 16.1

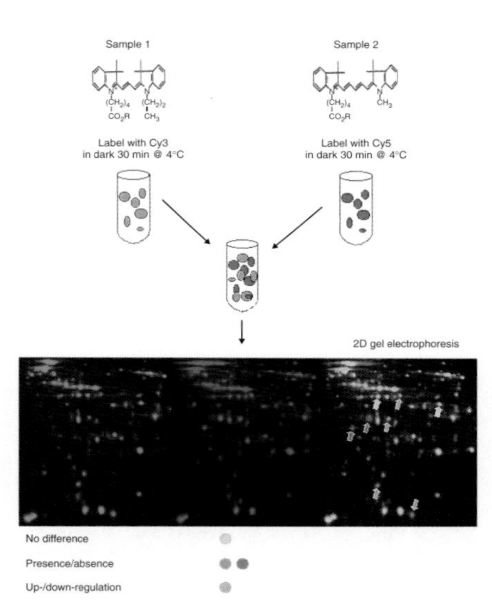

Figure 17.2

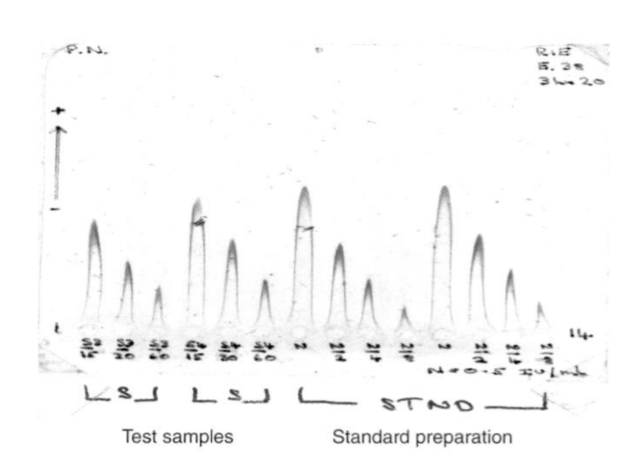

Figure 16.2

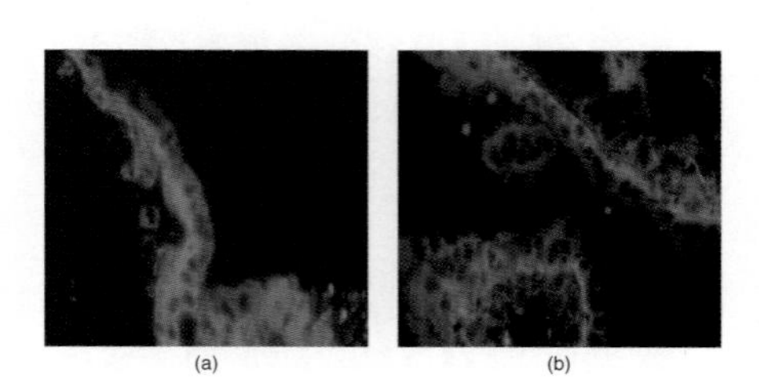

Figure 16.3

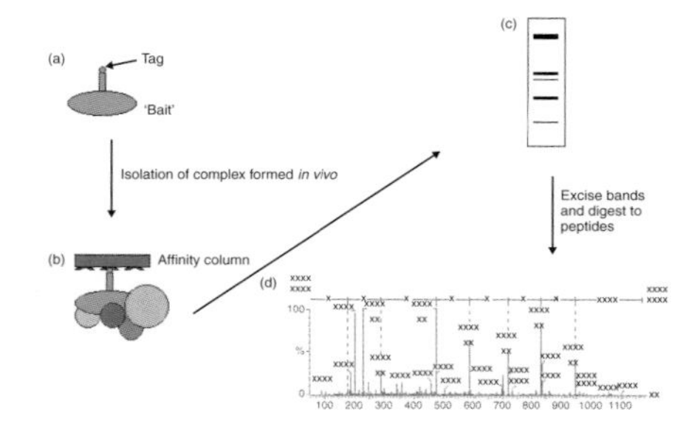

Figure 17.5

4

Interphase Cytogenetics

Sara A. Dyer and **Jonathan J. Waters**

4.1 Introduction

Chromosome Abnormalities in Clinical Medicine

Cytogenetics became of clinical interest in the late 1950s, when the association between particular syndromes and chromosomal aneuploidy (e.g. Down syndrome and trisomy 21) was established. A decade later banding techniques were developed that allowed the unequivocal identification of individual chromosomes and their subdivision into visually recognisable zones or 'bands'. More recently, the development of FISH (fluorescence *in situ* hybridisation) has allowed chromosome structure to be examined in even finer detail. One outcome has been the delineation of a group of syndromes that are characterised by submicroscopic chromosome deletions (e.g. Williams syndrome and microdeletion of chromosome 7) which can be delineated by specific, suitably labelled DNA probes. Cytogenetics is now used as a diagnostic tool for a variety of clinical disciplines and increasingly in the diagnosis and management of cancer.

Chromosome Abnormalities and Cancer

The study of chromosomes in human cancer was given impetus in the early 1960s by the observation that metaphases from bone marrow cells obtained from patients with chronic myeloid leukaemia contained a small marker chromosome, named the 'Philadelphia' chromosome. This was the first association between a chromosome abnormality and a particular malignant disease. In this case the chromosomal change was acquired because it could be demonstrated to be restricted to certain haemopoietic cell lineages. The applicability of cytogenetics in highlighting the location of genes important in tumorigenesis and

Molecular Biology in Cellular Pathology. Edited by John Crocker and Paul G. Murray
 ISBN: 0-470-84475-2

in the diagnosis and management of malignant disease is now firmly established (reviewed by Roylance, 2001). This is particularly true of the leukaemias, but applications to other cancers are also emerging as part of routine clinical practice.

Limitations of Metaphase Chromosome Analysis

Metaphase chromosomes must be obtained from dividing cells for cytogenetic analysis to be performed. In practice the following constraints apply to the application of the technique:

1. Samples must be fresh and ideally processed within 72 h.
2. To obtain sufficient metaphases for analysis, *some* tissues may require lengthy culture times *in vitro*: typically 7–14 days for a skin biopsy or an amniotic fluid sample. Tissues with a high intrinsic mitotic activity (e.g. bone marrow, or chorion villous biopsy) may only require a short *in vitro* incubation period; typically 6–18 h.
3. The cell population grown *in vitro* may not be representative of the original sample. For example, overgrowth of normal stromal cells which are not part of the malignant clone is a common occurrence in the monolayer culture of solid tumours. Metaphase chromosomes may be of poor quality and therefore difficult to band and characterise.

4.2 Interphase cytogenetics

Given the limitations of metaphase analysis, reliable detection of abnormalities in chromosome copy number and structure in nuclei without the need to directly visualise metaphase chromosomes is clearly desirable. The term 'interphase cytogenetics' was coined by Cremer and his co-workers to describe this approach. They developed radio-isotopically or enzymatically labelled probes for specific chromosome targets in interphase nuclei (Cremer *et al.*, 1986). Continuing developments in interphase FISH, together with the increasing availability of gene-specific probes, has considerably extended the application of interphase cytogenetics to a wider range of fresh and archival tissues.

Sample Type and Preparation

FISH on interphase nuclei is possible on a wide range of specimen types. Early interphase cytogenetic work concentrated largely on fixed cells, but the technique is equally applicable to tissue imprints from tumour or tissue samples. For the latter approach, a cut section of fresh or frozen tissue is dabbed onto a clean microscope slide allowing a thin layer of cells to be transferred onto the slide.

This 'no culture' method provides an immediate source of target material for interphase cytogenetics making it an attractive option in cases where a rapid clinical result is required.

Unlike metaphase FISH which requires dividing cells from fresh tissue samples, interphase FISH may be performed on both fresh/frozen tissue and archival specimens including formalin-fixed paraffin embedded tissues. Prior to FISH, paraffin embedded material is cut, placed on slides, de-waxed and then pretreated with enzymes to remove stromal tissue from around the target cells. For some applications, de-waxed cells are dissociated into a cell suspension, while for others it is possible to carry out FISH *in situ*. The latter approach allows distinct histological regions of a tissue to be examined for potentially different genetic changes. For technical information regarding FISH on paraffin sections, see the review by Werner and colleagues (Werner *et al.*, 1997).

Probe Preparation

Probes for *in situ* hybridisation are, for the most part, cloned in vectors such as bacterial artificial chromosomes (BACs), yeast artificial chromosomes (YACs) and cosmids which allow DNA fragments of several hundreds of kilobases of DNA to be amplified. Probe production by the polymerase chain reaction (PCR) has also become commonplace in recent years. Probes are either directly labelled with a fluorescent marker or indirectly labelled with a reporter molecule which is detected post-hybridisation with a fluorescent molecule. By using differentially labelled probes it is possible to apply several different probes simultaneously. Many different probes are now commercially available in a pre-labelled 'ready-to-use' form making the FISH technique a simple and accessible technique for most clinical and research laboratories.

Probe Types

FISH probes commonly used to interrogate interphase preparations fall into three main categories: repetitive sequence probes, locus-specific probes, and chromosome libraries (Figure 4.1). Centromere-specific repetitive sequence alpha satellite DNA has now been isolated from nearly all human chromosomes and forms the basis of chromosome-specific probes or chromosome enumeration probes. Application of these alphoid probes to interphase nuclei allows the rapid detection of chromosome copy number in non-dividing cells. Furthermore, by using differentially labelled probes it is possible to determine the copy number for several chromosomes simultaneously.

Locus-specific probes are designed to hybridise to a unique sequence or gene region within the genome and their use allows not only the detection of copy number changes, but also the detection of structural rearrangements involving these loci.

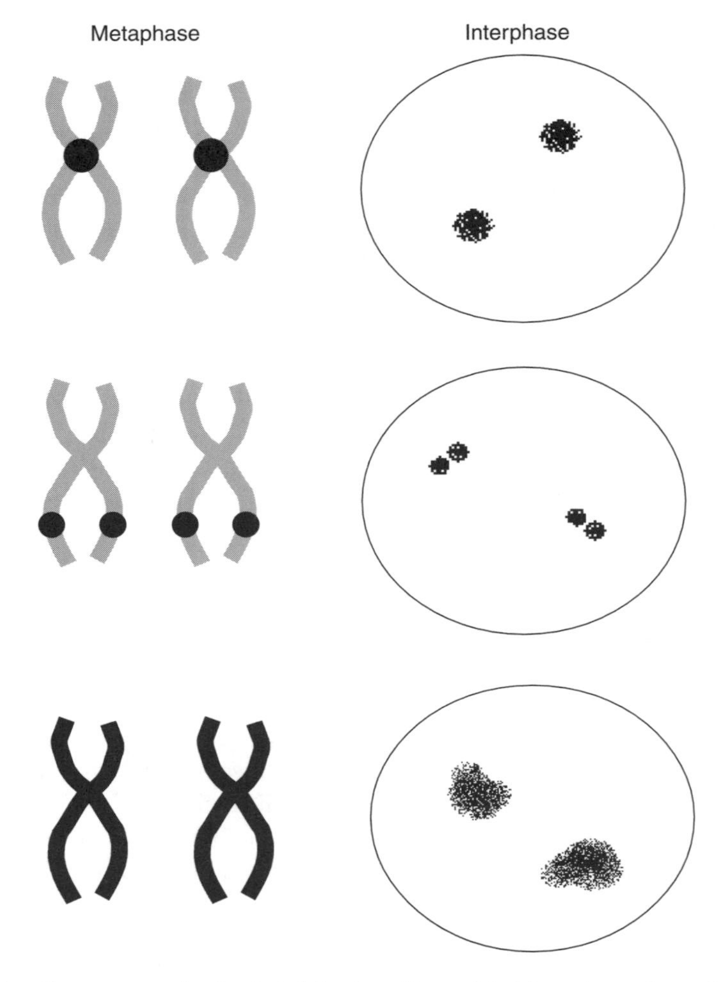

Figure 4.1 Fluorescence *in situ* hybridisation. Top: alphoid centromeric probe; middle: locus-specific probe; bottom: whole chromosome library

Whole chromosome 'paints' or 'libraries' are collections of composite probes containing DNA sequences from a given individual chromosome. DNA from specific chromosomes is initially obtained using techniques such as flow-sorting and subsequently digested into DNA fragments and cloned using vectors (e.g. plamids, cosmids) or PCR technology. Repetitive sequences within the DNA fragments are blocked prior to hybridisation leaving the unique sequences to hybridise along the length of one specific chromosome pair. This has been termed 'chromosome painting' (Pinkel *et al*., 1988) and is most useful in metaphase chromosomes, although the chromosome library can also be visualised in interphase nuclei where the chromosomes are spatially arranged in individual 'territories' (Figure 4.1 and Figure 4.2(a)).

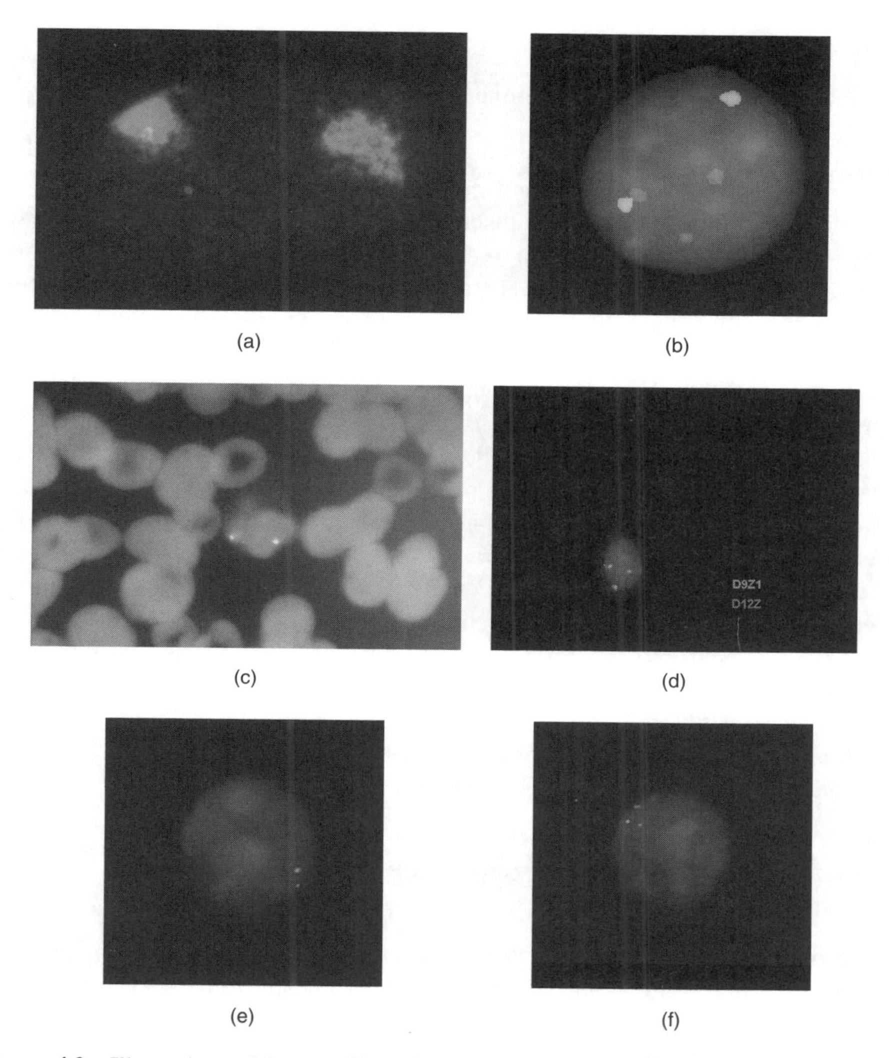

Figure 4.2 Illustrations of the use of interphase cytogenetics in the analysis of constitutional genetic disorders. (a) FISH analysis of large-scale chromatin organisation in a fibroblast cell. A signal for an MHC class II probe (blue) at band 6p21.3 is a long distance away from the chromosome 6 territory (green) on the left side of the cell. The chromosome 6 centromere is labelled in red. This configuration indicates the presence of a large chromatin loop extending from the surface of the chromosome territory; courtesy of Dr D. Sheer. (b) Uncultured amniocyte preparation with Vysis probe set LSI 13/31 showing three chromosome 21 signals in red and two chromosome 13 signals in green. (c) A blood smear showing a polymorphonucleocyte with two signals (yellow) with the probe D12Z specific for chromosome 12. (d) A squamous buccal mucosal cell with three green signals with the probe D9Z1 and two red signals with the control probe D12Z. (e) and (f) Lymphocytes from a control (male) and a male patient with Pelizaeus–Merzbacher disease with the probe PLP 125a1 (green) which maps to Xq22 and an X centromere-specific probe (red). Note that in the patient the PLP target sequence is duplicated and two discrete signals can be visualized; courtesy of Mr R. Palmer. A colour version of this figure appears in the colour plate section

In situ Hybridisation

The techniques are essentially similar to those described in Chapter 2.

Organisation of Chromosomes

Interphase chromosomes occupy discrete 'territories' in the nucleus (reviewed in Chevret *et al.*, 2000). In turn these territories contain relatively compact chromatin-rich 'domains' which are associated with active gene transcription. Transcriptionally active DNA appears to be markedly compartmentalised at or near the surface of these domains (Verschure *et al.*, 1999). For example, use of FISH, with multiple probes for the major histocompatibility complex (MHC) on chromosome 6, has allowed direct visualisation of gene transcriptional activity as judged by the presence of the HLA gene cluster on large chromatin loops extending out from chromatin-rich domains. The observation of these loops is cell-type-dependent and can be enhanced by appropriate external stimuli (e.g. interferon) (Volpi *et al.*, 2000) (Figure 4.2(a)). More generally, there appears to be a correlation between the gene density of a particular chromosome and its internal nucleus position: gene-rich chromosomes (e.g. 19) are located centrally and gene-poor chromosomes (e.g. 18) are located peripherally. This property is evolutionarily conserved and suggests a functional role for such positioning (Tanabe *et al.*, 2002). The pathological consequences of 'repositioning' of chromatin as a result, for example, of chromosomal translocation, is not yet known but may turn out to be of clinical interest.

Behaviour of Chromosomes: Replication Asynchrony, Spacial Association and Imprinting

The proximal region of the long arm of chromosome 15 (15q11–q13) is subject to genomic imprinting. That is, each region of chromosome 15q11–q13 (one paternal and one maternal) shows a pattern of gene expression which is determined by the parent of origin of that region of the chromosome. Interphase FISH studies have demonstrated that the two 15 homologues replicate asynchronously within the imprinted region. Patients with Prader–Willi or Angelman syndrome with deletions or mutations within the imprinted region, but who demonstrate biparental inheritance of chromosome 15, also show asynchronous replication. In contrast, those patients with Prader–Willi syndrome who show maternal uniparental disomy (for chromosome 15) and those with Angelman syndrome who show paternal uniparental disomy, show synchronous replication (Knoll *et al.*, 1994). White *et al.* (1996) were able to show that this pattern of synchronicity could form the basis of a diagnostic test for uniparental disomy at least for chromosome 15. Further studies demonstrated a spatial and temporal association at interphase between oppositely imprinted regions of 15q11–q13 in normal individuals. This association occurred specifically at the imprinted 15q11–q13 region and only during late S-phase of the cell cycle (LaSalle and Lalande, 1996). The clinical value of these observations remains to be determined.

4.3 Applications

Prenatal Genetic Diagnosis

Interphase FISH provides a powerful technique both to exploit classical sampling methods (i.e. chorion villous biopsy and amniocentesis) and in the development of new non-invasive approaches to prenatal diagnosis.

Rapid screening for aneuploidy

Detection of trisomy 21 (Down syndrome) is the most common indication for prenatal diagnosis in the United Kingdom, accounting for approximately 37 000 procedures per annum. Rapid interphase FISH on uncultured cells from an amniotic fluid sample allows a result for chromosome 21 copy number to be obtained in 24–48 h rather than the 7–14 days required for conventional chromosome analysis. The screen can also be set up to co-detect aneuploidy for 13, 18, X and Y (Figure 4.2(b)). In addition, enumerators for the sex chromosomes can allow for rapid fetal sexing, for instance in cases of sex-linked genetic disorders.

Numerous studies have shown that the sensitivity of interphase FISH on uncultured amniocytes approaches that of conventional analyses in appropriate samples (Ward *et al.*, 1993), even though most of the cells (>80%) in the mid-trimester amniotic fluid sample are degenerate squamous epithelial cells that are unsuitable targets for FISH. Maternal cell contamination may confound interpretation in a small proportion of cases (Witters *et al.*, 2002). In modern diagnostic practice the technique usually is used as an adjunct to conventional analysis in high-risk cases. For this particular application, PCR-based methods are likely to prove more amenable to a high volume, automated approach.

Non-invasive methods of prenatal diagnosis

Invasive procedures carry the risk of associated fetal loss. Alternative 'non-invasive' techniques are being explored to try to remove or minimise this risk. Approaches being investigated include retrieval of fetal cells from the uterine cavity by lavage and isolation of fetal cells from the maternal circulation. Interphase FISH provides a powerful method for investigating these possibilities by allowing direct access to small numbers of nuclei that can be interrogated for chromosome copy number at the single cell level.

Intrauterine lavage

Trophoblast cells are shed towards the cervix until weeks 13–15 of pregnancy when the uterine cavity is obliterated. Cioni *et al.* (2002) used cells retrieved by intrauterine lavage from women electing to undergo termination of pregnancy between 7 and 12 weeks gestation. Using FISH they were able to demonstrate a male signal pattern (X/Y) in 32/40 (80%) of male fetuses. PCR methods showed a similar level of efficiency. Questions concerning the efficiency of the technique and its safety remain to be resolved.

Fetal cells in maternal circulation

The ability to reliably and efficiently access fetal cells (or fetal DNA) from the maternal circulation for genetic diagnoses would clearly be of considerable clinical value. FISH has been used to successfully determine chromosome copy number after isolation of suitable target cells [e.g. nucleated red blood cells (NRBCs)]. Typically, methods involve one or more enrichment steps (e.g. fluorescent-activated cell sorting using fetal-specific cell surface markers) followed by identification of individual fetal cells prior to FISH or PCR by, for example, cytological observation, where large nuclear/cytoplasm ratio, compact mononuclear shape and symmetrical chromatin condensation are characteristics screened for (Samura *et al.*, 2001). In prospect is the possibility that isolated intact cells may not be required. Free-fetal DNA (ffDNA) is present in the maternal circulation during pregnancy, and using real-time PCR analysis, Lo *et al.* (1999) were able to consistently measure gene dosage levels consistent with trisomy 21 in blood samples from women between 12 and 21 weeks gestation.

Preimplantation Genetic Diagnosis

Preimplantation genetic diagnosis (PGD) by embryo biopsy represents a special form of prenatal diagnosis which is usually performed at the cleavage stage at day 3 of development. FISH or PCR-based assays can be performed on a single cell and a diagnosis obtained, although chromosomal mosaicism (FISH) and allelic drop-out (PCR) may present technical difficulties (Harper and Delhanty, 2000). A further refinement involves FISH to detect and distinguish between the products of balanced structural rearrangements in preimplantation embryos (Scriven *et al.*, 1998). FISH has also been used to provide insights into human fetal wastage: even normal preimplantation embryos from fertile patients show frequent chromosomal mosaicism or a 'chaotic' chromosomal picture (e.g. diploid/triploid cells with accompanying chromosome monosomy) (Delhanty *et al.*, 1997).

Postnatal Diagnosis

Screening for aneuploidy

Rapid confirmation of chromosome abnormality sometimes may be required in the neonatal period in the presence of congenital malformations such as tracheo-oesophageal fistula (possible Edward syndrome) or particular heart defects (possible Down syndrome). Interphase FISH can be performed on uncultured blood smears (Figure 4.2(c)) or cells grown in short-term *in vitro* culture and a result obtained within 24 h (McKeown *et al.*, 1992). This approach can also be used for rapid confirmation of chromosomal sex in newborns with ambiguous genitalia.

Screening for submicroscopic deletions and duplications

FISH probes may be used to detect submicroscopic deletions on metaphase chromosomes where the binding or otherwise of a probe specific to a particular chromosomal region can be observed and recorded using the non-deleted chromosome pair as a control. A number of microdeletion syndromes have been described and their diagnosis using metaphase FISH analysis is well established. A much smaller number of syndromes have been described which are associated with a submicroscopic *duplication* of a particular chromosomal region. Examples are a peripheral neuropathy, Charcot–Marie–Tooth disease Type 1A (CMT1A) and Pelizaeus–Merzbacher disease (PMD). PMD is an X-linked recessive dysmyelinating disorder of the central nervous system. Submicroscopic duplication of the proteolipid protein gene (PLP), which maps to the long arm of the X chromosome at Xq22 are frequently observed in this disease (Woodward *et al.*, 1998). The PLP duplications associated with the disease cannot be reliably detected using metaphase FISH and interphase analysis is required for their detection (Woodward *et al.*, 1999) [Figure 4.2(e) and 4.2(f)]. This is because chromatin in metaphase chromosomes is more compact and does not allow the resolution by FISH of sequences separated by less that 1 Mbp. In contrast in interphase nuclei, a resolution of 100 kbp can be achieved (Trask *et al.*, 1993). Such microduplication syndromes may be under-reported as, on theoretical grounds, they might be expected to be at least as common as microdeletion syndromes (Ji *et al.*, 2000).

Detection of mosaicism

Screening for chromosomal mosaicism is time consuming using banded metaphases spreads and FISH may provide a useful alternative approach. Metaphases and interphase nuclei may be screened simultaneously if a suitable probe is used. Differences in levels of mosaicism may be observed, reflecting the fact that the metaphase cells may not be representative of the total cell population in that sample, as illustrated in the following case. Cytogenetic analysis of a sample from an eight-month old female, who presented with dysmorphic features, revealed that she had three constitutional cell lines in blood lymphocytes, as shown in Table 4.1. Note that the proportion of each cell line differed considerably both with respect to culture time (48 h and 72 h metaphase analysis) and between metaphase and interphase analysis. FISH (using a whole chromosome library, wcp 9), also allowed the marker chromosome to be identified as chromosome

Table 4.1 Screening for chromosome mosaicism in a dysmorphic child

	Lymphocytes (48 h culture)	Lymphocytes (72 h culture)	Buccal mucosal cells
Metaphase	47,XX,+9[3] 47,XX,+mar[20] 46,XX[7]	47,XX,+mar[51] 46,XX[9]	
Interphase	18/80–3 signals 62/80–2 signals	14/80–3 signals 64/80–2 signals	12/30–3 signals 18/30–2 signals

Metaphases were scored using conventional cytogenetic analysis. mar = marker chromosome; numbers in brackets indicates number of metaphases observed.
Interphase cells were scored from the same samples using FISH with the probe D9Z1 which recognises the additional chromosome 9 but not the 9-derived marker chromosome. Note that the interphase cells in all three samples show a consistent signal pattern which is discrepant with the proportion of trisomy 9 cells seen at metaphase.

9-derived. Interphase FISH also allowed the identification of the trisomy 9 cell line in buccal mucosal cells (Figure 4.2(d)). The phenotype was consistent with mosaicism for trisomy 9.

Interphase Cytogenetics of Non-malignant Lesions

Trisomy 7 is a common chromosomal gain in a variety of tumours. More surprisingly, trisomy 7 mosaicism has been observed in a number of non-malignant lesions, e.g. osteoarthritis, rheumatoid arthritis and pigmented villonodular synovitis (PVNS) as well as in apparently normal tissue. Broberg *et al.* (2001) provided evidence that the acquisition of trisomy 7 in synovial samples may be age-related rather than a product of the disease process itself. Other studies have shown that the gain of chromosome 7 may occur independently in a number of cells within the same tissue (Broberg *et al.*, 1998). Furthermore, Dahlen *et al.* (2001) in a study involving patients affected by osteoarthritis, PVNS, and other forms of synovitis were able to demonstrate, using FISH, that in some cases trisomic cells were confined to proliferating synoviocytes within villous extensions of the synovial membrane.

Management of Haematological Malignancies

Many haematological malignancies are associated with well-defined genetic abnormalities including chromosome rearrangements and aneuploidy. Although these abnormalities are often detectable by the examination of metaphase chromosomes using a conventional cytogenetic approach, it is often the case that limited dividing cells, or only cells of inferior quality, are available for analysis. Interphase cytogenetics allows genetic abnormalities to be detected without the need for dividing cells and has thus revolutionised the management of haematological malignancies contributing not only diagnostic and prognostic information, but also enabling treatment response to be monitored.

Diagnosis and prognosis

Chronic myeloid leukaemia

Reciprocal translocations in haematological malignancies are relatively common and the consequences of these rearrangements at the molecular level are well understood. The classic example of such a chromosome translocation is the t(9;22)(q34;q11), diagnostic of chronic myeloid leukaemia (CML) and seen in a proportion of acute myeloid and acute lymphoid leukaemias. This rearrangement fuses the *ABL* oncogene normally located on the long arm of chromosome 9 with the *BCR* region of chromosome 22 creating a novel fusion gene that promotes increased tyrosine kinase activity. *BCR/ABL* fusion can be detected in interphase nuclei using differentially labelled 'single-fusion' locus-specific probes located immediately proximal to the *BCR* breakpoint and immediately distal to the *ABL* breakpoint. A nucleus that has a *BCR/ABL* rearrangement will in most cases show one overlapping *BCR/ABL* signal representing the derivative chromosome 22 together with one native *BCR* signal and one native *ABL* signal representing the normal chromosome 22 and normal chromosome 9, respectively. In contrast, a nucleus lacking the 9;22 rearrangement will show two native *ABL* and two native *BCR* probe signals (Figure 4.3(a) and (b)).

Further examples of translocations detectable by interphase cytogenetics include *PML/RARα* fusion associated with the t(15;17)(q21;q21) in acute myeloid leukaemia, subtype M3 (AML-M3), *TEL/AML1* fusion associated with t(12;21)(p13;q21) in paediatric acute lymphoblastic leukaemia (ALL) and *IGH/MYC* associated with t(8;14)(q24;q32) in chronic lymphocytic leukaemia (CLL). The advent of interphase cytogenetics using translocation-specific probes has meant that it is now possible to make an accurate and rapid diagnosis of many haematological malignancies, even if no dividing cells are available, thereby allowing the most appropriate treatment regimes to be initiated.

Mixed lineage leukaemia

Some genetic rearrangements implicated in haematological malignancies involve the fusion of a specific oncogene with one of various potential partner chromosomes; one such oncogene is mixed lineage leukaemia (MLL) which is normally located at 11q23. By using differentially labelled probes that flank the MLL locus, it is possible to detect an MLL rearrangement regardless of the partner chromosome involved. Interphases in which MLL is rearranged will show one pair of differentially labelled flanking probes separated from each other while cells with no rearrangement will show both pairs of flanking probes juxtaposed (Figure 4.4).

Acute lymphoblastic leukaemia

It is not only locus-specific probes that may be used to provide diagnostic and prognostic information in haematological malignancies. Differentially labelled

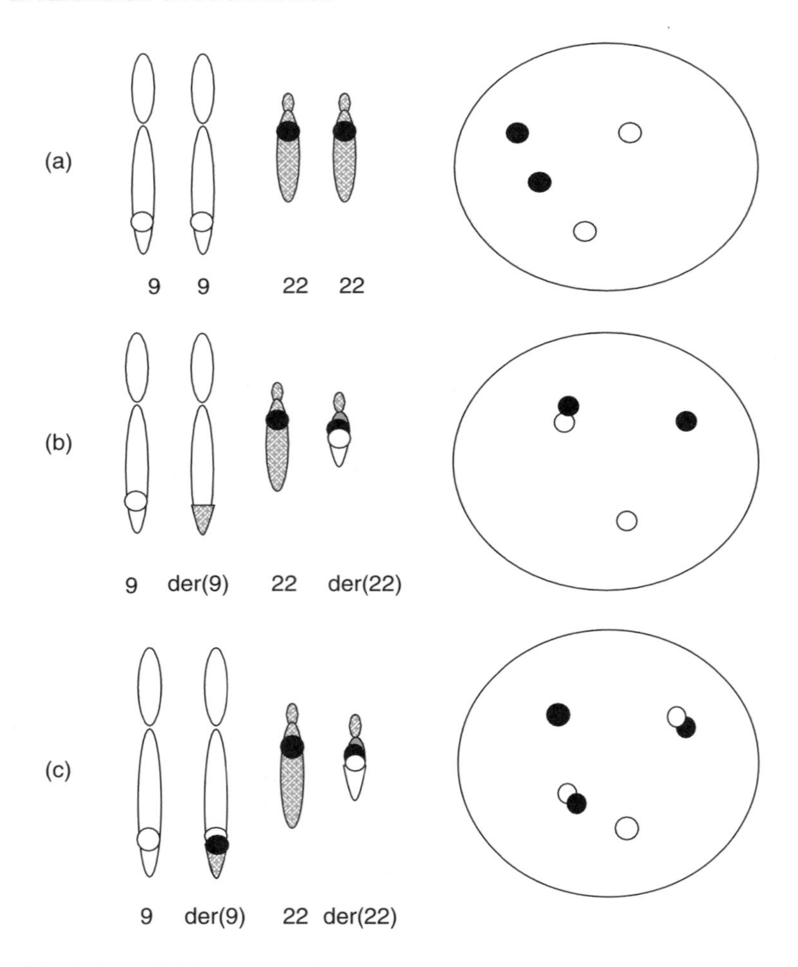

Figure 4.3 Schematic illustration of *BCR/ABL* locus-specific probe sets to detect translocations at metaphase and interphase: (a) no translocation present; (b) translocation as detected using 'single-fusion' probe set; (c) translocations as detected using 'dual-fusion' probe set

centromeric probes can be used to detect gain of specific whole chromosomes (hyperdiploidy) in some forms of childhood acute lymphoblastic leukaemia. As these patients respond well to certain forms of treatment, the rapid detection of hyperdiploidy using interphase cytogenetics can be extremely valuable prior to the commencement of treatment.

Disease monitoring

The single-fusion probes discussed above provide an excellent option for the diagnosis of diseases such as CML where the proportion of abnormal cells is likely to approach 100%. However, where levels of abnormal cells are lower,

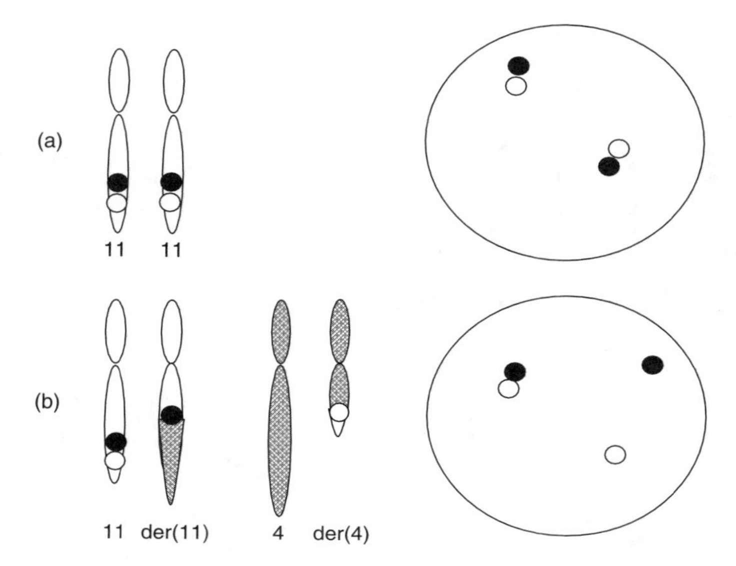

Figure 4.4 Schematic illustration of MLL locus-specific 'break-apart' probe set to detect translocations at metaphase and interphase: (a) no translocation present; (b) translocation between chromosomes 4 and 11 as seen in paediatric acute lymphoblastic leukaemia (ALL); probes flanking the MLL locus are separated as a result of the chromosome rearrangement

as would be expected post-treatment, the single-fusion probes are less useful as random juxtaposition of native *BCR* and *ABL* signals is possible in up to 10% of normal interphase cells. By using larger 'dual-fusion' probes that actually span the *BCR* and *ABL* breakpoints, the sensitivity of these probes can be increased to the detection of less than 1% of affected cells (Dewald *et al.*, 1998). A positive result using these larger probes shows two fusion signals representing not only the derived chromosome 22, but also the derived chromosome 9 (Figures 4.3(c), 4.5(a)).

Post bone marrow transplantation

Allogeneic bone marrow transplantation (BMT) is the only hope of cure for some malignant haematological conditions. When the donor and recipient are of opposite sex, interphase cytogenetics using centromeric probes specific for the X and Y chromosomes can be used to assess the degree of mixed chimaerism and therefore provide prognostic information regarding the success of the graft (Figure 4.5(b)). Since the hybridisation and analysis of X and Y probes to blood or bone marrow preparations can be rapidly and relatively cheaply performed, it is an attractive and cost-effective method for sequential chimaerism studies post-transplantation.

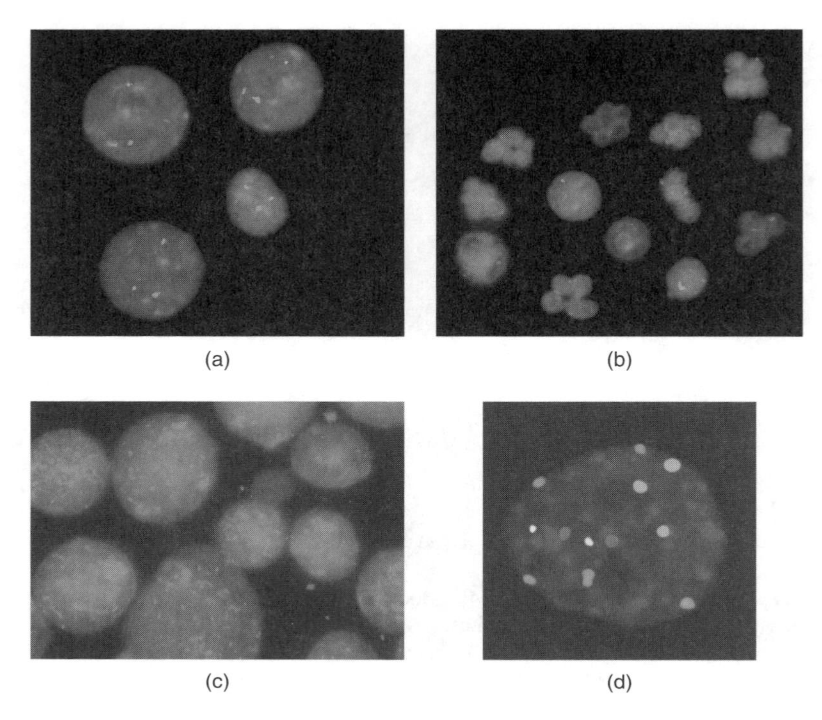

Figure 4.5 Illustrations of the use of interphase cytogenetics in the analysis of haematological malignancies and solid tumours. (a) *BCR/ABL* rearrangement in fixed bone marrow cell suspension. *BCR/ABL* dual-fusion probe (×750). (b) Mixed cell chimaerism in fixed bone marrow cell suspension. Alphoid X centromeric probe directly labelled with red fluorophore and alphoid Y centromeric probe directly labelled with green fluorophore (×750). (c) MYCN amplification in tumour imprint prepared from Stage 4 paediatric neuroblastoma. MYCN probe specific for chromosome 2p23–24 directly labelled with red fluorophore (×750). (d) Aneusomic bladder cancer cell obtained from a urine sample showing two copies of chromosome 3 (red), six copies of chromosome 7 (green), three copies of chromosome 17 (aqua) and two copies of p16 gene (gold); picture courtesy of Abbott Laboratories Ltd, UK, and hybridised with Vysis® UroVysion probe set (×750). A colour version of this figure appears in the colour plate section

Management of Solid Tumours

Progress in our understanding of the chromosomal abnormalities in malignant solid tumours has lagged behind that of haematological malignancies for a number of reasons: primary problems involve tissue inaccessibility; a lack of dividing cells; and overgrowth in the culture of normal stromal tissue. However, it has become clear in recent years that specific genetic abnormalities characterise certain tumour types and, perhaps more importantly, provide vital prognostic information enabling a more judicious use of adjuvant therapies to be developed in some cancers. Interphase cytogenetics using probes specific for known abnormalities negates the need for dividing cells and is therefore an extremely valuable tool in the clinical management of solid tumours.

Diagnosis and prognosis

Neuroblastoma

Neuroblastoma is the most common extra-cranial solid tumour in children with a very variable disease history; some children die extremely rapidly from tumour progression while in other individuals, spontaneous tumour regression occurs in the absence of any surgery or adjuvant therapy. Over the last 10 years, the genetics of neuroblastoma have been well studied and different prognostic groups have been shown to be linked with specific genetic abnormalities (Lastowska *et al.*, 2001). Tumours with amplification of the *MYCN* oncogene invariably progress rapidly and need aggressive clinical management, while tumours with a near-triploid karyotype and without *MYCN* amplification are much less progressive and benefit from reduced adjuvant therapies and their associated cytotoxic side-effects. Interphase cytogenetic studies, including the use of probes specific for the *MYCN* region of chromosome 2, have become an integral part of the clinical management of neuroblastomas and results of these studies are used directly in the treatment decisions for individual patients. In most Centres, an interphase cytogenetic study, using a battery of probes including *MYCN*, is initiated as soon as a neuroblastoma is excised. Patients whose tumours show *MYCN* amplification (Figure 4.5(c)) are stratified into an aggressive treatment regime immediately.

Breast cancer

The proto-oncogene c-*erbB2 (Her2/Neu)* located within the long arm of chromosome 17 is a trans-membrane growth factor receptor found in normal and malignant breast epithelial cells. C-*erbB2* is amplified in approximately 20% of invasive breast cancers and is associated with lymph node metastases, poor survival, decreased response to conventional therapies but an enhanced survival benefit from c-*erbB2*-targeted therapies such as Herceptin (Genentech, Inc., South San Francisco, CA) (Lakhani, 2001). Testing for c-*erbB2* amplification is therefore a vital part of patient management to determine the best treatment strategy for individual patients (Hanna, 2001). This testing is possible, accurately and rapidly, using an interphase cytogenetic approach with a locus-specific probe for c-*erbB2*. Tumours with c-*erbB2* amplification show multiple copies of the c-*erbB2* probe signal, while tumours lacking amplification show two signals representing two copies of chromosome 17.

Bladder carcinoma

Recent evidence suggests that genetic markers in bladder cancer may be a valuable predictor of tumour recurrence. Watters and colleagues (Watters *et al.*, 2000) examined interphases from 129 transitional cell carcinomas of the urinary bladder excised from 52 patients using centromeric FISH probes specific for chromosomes 7 and 17. Aneusomy for these chromosomes was found in the index primary tumours of 31% of patients with recurrent disease but was not present in patients without detectable tumour recurrence. From this data they conclude that the measurement of aneusomy by FISH for chromosomes 7 and 17 predicts recurrence in a sub-group of transitional bladder cancers and may offer a new objective and quantitative test for patients destined to recur.

Emerging Applications

Tissue microarrays

Although interphase FISH is a rapid technique for individual samples, it can be labour intensive to test large numbers of samples over a short time period. Tissue microarrays (TMAs) allow large numbers of samples to be studied by FISH in a single hybridisation experiment under identical conditions. (See Wang *et al*., 2002, for a review of TMAs.) To construct a TMA, cores of tissue from up to several hundred tissue samples are embedded in a single recipient paraffin block at defined co-ordinates. TMA sections cut from the recipient block can subsequently be analysed at the DNA or protein levels using FISH or immunohistochemistry techniques, respectively.

Sequential immunophenotyping and molecular cytogenetics

Tumour cells circulating in the haemopoietic system may be valuable in assessing marrow metastases in patients with solid tumours. The major difficulty in this approach lies in the detection and verification of scarce tumour cells amongst vast numbers of bone marrow cells. To overcome this problem a sequential immunological and molecular cytogenetic approach has been developed (Mehes *et al*., 2001). Tumour cells are first identified using immunological markers and then their genetic profiles are characterised using specific FISH probes. Mehes and colleagues report that these sequential FISH analyses of marker-positive cells improve both the specificity and sensitivity of the detection of microscopic minimal residual disease in neuroblastoma patients (Mehes *et al*., 2001).

4.4 Conclusion

Interphase cytogenetics using the FISH technique can provide precise, rapid and biologically accurate information about a particular population of cells. It offers considerable advantages over conventional metaphase chromosome analysis in a number of ways which include the variety of cells or tissues that can be examined. This chapter has given examples of the way in which interphase cytogenetics is currently applied in genetic diagnosis and how other applications, particularly in cancer genetics, may develop.

4.5 References

Broberg K, Hoglund M, Lindstrand A *et al.* (1998) Polyclonal expansion of cells with trisomy 7 in synovia from patients with osteoarthritis. *Cytogenet. Cell Genet.* **83**: 30–4.

Broberg K, Toksvig-Larsen S, Lindstrand A *et al.* (2001) Trisomy 7 accumulates with age in solid tumours and non-neoplastic synovia. *Genes Chromosomes and Cancer* **30**: 310–5.

Chevret E, Volpi EV and Sheer D (2000) Minireview: form and function in the human interphase chromosome. *Cytogenet. Cell Genet.* **90**: 13–21.

Cioni R, Bussani C, Scarselli B *et al.* (2002) Detection of fetal cells in intrauterine lavage samples collected in the first trimester of pregnancy. *Prenat. Diagn.* **22**: 52–5.

Cremer T, Landegent J, Bruckner A *et al.* (1986) Detection of chromosome aberration in the human interphase nucleus by visualization of specific target DNAs with radioactive and non-radioactive *in situ* hybridisation techniques: diagnosis of trisomy 18 with probe L1.84. *Hum. Genet.* **74**: 346–51.

Dahlen A, Broberg K, Domanski HA *et al.* (2001) Analysis of the distribution and frequency of trisomy 7 *in vivo* in synovia from patients with osteoarthritis and pigmented villonodular synovitis. *Cancer Genet. Cytogenet.* **131**: 19–24.

Delhanty JDA, Harper JC, Ao A *et al.* (1997) Multicolour FISH detects frequent chromosomal mosaicism and chaotic division in normal pre-implantation embryos from fertile patients. *Hum. Genet.* **99**: 755–60.

Dewald G, Wyatt W, Juneau A *et al.* (1998) Highly sensitive fluorescence *in situ* hybridisation method to detect double BCR/ABL fusion and marrow response to therapy in chronic myeloid leukaemia. *Blood* **91**(9): 3357–65.

Hanna W (2001) Testing for Her2 status. *Oncology* **61** Suppl 2: 22–30.

Harper JC and Delhanty JD (2000) Preimplantation genetic diagnosis. *Curr. Opin. Obstet. Gynaecol.* **12**: 67–72.

Ji Y, Eicher EE, Scwartz S *et al.* (2000) Structure of chromosomal duplicons and their role in mediating human genomic disorders. *Genome Res.* **10**: 597–610.

Knoll JH, Cheng SD and Lalande M (1994) Allele specificity of DNA replication timing in the Angelman/Prader Willi syndrome imprinted chromosomal region. *Nat. Genet.* **6**: 41–6.

Lakhani S (2001) Molecular genetics of solid tumours: translating research into clinical practice. What could we do now: breast cancer. *J. Clin. Pathol.: Mol. Pathol.* **54**: 281–4.

LaSalle JM and Lalande M (1996) Homologous association of oppositely imprinted chromosomal domains. *Science* **272**: 725–8.

Lastowska M, Cullinane C, Variend S *et al.* (2001) Comprehensive genetic and histopathological study reveals three types of neuroblastoma tumours. *J. Clin. Oncol.* **15**(12): 3080–90.

Lo YMD, Lau TK, Zhang J *et al.* (1999) Increased fetal concentrations in the plasma of pregnant women carrying fetuses with trisomy 21. *Clin. Chem.* **45**: 1747–51.

McKeown CME, Waters JJ, Stacey M *et al.* (1992) Rapid interphase FISH for the diagnosis of trisomy 18 on blood smears. *Lancet* **340**: 495.

Mehes G, Luegmayr A, Hattinger CM *et al.* (2001) Automatic detection and genetic profiling of disseminated neuroblastoma cells. *Med. Paediatr. Oncol.* **36**: 205–9.

Pinkel D, Landegent J, Collins C *et al.* (1988) Fluorescence *in situ* hybridisation with human chromosome specific libraries: detection of trisomy 21 and translocations of chromosome 4. *Proc. Natl. Acad. Sci. USA* **85**: 9138–42.

Roylance R (2001) Methods of molecular analysis: assessing losses and gains in tumours. *J. Clin. Pathol.: Mol. Pathol.* **55**: 25–8.

Samura O, Sohda S, Johnson KL *et al.* (2001) Diagnosis of trisomy 21 in fetal nucleated erythrocytes from maternal blood by use of short tandem repeat sequences. *Clin. Chem.* **47**: 1622–26.

Scriven PN, Handyside AH, Mackie Ogilvie C (1998) Chromosome translocations: segregation modes and strategies for preimplantation diagnosis. *Prenat. Diagn.* **18**: 1437–49.

Tanabe H, Muller S, Neusser M *et al.* (2002) Evolutionary conservation of chromosome territory arrangements in cell nuclei from higher primates. *Proc. Natl. Acad. Sci.* **99**: 4424–9.

Trask BJ, Allen S, Massa H *et al.* (1993) Studies of metaphase and interphase chromosomes using fluorescence *in situ* hybridisation. Cold Spring Harbour Symposium. *Quant. Biol.* **53**: 767–75.

Verschure PJ, van der Kraan I, Manders EM *et al.* (1999) Spatial relationship between transcription sites and chromosome territories. *J. Cell Biol.* **147**: 13–24.

Volpi EV, Chevret E, Jones T *et al.* (2000) Large scale chromatin organization of the MHC and other regions of human chromosome 6 and its response to interferon in interphase nuclei. *J. Cell Sci.* **113**: 1565–76.

Wang H, Wang H, Zhang W *et al.* (2002) Tissue microarrays: application in neuropathology research, diagnosis and education. *Brain Pathol.* **12**: 95–107.

Ward BE, Gersen SL, Carelli MP *et al.* (1993) Rapid prenatal diagnosis of chromosomal aneuploides by FISH: clinical experience with 4500 specimens. *Am. J. Hum. Genet.* **52**: 854–65.

Watters AD, Ballantyne SA, Going JJ *et al.* (2000) Aneusomy for chromosomes 7 and 17 predicts the recurrence of transitional cell carcinoma of the urinary bladder. *BJU Int.* **85**(1): 42–7.

Werner M, Wilkens L, Aubele M *et al.* (1997) Interphase cytogenetics in pathology: principles, methods, and applications of fluorescence *in situ* hybridisation. *Histochem. Cell Biol.* **108**: 381–90.

White LM, Rogan PK, Nicholls RD *et al.* (1996) Allele-specific replication of 15q11–q13 loci: a diagnostic test for detection of uniparental disomy. *Am. J. Hum. Genet.* **59**: 423–30.

Witters I, Devriendt K, Legius E *et al.* (2002) Rapid prenatal diagnosis of trisomy 21 in 5049 consecutive uncultured amniotic fluid samples in fluorescence *in situ* hybridisation (FISH). *Prenat. Diagn.* **22**: 29–33.

Woodward K, Kendall E, Vetrie D *et al.* (1998) Pelizaeus–Merzbacher disease: identification of Xq22 proteolipid–protein duplications and characterization of breakpoints by interphase FISH. *Am. J. Hum. Genet.* **63**: 207–17.

Woodward K, Palmer R, Rao K *et al.* (1999) Prenatal diagnosis by FISH in a family with Pelizaeus–Merzbacher disease caused by duplication of the PLP gene. *Prenat. Diagn.* **19**: 266–8.

5

Oncogenes

Fiona Macdonald

5.1 Introduction

Cancer occurs as a consequence of the accumulation of mutations in genes whose protein products are important in the control of cell proliferation, differentiation and apoptosis. Usually, several cancer-promoting factors have to occur before a person will develop a tumour and with a few exceptions, no one mutation alone is sufficient. The predominant mechanisms for the development of cancer involve (a) impairment of a DNA repair pathway, (b) the transformation of a normal gene (the proto-oncogene) into an oncogene, (c) the mutation of a tumour suppressor gene or (d) mutations in genes involved in apoptotic control. Over the last 20 years considerable advances have been made in our understanding of the genes involved in these processes. This chapter will describe how one group of genes, the oncogenes, were identified, how they function, and will give examples of the role of some oncogenes in some of the commoner cancers.

5.2 Identification of the oncogenes

Some of the first evidence for the existence of oncogenes came from the study of the group of RNA viruses known as retroviruses (Stehelin *et al*., 1976). These are relatively simple viruses consisting of three genes, *GAG, POL* and *ENV*, which produce a core protein, a reverse transcriptase and an envelope protein, respectively. The clue as to how these viruses caused cancer came when one virus, associated with malignancies in chickens, was being studied. The causative virus, the Rous sarcoma virus, was shown to carry a fourth gene, the *SRC* gene, which was mammalian in origin. The virus was shown to be able to integrate reversibly into the mammalian genome using its reverse transcriptase

Molecular Biology in Cellular Pathology. Edited by John Crocker and Paul G. Murray
 ISBN: 0-470-84475-2

to make a DNA copy. Recombination then allowed the virus to exit from the host taking a mammalian gene with it that consequently became part of the viral genome. This process is known as transduction. The mammalian gene was expressed when the virus re-infected a cell and was oncogenic in nature either because (a) it was altered during the transduction process leading to production of an aberrant protein or (b) it was abnormally expressed by the viral promoters under whose control it had been placed by the process of transduction. Although this mechanism of activation has been implicated as the cause of a number of animal tumours it has never been shown to be the cause of human malignancies. However, subsequent studies detected homology between the genes identified in retroviruses and genes isolated from human tumours and this whole field advanced rapidly.

More direct methods were developed to identify genes in human tumours with the potential for transformation. The first assay used was a DNA transfection assay (Shih *et al.*, 1981). DNA was extracted from a human tumour and transfected into a normal cell line. Occasionally a transformed cell was produced which could be recognised because contact inhibition was lost causing cells to pile up rather than form a monolayer in culture. These transformed cells were capable of forming tumours when injected into 'nude' mice. Analysis of the genome of these transformed cells identified the presence of an oncogene similar in many cases to those identified in the retroviruses.

Further oncogenes were identified through the analysis of breakpoints of chromosome translocations commonly found in human leukemias. The best known of these is the translocation between chromosomes 9 and 22 found in chronic myeloid leukaemia (CML). An oncogene previously found associated with retroviruses, the *ABL* oncogene, was localised to the breakpoint on chromosome 9. Similarly, the *MYC* oncogene was identified at the breakpoint on chromosome 8 associated with the 8;14 translocation in Burkitt's lymphoma. Overall 19 oncogenes were identified at the position of translocation breakpoints (Heim and Mittleman, 1987). Analysis of other chromosome abnormalities, homogeneous staining regions (HSRs) and double minutes (DMs) led to the identification of further oncogenes such as *MYCL* and *MYCN*.

Altogether close to 200 different oncogenes that have been implicated in human tumours have been identified (Futreal *et al.*, 2001). In normal cells, the non-transformed variant of the oncogenes – the proto-oncogenes – are all highly conserved, have important roles in the cell and their expression are tightly controlled. Activation of these genes by a number of processes described below leads to their activation and hence to inappropriate expression.

5.3 Functions of the proto-oncogenes

Proto-oncogenes regulate the normal processes of the cell including cell proliferation, differentiation, adhesion, cell cycle control and apoptosis. Not unexpectedly, therefore, any mutations in these important genes will lead to abnormal expression resulting in malignant transformation. The products of proto-oncogenes fall into five main groups with some genes having more than one role:

1. growth factors,
2. growth factor receptors,
3. signal transducers,
4. nuclear genes and transcription factors and
5. genes that are involved with control of the cell cycle.

Growth Factors

Growth factors play a major role in the development of tissues and, if limited, result in cell death. In general, normal cells require growth factors to remain viable but tumour cells circumvent this process. Growth factors bind to their own specific receptors or, in a few cases, a group of receptors on the cell surface. Once bound to the receptor they regulate major cellular processes via a number of intracellular signalling pathways. The end-point of this process is to change expression of genes including those involved in cell cycle regulation and differentiation. Growth factors can function either in an autocrine or paracrine manner. In the former, binding of the growth factor to its receptor leads to increased secretion of more growth factor from the same cell. In the paracrine model, growth factor released from one cell binds to receptors on neighbouring cells releasing factors that increase production from the first cell. Growth factors function in a number of different ways. Following binding of epidermal growth factor (EGF) and platelet derived growth factor (PDGF) to their receptors, cell division is stimulated, advancing the cell from the stationary phase into the G1 phase of the cycle (Normano *et al.*, 2001). Insulin growth factors (IGF-I and IGF-II) are responsible for assisting in the passage of cells through G_1 of the cell cycle. They have endocrine functions, e.g. in mediating the role of growth factors, but also have both autocrine and paracrine functions during normal development and growth as well as in cancer development (Sachdev and Yee, 2001). Transforming growth factor β (TGFβ) is considered both as a tumour suppressor and as a promoter of progression and invasion (Pasche, 2001). It is a potent inhibitor of normal stromal, haematopoietic and epithelial cell growth. Through its ability to inhibit cell proliferation it suppresses tumour development in its early stages. However, once tumours develop, they often become resistant to TGFβ growth inhibition and in the later stages of development of a cancer the tumour cells themselves secrete large amounts of the growth factor thereby enhancing invasion and metastases.

Growth Factor Receptors

A second group of proto-oncogenes are the growth factor receptors or their functional homologues that link signals from growth factors to the intracellular signalling pathways. These genes, when mutated, make up a large proportion of all known oncogenes. The most important group of these receptors with respect to tumour development is the transmembrane receptor protein tyrosine kinase family (Figure 5.1). These receptors have an extracellular domain to which the growth factors bind, a transmembrane domain and one or two intracellular tyrosine kinase domains. At least 58 different transmembrane receptor tyrosine kinases have been recognised (Blume-Jensen and Hunter, 2000; Robinson *et al.*, 2000). Binding of the growth factor results in oligomerisation of the receptors. This in turn leads to tyrosine autophosphorylation of the receptor which both activates its catalytic activity and provides phosphorylated residues that mediate the specific binding of cytoplasmic signalling proteins. This activated receptor acts as a centre for the assembly of a 'signal particle' on the inner surface of the membrane which transmits the signal to the nucleus via the intracellular signalling pathways (see below). As these receptors are important in the development and progression of cancers they have become the focus of interest for the development of novel targets for anti-cancer therapy (Zwick *et al.*, 2001).

Over 20 years ago, *EGFR* was shown to be the proto-oncogene of the mutant constitutively active viral oncogene v-*erbB*. The epidermal growth factor receptor family, which consists of four receptors, epidermal growth factor receptor (EGFR) itself (also known as HER1) and HER2-4, interact with a wide range of downstream signalling pathways to regulate cell growth, differentiation and survival (Yardin, 2001). Over-expression of these genes has been implicated in a wide range of tumours and the mechanism of signal transduction by this family have revealed major therapeutic targets. An antibody to HER2 (herceptin) has been widely evaluated for the treatment of breast cancer (Rubin and Yardin, 2001).

Platelet derived growth factor receptor (PDGF-R) was also one of the earliest growth factor/receptor pathways to be identified and shown to constitute a family of ligands and receptors, primarily PDGF-R alpha and beta. Again these receptors have been shown to be novel therapeutic targets and the tyrosine kinase inhibitor Imatinib, originally developed for the treatment of chronic myeloid leukemia by blocking the activity of the BCR–ABL oncoprotein, has also been shown to inhibit PDGF-R.

Binding of IGF-I and II to their receptors (IGF-IR and IIR) mediates pleiotropic effects in cancers. The effect is to relay signals via lipid and protein kinases which ultimately result in increased translation of specific types of RNAs.

Transforming growth factor receptors are serine/threonine kinases rather than tyrosine kinases. There are two recognised receptors, Type 1 and II, which operate via an unusual mechanism. TGFβ binds to TGFβ-RII and this complex then rapidly leads to phosphorylation of TGFβ-RI. TGFβ-RI in turn phosphorylates

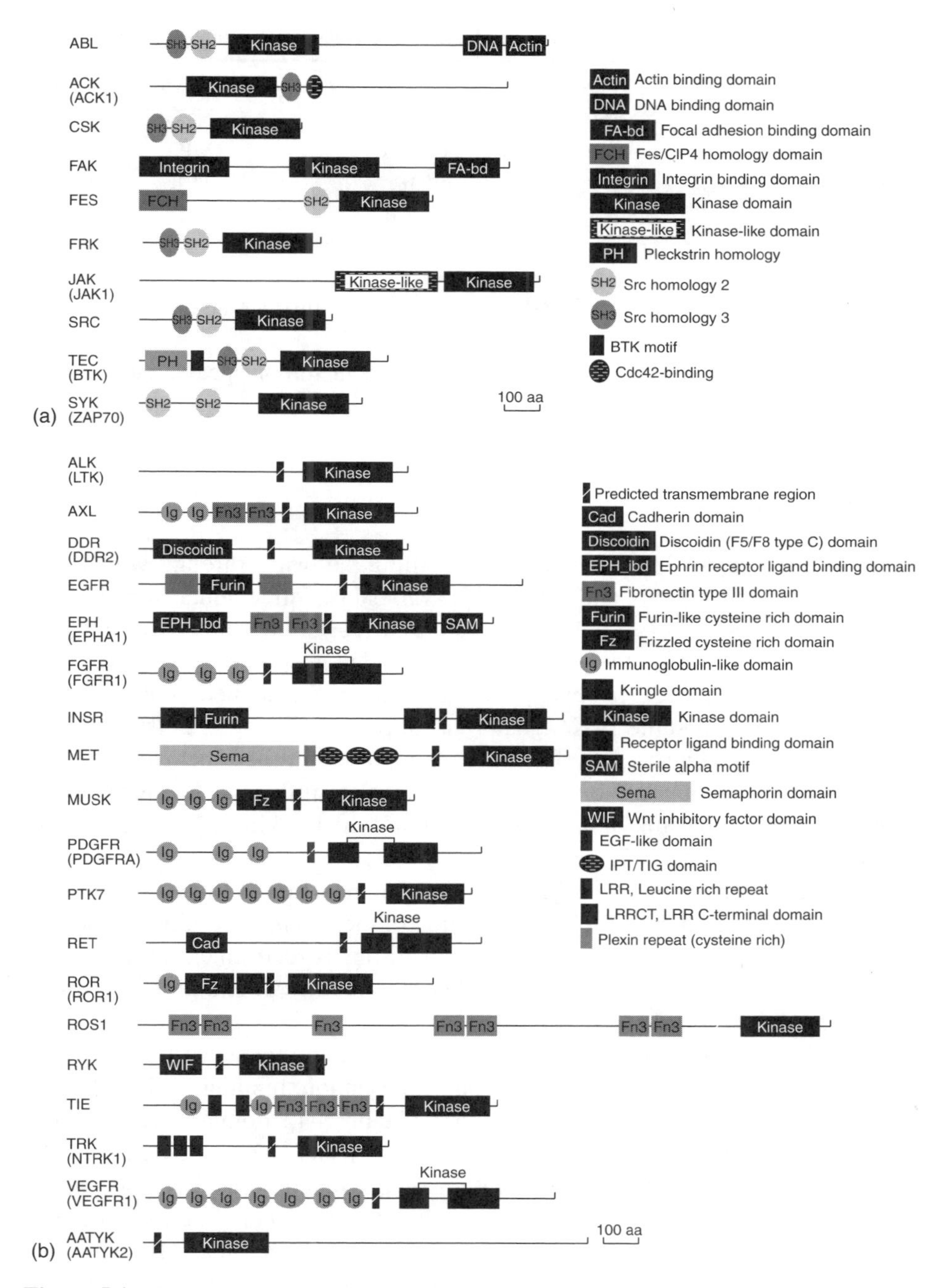

Figure 5.1 Domain structure of the receptor (a) and non-receptor (b) tyrosine kinases. From Robinson *et al.* (2000). Reproduced by permission of Nature Publishing Group

Table 5.1 Receptor protein tyrosine kinases associated with human tumours

EGFR	HER2-4	IGF-IR/IGF-IIR	PDGFR – α and β
CSF-1R	KIT	FLK2/FLT3	VEGFR1-3
FGFR1-4	TRKA/TRKC	RON	MET
EPHA2/EPHB2/EPHB4	AXL	TIE	TEK
RET	ROS	ALK	

and activates members of the Smad proteins which transduce signals from cytoplasm to nucleus (Kloos *et al.*, 2002).

Other growth factor receptors that play a role in tumourigenesis are shown in Table 5.1.

Signal Transducers

Many of the receptors utilise common signalling pathways through which they can transmit their signals and although a variety of intracellular signal transduction pathways exist (see `http://biocarta1.epangea.net/` for diagrams of signalling pathways), three major pathways are recognised. These are the PI3-kinase (PI3-K)/AKT pathway, the RAS/mitogen activated protein kinase (MAPK) pathway, and the JAK/signal transducers and activators of transcription (STATs) pathway (Figure 5.2). The proteins involved in these pathways belong to the third group of proto-oncogenes, the signal transducers. Included in this group are the cytoplasmic non-receptor protein tyrosine kinases of which there are at least 32 members (Table 5.2).

The initial step in both the PI3-K and RAS mediated pathways, following growth factor binding, is autophosphorylation of the receptor, resulting in activation of the kinase function. The consequence of this is recruitment of a group of cytoplasmic signalling proteins with SH2 (src homology 2) and protein tyrosine binding domains.

PI3-K is composed of two subunits – the p110 catalytic subunit and the p85 regulatory subunit. Following receptor phosphorylation, binding of the p85 subunit to the growth factor receptor occurs via its SH2 domain. This in turn activates the p110 subunit leading to an increase in its enzymatic activity. The immediate result of this is the generation of phosphatidylinositol phosphates. These lipid messengers are then binding sites for a further group of proteins, which in PI3-K signalling, primarily involves AKT (protein kinase B). This protein is then translocated to the plasma membrane and is itself phosphorylated to activate it. At least 13 substrates for AKT have been recognised (Blume-Jensen and Hunter, 2000) which fall into two main groups: (a) regulators of apoptosis such as BAD and the Forkhead transcription factors, and (b) regulators of cell growth including protein synthesis and glycogen metabolism and cell

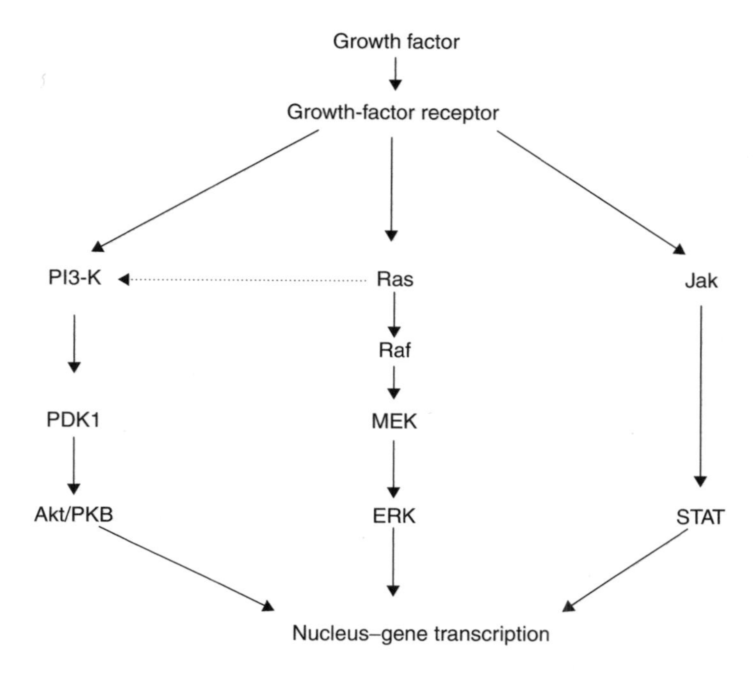

Figure 5.2 Three common cell signalling pathways: (a) PI3-kinase pathway, (b) the RAS/MAPK pathway, and (c) the JAK/STAT pathway

Table 5.2 Cytoplasmic non-receptor protein tyrosine kinases associated with human tumours

FGR	SRC	YES1
ABL1	JAK1-3	FAK
PYK2	FES	BRK
SYK	LCK	

cycle regulation (Talapatra and Thompson, 2001). The end result of this signalling pathway is therefore the promotion of cell survival and opposition of apoptosis.

Many of the growth factors that activate the PI3-K pathway also activate the RAS/MAPK pathway. The *RAS* gene is one of the best studied of all of the oncogenes and the family consists of three members, *HRAS*, *KRAS* and *NRAS*. *RAS* encodes the G protein, p21, which cycles between a GDP bound inactive protein and a GTP bound active protein. Conversion between the active and inactive states is a tightly regulated event. The exchange of GDP for GTP to activate RAS is promoted by the nucleotide exchange factor SOS (Jones and Kazlauskas, 2000). Inactivation occurs by hydrolysis of GTP to GDP and is promoted by GTPase activating proteins called GAPs. Activation of RAS occurs following growth factor or cytokine interaction with cell surface receptors, as described above. As SOS is a cytoplasmic protein, activation of RAS p21 requires firstly that SOS is translocated to the cell membrane, a process mediated by the adapter protein Grb2. The signalling process therefore involves tyrosine phosphorylation of the growth factor receptor, binding of Grb2 to the receptor via its SH2 domain and translocation of SOS to the membrane. Interaction of inactive RAS with SOS leads to GDP/GTP exchange, activating RAS. An alternative scenario involves a third adaptor protein, Shc, which is phosphorylated by growth factors leading to its association with the Grb2/SOS complex. The functional consequences of this process are then the same as when Grb2/SOS interacts directly with the receptor. The next stage of the pathway involves the MAP kinases (MAPK). Active RAS recruits the serine/threonine kinase, RAF, to the plasma membrane. In turn it then phosphorylates and activates its substrate MEK which then activates a further kinase, ERK, leading to activation of transcription factors such as MYC and FOS or components of the cell cycle machinery such as cyclin D1 (Frame and Balmain, 2000). In addition several studies have attributed a role in cell survival to this pathway. RAS is also active in other signalling pathways and in particular has been shown to activate the PI3-K pathway by binding to the catalytic subunit of PI3-K.

The Janus protein tyrosine kinases (JAKs) are the main mediators of cytokine signalling pathways, such as the interleukin 3 pathway, via the STAT family of transcription factors (Reddy *et al.*, 2000). Cytokine receptors are phosphorylated by JAKs following ligand binding. These phosphorylated residues bind STATs through their SH2 domains leading to STAT oligomerisation. Dimeric STATs are released from the receptor and translocate to the nucleus where they activate transcription. As before, there is however cross-talk between signalling pathways. PI3-K also interacts with STATs and growth factors such as EGF, and PDGF will also activate the pathway following interaction with its own receptors. Aberrations in the JAK/STAT pathway are primarily associated with the development of some leukaemias and B cell malignancies.

The cytoplasmic protein–tyrosine kinases are also important in receptor signalling pathways. SRC is a non-receptor protein kinase which transduces signals involved in processes such as proliferation, differentiation, motility and adhesion (Bjorge *et al.*, 2000). The activated form of the SRC cytoplasmic tyrosine kinase was one of the first oncogenes to be identified. SRC is maintained in its inactive form by suppression of its kinase activity through phosphorylation of a specific tyrosine residue (Tyr530) in the carboxy terminus of the protein and interaction of this residue with the SH2 domain. A second tyrosine (Tyr419) is also important in the regulation of SRC and, when phosphorylated, is a positive regulator of SRC activity. Binding of ligands to receptor tyrosine kinases such as the PDGFR or dephosphorylation of Tyr530 by protein tyrosine phosphatases relieves the inhibitory constraints on the kinase (Figure 5.3). This

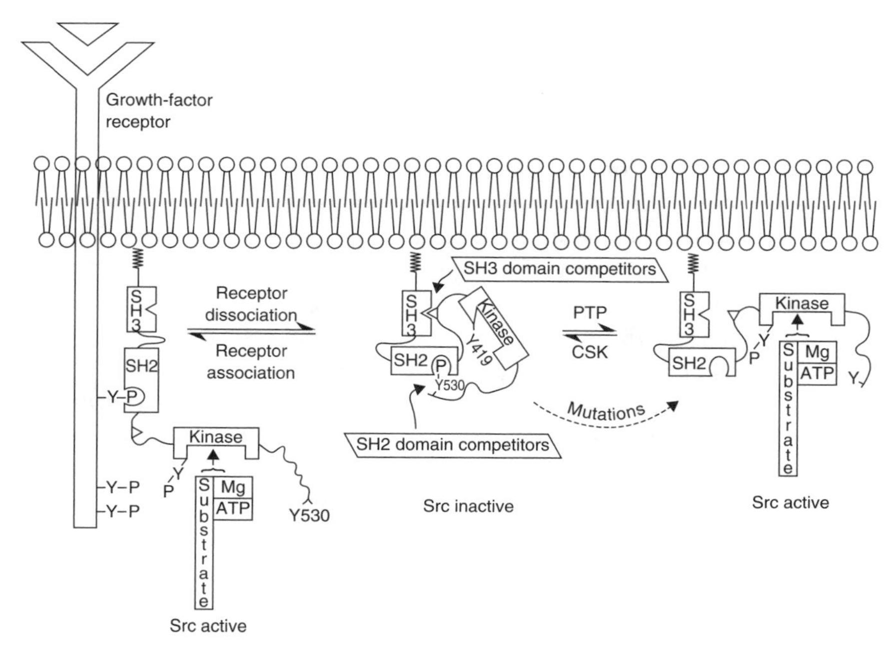

Figure 5.3 Mechanism of Src activation. Src is maintained in its inactive state as shown in the centre with the SH2 domain engaged with Tyr530, the SH3 domain engaged with the SH2-kinase linker and Tyr419 dephosphorylated. Binding of Src to growth-factor receptors via its SH2 domain results in displacement of Tyr530 allowing Tyr419 phosphorylation and Src activation. From Bjorge *et al.* (2000). Reproduced by permission of Nature Publishing Group

results in activation of downstream signalling pathways such as RAS/MAPK or PI3-K.

The product of the *ABL* gene is also a cytoplasmic protein tyrosine kinase and the gene was one of the first to be identified through a consistent chromosomal translocation seen in chronic myeloid leukaemia. The product of the *ABL* proto-oncogene has a number of features in common with that of the *SRC* gene in that it has both SH2 and SH3 domains, a kinase domain and two C terminal domains – a DNA binding domain and a domain through which it can bind to actin. Evidence suggests that ABL has a role in DNA damage induced apoptosis. The result of the chromosome translocation seen in leukaemias is the production of the *BCR–ABL* fusion gene (see below). Activation of the kinase activity of the fusion protein is mediated by dimerisation of the protein through the BCR domain leading to phosphorylation of the tyrosine residues of the BCR–ABL protein. These residues then form docking sites for adaptor molecules and other enzymes which activate signal transduction pathways. All three of the signalling pathways described above are involved in BCR–ABL signalling (Thijsen *et al.*, 1999).

Nuclear Proto-oncogenes

The final group of proto-oncogenes are those involved in the control of gene expression by binding of their protein products directly to DNA to control transcription. Examples of this group include *MYC*, *FOS* and *JUN*.

MYC was initially identified as the cellular homologue of the v-*myc* gene. The *MYC* family consists of a number of genes, three of which, *MYC*, *MYCL* and *MYCN*, have been associated with human cancers. *MYC* has a role in differentiation, in regulating cell proliferation and in apoptosis (Dang, 1999). It can drive quiescent cells into continuous cycling and can prevent cells from exiting the cell cycle (Pelengaris *et al.*, 2000). It appears to sensitise cells to a variety of apoptotic triggers rather than directly inducing apoptosis (Prendergast, 1999). The MYC protein, p62, contains two domains, an N-terminal transactivation domain and a C-terminal domain with the ability to bind to DNA and to dimerise. MYC has been shown to act in conjunction with a second protein, MAX, and these heterodimers bind to target sites to transactivate genes. Binding occurs at a specific DNA sequence motif (CACGTG) found in a number of genes shown to be regulated by MYC. However, the model is complicated by the existence of two further proteins, MAD and MXI-1, which can also dimerise with MAX. These dimers bind to the same motif as MYC/MAX dimers. Changes in the level of expression of the four genes in response to growth stimuli lead to alterations in the composition of the two types of dimer which have opposite effects on regulation of expression. MYC/MAX dimers bind to DNA and MYC interacts with the TATA box binding protein TBP and transcriptional machinery leading to activation of target genes. In contrast, following DNA binding, MAX/MAD dimers recruit a number of proteins, including those with histone deacetylation activity. This deacetylation process locks up the DNA in nucleosomes thereby preventing transcription. Many genes have been implicated as the targets transactivated by MYC. Some, such as the cyclins and cyclin-dependent kinases, are involved in cell growth (Hermeking *et al.*, 2000); others, such as LDH-A, are involved in growth and metabolism. The more recent finding is the association of MYC with apoptosis and it is assumed that MYC affects transcription of genes in the apoptotic pathway. In addition, MYC has been identified as an inducer of telomerase activity (Dang, 1999).

FOS and JUN are also transcription factors belonging to the AP-1 transcription family which dimerise and regulate transcription from promoters containing AP-1 DNA binding sites found in genes associated with proliferation and differentiation (Bannister and Kouzarides, 1997). Dimerisation occurs via a region known as the leucine zipper region and these proteins can then bind to DNA through a region rich in basic amino acids in a sequence-specific manner. Both FOS and JUN also contain a number of activation domains. Activity of FOS and JUN is under tight control in the cell and they are regulated by phosphorylation of specific amino acids.

5.4 Mechanism of oncogene activation

In the normal cell, expression of each of the proto-oncogenes is tightly controlled. Any disruption of this tight control leads to deregulation and the subsequent eventual transformation of the cell. Activation of the proto-oncogenes occurs through a number of mechanisms that are not unique to any one gene. These mechanisms fall broadly into three categories: alteration in the protein itself; over-expression of the protein, or loss of control mechanisms. Examples of some of the best studied of these are described below. There is, however, no single consistent mechanism of activation of any one oncogene and many oncogenes can be activated by a number of oncogenes.

Structural Alteration

Alterations in the proteins can occur in a number of different ways. The simplest case is through point mutation of the gene, as described in activation of the *RAS* gene where it has been studied extensively. Point mutations, resulting in amino acid substitutions, occur at a number of positions within the gene, notably at codons 12, 13 and 61. These alterations have been seen in approximately 30% of human tumours. Mutations in *KRAS* are found widely in colorectal and pancreatic cancers, and non-small-cell lung tumours. Mutations in *HRAS* are found in bladder and kidney tumours and *NRAS* mutations have been associated with haematological malignancy and hepatocellular cancers. Mutant RAS is locked in its activated state bound to GTP circumventing the need for growth factor stimulation and resulting in constitutive signalling.

Point mutations are also seen in the receptor protein tyrosine kinase genes such as *MET* as found in renal cancers and, classically, in the *RET* gene in the inherited disorders, familial medullary thyroid cancers (FMTC) and multiple endocrine neoplasia types 2A (MEN2A) and 2B (MEN2B) (Blume-Jensen and Hunter, 2000). In FMTC and MEN2A, mutations in the extracellular domains at conserved cysteine residues cause formation of intermolecular disulphide bonds between RET molecules leading to constitutive dimerisation and activation. In MEN2B, the common methionine to threonine mutation at codon 918 results in increased kinase activity of the protein without the need for dimerisation.

Larger structural alterations are found following chromosomal translocations as described in many lymphomas and leukemias. One of the best studied of these is the 9,22 translocation found in CML which places the *ABL* oncogene on chromosome 9 next to the breakpoint cluster region (*BCR*) of the Philadelphia gene on chromosome 22. The fusion gene produced in this process produces a fusion protein with constitutive protein tyrosine kinase activity (Figure 5.4).

Truncation of a proto-oncogene resulting in a shorter protein product can also result in constitutive activation. Deletion of the 3′ end of the *SRC* gene has been described which leads to removal of part of the protein that binds to the SH2 domain required for maintenance of the inactive state.

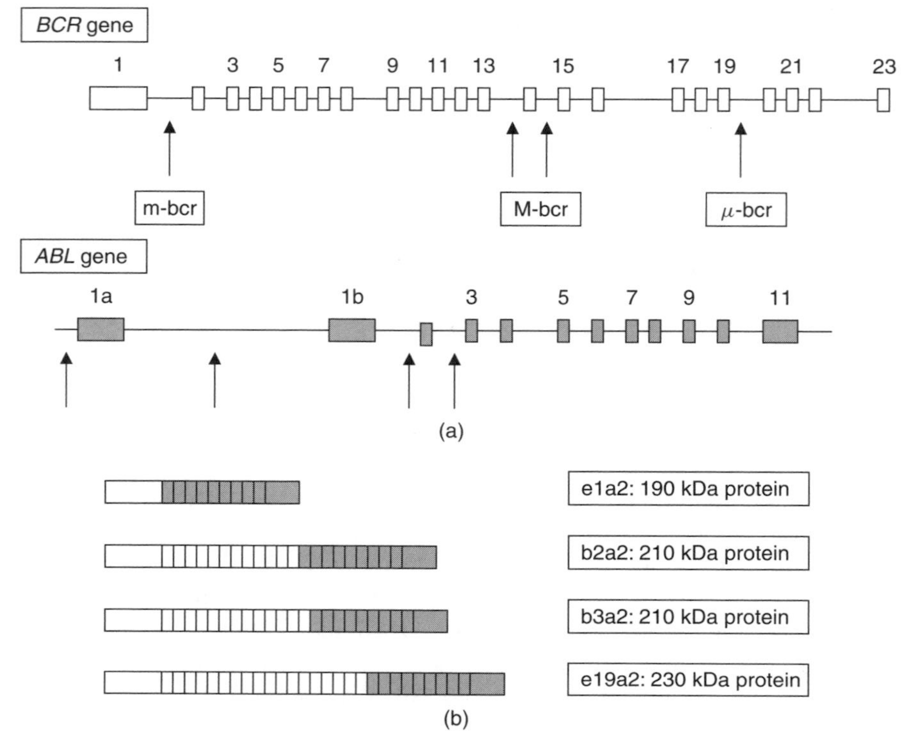

Figure 5.4 *BCR* and *ABL* genes, the position of breakpoints associated with leukemias and the resulting fusion proteins. (a) Exons are indicated by light and shaded boxes and are numbered above the boxes. Breakpoints are indicated by arrows. (b) Fusion proteins produced by each of the rearrangements. b2 is equivalent to exon 13 (e13) and b3 to exon 14 (e14) in parallel nomenclature

Gene Amplification and Over-expression

Over-expression of normal proteins following amplification of a proto-oncogene is also involved with malignant transformation. Amplification of the *MYC* gene has been associated with a number of tumours and can be seen microscopically as DMs or of HSRs. *NMYC* is amplified in late stage neuroblastomas. Amplification of the *HER2* gene is an important prognostic indicator in breast cancer (Kaptain *et al.*, 2001). A meta-analysis of over-expression of EGFR showed that it was a strong prognostic indicator in head and neck cancer, ovarian, cervical, bladder and oesophageal cancers (Nicholson *et al.*, 2001).

De-regulated Expression

This mechanism of activation is typified by activation of the *MYC* oncogene in Burkitt's lymphomas. Expression of *MYC* is normally under very fine control. In Burkitt's lymphoma the *MYC* gene is translocated to one of the immunoglobulin loci in virtually every case of the disease. The translocation between chromosomes 8 and 14 involving *MYC* and the immunoglobulin heavy chain gene is seen in approximately 80% of cases of Burkitt's lymphoma (Boxer and Dang, 2001). In the remaining cases the translocation involves either the κ light chain gene on chromosome 2 or the λ light chain gene on chromosome 22. Breakpoints can occur either in the first intron or exon 1 of *MYC*, immediately 5′ to the *MYC* gene or distant to the gene by up to 100 kb. In all cases the two coding exons, exons 2 and 3 of *MYC*, are intact. The effect of the translocation appears therefore to involve the deregulation of *MYC* gene expression probably involving regulatory elements from the immunoglobulin loci resulting in over-expression of the *MYC* gene with continuous signalling and consequent cell proliferation. *MYC* is also involved in other haematological malignancies. In some cases of T-cell ALL, the *MYC* gene is translocated into one of the T-cell receptor loci. In multiple myeloma translocation of *MYC* also occurs into one of the immunoglobulin loci but here, unlike in Burkitt's lymphoma, the translocation is likely to be a secondary event.

The major reason for the enormous research into oncogenes is to identify them as possible diagnostic and prognostic indicators in cancers and as possible novel therapeutic targets. Examples of how oncogenes may be of use clinically in some of the commoner cancers are given in the remainder of this chapter. However, it must be remembered in all cases that tumours are not the result of a mutation in a single gene. Rather they are the end-point of a pathway of gene activation which involves the oncogenes but also other important genes such as the tumour suppressor genes and genes involved in cell cycle control, in addition to epigenetic phenomena such as DNA methylation.

5.5 Oncogenes in colorectal cancer

The stepwise progression to colorectal cancer is well recognised and involves a series of changes from normal epithelium through increasingly worsening degrees of dysplasia to carcinoma and metastases (Figure 5.5). The molecular changes that accompanied this pathway were described more than 10 years ago (Fearon and Vogelstein, 1990) which include mutations in both oncogenes and tumour suppressor genes (Figure 5.4). Initiation of colorectal tumourigenesis requires mutations in the *APC* gene, a gene which is not only found to be mutated in the germ line of individuals with familial adenomatous polyposis, but is also mutated in sporadic colorectal cancer. Progression from small to intermediate adenomas is associated with point mutations in the *KRAS* oncogene. Further adenoma progression then involves the deleted in colorectal cancer (*DCC*) tumour suppressor gene located on chromosome 18. Abnormalities in one of the best characterised tumour suppressor genes, *p53*, are associated with the development of early carcinomas. Although this pattern of changes is well recognised, it is the accumulation of mutations that is more important than

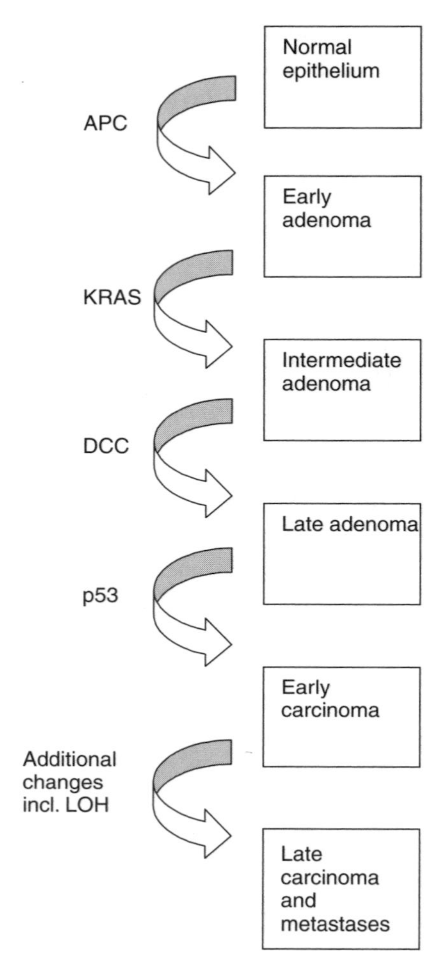

Figure 5.5 The genetic pathway to colorectal carcinogenesis

the specific order in which they occur. Other oncogenes and tumour suppressor genes are also important in colorectal cancers. The mismatch repair genes, *MSH2* and *MLH1*, may contribute to an increase in the mutation rate of genes such as *APC*. Consistent allele loss in colorectal tumours, such as seen on chromosomes 1, 8, 14 and 22, indicates the possible location of other tumour suppressor genes (Vogelstein *et al*., 1989).

Alterations in the *RAS* genes have been studied extensively in colorectal cancer. As described in the above pathway, mutations occur at an early stage of development. A PCR-based approach has been developed as a highly sensitive method of mutation detection and up to 60% of colorectal tumours and a similar proportion of adenomas have been shown to carry a mutation in *RAS* primarily occurring at codons 12, 13 or 61 of *KRAS* (Fearon and Vogelstein, 1990). *RAS* mutations have also been detected in the grossly normal mucosa of patients with colorectal cancer and this has been shown to be due to the presence of aberrant crypts in otherwise normal tissue (Yamashita *et al.*, 1995). Associations have been found between the nature of the amino acid substitutions and clinical features. In one study, distant metastases were associated with glycine to aspartic acid substitutions at codon 12, whereas tumours with valine substitutions at this position did not spread beyond the pericolonic and perirectal lymph nodes. Tumours with codon 13 mutations did not progress locally or distantly (Finkelstein *et al.*, 1993). A number of studies have associated the presence of *KRAS* mutations with decreased survival. A meta-analysis of over 2500 patients indicated that there was a statistically significant association between the presence of a *KRAS* mutation and increased risk of recurrence and death (Andreyev *et al.*, 1998). Mutations of guanine to thymine specifically increase the risk of recurrence and death adding to earlier studies that related this change in particular with metastatic behaviour (Moerkerk *et al.*, 1994). Meta-analysis also showed that only the codon 12 valine substitution was an independent indicator of increased risk of recurrence and death.

Since *RAS* mutations are an early event in the development of colorectal cancers and the number of specific mutations is limited, it has been proposed that screening of stool samples for mutations may be a useful test for colorectal cancer with the aim of early detection of disease (Sidransky *et al.*, 1992). Preliminary studies showed that up to 50% of patients with colorectal cancer had detectable mutations in their faecal samples (Smith Ravine *et al.*, 1995). These initial studies were however limited by technical problems which have recently been overcome by the use of a more reproducible method of DNA extraction and a more robust PCR method (Dong *et al.*, 2001). In addition, the inclusion of three genetic markers associated with colorectal cancer, *RAS*, p53 and a microsatellite instability marker BAT26, were shown to detect up to 70% of colorectal cancers. It now remains to be determined if this approach is sufficiently sensitive to be used routinely as a screening tool given that many of the tumours in these studies were advanced, with patients already symptomatic. Ideally, identification of *APC* gene mutations, which occur at an earlier stage of colorectal carcinogenesis than *RAS*, could prove to be more sensitive early markers. However, the wide number and spread of mutations in this gene limit the development of a simple test. Larger trials of normal individuals to evaluate the specificity of the three existing markers and studies of patients with only adenomas should prove the general usefulness of what appears to be a promising method of early detection of colorectal cancer.

The *SRC* gene has also been implicated in early colorectal cancer development with a 5–8-fold increase in activity in the majority of tumours (Irby and Yeatman, 2000). This level of increase is found even in premalignant lesions and in small adenomatous polyps as well as benign polyps containing villous change or severe dysplasia. *SRC* may also be involved in tumour progression with higher levels of expression found in tumours compared with polyps and still higher levels in distant metastases such as those in the liver.

Over-expression of the anti-apoptotic proto-oncogene *AKT* has been detected immunohistochemically in 57% of colorectal cancers as well as in adenomas again suggesting that this gene may be involved in early events in the development of this cancer (Roy *et al.*, 2002). This finding may relate to early inhibition of apoptosis during colorectal carcinogenesis.

High levels of *MYC* transcripts have been detected by Northern blotting in up to 70% of colorectal cancers and immunohistochemical studies of the expression of the MYC p62 protein have confirmed this. However, there are few diagnostic or prognostic implications associated with these findings. Elevated levels of MYC have been associated with tumours of the distal colon, a result that is consistent with the *APC* gene exerting its influence by deregulating *MYC*. Evidence suggests that activation of *MYC* is central to the way signals are transduced in the Wnt signalling pathway through APC to negatively regulate β-catenin expression (Dang, 1999).

5.6 Oncogenes in breast cancer

Abnormalities in a number of oncogenes have been associated with breast cancer including *HER2*, *SRC*, *MYC* and *RAS*.

HER2 (also known as *NEU* or *ERBB2*) is involved in the regulation of normal breast development and over-expression has been associated with breast cancer and subsequently with ovarian cancer (Slamon *et al.*, 1989). *HER2* amplification has been seen in 20%–30% of breast cancers. The clinical utility of *HER2* has received considerable attention over the last 15 years. Initial studies showed that amplification independently predicted a more aggressive disease and reduced overall survival and disease-free interval in node-positive patients (Slamon *et al.*, 1987, 1989). Subsequently many studies have been carried out to try to confirm these findings. Multivariate analysis of over 15 000 patients from 47 studies has shown that, in the majority of studies and the majority of patients, *HER2* amplification was associated with a worse prognosis in node-positive patients (Ross and Fletcher, 1998). Studies on HER2 status in node-negative individuals, however, remain conflicting (Ross and Fletcher, 1998). The American Society of Clinical Oncology has recommended that HER2 status of all primary breast cancers should be determined at diagnosis or recurrence (Bast *et al.*, 2001). HER2 testing has therefore become part of the routine work-up of breast cancer patients at many centres. A humanised monoclonal antibody (Herceptin, trastuzumab) to the HER2 protein has been developed which abolishes HER2 function and inhibits tumour growth, and determination of HER2 status is required for patients who are to undergo treatment with Herceptin. Studies have shown, perhaps not surprisingly, that the higher the level of *HER2* amplification or over-expression, the greater the response rate to the antibody.

SRC kinase activity in breast cancer has been found at 4–20 times the level of that seen in normal tissues. One reason for this increase appears to be related to increased phosphatase activity which dephosphorylates the SRC Tyr530 regulatory site resulting in SRC activation (Egan *et al.*, 1999). *SRC* primarily appears to be involved in the progression and metastasis of cancers and it has been suggested as a possible target for anti-cancer therapies. However, so far there are no diagnostic or prognostic implications for *SRC* activation.

Amplification of *MYC* is well recognised in breast cancers, although the frequency with which this occurs varies from one study to the next varying from 1% to 94%. A meta-analysis of 29 separate studies examined showed the average frequency of amplification to be 15.5% (Deming *et al.*, 2000). This analysis showed that amplification had significant but weak associations with tumour grade, lymph node metastasis, post-menopausal status and negative progesterone receptor status, although the only feature reaching statistical significance was the last. Amplification was significantly associated with the risk of relapse and death. However, reports on the prognostic value of amplification and prognosis of *MYC* amplification are conflicting. The major reasons for the wide differences in findings are due to the variation in the sensitivities of the technology used, the small size of studies and to variation in the tumour grades studied. A review of *MYC* amplification in breast cancer highlights these problems (Liao and Dickson, 2000).

Point mutations in the *RAS* family are not commonly associated with breast cancer. However, over-expression of the RAS p21 protein has been found in up to 83% of cases. Several immunohistochemical studies have shown a correlation between over-expression of RAS and a better prognosis but only in the group of node-negative patients (Gohring *et al.*, 1999; Schondorf *et al.*, 2002). However, as with *MYC* gene abnormalities, reports on the role of *RAS* abnormalities are conflicting and some studies have associated over-expression of RAS with progression and poor prognosis. Again the reasons for these differences are almost certainly due to variations in methodology, studies involving small patient numbers and variations in the patient groups investigated.

Polymorphisms of the *HRAS* microsatellite located downstream of the *HRAS* gene have been investigated in breast cancer. Meta-analyses have shown that women with a single copy of one of the rare alleles of this polymorphism have a 1.7-fold increased risk of breast cancer compared with the general population, whilst those who are homozygous for the rare allele are at a 4.6-fold increased risk (Krontiris *et al.*, 1993).

5.7 Oncogenes in lung cancer

There are a series of morphologically distinct changes that occur in the bronchial epithelium before the appearance of a clinically overt lung tumour including hyperplasia, dysplasia and carcinoma *in situ*. A number of mutations in oncogenes have been associated with these early stages as well as with the cancers themselves and may be of use as diagnostic or prognostic tools. Lung cancers can be divided broadly into two groups – the small-cell lung cancers (SCLCs) and the non-small-cell lung cancers (NSCLCs). The latter group can be subdivided into adenocarcinomas, squamous cell carcinomas (SQCs) and large-cell carcinomas with different genes being associated with each group.

Over-expression of the growth factor receptor group of oncogenes is a common abnormality in lung cancers. Over-expression of EGFR is consistently seen, primarily in NSCLCs of the SQC type. One study showed that over-expression was not associated with a survival difference (Rusch *et al.*, 1995) but suggested that this growth factor/receptor loop was more important for lung tumour formation than for tumour progression. However, others have found an association between over-expression of the receptor and poor prognosis (Brabender *et al.*, 2001; Pastorino *et al.*, 1997). *HER2* amplification is seen in approximately a third of NSCLC, particularly adenocarcinomas, and over-expression has been correlated with shorter survival (Brabender *et al.*, 2001; Han *et al.*, 2002). Given these findings prospective studies on the use of the anti-HER2 monoclonal antibody, Herceptin, are underway (Hirsch *et al.*, 2002). The *KIT* oncogene has been shown to be expressed in over 70% of SCLCs and has been suggested as a possible target for therapy with the tyrosine kinase inhibitor STI571 (Heinrich *et al.*, 2002). *MET* as been shown to be consistently activated in lung cancers where it is believed to contribute to tumour progression.

RAS mutations are found in 30% of NSCLCs especially in adenocarcinomas, though *RAS* is rarely mutated in SCLCs (Mao *et al.*, 1994). Mutations occur primarily in *KRAS* and mutations at codon 12 are those most frequently found. Most mutations are G to T transversions which may be a reflection of DNA damage by nitrosamines in cigarette smoke (Ahrendt *et al.*, 2001). *KRAS* mutations were initially believed to be poor prognostic indicators, but more recent studies have not upheld this observation (Niklinski *et al.*, 2001). However, a combination of *KRAS* mutations plus alterations in other markers such as *p53* and *ERBB2* have been demonstrated to have an improved prognostic value (Schneider *et al.*, 2000).

Of all the characterised members of the *MYC* family of genes, only *MYC* gene amplification or over-expression due to transcriptional deregulation is common to both NSCLC and SCLC, whereas abnormalities of two others, *MYCN* and *MYCL*, are specifically associated with SCLC. In general, abnormalities of *MYC* are associated with an adverse outcome. *MYC* deregulation has been found in preneoplastic lesions associated with NSCLCs suggesting an early role for *MYC* in lung cancer development (Broers *et al.*, 1993).

5.8 Oncogenes in haematological malignancies

Given the frequency with which chromosome translocations are seen in leukemias and lymphomas, cytogenetic analysis and molecular diagnostic testing of these abnormalities plays an important role in their diagnosis and monitoring.

One of the first chromosome translocations to be associated with leukemias was the t(9;22) rearrangement which results in the production of the Philadelphia (Ph) chromosome. This translocation is found in up to 90% of cases of CML. This translocation is not diagnostic of CML as it is also seen in 25% of adult acute lymphocytic leukaemia (ALL), in 2%–10% of childhood ALL and in a few cases of acute myeloid leukaemia (AML). The breakpoints on chromosome 9 usually occur in the first two introns of the *ABL* gene. Breakpoints on chromosome 22 occur in the *BCR* gene in one of three possible breakpoint cluster regions: the major (M) bcr, the minor (m) bcr or the micro (μ) bcr (Thijsen *et al.*, 1999) (Figure 5.4). In the majority of case of CML, breakpoints occur in intron 1 or 2 of *ABL* which is connected to exon b2 or b3 of the *BCR* gene giving rise to either the b2a2 or b3a2 variants, both of which are translated into the p210 fusion protein. Conflicting results have been published concerning the implications for clinical heterogeneity in CML of the presence of exon b3. Less common breakpoints are those in the minor bcr region leading to an e1a2 fusion gene and a fusion protein of 190 kDa which is seen in most cases of Ph positive ALL patients. The p190 protein has an increased tyrosine kinase activity compared with p210 and is associated with a more aggressive leukaemia. Micro bcr breakpoints resulting in the a19b2 fusion gene leading to the p230 protein are also occasionally found. Two to three percent of patients show no cytogenetic evidence of a rearrangement but this can be detected molecularly by a sensitive reverse transcriptase PCR (RT-PCR) based method (Cross *et al.*, 1994). As this can detect cases that cytogenetically appear to be normal, it is a useful diagnostic test (Figure 5.6). Given its sensitivity, the test can be used routinely to detect minimal residual disease following bone marrow transplantation and can detect 1 malignant cell amongst 10^6 normal ones (van Dongen *et al.*, 1999). However, qualitative PCR is not a good predictor of patients who will relapse. Repeated post-transplant positivity, and positivity six months post-transplant, has been shown to correlate with relapse. However, positive PCR results 10 years after transplant with no sign of relapse in the patient, have also been detected (Thijsen *et al.*, 1999). Quantitative assays have therefore been developed to estimate the level of residual disease. The use of real time PCR, using TaqMan® or LightCycler® systems, has gradually been introduced to enable a more quantitative assessment and has been shown to be a robust method on which clinical decisions can be made (Hochhaus *et al.*, 2000).

Although *RAS* mutations are rare in chronic phase CML, the RAS/MAPK pathway is important for the transforming activity of BCR–ABL with *RAS* being directly activated by the fusion protein. There is therefore ongoing research to identify the downstream targets of this pathway which include *BCL2*, a member of the family of apoptotic regulators. *RAS* deregulation is generally thought to be of importance in leukemias and is accepted as contributing to disease aggression and reduced survival rates. The importance of its deregulation is reflected in the interest in the development of anti-RAS drugs. Mutations in *RAS* have been

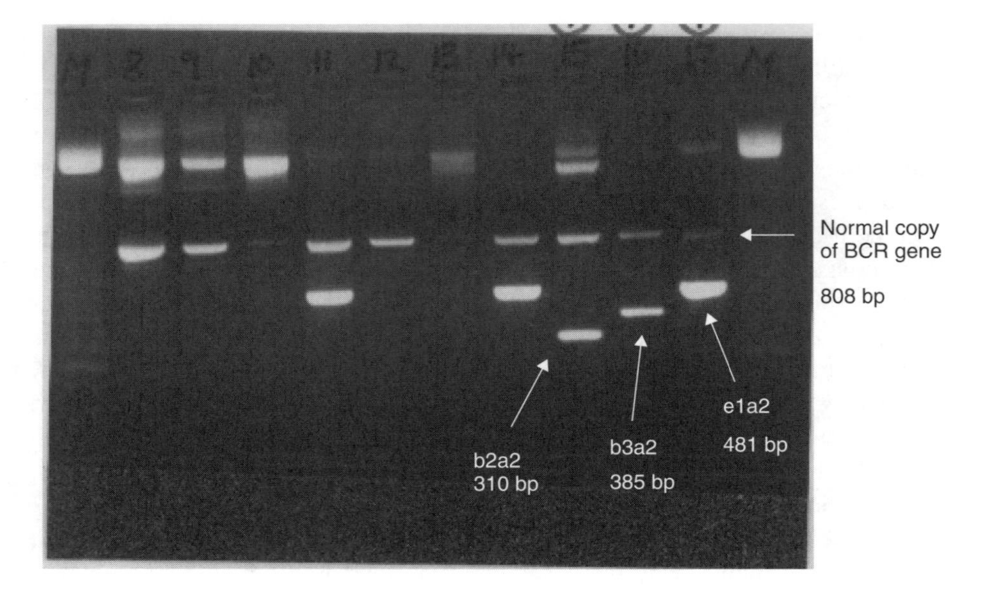

Figure 5.6 RT-PCR for BCR–ABL rearrangements. The e1a2, b2a2 and b3a2 rearrangements are arrowed, as is the position of the normal copy of the BCR gene included as a control (Cross *et al.*, 1994). Photograph courtesy of the West Midlands Regional Genetics Laboratory

found in 20%–30% of cases of AML and myelodysplastic syndrome (MDS). Several studies have shown that cases of MDS with a mutation are those most likely to progress to AML and hence have a worse prognosis.

The BCL2 protein has been shown to be over-expressed in B-cell lymphomas as a result of a translocation involving chromosome 14 and the *BCL2* gene on chromosome 18. It is also involved in AML where it is expressed in approximately 87% of cases at presentation rising to 100% at relapse. It is an indicator of poor prognosis related to unsuccessful drug treatment, poor patient survival and clinical outcome (O'Gorman and Cotter, 2001).

Two subunits of the core binding factor (CBF), AML1 and CBFβ, key regulators of early haematopoiesis, are involved in several translocations associated with acute leukemias. The fusion proteins CBFβ–SMMHC, AML1–ETO and AML1–MDS1/EVI1 are commonly found in subsets of acute myeloid leukaemia and TEL–AML1 is associated with acute lymphocytic leukaemia. The t(8;21) translocation is seen in approximately 12% of AML and involves the *AML1* gene on chromosome 8 and the *ETO* gene on chromosome 21. Expression of this fusion gene appears to be an early event in the development of the leukaemia (Friedman, 1999). The inv(16) abnormality fuses the *CBF*β gene with the smooth muscle myosin heavy chain gene (*SMMHC*) and is seen in 10% of AML where it is associated with a good prognosis. The t(12;21) translocation, which places the *AML1* gene next to the *TEL* gene, is common in ALL being found in around 20%–25% of cases of paediatric ALL and 3% of cases of adult disease. Children with this translocation have an excellent prognosis with a 10-year relapse rate of less than 10%. All three rearrangements can be detected in the laboratory by RT-PCR based techniques and given the prognostic significance of the presence of the rearrangement are useful diagnostic tests in the clinical setting.

Finally, the t(15;17) translocation found in acute promyelocytic leukaemia places the *PML* gene from chromosome 15 next to the retinoic acid receptor on chromosome 17. It is seen in acute myeloid leukaemia and is associated with a good prognosis. Again RT-PCR detection of the rearrangement is a helpful tool in the diagnosis of the condition.

5.9 Other cancers

Oncogenes have been associated with the majority of other cancers, some of which have diagnostic or prognostic roles. Space limitations do not allow them all to be mentioned here, but there are a few important cases that are worthy of specific mention.

Familial cancer syndromes are not generally associated with germ line mutations in the oncogenes, the majority being associated with tumour suppressor gene mutations. One exception to this is the *RET* oncogene which is mutated in familial medullary thyroid cancer (FMTC) and multiple endocrine neoplasia types 2A and B (MEN2A/MEN2B).

In MEN2A, mutations at 5 cysteine residues of the extra cellular domain of RET have been found in 93% of cases, with mutations at codon 634 being present in 83% (Mak and Ponder, 1996). A strong genotype–phenotype correlation exists between the mutation at codon 634, particularly a cysteine to arginine change, and the presence of pheochromocytoma. A single missense mutation at codon 918, in which methionine is replaced by threonine, is found in over 90% of cases of MEN2B. FMTC is associated with the same spread of mutations as in MEN2A, but the incidence of mutations is more evenly spread across the five cysteines. In all three conditions the effect of the mutations is to confer a gain of function to the protein. Interestingly a loss of function mutation is also seen in *RET* in which the gene is inactivated or there is abrogation of the function of the RET protein. These mutations cause a completely unrelated disorder, Hirschprungs disease, in which there is an abnormality of the hindgut characterised by the absence of enteric autonomic ganglia.

There is considerable clinical heterogeneity in paediatric neuroblastomas ranging from benign tumours requiring little or no treatment to highly aggressive tumours. Amplification of the *MYCN* oncogene in neuroblastomas almost always correlates with rapid tumour progression and poor outcome irrespective of tumour stage, and many studies have shown that *MYC* amplification is an independent prognostic indicator (George *et al.*, 2001). Patients with stage 4s neuroblastomas are those who frequently show spontaneous remission and this is strongly associated with lack of *MYCN* amplification (Ikeda *et al.*, 2002). Measurement of *MYCN* amplification is therefore an essential component of the routine diagnostic evaluation of new neuroblastoma patients (Favrot *et al.*, 1996). Amplification can be detected by fluorescence *in situ* hybridisation (FISH) and PCR and more recently by real time PCR allowing accurate quantification.

Mutations at codon 12 of the *KRAS* gene have been identified in up to 95% of pancreatic cancers and are believed to be early events in pancreatic carcinogenesis (Longnecker, 1999). Although there appears to be little correlation with survival, an association between the type of mutation and survival has been seen (Kawesha *et al.*, 2000). PCR-based assays have been used to detect *RAS* mutations in pancreatic juice, bile and pancreatic duct brushings obtained at ERCP as well as in serum and faecal specimens (Longnecker, 1999; Theodor *et al.*, 2000) and have all been suggested as useful diagnostic tests. However, there is a lack of specificity in the testing as positive results have been obtained in the absence of carcinoma (Hruban *et al.*, 1998). In addition, *RAS* mutations are frequently found in chronic pancreatitis reducing the sensitivity of *RAS* as a tumour marker.

5.10 Conclusion

The identification of the oncogenes has led to a much better understanding of the involvement of this group of genes in the development of tumours. It must however be remembered that cancers are caused by an accumulation of changes and abnormalities in one gene are insufficient for tumour development. In some types of tumours the oncogenes have proven to be helpful diagnostic or prognostic tools. Importantly, in recent years the products of these genes have started to become novel targets for therapeutic intervention and some therapies, such as Herceptin and STI571, have shown real clinical usefulness. Future work in this area is likely to lead to the development of other treatment strategies to increase the arsenal of drugs available for the treatment of cancers.

5.11 References

Ahrendt SA, Decker PA, Alawi EA *et al.* (2001) Cigarette smoking is strongly associated with mutation of the K-ras gene in patients with primary adenocarcinoma of the lung. *Cancer* **92**: 1525–30.

Andreyev HJ, Norman AR, Cunningham D *et al.* (1998) Kirsten ras mutations in patients with colorectal cancer: the multicentre RASCAL study. *J. Natl Cancer Inst.* **90**: 675–84.

Bannister AJ and Kouzarides T (1997) Structure/function and oncogenic conversion of Fos and Jun, in: *Oncogenes as Transcriptional Regulators* (eds Yaniv M and Ghysdael J). Birkhauser Verlag, Basel.

Bast RC, Ravdin P, Hayes DF *et al.* for the American Society of Clinical Oncology Tumor Markers expert panel (2001) 2000 update of recommendations for the use of tumor markers in beast and colorectal cancer: clinical practice guidelines of the American Society of Clinical Oncology. *J. Clin. Oncol.* **19**: 1865–78.

Bjorge JD, Jakymiw A and Fujita DH (2000) Selected glimpses into the activation and function of Src Kinase. *Oncogene* **19**: 5620–35.

Blume-Jensen P and Hunter T (2000) Oncogenic kinase signalling. *Nature* **411**: 355–65.

Boxer LM and Dang CV (2001) Translocations involving c-myc and c-myc function. *Oncogene* **20**: 5595–610.

Brabender J, Danenberg KD, Metzger R *et al.* (2001) Epidermal growth factor receptor and Her2-neu mRNA expression in non-small cell lung cancer is correlated with survival. *Clin. Chem. Res.* **7**: 1850–5.

Broers JL, Viallet J, Jensen S *et al.* (1993) Expression of c-myc in progenitor cells of the bronchopulmonary epithelium and in a large number of non-small cell lung cancers. *Am. J. Resp. Cell Mol.* **9**: 33–43.

Cross NCP, Melo JV, Feng L and Goldman JM (1994) An optimised multiplex polymerase chain reaction (PCR) for the detection of BCR–ABL fusion RNA in haematological disorders. *Leukemia* **8**: 86–189.

Dang CV (1999) C-myc target genes involved in cell growth, apoptosis and metabolism. *Mol. Cell Biol.* **19**: 1–11.

Deming SL, Nass SJ, Dickson RB and Trock BJ (2000) C-myc amplification in breast cancer: a meta-analysis of its occurrence and prognostic relevance. *Brit. J. Cancer* **83**: 1688–95.

Dong S, Traverso G, Johnson C *et al.* (2001) Detecting colorectal cancer in stool with the use of multiple genetic targets. *J. Natl. Cancer Inst.* **93**: 858–65.

Egan C, Pang A, Durda D *et al.* (1999) Activation of Src in human breast tumor cell lines: elevated levels of phosphotyrosine phosphatase activity that preferentially recognizes the Src carboxy terminal negative regulatory tyrosine 530. *Oncogene* **18**: 1227–1237.

Favrot MC, Ambros P, Schilling F *et al.* (1996) Comparison of the diagnostic and prognostic value of biological markers in neuroblastoma – proposal for a common methodology. *Ann. Oncol.* **7**: 607–11.

Fearon ER and Vogelstein B (1990) A genetic model for colorectal tumorigenesis. *Cell* **61**: 759–67.

Finkelstein SD, Sayegh R, Christensen S *et al.* (1993) Genotypic classification of colorectal adenocarcinoma. Biologic behaviour correlates K ras 2 mutation type. *Cancer* **71**: 3827–38.

Frame S and Balmain A (2000) Integration of positive and negative growth signals during ras pathway activation *in vivo*. *Curr. Opinion Genet. Dev.* **10**: 106–13.

Friedman AD (1999) Leukemogenesis in CBF oncoproteins. *Leukemia* **13**: 1932–42.

Futreal PA, Kasprzyk A, Birmey E, Mullikin, Wooster R and Stratton MR (2001) Cancer and genomics. *Nature* **409**: 850–2.

George RE, Variend S, Cullinane C *et al.* (2001) Relationship between histopathological features, MYCN amplification and prognosis: a UKCCSG study. United Kingdom Children Cancer Study Group. *Med. Ped. Oncol.* **36**: 169–76.

Gohring UJ, Schondorf T, Kiecker VR, Becker M, Kurbacher C and Scharl A (1999) Immunohistochemical detection of H-Ras proto-oncogene p21 indicates favourable prognosis in node negative breast cancer patients. *Tumour Biol.* **20**: 173–83.

Han H, Landreneau RJ and Santucci TS (2002) Prognostic significance of immunohistochemical expressions of p53, Her2/neu and bcl-2 in stage 1 non small cell lung cancer. *Hum. Pathol.* **33**: 105–10.

Heim S and Mittleman F (1987) Nineteen of 26 cellular oncogenes precisely localized in the human genome map to one of the 83 bands involved in primary cancer-specific rearrangements. *Human Genetics* **75**: 70–2.

Heinrich MC, Blanke CD, Druker BJ and Corless (2002) Inhibition of KIT tyrosine kinase activity: a novel molecular approach to the treatment of KIT positive malignancies. *J. Clin. Oncol.* **20**: 1692–703.

Hermeking H, Rago C, Schuhmacher M *et al.* (2000) Identification of CDK4 as a target for c-MYC. *Proc. Natl. Acad. Sci. USA* **97**: 2229–34.

Hirsch FR, Franklin WA, Veve R, Varella-Garcia M and Nunn PA (2002) Her2/neu expression in malignant lung tumors. *Sem. Oncol.* **29**: 51–8.
Hochhaus A, Weisser A, La Rosee P *et al.* (2000) Detection and quantification of residual disease in chronic myelogenous leukemia. *Leukemia* **14**: 998–1005.
Hruban RH, Petersen G and Kern SE (1998) Genetics of pancreatic cancer. From genes to families. *Surg. Oncol. Clin. North Am.* **7**: 1–23.
Ikeda N, Iehara T, Tsuchida Y *et al.* (2002) Experience with international neuroblastoma staging system and pathology classification. *Brit. J. Cancer* **86**: 1110–6.
Irby RB and Yeatman TJ (2000) Role of Src expression and activation in human cancer. *Oncogene* **19**: 5636–42.
Jones SM and Kazlauskas A (2000) Connecting signaling and cell cycle progression in growth-factor stimulated cells. *Oncogene* **19**: 5558–67.
Kaptain S, Tan LK and Chen B (2001) Her2/neu and breast cancer. *Diagn. Mol. Pathol.* **10**: 139–52.
Kawesha A, Ghaneh P, Andren-Sanberg A *et al.* (2000) K-ras oncogene subtype mutations are associated with survival but not expression of p63, p16 (INK4A), p21 (WAF-1), cyclin D1, erbB-2 and erbB-3 in resected pancreatic ductal adenocarcinoma. *Int. J. Cancer* **20**: 469–74.
Kloos DU, Choi C and Wingender E (2002) The TGF-beta-Smad network: introducing bioinformatic networks. *Trends Genet.* **18**: 96–103.
Krontiris TG, Devlin B, Karp DD, Robert NJ and Risch N (1993) An association between the risk of cancer and mutations in the HRAS1 minisatellite locus. *New Eng. J. Med.* **329**: 517–23.
Liao DJ and Dickson RB (2000) C-myc in breast cancer. *Endo. related cancer* **7**: 143–64.
Longnecker D (1999) Molecular pathology of invasive carcinoma. *Ann. NY Acad. Sci.* **880**: 74–82.
Mak YF and Ponder BAJ (1996) RET oncogene. *Curr. Opinion Gen. Dev.* **6**: 82–6.
Mao L, Hruban RH, Boyle JO, Tockman M and Sidransky D (1994) Detection of oncogene mutations in sputum precedes diagnosis of lung cancer. *Cancer Res.* **54**: 1634–7.
Moerkerk P, Arends JW, van Driel M *et al.* (1994) Type and number of Ki ras point mutations relate to stage of human colorectal cancer. *Cancer Res.* **54**: 3376–8.
Nicholson RI, Gee JM and Harper ME (2001) EGFR and cancer prognosis. *Eur. J. Cancer* **37** Suppl 4: S9–15.
Niklinski J, Niklinski W, Laudanski J, Chyczewska E and Chyczewska L (2001) Prognostic molecular markers in non-small cell lung cancer. *Lung Cancer* **34**: S53–S58.
Normano N, Bianco C, De Luca A and Salomon DS (2001) The role of EGF-related peptides in tumor growth. *Front Biosci.* **6**: D685–707.
O'Gorman DM and Cotter TG (2001) Molecular signals in anti-apoptotic survival pathways. *Leukemia* **15**: 21–34.
Pasche B (2001) Role of transforming growth factor beta in cancer. *J. Cell Physiol.* **186**: 153–68.
Pastorino U, Andreola S, Tagliabue E *et al.* (1997) Immunocytochemical markers in stage 1 lung cancer: relevance to prognosis. *J. Clin. Oncol.* **15**: 2858–65.
Pelengaris S, Rudolph B and Littlewood T (2000) Action of Myc *in vivo* – proliferation and apoptosis. *Curr. Opinion Genet. Dev.* **10**: 100–5.
Prendergast GC (1999) Mechanism of apoptosis by c-Myc. *Oncogene* **18**: 2967–87.
Reddy EP, Korapati A, Chaturvedi P and Rane S (2000) IL-3 signaling and the role of Src kinases, JAKs and STATs: a covert liaison unveiled. *Oncogene* **19**: 2532–47.
Robinson D, Wu Y-M and Lin S-F (2000) The protein tyrosine kinase family of the human genome. *Oncogene* **19**: 5548–57.
Ross JS and Fletcher JA (1998) The Her2/neu oncogene in breast cancer: prognostic factor, predictive factor and target for therapy. *Stem Cells* **16**: 413–28.
Roy HK, Olusola BF, Clemens DL *et al.* (2002) AKT proto-oncogene over-expression is an early event during sporadic colon carcinogenesis. *Carcinogenesis* **23**: 201–5.

Rubin I and Yardin Y (2001) The basic biology of HER2. *Ann. Oncol.* **12**: S3–8.

Rusch V, Klimsta D, Linkov E *et al.* (1995) Aberrant expression of p53 or the epidermal growth factor receptor is frequent in early bronchial neoplasia and co-expression precedes squamous cell carcinoma development. *Cancer Res.* **55**: 1365–72.

Sachdev D and Yee D (2001) The IGF system and breast cancer. *Endocrine related Cancer* **8**: 197–209.

Schneider PM, Praeuer HW, Stoeltzing O *et al.* (2000) Multiple molecular marker testing (p53, C-Ki-ras, c-erbB-2) improves estimation of prognosis in potentially curative resected non-small cell lung cancer. *Brit. J. Cancer* **83**: 473–9.

Schondorf T, Rutzel S, Andrack A, Becker M, Hoopmann M, Breidenbach M and Gohring UK (2002) Immunohistochemical analysis reveals a protective effect of H-ras expression mediated via apoptosis in node negative breast cancer patients. *Int. J. Oncol.* **20**: 273–7.

Shih C, Padhy LC, Murray M and Weinberg RA (1981) Transforming genes of carcinomas and neuroblastomas introduced into mouse fibroblasts. *Nature* **290**: 261–4.

Sidransky D, Tokino T, Hamilton SR, Kinsler KW *et al.* (1992) Identification of ras mutations in the stool of patients with curable colorectal cancers. *Science* **256**: 102–5.

Slamon DJ, Clark GM, Wong SG, Levin WJ, Ullrich A and McGuire WL (1987) Human breast cancer: correlation of relapse and survival with amplification of the Her2/neu oncogene. *Science* **235**: 177–82.

Slamon DJ, Godolphin W, Jones LA *et al.* (1989) Studies of the her2/neu proto-oncogene in human breast and ovarian cancer. *Science* **244**: 707–12.

Smith Ravine J, England J, Talbot IC and Bodmer W (1995) Detection of c-Ki-ras mutations in faecal samples from sporadic colorectal cancer. *Gut* **36**: 81–6.

Stehelin D, Varmus HV, Bishop JM and Vogt PK (1976) DNA related to the transforming gene(s) of avian sarcoma viruses is present in normal avian DNA. *Nature* **260**: 170–3.

Talapatra S and Thompson C (2001) Growth factor signalling in cell survival: implications for cancer treatment. *J. Pharmacol. Exp. Ther.* **298**: 873–8.

Theodor L, Melzer E, Sologov M and Bar-Meir S (2000) Diagnostic value of K-ras mutations in the serum of pancreatic cancer patients. *Ann. NY Acad. Sci.* **906**: 19–24.

Thijsen SFT, Schuurhuis GJ and van Oostveen JW (1999) Chronic myeloid leukemia from basics to bedside. *Leukemia* **13**: 1646–74.

van Dongen JJM, Macintyre EA, Gabert JA *et al.* (1999) Standardized RT-PCR analysis of fusion gene transcripts for chromosome aberrations in acute leukemia for detection of minimal residual disease. *Leukemia* **12**: 1901–28.

Vogelstein B, Fearon ER, Kern SE *et al.* (1989) Allelotype of colorectal carcinomas. *Science* **24**: 207–11.

Yamshita N, Minamoto T, Ochiai A, Onda M and Esumi H (1995) Frequent and characteristic K-ras activation in aberrant crypt foci of colon. Is there preference among K-ras mutants for malignant progression? *Cancer* **75**: 1527–33.

Yardin Y (2001) Biology of HER2 and its importance in breast cancer. *Oncogene* **61** Suppl 2: 1–13.

Zwick E, Bange J and Ullrich A (2001) Receptor tyrosine kinase signalling as a target for cancer intervention strategies. *Endocr. Relat. Cancer* **8**: 161–73.

6

Molecular and Immunological Aspects of Cell Proliferation

Karl Baumforth and **John Crocker**

6.1 The cell cycle and its importance in clinical pathology

In recent years, it has become increasingly apparent to histopathologists that a simple description of the morphological appearance of a section of tissue cannot give sufficient idea of its behavioural status. This is most unfortunate, since the accurate diagnosis of a specimen is linked to effective treatment of the patient and further management. It has also become clear to us that in this context the description of individual tissues (notably cancers), in terms of their histogenetic type, is insufficient. Modern diagnosis of many biopsy specimens depends upon various changes which are directly or indirectly related to cell replication. This is particularly true in the field of cancer, where cell division is known to be disordered. In general, such assessments are largely subjective and it is therefore highly desirable that more objective techniques become available. An example of a 'problem tumour' lies in leiomyosarcoma of the stomach; this is a neoplasm encountered by surgeons on occasion and which has a very variable prognosis. This latter is determined, presumably, by various factors in the biology of each case; however, variables as divergent as tumour mass and mitotic counts have been promulgated as indicators of behaviour. Nonetheless, it would seem to be logical that the level of cell proliferation in a tumour might be one of the more important features governing its future behaviour. It must never escape our understanding that factors including cell motility and adhesion, destruction of the basement membrane by proteinases and escape

Molecular Biology in Cellular Pathology. Edited by John Crocker and Paul G. Murray
 ISBN: 0-470-84475-2

from the immune system may be at least as important as cell proliferation in the natural history of a malignant tumour. None of us should therefore be in the least surprised if measurement of the proliferative ability of a specimen or series of specimens of cancer does not relate well to progression or clinical parameters of survival. Notwithstanding this caveat, the assessment of cell proliferation has proved to be most useful in many contexts in histopathology and it is timely that the immunological means available for this are described later.

Mitosis

The phases and mechanisms of mitosis (McIntosh and Koonce, 1989) are outlined in brief below:

1. In prophase, the chromosomes condense, ready for transport; this is effected by means of microtubules, under the control of the centrosome, which divides in mitosis and increases and rearranges the microtubules. In addition, protein synthetic levels decline and RNA synthesis ceases.
2. In prometaphase, one copy of each chromosome aligns with one end of the cell, by means of the 'metaphase plate'; this is organized by means of the centromere of the chromosome, which forms the kinetochore with associated, bound proteins. These structures drag the chromosomes into the correct orientation at the metaphase plate. As this is occurring, the nuclear membrane breaks down, as described above.
3. In anaphase, the two copies of the chromosome separate and migrate to the ends of the cell.
4. In telophase, the cell reverts to its interphase configuration, with decondensation of the chromosomes. After this, the cell itself divides, forming two daughter cells. This is an active metabolic event and requires the presence of an actomyosin-dependent contractile ring.

The cell cycle is divided into four stages, G1, S, G2, and M (Table 6.1). These are defined below and are also shown schematically in Figure 6.1. Cells

Table 6.1 Summary of cell cycle phases

Phase	Description
M (mitotic)	Segregation of one of the two sister chromatids to each daughter cell. Cell constituents are also divided, e.g. Golgi and ER
G1 (first gap)	Post-mitotic pre-synthetic. Cell becomes committed to cell cycle progression. Cells can leave the cycle here and enter G0
S (synthetic)	Chromosomes are replicated
G2 (second gap)	Post-synthetic pre-mitotic. Cells can repair errors in replicated DNA prior to cell division

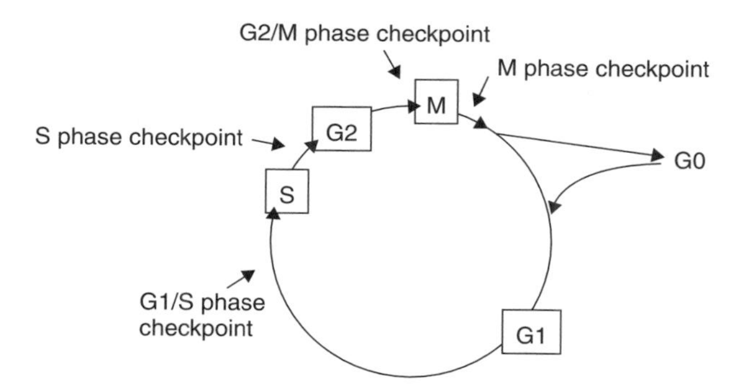

Figure 6.1 Phases and checkpoints of the cell cycle

may also exit the cycle into a non-dividing quiescent state (G0) from which they can either re-enter the cell cycle or differentiate into a committed cell. The frequency with which cells enter the cycle varies with the cell type; for example, lymph nodal follicular centroblasts have a high frequency of cycling, whereas mature neurons do not divide. In cancer cells, a great range of cell turnover rate is observed, thus high-grade malignancies have more cells in cycle at a particular time than do low-grade tumours.

- S Phase: The 'S' (synthetic) phase is when the chromosomes are replicated.
- M Phase: The 'M' (mitotic) phase is the segregation of one of the two sister chromatids to each daughter cell. This can be subdivided into prophase, anaphase, metaphase and telophase. Cell constituents for example, Golgi and endoplasmic reticulum, are also divided.

These two phases are separated by the 'gap' phases, G1 and G2.

- G1 Phase: The G1 (first gap) phase is a post-mitotic but pre-synthetic period, during which the cell becomes committed to cell cycle progression. Cells can also leave the cell cycle during this phase and enter G0.
- G2 Phase: The G2 (second gap) phase is a post-synthetic but pre-mitotic period during which cells can repair errors in the replicated DNA prior to cell division.

In rapidly dividing human cells this whole process takes 24 hours (M $\cong$ 30 min, G1 $\cong$ 9 h, S $\cong$ 10 h, and G2 $\cong$ 4.5 h, compared with about 90 min for the whole process in yeast cells).

Such a complex procedure is subject to stringent regulation by biochemical mechanisms, which are phylogenetically conserved between organisms as diverse as yeast, amphibian oocytes and mammalian cells. Cancer cells become able to override certain aspects of this regulation.

6.2 Molecular control of the cell cycle

There are three main groups of regulatory proteins which control the cell cycle. These are (a) cyclin-dependent kinases (CDKs), (b) cyclins and (c) CDK inhibitors (CDKIs). The groups of proteins and their interactions are described below and shown in Figure 6.2.

Cyclin Dependent Kinases (CDKs)

There are currently nine recognised members of this family (CDK1-9), all of which phosphorylate other proteins and allow the cell cycle to progress (Ekholm and Reed, 2000; Morgan, 1997). The CDKs are entirely reliant upon the presence of a cyclin partner for their activity, e.g. cyclin E/CDK2. CDKs are activated only when their cyclin partner reaches a critical concentration. Conversely, they are deactivated when the cyclin concentration decreases. Some CDKs may be activated by more than one cyclin (for example, CDK2 can be activated by cyclin D2, cyclin E and cyclin A, depending upon the phase of the cell cycle). Therefore, the activity of CDK2, for example, is maintained through several phases of the cell cycle (see Figure 6.2). The CDKs show 75% sequence homology but possess unique cyclin-binding sites. To become fully activated, the CDKs also require the phosphorylation and dephosphorylation of certain amino acid residues. This is performed by the CDK-activating kinase complex (CAK), which is composed of a CDK, CDK7, and a cyclin, namely cyclin H (Fisher and Morgan, 1994).

Cyclins

The cyclins are activators of CDKs. As their name suggests, their concentrations oscillate in a cyclical manner as the cell cycle progresses (Evans *et al.*, 1983).

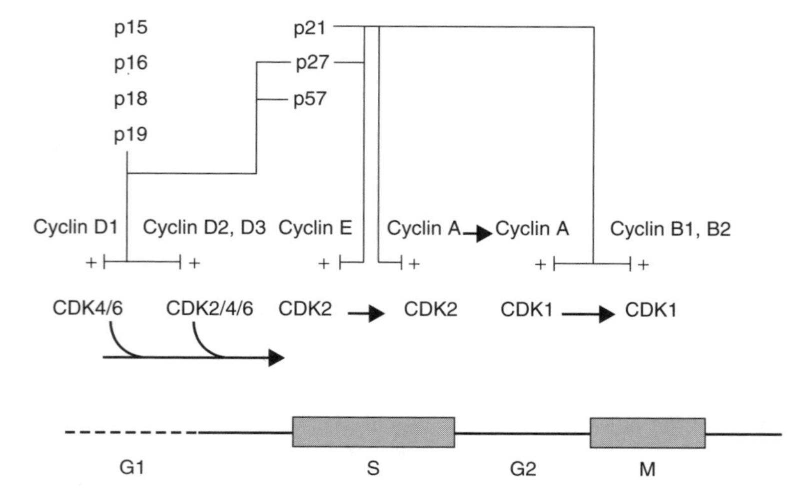

Figure 6.2 The interactions between the cyclins, CDKs and CDKIs during the cell cycle

There are currently 15 members of the cyclin family (cyclin A-T), although the functions of some of them have not yet been resolved. The cyclins have a conserved 100 amino acid region known as the cyclin box (Lees and Harlow, 1993) and some have a conserved destruction box near their N-terminus, which is involved in their degradation. The cyclin expression level is rate-limiting for CDK activity and the amount of a cyclin present is controlled, both at the transcriptional level and degradation, via the ubiquitin proteasome system (Murray, 1995).

Cyclin-dependent Kinase Inhibitors (CDKIs)

In order to prevent perpetual cell reproduction, the activation of CDKs by their partner cyclins is itself controlled by the CDKIs (Chellappan *et al.*, 1998). The CDKIs control the transition of the cell cycle from one phase to another [i.e. they act at cell cycle checkpoints (discussed later)]. CDKIs can block the activity of the CDKs, retaining the cell in a particular phase of the cell cycle, until conditions become favourable for cell proliferation to continue or that a cell is directed towards apoptosis.

There are two main classes of CDKIs within mammalian cells and these are the p21 and INK4 families. The p21 (or Cip/Kip) family consists of three members, namely p21, p27 and p57. They are universal CDK inhibitors with multiple roles, e.g. over-expression leads to G1 arrest. The INK4 family consists of $p15^{INK4B}$, $p16^{INK4A}$, $p18^{INK4C}$ and $p19^{INK4D}$ and these are far more specific, inhibiting CDK4 and CDK6 during G1 of the cell cycle. A related protein, $p19^{ARF}$, has been shown to play a role in increasing p53 stability and the onset of cell cycle arrest (Pomerantz *et al.*, 1998).

Cell Cycle Initiation

Normal cells are stimulated to divide by numerous external signals (mitogens), such as hormones and growth factors. These ligands interact with receptors, either on the cell surface or within the cytoplasm, which signal via secondary messengers. These eventually activate transcription factors within the nucleus leading to increased expression of the G1 cyclins. Cells in G0 or those immediately after mitosis in early G1 require growth signals to allow the progression through G1 and into the S phase. However, at a point in G1 these external signals become unimportant and the cell cycle progresses whatever. This point is known as the restriction point 'R' (Pardee, 1974, 1989) (START in yeast). Tumour cells characteristically lose 'R' point control and become mitogen-independent (see Section 6.4).

Cell stimulation leads to increased expression of the D-type cyclins, D1, D2, and D3, (Scherr, 1995) which form complexes with their partner CDKs (see Figure 6.2) i.e. cyclin D1 with CDK4 and CDK6, and cyclins D2 and D3 with CDK2, CDK4 or CDK6. This ultimately leads to the phosphorylation of the Rb protein family members, Rb, p107 and p130. In early G1, Rb is hypophosphorylated and binds to and inactivates members of the E2F transcription factor family. E2F transcription factors are involved in the onset of DNA replication. Until Rb is fully phosphorylated, the E2F cannot be released and DNA synthesis is blocked (Weinberg, 1995).

The phosphorylation of Rb is started by activation of CDK2, CDK4 and CDK6 by the D-type cyclins. The concentrations of the D-type cyclins increase through G1, peaking just prior to the S phase. Unfortunately, at this stage the Rb protein is not phosphorylated enough to release the E2F. Therefore, cyclin E/CDK2 takes over and completes the phosphorylation of Rb. Cyclin E levels have begun to increase from mid-to late G1 and peak at the G1/S phase transition, where the cyclin E/CDK2 complex has peak activity. The fully phosphorylated Rb can now release E2F, and DNA synthesis (the S phase) can now proceed. Therefore, cyclin E/CDK2 activity triggers the onset of the S phase. The G1/S phase transition is therefore a point in the cell cycle where control can occur, a checkpoint, to determine whether the cell cycle proceeds or arrests (see Section 6.3).

DNA Synthesis

At the onset of the S phase, cyclin E is degraded rapidly and the cyclin A/CDK2 complex becomes necessary for the continuation of DNA synthesis. Cyclin A levels have increased during the latter stages of G1. Cyclin A/CDK2 complexes activate the proteins involved with the origins of replication upon the DNA (chromatids) and hence DNA synthesis occurs (Stillman, 1996). Towards the end of the S phase, cyclin A starts to activate CDK1; this signals the onset of G2, which provides a break between DNA synthesis and mitosis. During this time the cell can check for the accuracy and completeness of DNA synthesis and if necessary be directed towards apoptosis. The G2/M transition is therefore another cell cycle control checkpoint (see Section 6.3).

Mitosis

The onset of mitosis is signalled by a group of proteins, initially known as maturation-promoting factor (see the first edition of this book) (Lohka *et al.*, 1988) but know referred to as M phase promoting factor (MPF). In reality, MPF consists of CDK1 complexed to either cyclin A or cyclin B. Cyclin A/CDK1 is involved in the completion of the S phase and preparation for mitosis, whilst cyclin B/CDK1 controls the onset, sequence and completion of mitosis. The concentrations of the B cyclins (B1 and B2) is very low in G1, they increase gradually during the latter stages of the S phase and through G2 to peak in the M phase. Although the B cyclins appear to perform the same functions, their sub-cellular localization differs, thus cyclin B1 localizes to microtubules and cyclin B2 to the Golgi apparatus (Jackman *et al.*, 1995).

Division occurs only when the cyclin B/CDK1 complex has been activated (Morgan, 1995) by a sequential series of events:

1. Phosphorylated on threonine 14 and tyrosine 15 by WEE1.
2. Phosphorylated on threonine 16 by CAK.
3. De-phosphorylated on threonine 14 and tyrosine 15 by CDC25.

This activation leads to mitotic events (McIntosh and Koonce, 1989) such as chromosome condensation, nuclear envelope breakdown and chromatid separation via the action of cyclin B/CDK1 upon histone H1, nuclear lamins, vimentin and so on. Therefore, the nucleus embarks upon mitosis and cell division occurs. Upon completion of cell division cyclin B is degraded, therefore CDK1 is no longer active and the cell enters G1 or G0.

6.3 Cell cycle control

The complicated processes undertaken during the cell cycle require a plethora of mechanisms to keep the whole system under control. The individual phases have to progress in the correct order and there has to be flexibility to allow repair mechanisms to be undertaken or for the cell to be directed to undergo apoptosis. The preceding sections have shown how varying the concentration of the major activating molecules imparts a certain degree of control upon the cell cycle. However, there are many other control mechanisms, a few of which are described here.

Prior to the onset of the cell cycle, that is, in non-dividing cells, p27 is present at high concentration, therefore preventing the activation of CDK4 or CDK6 by cyclins D1, D2 and D3. The cell therefore remains in G0 or G1. Mitogenic stimulation causes a decrease in the levels of p27 and the increasing levels of cyclin D activate their partner CDKs, leading to the onset of Rb phosphorylation. Rb needs to be phosphorylated in order to release E2F and allow cell cycle progression (Weinberg, 1995, 1996). Therefore, active Rb regulates the cell cycle. During G1, p27 levels remain high and a sudden removal of the mitogenic stimulus increases its concentration, further stopping the cell cycle (Hengst and Reed, 1996). The primary function of p27 during G0 and G1 is to inhibit CDK2, CDK4 and CDK6. The p15 and p16 proteins are tumour suppressors which are regulated by Rb; they, together with p18 and p19, can inhibit CDKs 4 and 6, either by displacing the cyclin from the CDK or by destabilizing its association with the CDK.

In the early stages of G1, any DNA damage, e.g. DNA damage caused by UV irradiation, can be repaired by delaying the cell in G1. This is a p53-mediated response to DNA damage; p53 induces p21 expression, which in turn acts to inhibit CDK2 activation. The cell therefore arrests in late G1 or early S, allowing repair or apoptosis (El-Deiry *et al.*, 1994).

The p57 protein can bind to and inactivate CDK2, CDK3 and CDK4 early in the cell cycle, although the control mechanisms for p57 expression are not fully understood. Another checkpoint occurs at the G2/M transition. Again, p53 mediates a response to DNA damage by up-regulating p21. Under these conditions, the p21 inhibits the cyclin B/CDK1 complex. Mitosis itself is controlled by an intricate series of checkpoints, which monitor chromosome condensation, mitotic apparatus assembly, chromosome alignment, movement and, finally, cytokinesis (Cahill *et al.*, 1998).

The transitions between phases of the cell cycle can also be controlled by the proteolysis of the components involved via eliminating proteins involved in the last phase and removing those that inhibit progression to the next phase (King *et al.*, 1996). An example is cyclin E, which is degraded by ubiquitin-dependent proteolysis. Cyclin E is phosphorylated by its own CDK and this phosphorylation triggers the ubiquitination of cyclin E and its eventual destruction. Therefore, cyclin E/CDK2 is self-limiting. Antiproliferative affects such as those involved in DNA damage or cAMP and TGF-β (see Section 6.4) can also help control a cell's progression through the cycle.

6.4 The cell cycle and cancer

Tumours most often originate in adult tissues, within which many cells are quiescent, i.e. in G0. However, tumour cells undergo uncontrolled proliferation and must therefore be able to bypass quiescence. All cells, apart from those that are terminally differentiated, are able to re-enter the cell cycle. Mutations and alterations commonly seen in cancer include alterations to mitogenic cell signalling pathways *via* growth factors, receptors and their associated downstream effectors, inactivation of apoptotic pathways (therefore introducing an imbalance in cell survival vs. cell death), induction of genomic instability and the promotion of angiogenesis. These alterations in turn can lead to changes in the mechanisms that control the cell cycle.

Cells undergo a period of mitogen dependence before they can enter the cell cycle; this transition in G1 is known as the restriction point 'R' described as START in the first edition of this book. Once past this point, the cells are committed to entering the cell cycle. It is postulated that the loss of regulation of 'R' is critical in cancer (Dannenberg *et al.*, 2000; Sage *et al.*, 2000). Loosening of control of 'R' resulting from mutations in the regulators of G1, for example, will allow cells to enter the cycle in the absence of sufficient mitogenic signalling and lead to cell proliferation. The cyclins, cyclin-dependent kinases (CDKs) and the Rb protein are all involved in the control of passage through the restriction point.

Table 6.2 Mutations of G1/S regulators in human cancers

Cancer type	Alterations	Cell cycle regulators altered: Genetic or epigenetic alteration defined	Cell cycle regulators altered: Alterations with no mechanistic explanation
Head and neck	>90%	p130, CDK4, D1[a], $p16^{INK4A}$	$p27^{KIP1a}$
Bone marrow (leukemia)	>90%	$p16^{INK4A}$, $p15^{INK4B}$, D1, CDK4, Rb[a]	E1, $p27^{KIP1a}$
Liver	>90%	Rb, CDK4, D1, $p16^{INK4A}$, $p15^{INK4B}$, p15, CDK2	E1[a], $p27^{KIP1a}$
Lung	>90%	$p16^{INK4Aa}$, $p15^{INK4B}$, D1, CDK4, p130[a], Rb[a]	E1[a], $p27^{KIP1a}$
Lymphoma	>90%	Rb, CDK6, D1, $p16^{INK4A}$, $p15^{INK4B}$	D2, D3, E1[a], $p27^{KIP1a}$
Breast	>80%	$p16^{INK4A}$, D1, CDK4, p130, Rb	E1[a], $p27^{KIP1a}$
Pituitary	>80%	Rb, D1, $p16^{INK4A}$	D3, $p27^{KIP1a}$
Bladder	>70%	$p16^{INK4A}$, D1, Rb[a]	E1[a], $p27^{KIP1a}$
Prostate	>70%	$p16^{INK4A}$, D1, Rb[a]	E1, $p27^{KIP1a}$
Melanoma	>20%	p130[a], D1, p16[a]	E1

[a]Represents an alteration which is relevant for tumour progression.
Table adapted from Figure 6.2 in Malumbres M and Barbacid M (2001) To cycle or not to cycle: a critical decision in cancer. *Nature* Rev. *Cancer* **1**: 222–31.

Cell cycle regulators are frequently mutated in human cancers (see Figure 6.2 or Table 6.2) (Easton *et al*., 1998; Wolfel *et al*., 1995). Over-expression of proteins involved in cell cycle control commonly occur. Examples include over-expression of the cyclins D1 and E1 and cyclin-dependent kinases CDK4 and CDK6. Under-expression/loss of proteins also occurs, for example of the cyclin kinase inhibitors, $p16^{INK4A}$, $p15^{INK4B}$, and $p27^{KIP1}$ and Rb. These changes occur as a result of chromosomal alterations in the cases of cyclin D1, CDK4, CDK6, $p16^{INK4A}$, $p15^{INK4B}$ and Rb or epigenetic inactivation via methylation of the Rb or INK4 promoters.

Cyclin D1 Over-expression

Cyclin D1 is over-expressed in many human cancers, including that of breast, as a result of gene amplification or a translocation, which places the gene under the control of a stronger promoter (Sicinski *et al*., 1995). In B-cell tumours, cyclin D1 is placed under the control of an antibody gene-enhancer, leading to increased cyclin D1 production throughout the cell cycle. This is analogous to the c-myc translocation seen in Burkitt's Lymphoma. Human tumour viruses can also express genes with homology to human cyclin genes (for example, KSHV expresses v-cyclin-D, which activates CDK6, in turn phosphorylating the Rb protein and releasing the cell from G1 arrest).

Loss of INK4A (p16) Function

The INK4A protein p16 usually functions as a tumour suppressor; it belongs to a family of proteins that inhibit cyclin kinases and hence help to regulate the cell cycle. Cyclin D-dependent kinase activity is inhibited by $p16^{INK4A}$. Mutations that inactivate this gene are common in some human cancers (Easton *et al*., 1998; Wolfel *et al*., 1995). Loss of this gene mimics the effects of cyclin D1 over-expression and Rb becomes hyperphosphorylated, releasing active E2F transcription factors, which in turn activate genes required for entry into the S phase. The genes activated by E2F transcription factors include those for CDK2, cyclins A and E, and genes encoding proteins required for DNA replication.

Loss of Rb Function

The loss of Rb function leads to many types of cancer, most notably childhood retinoblastoma, a condition caused by a mutation, either inherited (autosomal dominant) or somatic, in the retinoblastoma (Rb) gene. Loss of Rb function also occurs in other tumours, e.g. osteosarcomas. In most tumours with inactivated Rb there are usually both normal levels of cyclin D1 and p16, whereas in tumours that over-express cyclin D1 or have lost p16 there is wild-type Rb protein. Therefore, an alteration to one of these proteins can cause the cell to pass through the 'R' point and into the S phase. Hypophosphorylated Rb protein inhibits E2F transcription factors (during early G1 and G0). Rb becomes hyperphosphorylated, initially by cyclin D1/CDK4, cyclin D1/CDK6 complexes in mid G1, cyclin E/CDK2 in late G1 (Ezhevsky *et al*., 2001) and CDK2, CDK1 cyclin complexes through the S, G2 and M phases. This hyperphosphorylation of Rb leads to the activation of E2F and progression through the cell cycle. Normal Rb is dephosphorylated by phosphatases during early G1, resulting from a lack of cyclin/CDK complexes and hence hypophosphorylated Rb is able to inhibit the E2F transcription factors (Harbour and Dean, 2000).

Loss of TGF-β Signalling

Transforming growth factor β (TGF-β) inhibits cell proliferation at G1; it is secreted by most cells and has a wide variety of functions. Normally TGF-β binds to cell surface receptors (which have intrinsic serine threonine kinase activity) causing the formation of receptor complexes, which phosphorylate the Smad proteins (either Smad2 or Smad3). These then form complexes with Smad4, translocate to the nucleus and transcribe various genes including p15.

The G1 cyclin kinase inhibitor, $p15^{INK4B}$, displaces p27 from cyclinD-CDK4 complexes allowing p27 to inhibit cyclinE-CDK2 complexes and preventing entry into the S phase. The net result is that cells arrest in G1.

Many tumours possess inactivating mutations in either the Smad proteins or TGF-β receptors. For example, gastric cancer, hepatoma and retinoblastomas all show loss of TGF-β receptors, whereas many pancreatic cancers have a deletion within the Smad4 gene. The TGF-β pathway also induces expression of genes which encode the proteins of the extracellular matrix (ECM). Therefore, an inability correctly to synthesize the ECM may contribute to the metastatic potential of some tumours.

6.5 Immunocytochemical markers of proliferating cells

General Considerations

It will be apparent from the foregoing that there are several, if not many, candidates for 'markers' of cells which are proliferating. These can be expressed throughout much or all of the replicative phases of the cell cycle or be highly restricted in expression to only a very limited portion of the cycle. Furthermore, some of the antigens defined by the available antibodies may be of unknown location and chemical conformation; in these cases, it has often been the case that their value has been found on an empirical basis only. Examples of this type of preparation lie in some of the autoantibodies to human nuclear and nucleolar molecules. Conversely, in more recent years, the chemical nature and ultrastructural localization have been determined for some of the epitopes or molecules binding to certain antibodies. Our greater understanding of the molecular events and control of the cell cycle should enable us to label specific phases of the cycle in cells or tissue sections, although it could be argued that for 'routine' diagnostic purposes such 'fine tuning' is at present unnecessary (or, at least, of unknown value!).

Certain general comments should be considered with regard to the practical aspects of the investigation of proliferative status by means of immunocytochemistry. Firstly, in general it is the nucleus which is labelled in these methods and it is then necessary to count the numbers of positively stained nuclei in relation to a certain number of overall cells per nuclei. The 'lazy' technique of enumerating cells per nuclei per high-power field is to be eschewed at all costs, since not only does the area of such a field vary considerably from microscope to microscope, but also the measurement becomes meaningless since cell size varies from specimen to specimen (Figure 6.3). Accordingly, the score of 'events' per area tells us little of the frequency of such events per number of cells, which is of much greater biological significance. The minimum number of total cells to be included in the counting procedure should always be derived by the standard continuous mean method.

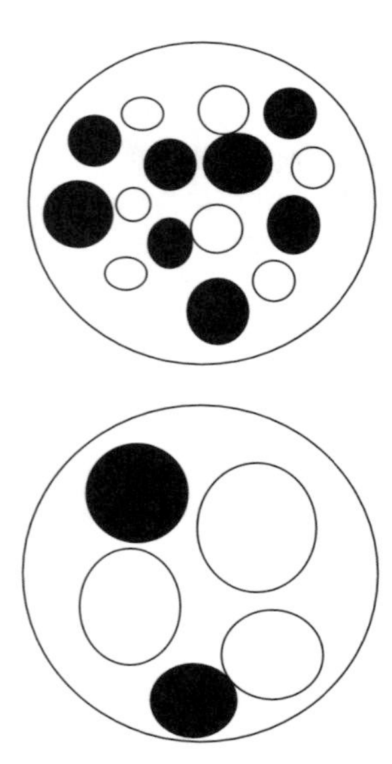

Figure 6.3 The problems encountered on counting 'events per microscope field'. Depending on the overall cell size, counts for the positive structures (black) may seem more or less frequent relative to the negative structures (white). It is more satisfactory to express results as 'positive cells per *n* negative cells'

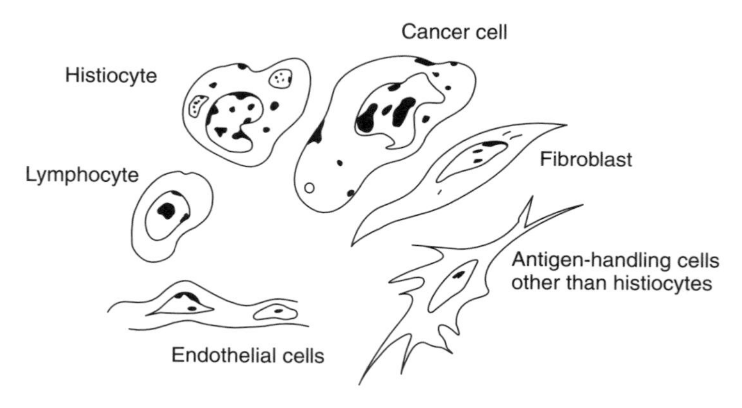

Figure 6.4 Another problem encountered in enumerating positively stained cells in a malignant neoplasm. Many of the cells present may not be neoplastic and yet may express molecules leading them to be included in the score

The next consideration is that of tissue heterogeneity. This is a particular problem in some neoplasms, where some of the cells may be residual from the original tissue or may be 'reactive' (for example, inflammatory or endothelial cells) (Figure 6.4). A rather arbitrary correction factor has been proposed by some workers to deal with this problem; however, the most stringent (and often the least practical) method is to use dual labelling of the tissue under investigation, with one or more other antibodies to highlight the cells of interest or to exclude others. This latter approach, although theoretically highly desirable, is very labour-intensive. The next practical point to be considered is that of the suitability of the tissue for study with a particular antibody. Unfortunately, many of the antibodies available for labelling proliferating cells will only succeed with frozen tissue, and routinely processed paraffin wax-embedded tissue is not suitable in this context. This can cause problems in two ways; firstly, paraffin wax-embedded archival material used in most studies in histopathology laboratories cannot be accessed; secondly, the morphology attained with frozen sections is never as clear as that available with a paraffin section. Thus, the cell types reacting with an antibody cannot be recognized with certainty on morphological grounds.

It has become apparent that the data derived from studies of, for example, neoplastic tissues using a particular antibody may not correlate well with those obtained from the use of other antibodies or methods. For example, staining with antibodies such as Ki-67 (see below) usually gives results in accord with those afforded by 'AgNOR' enumeration; conversely, other parameters may correlate poorly. The reasons for these observations are complex but probably lie in the fact that most of the antibodies in use label more than one phase of the cell cycle, unlike the uptake of thymidine or its analogues, which is restricted to the S phase. Furthermore, the half-life of some of the marker proteins may be relatively lengthy and, although they may not still be metabolically active, they may yet be detectable immunologically. Under ideal conditions, perhaps, any specimen to be evaluated for proliferation status should be subjected to more than one type of method. In reality, however, this cannot always be practical.

The 'classical' method for the assessment of proliferation in tissues or cells was to subject them to labelling with [^{3}H]thymidine, which was incorporated into the DNA of S phase nuclei; however, the method required the availability of fresh tissue and involved cumbersome autoradiography, which was also time-consuming. Accordingly, thymidine labelling is rarely used today, having been largely supplanted by newer methodologies.

The following part of this chapter details the various antibodies available for the assessment of cellular proliferation in tissue sections. It should become apparent to the reader that although advances have been made in terms of application to paraffin sections, many problems still exist in the practical side of this field.

Specimen Handling and Preparation

For most immunohistochemical purposes tissue can be fixed in 10% formalin and processed in the conventional manner. To optimize fixation, it is recommended that the tissue is obtained fresh, as below, and that thin (2 mm) slices are placed in the formalin. However, for many markers of cell proliferation it is necessary to use frozen sections and it is therefore necessary either to stain freshly cut frozen sections or to maintain tissue at a low temperature until required for sectioning. The latter is the usual case, so it must be arranged with the surgeon that fresh tissue is available for collection from the operating theatres to enable appropriate handling. Slices of tissue, about 2–3 mm thick, should be dropped into liquid nitrogen, then stored in the same. In the author's experience, there is no need to make use of cooled isopentane to freeze tissues; this merely adds complexity to the process. Whether it is necessary to use frozen sections or whether paraffin sections are amenable, the immunohistochemical labelling process to be applied should be considered. Whenever possible, a chromogenic system which gives a permanent reaction product is preferable. Such a system is the avidin–biotin complex method, which has found widespread routine use. However, there are occasions when the alkaline phosphatase–anti-alkaline phosphatase technique can be invaluable, especially when high amplification of the 'signal' is necessary. This latter method may also be very helpful when double-staining (i.e. staining for two antigens or staining for one antigen plus *in situ* hybridization) is important. For details of the methodology, the reader should refer to one of the several practical texts in this field.

'Traditional' Antibodies

Antibodies to transferrin receptor

Transferrin receptor (TfR) is a substance which, like its ligand transferrin, is essential to the life of many (if not most) living organisms. The uptake of iron is as necessary for life-forms as lowly as bacteria as it is to mammals, and the regulation of its usage is predictably quite highly controlled. The availability of iron binding is of great importance, since in the native state iron is in its ferric (Fe^{3+}) form and thus tends to be hydrolysed to insoluble $Fe(OH)_3$. Iron-binding molecules such as transferrin 'protect' iron from this process and each molecule of transferrin can bind up to two atoms of iron. The transferrin binds to TfR on the cell surface and the complex is then internalized via coated vesicles (Figure 6.5). The highest rate of such iron uptake is observed in haemoglobin-synthesizing reticulocytes and in the placental trophoblast; in the former, each cell has up to 3.10^5 TfR sites and the entire sequence of iron/transferrin uptake, release of iron and return of apotransferrin to the exterior takes up to only 30 s. It is likely that all human cells possess TfR sites but at greatly divergent densities. Thus, to give an assessment of TfR positivity for a particular cell species is really a statement describing the ability of the antibody used to detect a certain amount of surface TfR. Monoclonal antibodies to TfR have been available for some years; two such antibodies are B3/25, which was raised against a human erythroid leukaemic cell line, and OKT9, which is also reactive with some T lymphocytes.

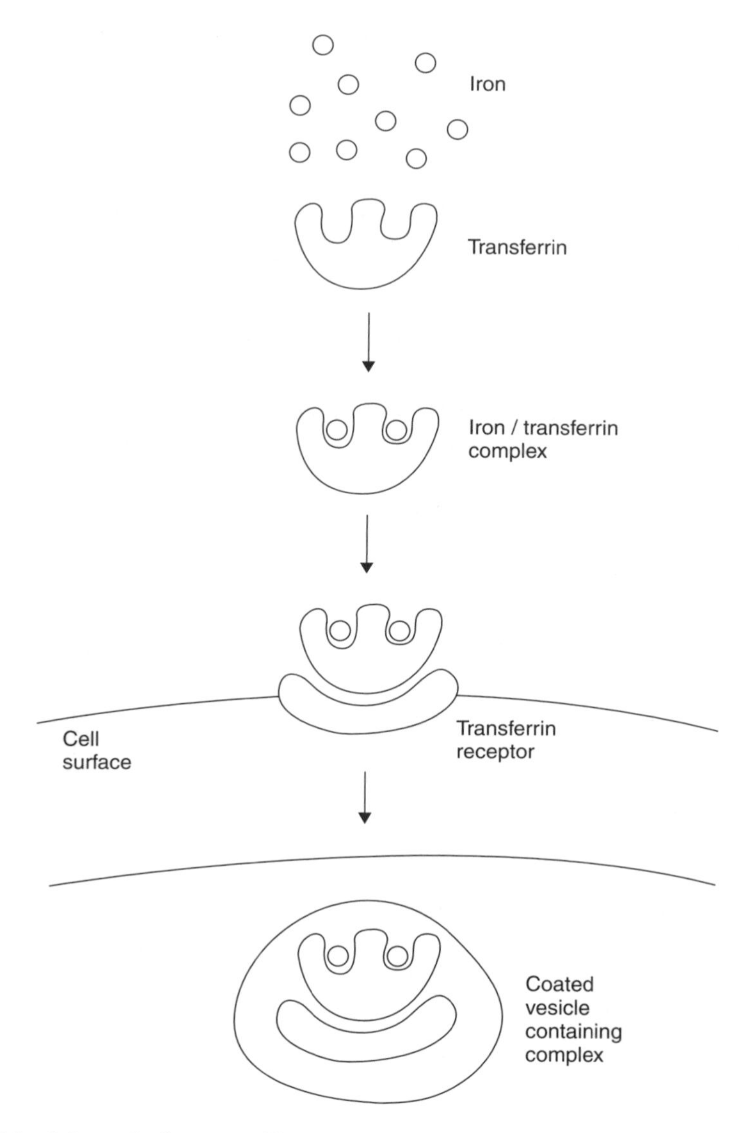

Figure 6.5 Schematic diagram to illustrate the uptake of iron and transferrin by an active cell

Biochemical analysis of TfR has been facilitated by the use of monoclonal antibodies against transferrin itself, to purify the transferrin/TfR complex or, more directly by means of OKT9, directed against TfR itself. It has been shown that the receptor is composed of two 90 kDa subunits, joined by disulphide linkage. The evidence is that TfR is a transmembrane molecule, with covalently bound fatty acid and phosphoryl serine residues. There is also a transmembrane 'tail' of 5 kDa size, as demonstrated by the protease treatment of microsomes. It also appears that each 90 kDa subunit of TfR binds to a single transferrin molecule. Recently, chromosome-mapping techniques have been applied to human–murine cell hybrids, to localize the gene encoding for TfR; these experiments have made use of the fact that the OKT9 antibody is species-specific for humans and that only the hybrid cells retaining the appropriate human chromosome will express TfR binding to this antibody. By this method, it has been shown that chromosome 3 is the location of the receptor gene, lying at the 3q26.2 position. (Interestingly, the location of the gene for transferrin itself lies close by, at 3q21. Furthermore, there is a protein gp97, associated with malignant melanoma cells and also capable of binding iron, which is encoded by a gene at 3q29. This suggests that the inheritance of these genes may be linked as a family.)

It was noted in the early 1980s that immunostaining for TfR might prove to be a useful method for the assessment of numbers of proliferating cells in a tissue sample. When fluorescent labelling with OKT9 was applied to leukaemic cells, with double tagging for DNA with mithramycin, followed by cell sorting, there was good correlation between immunolabelling and proliferation status. Induction of differentiation, with lowering of proliferation, was associated with a decrease in TfR expression. Furthermore, when the OKT9 antibody was applied to cell suspensions of cells from high- and low-grade lymphomas, there was a much greater level of labelling of the former (range 3%–57%; mean 22.5%) than the latter (range <1%–22%; mean 2.5%) (Habeshaw *et al.*, 1983). Similar results have also been reported for carcinoma of the breast. Antibodies to TfR have also been applied to frozen sections of human tissue, although paraffin wax sections are not suitable. However, the most unfortunate limitation to the use of anti-TfR antibodies in the analysis of histological material, especially when it is neoplastic, is that they bind strongly to macrophages/histiocytes. This is an especially great problem, since most tumours, especially those of high malignancy, contain these cells. Although double immunostaining with anti-TfR and anti-macrophage antibodies could help in this context, it would be highly labour-intensive to perform this procedure and it must be admitted that the use of antibodies to TfR should be regarded now as being of academic rather than practical value in the field of histopathology.

Antibody BK 19.9

This antibody has found only limited use in the demonstration of proliferating cells, probably because it has not been fully characterized and because it can only be applied successfully to frozen section material (Figure 6.6) (having been overtaken by the antibody Ki-67, as described below). BK 19.9 was raised against myeloid leukaemia cells and was shown by means of binding of anti-TfR and transferrin itself to co-cap (that is to say, to move to one pole of the cell) with TfR. It was therefore concluded that the antibody was directed against transferrin receptor. However, subsequent evaluation (Gatter *et al.*, 1983)

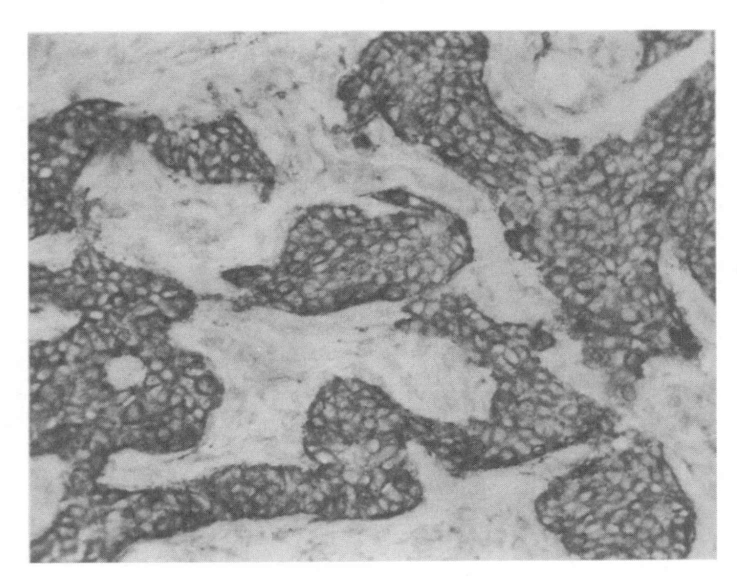

Figure 6.6 A frozen section of a carcinoma of breast, showing extensive positive labelling with the antibody BK 19.9. Photomicrograph courtesy of Prof Rosemary Walker, Department of Pathology, University of Leicester

demonstrated that this may not be the case. For example, hepatocytes are BK 19.9^{-} but $B3/25^{+}$ and Kupffer cells are BK 19.9^{+}, although both cells are known to express TfR. Anomalous staining with BK 19.9 of pancreatic islets is also seen. Furthermore, expression of the BK 19.9 antigen precedes that of TfR during the entry of G_0B lymphocytes into the cell cycle.

Antibodies to 5-bromodeoxyuridine

Antibodies to 5-bromodeoxyuridine (BrdUrd) and 5-iododeoxyuridine (IrdUrd) have been available for over a decade (Gratzner, 1982). As cells replicate, they incorporate BrdUrd into their DNA (since it is an analogue of thymidine) and this can then be detected by means of anti-BrdUrd antibody. Earlier studies with anti-BrdUrd made use of fluorescence-activation cell sorting (FACS) and, for example, in non-Hodgkin's lymphomas, gave results equivalent to those obtained with [^{3}H]thymidine uptake. Subsequently, the antibody was applied to cell suspensions from tumours, with more cells binding BrdUrd in high-malignancy specimens than in those with low aggressiveness. An important advance lay in the discovery that fresh tissue slices can be incubated in a solution of BrdUrd, cut as frozen sections or processed to paraffin wax, then reacted with anti-BrdUrd antibody. To optimize the reaction, it is necessary to run the BrdUrd-binding reaction at a high oxygen tension, to 'drive' the metabolic uptake of this substrate. This can be done either by bubbling oxygen through the incubation medium or by increasing the pressure to at least 3 atmospheres. In addition,

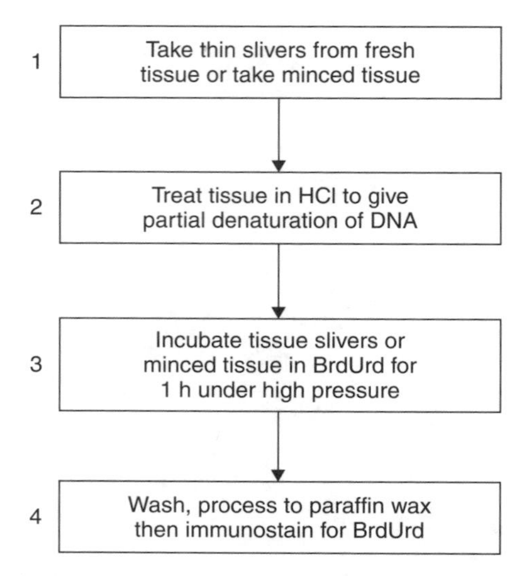

Figure 6.7 Scheme for the labelling of fresh tissue slice with BrdUrd prior to immunostaining

it is necessary to pre-treat the sections with hydrochloric acid to denature or unwind the DNA, to allow access of the antibody to the incorporated BrdUrd (Figure 6.7). It must be emphasized that this is not a true 'paraffin method' since it requires the availability of fresh tissue. Furthermore, diffusion of the reaction mixture into the tissue sample is often limited and a good reaction may be seen only at the periphery of the section (Figure 6.8).

Ki-67 antibody

In the early 1980s a novel antibody was described by a German group, having been raised against the L 428 Hodgkin's disease cell line (Gerdes *et al.*, 1984). It was found that Ki-67 reacted with cells known to be proliferating and could be applied to tissue sections (Figure 6.9), although a constraint was that paraffin wax-embedded material was not amenable to staining with the antibody. There have been many studies of wide-ranging tissues and these have generally shown a good correlation between Ki-67^+ cell numbers and tumour grade; furthermore, in some instances the Ki-67 score has been related closely to survival or prognosis. The Ki-67 antigen is expressed by cells in all phases of the cycle other than G_0 and early G_1 and is maximal in amount in G_2 and M phases. In general, there is a good correlation between the Ki-67 index and other measures of cell proliferation such as [^{3}H]thymidine and BrdUrd uptake, DNA flow cytometric analysis and interphase AgNOR scores. There have been two recent significant advances in relation to Ki-67; firstly, the antigen has been much better characterized than before and, secondly, an antibody with the same reactivity has been

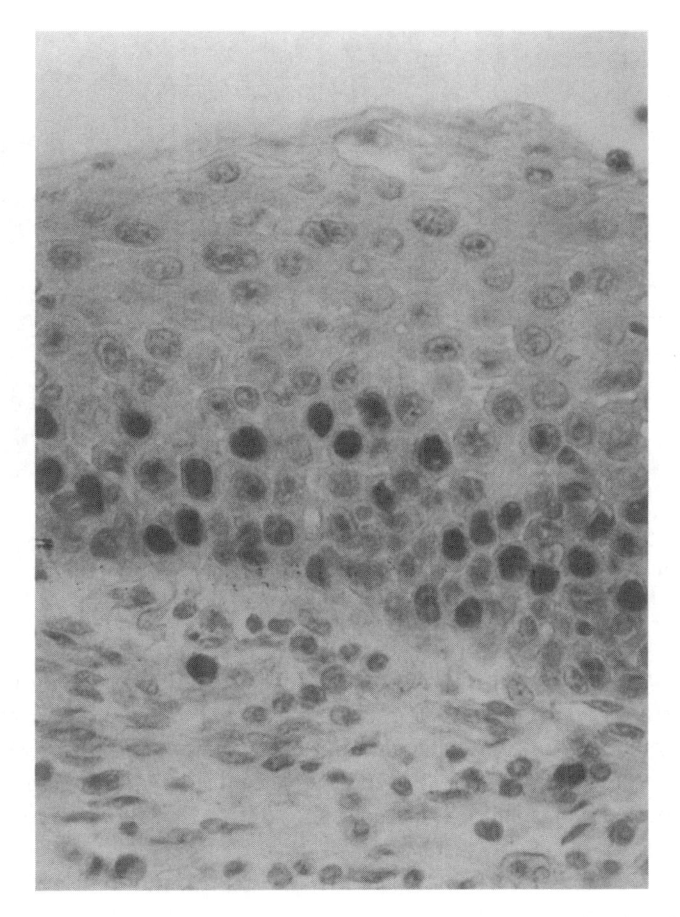

Figure 6.8 Tonsillar pharyngeal squamous epithelium which has been incubated with BrdUrd then immunostained with anti-BrdUrd antibody. As would be expected, most of the nuclei in the suprabasal layer are positively stained

produced which can be applied to paraffin wax-embedded tissue sections. These developments are described below. It has been shown that on western blotting of proliferating (but not resting) cells there are two very large polypeptides, of 345 and 395 kDa sizes, which react with Ki-67. When immunoscreening of human λT11 cDNA libraries was performed, 10 Ki-67^{+} clones were found, of which eight had identical DNA sequences with respect to the Ki-67 antigen. It was also observed that all of these Ki-67 clones had 65%–100% homology, which is remarkably high. When the Ki-67 cDNA was cloned into a bacterial expression vector, the antibody reacted with fusion proteins prepared only from bacteria where the insert was in the correct reading frame. Conversely, when it was present in the wrong reading frame, there was no binding. It can probably be safely assumed that the repetitious 62 bp unit encodes the Ki-67 epitope and, together with isolated cDNA clones, it has been used to characterize the Ki-67

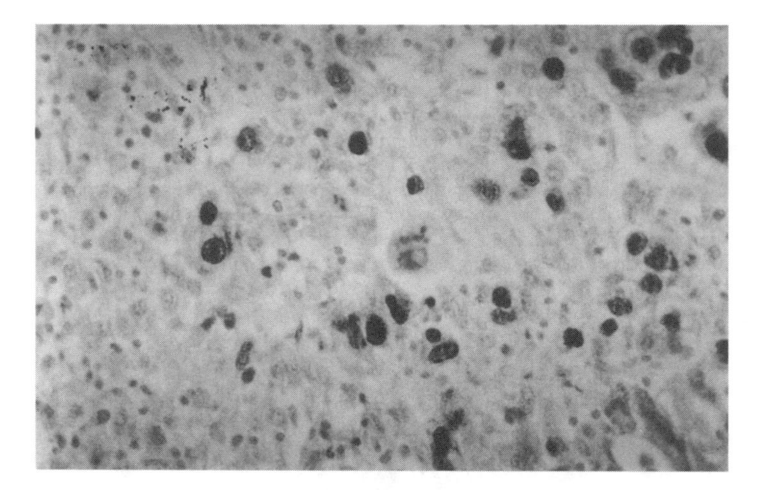

Figure 6.9 Frozen section of a high-grade non-Hodgkin's lymphoma stained with the monoclonal antibody Ki-67, with labelling of the nuclei of proliferating cells. The proliferating nuclei are stained intensely but the morphology is rather poor. Preparation courtesy of Mrs Jane Starczynski, Cellular Pathology, Birmingham Heartland, Hospital

gene. Both of these probes hybridize with mRNA (11.5 kbp or more) prepared from proliferating but not resting cells, and the polymerase chain reaction has subsequently been applied to the cloning and sequencing of a 6.3 kbp portion of the gene. Interestingly, this sequence appears to be unique and contains 16 repeats of the characteristic 62 bp sequence referred to above. The entire exon is of 6845 bp size.

The gene encoding Ki-67 appears to be located on the long arm of chromosome 10 (10q25), as has been shown by means of *in situ* hybridization using the 1095 bp sequence as a probe. Furthermore, the Ki-67 antigen has been localized at the ultrastructural level in the interphase of proliferating cells. The antigen is present in the outer parts of the nucleolus, especially in the granular component. When mitotic prophase commences, the antigen is seen in condensed chromatin and in metaphase on the chromatids. Later, it lies in the nucleoplasm. Thus, the localization differs from proliferating cell nuclear antigen and the major proteins associated with nucleolar organizer regions (see below), although numatrin (B23 protein) is also observed in the periribosomal zone.

Despite the acquisition of so much information regarding the Ki-67 antigen, the antibody generally available had suffered from a major drawback, namely the extreme lability of its antigen in the face of 'routine' processing to paraffin wax-embedded sections. However, there was a major advance in the production of a novel antibody, designated MIB1, which was made by the immunization of mice with the central repeat area of Ki-67 antigen, expressed in *E. coli*. Prior to immunostaining, exposure of the sections to microwave irradiation is necessary (in citrate buffer) but morphological preservation was much better than that

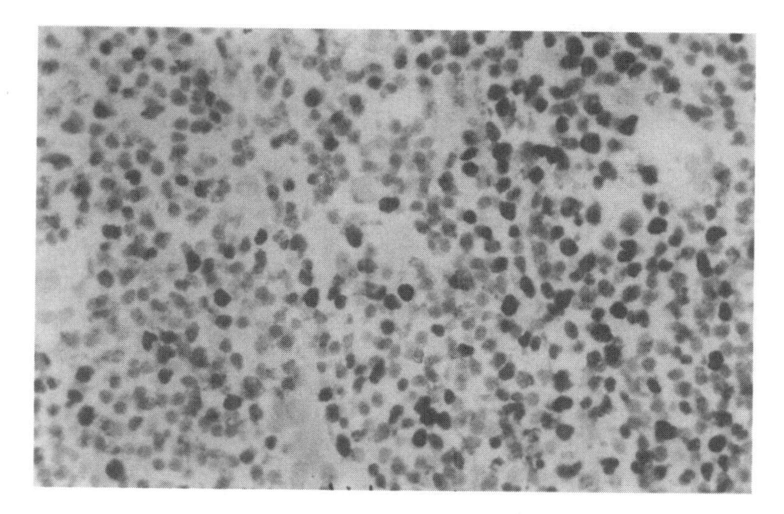

Figure 6.10 Paraffin-wax section of a high-grade breast carcinoma, showing labelling of most nuclei with the antibody Ki-67 after prior microwave irradiation of the section. Note that the morphology is very well preserved. Preparation courtesy of Mrs Jane Starczynski, Cellular Pathology, Birmingham Heartlands Hospital

in frozen sections and punctate areas of activity could be seen in proliferating nuclei. It is now also possible to obtain highly satisfactory staining with the original Ki-67 antibody in this way (Figure 6.10) and this is now the method of preference.

Recently, it has been shown that the Ki-67 protein binds to members of the heterochromatin protein 1 (HP1) family (Schölzen *et al.*, 2002). This has been demonstrated both *in vivo* and *in vitro*. The implication of this is that Ki-67 protein may be able to control higher-order chromatin structure.

Proliferating cell nuclear antibody

Proliferating cell nuclear antibody (PCNA) was first detected by means of human autoantibodies from patients with systemic lupus erythematosus. These express PCNA and the 'PCNA score' presumably overestimates the numbers of proliferating cells in tissue sections (Figure 6.11).

As a result of the above observations, it is apparent that the interpretation of PCNA scores must be approached with considerable caution, since there are so many important variables to be considered. These include the method and duration of fixation, the clone of antibody used, the half-life of the antigen and the effects of growth factors. It is perhaps not in the least surprising that there are so many conflicting results with regard to staining with anti-PCNA antibodies and other measures of cell proliferation and in the context of studies of different tissues from different groups of researchers.

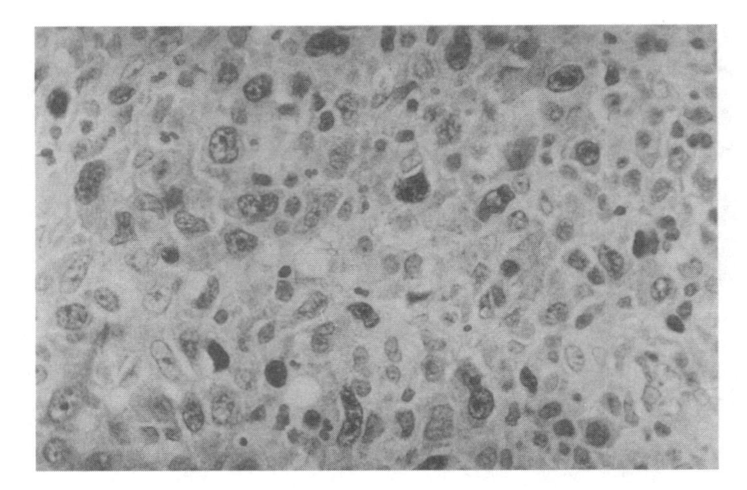

Figure 6.11 A section of high-grade carcinoma of the breast stained with the anti-PCNA antibody PC10. Most nuclei are reactive. Preparation courtesy of Mrs Jane Starczynski, Cellular Pathology, Birmingham Heartlands Hospital

Other Antibodies

Antibody to p105 antigen

Antibody to p105 in fact reacts with two proteins, with molecular weights of 105 and 41 kDa. The interpretation of this observation may be that there are monomeric and dimeric forms of the protein or that the smaller fragment is formed by partial proteolysis of the larger molecule. p105 is important in RNA synthesis and transport and in the regulation of the cell cycle. Teleologically, then, it is perhaps not surprising that it can be localized by means of antibodies were found to react with the nuclei of proliferating cells and we now have an increasing knowledge of the molecular significance of PCNA. The antigen is of 36 kDa size and is an acidic non-histone protein. Cells cannot divide in its absence and it is an auxiliary protein for DNA polymerase δ. There are two populations of PCNA during the S phase of the cell cycle. One of these is nucleoplasmic in distribution, is present in low levels in resting cells and is readily extracted by detergents and organic solvents; the other is present in replication sites and is detergent-resistant. The latter co-localizes in space and time with incorporated BrdUrd and its level is lowered by the application of anti-sense oligonucleotides. It is this latter type of the protein which shows a large *proportional* rise of level during the cell cycle, even though there is only a two- to three-fold increase in *total* PCNA. The seminal importance of PCNA in DNA synthesis is underlined by the high level of evolutionary structural conservation of the protein. Thus, rat and human PCNA differ by only four amino acids, and even *Drosophila* PCNA has about 70% homology with the human protein.

Whether proliferating or not, the gene encoding for PCNA is transcribed; however, its mRNA accumulates only in proliferating cells. Growth factors, especially platelet-derived growth factor (PDGF), are of great importance in this context, since they stabilize PCNA mRNA and thus encourage translation, as the mRNA is otherwise unstable. This enables rapid response to stimulation and seems to be the result of splicing of intron 4. Oncogenes may also be involved in the control of PCNA mRNA levels. It is therefore apparent that control of PCNA production is under both transcriptional and post-transcriptional control.

Although autoantibodies were first used to detect PCNA, immunization of mice with *E. coli* bearing genetically engineered PCNA has led to the production of monoclonal antibodies to the protein. These include PC10, 19A2 and 19F4, which recognize different epitopes and demonstrate PCNA in routinely processed paraffin wax-embedded tissues (Hall *et al.*, 1990). It is of interest that these have rather different staining patterns to those seen with human autoantibodies and different sensitivities to histological processing. Variables such as tissue block size and type and duration are of importance, and sections must be produced and handled with great care. Two further problems may arise; firstly, PCNA may be seen in large amounts in normal or non-neoplastic cells adjacent to malignant tumours. This is the result of the formation of growth factors in the latter which induce the formation of PCNA in the former. This phenomenon has been shown experimentally in nude mice, when human cancer cells were injected into their kidneys or livers; there was a rise in the numbers of $PCNA^+$ cells in the adjacent hepatic and renal tissues without any increase in the numbers of S phase (i.e. genuinely proliferating) cells. The same effect can be shown by giving parenteral growth factors to rats. Secondly, difficulties may arise as a result of the relatively long half-life (>20 h) of the protein. Thus, cells which have exited from the cell cycle may still show, by means of immuno-electron microscopy of the interchromatin granules of the nucleus, the areas where RNA synthesis occurs. In the early phases of the cell cycle, p105 cannot be detected but there is a great rise in concentration during G_2 and mitosis itself. There is a concurrent change in the distribution of the protein during the cell cycle; thus, in interphase and early prophase p105 is seen in the interchromatin areas of the nucleus, but as the cycle progresses it is also present in the cytoplasm, after the nuclear membrane lyses. By means of immuno-cytochemistry it can readily be shown that p105 is present in proliferating but not in non-cycling cells. Furthermore, there is much greater expression in malignant cells, especially in those of high-grade malignancy. The antibody to p105 which is available commercially can be applied satisfactorily to 'routine' paraffin wax-embedded sections.

Antibody to p125 antigen

A protein, p125, so designated because of its 125 kDa size, has been shown in the nuclear matrix by the application of a monoclonal antibody. Increased levels of p125 can be seen in mitotic cells, being maximal at metaphase and anaphase, and its formation can be induced in lymphocytes by means of phytohaemagglutinin. In neoplasms and normal tissues, $p125^+$ cell numbers correspond approximately to those in S phase, although it may be that some non-proliferating cells may also express this protein. The antibody to p125 has yet to be fully evaluated in histopathology but can be applied successfully to paraffin sections.

Antibody to p145 antigen

A monoclonal antibody has been produced which reacts with a nucleolar protein in malignant cells but not with the cells of normal tissues. The protein is of 145 kDa size and the antibody was raised by immunizing mice with nucleolar extracts from HeLa cells. It would appear that the antibody has only been applied to cell preparations and to frozen tissue sections with immunofluorescent labelling.

Antibodies to DNA polymerase α and δ

DNA polymerase δ, which is an enzyme with fluctuating levels through the cell cycle, has peaks in the G_2 and M phases and tends to be associated with DNA polymerse α in its occurrence. It has 3′ to 5′ exonuclease activity and might be expected to reflect cell proliferation. DNA polymerase α is observed to peak, together with mRNA levels, at the transition from G_0 to G_1, just before the peak of DNA synthesis. Monoclonal antibodies to both enzymes, which are applicable only to frozen sections, have been described, and when antibodies to DNA polymerase α are applied to tumours proliferating cells are demonstrated. With antibodies to DNA polymerase δ, the picture is more complex, with staining of the nucleus, which spreads to the entire cell on mitosis. Unfortunately, these antibodies have yet to be fully investigated.

Antibody C_5F_{10}

A murine monoclonal antibody, C_5F_{10}, has been described and was produced by immunizing mice against 'tumour polysaccharide substance (TPS)-1–28', extracted from a human pulmonary carcinoma. The antibody demonstrates proliferating cells in paraffin wax-embedded sections and can be shown at the ultrastructural level to bind to multiple nuclear sites and sometimes to lysosomes. (The latter finding may explain the recent observation that the antibody may react with non-cycling cells.) It has been shown that the antibody does not react at the same sites as anti-PCNA or anti-tubulin. This is of importance, since tubulin (and vimentin) are known to be affected in their distribution by neoplastic transformation; similarly, so is the aggregation of tubules into microtubules with the occurrence of mitosis.

Antibody to protein p40

Antibody has been raised against a protein, *M*r 40 000, designated p40, which is associated with nucleoli. It appears to be distinct from the major nucleolar organizer region-associated proteins and has been shown to be absent from normal human tissues but present in a range of malignant tissues, where it has a nucleolar distribution different from that of PCNA.

Antibody JC1

Recently a novel antibody, JC1, has been produced. It reacts with two antigenic components, of 123 and 212 kDa molecular weights, quite distinct from the proteins recognized by Ki-67. Furthermore, JC1 does not react with recombinant Ki-67 protein. The antibody is suitable for use with frozen sections only and has a similar but not identical staining pattern to that obtained with Ki-67.

Antibody KiS1

This novel antibody was raised against nuclear extracts from the promonocytic leukaemic cell line U937. It runs satisfactorily with paraffin sections and it has been related to prognosis in breast carcinoma. It reacts with a 160 kDa nuclear protein with a minor, 140 kDa element. Flow cytometric analysis shows that the KiS1 antigen increases in amount during the G_1 and S phases of the cell cycle, reaching levels four times greater than at G_2/M.

Antibody IND.64

This monoclonal antibody was produced by immunizing splenic cells from athymic nude mice grafted with a human lymphoblastic leukaemic cell line. The antibody reacts with proliferating cells in frozen sections and recognizes two antigenic components, with molecular weights of 345 and 395 kDa, assessed by immunoblotting. The antigens appear in late G_1 phase through to S,G_2 and M and is absent from G_0 and early G_1.

Figure 6.12 summarizes the phases of the cell cycle at which some of the above molecules are expressed; clearly, there is considerable ‘overlap’ in the timing of their appearances. This may, in part, account for some of the apparent anomalies observed when these molecules are demonstrated in tissue sections.

Antibodies to Non-cycling Cells

A prospect for the future lies in the possibility of generating antibodies to molecules characteristic of cells which are *not* proliferating. A likely candidate for such a marker is *statin*, a 57 kDa protein present in the nucleus. An antibody, S-30, has been raised against statin by immunizing mice with a detergent-resistant preparation from senescent, non-dividing human fibroblasts

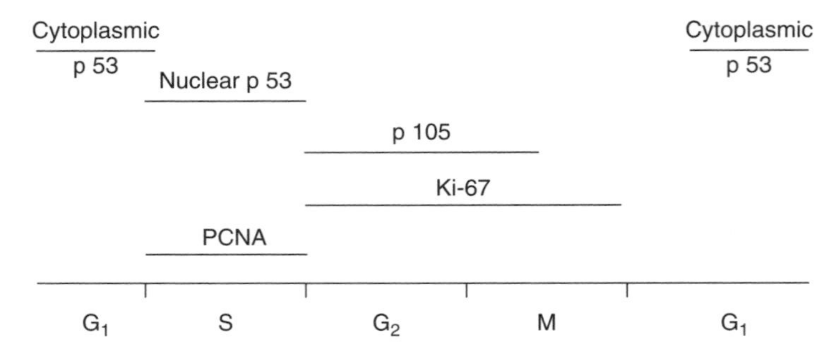

Figure 6.12 Schematic diagram of the cell cycle with *peak level* expression of various cycle-related antigens

and this has localized statin to the nuclear envelope. The antibody demonstrates non-proliferating cells in frozen sections but the picture is complicated by the discovery of two different forms of the molecule. One is detergent-soluble and is seen in both dividing and resting cells; the other is detergent-insoluble and is observed only in non-dividing cells. The latter form of the antigen is lost progressively when cells are stimulated by serum to pass from G_0 to S phase of the cycle. It seems likely that other antibodies of this sort will become available and enable extended studies of cell populations in malignant and other tissues. An example antigen is a 30 kDa protein called *prohibitin*, which is present in non-cycling cells. Furthermore, screening of clones may reveal antibodies which are found empirically to bind to resting cells although their target antigen is not characterized. Such an antibody is BU31, which may possibly be directed against lamin and can be shown by means of scanning confocal microscopy to bind in the region of the nuclear membrane. It can be applied with success only to frozen sections, where it binds to resting cells (Figures 6.13 and 6.14).

In situ Hybridization for Histone mRNA

Although not an immunocytochemical method as such, this novel approach has recently been used to demonstrate proliferating cells in paraffin sections (Figure 6.15). Digoxigenin-labelled oligonucleotide probes have been used to demonstrate the histone mRNAs which are up-regulated in the S phase of the cell cycle. It has been shown that the method gives a significantly higher labelling index in high-grade than low-grade lymphomas, although the scores were lower than those obtained with Ki-67, probably reflecting the fact that these mRNAs are present for a shorter part of the cycle than the Ki-67 epitope.

Correlation Between Immunological Markers of Cell Proliferation and Clinical Status

It is emphatically not the remit of a text such as this to describe the seemingly endless studies where immunohistochemistry for proliferation markers has been

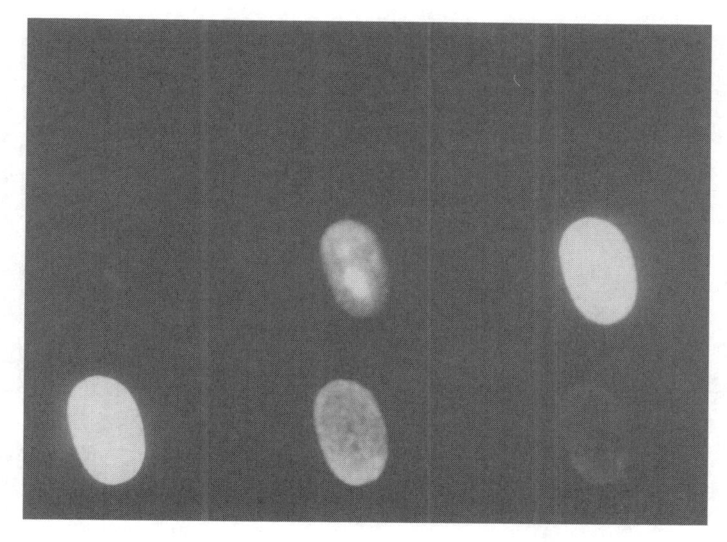

Figure 6.13 A confocal scanning laser microscope preparation of a resting fibroblast, with six optical 'slices' showing the nuclear membrane distribution of labelling with the antibody BU31. Photograph courtesy of Drs R. Dover and C.P. Gillmore, Histopathology Unit, ICRF, Lincoln's Inn Fields, London

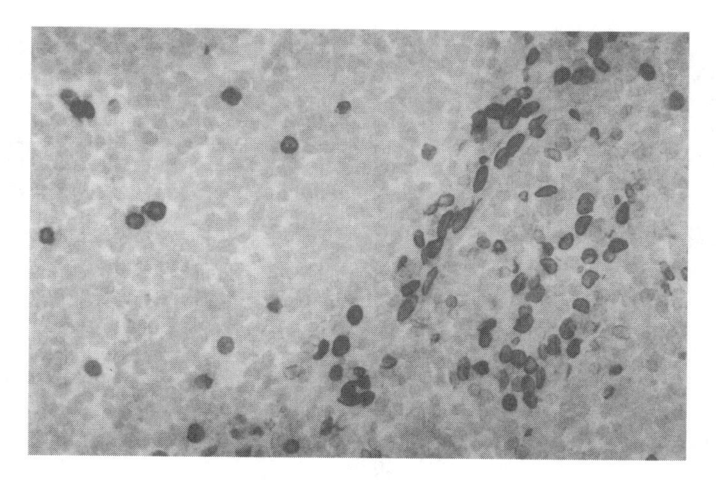

Figure 6.14 A photomicrograph of a tonsillar lymphoid follicle, reacted with the antibody BU31; the positively stained cells are largely external to the follicle centre and correspond to cells which are known to be non-proliferating. Preparation courtesy of Mr Jane Starczynski, Cellular Pathology, Birmingham Heartland, Hospital

applied to clinical or histopathological diagnostic problems. For example, the reader should be able to appreciate the magnitude of the available data by referring to the information in Table 6.3, which gives as an example some of the conclusions reached in relation to numerous studies of tumour labelling by Ki-67 (Brown and Gatter, 1990). A list of similar length could now be

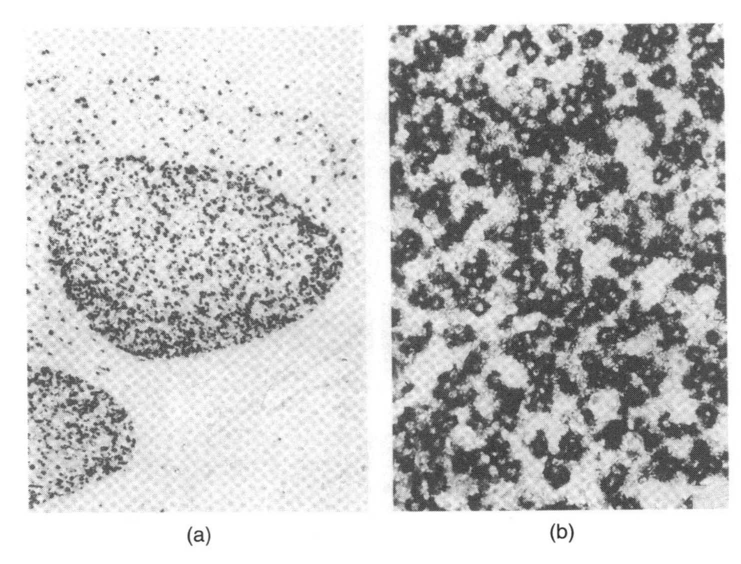

Figure 6.15 *In situ* hybridization used to demonstrate histones in (a) a 'reactive' lymphoid follicle, where the majority of basal cells are positive, and (b) a high-grade non-Hodgkin's lymphoma, with positive labelling of most cells. Preparations courtesy of Dr P.S. Colloby, formerly of Pathology Department, University of Leicester

Table 6.3 A few examples of the clinical application of Ki-67 in the assessment of malignant tumours

Tissue studied	Results and comments
Lymphoid	In general, high Ki-67 scores correlate with poor prognosis and high-grade malignancy in lymphomas. Good correlation with AgNOR scores
Breast	Ki-67 scores related well to recurrence after mastectomy, histological grade and mitotic activity. Most studies show no correlation with tumour size, lymph node involvement or oestrogen receptor status
CNS tumours	Histological grade in general reflected by Ki-67 counts although the latter did not always relate to survival data
Lung tumours	Ki-67 scores highest in oat cell carcinomas; lowest in carcinoids
Colorectal carcinomas	In general, poor correlation with histological type and other prognostic indices, including stage. High Ki-67 score in recurrent tumours
Hepatocellular carcinoma	Ki-67 score relates well with histological grade
Prostatic carcinoma	Ki-67 score related well to histological grade and clinical stage. With hormone therapy, the score decreases dramatically
Malignant melanoma	Tumour thickness correlates well with Ki-67 count
Uterine cervical carcinoma	No correlation between Ki-67 score and histological types and changes
Connective tissue tumours	Ki-67 index correlates well with histological grade and mitotic counts, as well as survival

given for studies with PCNA. Such investigations have sought to answer one or more of three main questions: (i) does a particular marker relate to grade or proliferative status of a specimen? (ii) does staining for the marker relate to survival/prognosis in neoplastic disease? (iii) how can the data obtained from the application of one marker be related to those derived from the use of others? Certain generalizations can be made. It would be naive to expect the measurement of any one aspect of tumour extension to give an overall assessment of survival; indeed, cell proliferation is only one of many aspects of the malignant phenotype. Thus, enumerating the proliferating compartment of a particular cancer will often correlate well with the tumour grade but may or may not give an indication of patient prognosis. Other factors, such as cell motility, immune escape and secretion of proteolytic enzymes all have roles to play in the overall process of malignancy. Another limitation of the techniques generally used for the assessment of cell proliferation is that they give us a measure of proliferation *state*, not *rate*. The latter could potentially tell us much more about a specimen, since it would indicate the flux of cells entering and leaving what is usually a relatively static dividing pool. Furthermore, we do not yet have the ability either to detect or measure the numbers of *clonogenic cells* in histological material. These cells, as their name implies, are those that give rise to a proliferating clone, and an ability to demonstrate and quantify them in histological material would be highly desirable.

The potential difficulties arising from the phenomenon of tumour cell heterogeneity have been outlined above and may, again, explain the lack of consistency of data from one study to another or between different specimens of the same tumour type. There may also be substantial variation in cell proliferation from areas which are, for example, at the growing edge of a neoplasm in relation to the deeper, relatively hypoxic areas. Tissue sampling, then, becomes a further consideration. The question of the relationship between data obtained from the use of different (immunocytochemical and non-immunocytochemical) techniques is complex. Certain 'markers', such as AgNOR scores (see Chapter 7) and Ki-67 counts generally correlate well, although conflicting results have recently been recorded between AgNOR numbers and PCNA counts. Likewise, there are papers describing high agreement between, say, S phase cells as assessed by DNA flow cytometry (Chapter 5) and AgNORs or Ki-67 labelling. When these data are further related to prognosis and survival, the picture becomes even more confused, doubtless for some or all of the reasons cited above. Nonetheless, the reader should not be discouraged from further studies of cell proliferation in tissues, since there is still much to be established and learned.

6.6 References

Cahill DP, Lengaeur C and Yu J *et al.* (1998) Mutations of mitotic checkpoint genes in human cancers. *Nature* **392**: 300–3.

Chellappan SP, Giordano A and Fisher PB (1998) Role of cyclin dependent kinases and their inhibitors in cellular differentiation and development, in: Vogt PK, SI Reed eds. *Cyclin Dependent Kinase (CDK) Inhibitors* (eds Vogt PK and Reed SI), pp. 57–103. Springer-Verlag, Berlin.

Dannenberg JH, van Rossum A, Schuiff L and te Riele H (2000) Ablation of the retinoblastoma gene family deregulates G1 control causing immortalization and increased cell turnover under growth-restricting conditions. *Genes Dev.* **14**: 3051–64.

Easton J, Wei T, Lahti JM and Kidd VJ (1998) Disruption of the cyclin D/cyclin-dependent kinase/INK4/retinoblastoma protein regulatory pathway in human neuroblastoma. *Cancer Res.* **58**: 2624–32.

Ekholm SV and Reed SI (2000) Regulation of G1 cyclin-dependent kinases in the mammalian cell cycle. *Curr. Opin. Cell Biol.* **12**: 676–84.

El-Deiry WS, Harper JW, O'Connor PM *et al.* (1994) WAF1/CIP1 is induced in p53-mediated G1 arrest and apoptosis. *Cancer Res.* **54**: 1169–74.

Evans T, Rosenthal ET, Youngblom J *et al.* (1983) Cyclin: a protein specified by maternal mRNA in sea urchin eggs is destroyed at each cleavage division. *Cell* **33**: 389–96.

Ezhevsky SA, Ho A, Becker-Hapak M, Davies PK and Dowdy SF (2001) Differential regulation of retinoblastoma tumor suppressor protein by G1 cyclin-dependent kinase complexes *in vivo*. *Mol. Cell. Biol.* **21**: 4773–84.

Fisher RP and Morgan DO (1994) A novel cyclin associates with MO15/CDK7 to form the CDK activating kinase. *Cell* **78**: 13–24.

Gatter KC, Brown NG, Trowbridge IS, Woolston RE and Mason DY (1983) Transferrin receptors in human tissues: their distribution and possible clinical relevance. *J. Clin. Pathol.* **36**: 539–545.

Gerdes J, Dallenbach F, Lennert K, Lemke H and Stein H (1984) Growth fractions in malignant non-Hodgkin's lymphomas (NHL) as determined *in situ* with the monoclonal antibody Ki67. *Hematol. Oncol.* **2**: 365–371.

Gratzner HG (1982) Monoclonal antibody to 5-bromo and 5-iodo-deoxyuridine: A new reagent for detection of DNA replication. *Science* **218**: 474–475.

Hall PA, Levison DA, Woods Al *et al.* (1990) Proliferating cell nuclear antigen (PCNA) immunolocalisation in paraffin sections: an index of cell proliferation with evidence of de-regulated expression in some neoplasms. *J. Pathol.* **162**: 285–294.

Harbour JW and Dean DC (2000) The pRb/E2F pathway: expanding roles and emerging paradigms. *Genes Dev.* **14**: 2393–409.

Hengst L and Reed SI (1996) Translational control of p27KIP1 accumulation during the cell cycle. *Science* **271**: 1861–4.

Jackman M, Firth M and Pines J (1995) Human cyclins B1 and B2 are localized to strikingly different structures: B1 to microtubules, B2 primarily to the Golgi apparatus. *Embo. J.* **14**: 1646–54.

King RW, Deshais RJ, Peters JM *et al.* (1996) How proteolysis drives the cell cycle. *Science* **274**: 1652–9.

Lees EM and Harlow E (1993) Sequences within the conserved cyclin box of human cyclin A are sufficient for binding to and inactivation of CDC2 kinase. *Mol. Cell. Biol.* **13**: 1194–1201.

Lohka MJ, Hayes MK and Maller LJ (1988) Purification of maturation promoting factor, an intracellular regulator of early mitotic events. *Proc. Natl. Acad. Sci. USA* **85**: 3009–13.

Luca FC and Ruderman JV (1989) Control of programmed cyclin destruction in a cell-free system. *J. Cell Biol.* **109**: 1895–1909.

Malumbres M and Barbacid M (2001) To cycle or not to cycle: a critical decision in cancer. *Nature Rev. Cancer* **1**: 222–31.

McIntosh JR and Koonce MP (1989) Mitosis. *Science* **246**: 622–8.

Morgan DO (1995) Principles of CDK regulation. *Nature* **374**: 131–4.

Morgan DO (1997) Cyclin-dependent kinases: engines, clocks and microprocessors. *Ann. Rev. Cell. Dev. Biol.* **13**: 261–91.

Murray A (1995) Cyclin ubiquitination: the destructive end of mitosis. *Cell* **81**: 149–52.
Pardee AB (1974) A restriction point for the control of normal animal cell replication. *Proc. Natl. Acad. Sci. USA* **71**: 1286–90.
Pardee AB (1989) G1 events and regulation of cell proliferation. *Science* **246**: 603–8.
Pomerantz J, Schreiber-Agus N, Liegeois J *et al.* (1998) The INK4A tumour suppressor gene product, $p19^{ARF}$, interacts with MDM2 and neutralizes MDM2's inhibition of p53. *Cell* **92**: 713–23.
Sage J, Mulligan GJ, Attardi LD *et al.* (2000). Targeted disruption of the three Rb-related genes leads to loss of G1 control and immortalization. *Genes Dev.* **14**: 3037–50.
Scherr CJ (1995) D-type cyclins. *Trends Biochem. Sci.* **20**: 187–90.
Schölzen T, EndL E, Wohlenberg C *et al.* (2000) The Ki67 protein interact in the member of the heterochromatin protein 1 (H/1) family – a potential role in the regulation of higher-order chromatin structure. *J. Pathol.* **196**: 135–444.
Sicinski P, Danaher JL, Parker SB *et al.* (1995) Cyclin D1 provides a link between development and oncogenesis in the retina and breast. *Cell* **82**: 621–30.
Stillman B (1996) Cell cycle control of DNA replication. *Science* **274**: 1659–64.
Weinberg RA (1995) The retinoblastoma protein and cell cycle control. *Cell* **81**: 323–30.
Weinberg RA (1996) The molecular basis of carcinogenesis: understanding the cell cycle clock. *Cytokines Mol. Ther.* **2**: 105–10.
Wolfel T, Hauer M, Schneider J *et al.* (1995) A $p16^{INK4A}$-insensitive CDK4 mutant targeted by cytolytic T-lymphocytes in a human melanoma. *Science* **269**: 1281–4.

6.7 Further reading

Brown DC and Gatter KC (1990) Monoclonal antibody Ki-67: its use in histopathology. *Histopathology* **17**: 489–503.
Brown DC and Gather KC (2002) Ki67 protein: the immaculate deception? *Histopathology* **40**: 2–11.
Ciba Foundation Symposium 170 (1992) *Regulation of the Eukaryotic Cell Cycle*. Wiley, Chichester.
Crocker J (1989) Review: proliferation indices in malignant lymphomas. *Clin. Exp. Immunol.* **77**: 299–308.
Crocker J (1993) *Cell Proliferation in Lymphomas*. Blackwell Scientific Publications, Oxford.
Hall PA, Levison DA and Wright NA (1992) *Assessment of Cell Proliferation in Clinical Practice*. Springer-Verlag, Berlin.
Young S (1992) Dangerous dance of the dividing cell. *New Scientist* **1824**: 23–7.

7

Interphase Nucleolar Organiser Regions in Tumour Pathology

Massimo Derenzini, Davide Treré, Marie-Françoise O'Donohue and **Dominique Ploton**

7.1 Introduction

The first work on interphase AgNORs in neoplastic pathology was carried out in 1986 by Ploton and co-workers (Ploton *et al.*, 1986). The authors showed that, in a small series of prostatic tumours, the malignant cells were characterised by greater numbers of interphase AgNORs than the benign tumours. That study and especially the one that followed it, carried out on a larger series of melanotic lesions (Crocker and Skilbeck, 1987) generated a great deal of enthusiasm among pathologists and a very positive editorial published in *The Lancet* (Anon., 1987). A series of investigations was then undertaken in order to demonstrate the possibility of distinguishing cancer from non-cancer cells on the basis of a different quantity of interphase AgNORs per cell nucleus. The subsequent observation of a close association between interphase AgNOR quantity and cell proliferation indices caused the interphase AgNOR parameter also to be applied to tumour prognosis. Up to now, more than a thousand papers have been published on the use of the interphase AgNOR measurement in tumour pathology. All of the common tumours have been widely and repeatedly investigated in order to evaluate the diagnostic and prognostic impact of interphase AgNOR assessment. Unfortunately, the results obtained appear to be quite conflicting, giving rise to two opposite parties: a smaller one in favour of and a larger one contrary to the value of interphase AgNORs. Consequently, the utilisation of interphase AgNORs in routine tumour pathology is now a rarity, as may have been expected according to the first data.

Molecular Biology in Cellular Pathology. Edited by John Crocker and Paul G. Murray
 ISBN: 0-470-84475-2

In our opinion, this scepticism results from inaccurate knowledge of the biological meaning of AgNORs. Indeed, many pathologists used the method to obtain information that it was not able to supply. Moreover, widespread methodological inaccuracy for the visualisation and quantification of interphase AgNORs is acknowledged.

Therefore, before discussing the applications of interphase AgNOR measurement in tumour pathology, the nature and function of interphase AgNORs will be outlined.

7.2 The AgNORs

In cell biology the abbreviation 'AgNOR' represents Nucleolar Organiser Regions (NORs) stained with silver (Ag). The NORs are those chromosomal regions around which nucleoli reorganise during telophase. These areas correspond to the secondary constrictions of metaphase chromosomes and, in man, are located in the acrocentric chromosomes (chromosomes 13, 14, 15, 21 and 22). The NORs contain the ribosomal genes. During interphase, the NORs are located in the nucleolus, where they constitute its functional heart for ribosome biogenesis.

Nucleolar Structure and Function

Each human cell contains at least one nucleolus. The nucleolus is the most prominent domain within the nucleus. Classical electron microscopic techniques allow the recognition of five main components, each characterised by different electron density, texture and shape (Figure 7.1). These are the fibrillar centres, dense fibrillar component, granular component, interstices and chromatin (peri- and intra-nucleolar) (Devenzini *et al.*, 1990). Fibrillar centres (FCs) are roundish,

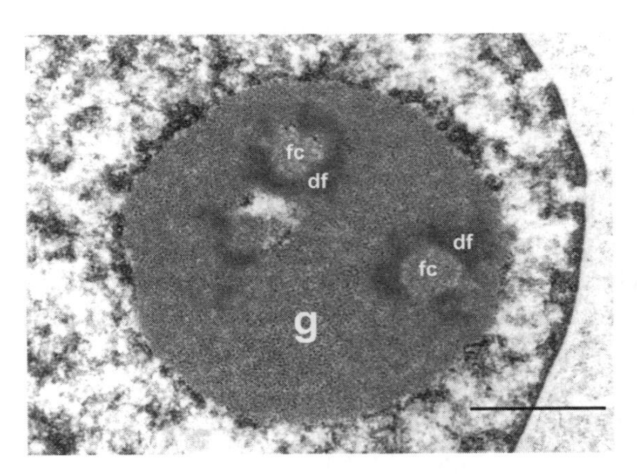

Figure 7.1 Electron microscopy of a cell nucleus. Two fibrillar centres (fc) surrounded by the dense fibrillar component (df) are visible; the granular component is indicated by ***g***. Bar, 0.3 μm

lightly electron-opaque structures 0.3–2 μm in diameter. Their numbers and sizes depend on the cell type but it is known that they are much more numerous and smaller in proliferating than in resting cells. The dense fibrillar component (DFC) is organised into fibrils, about 5 nm thick and mainly closely located round the FCs. The granular component (GC) is made up of granules 15–20 nm in diameter, surrounding the FCs and DFC. Interstices are small areas of different sizes, their density being similar to that of the nucleoplasm to which they are frequently connected. They always contain clumps of condensed chromatin (intra-nucleolar chromatin) which are linked to peri-nucleolar chromatin.

One of the main functions of the nucleolus, namely ribosomal biogenesis, is responsible for 60% of cellular transcription. The main steps in this complex process are: ribosomal DNA (rDNA) transcription; processing of ribosomal RNA (rRNA); association of the latter with numerous proteins to constitute ribosomal sub-units; and, finally, their assembly as mature cytoplasmic ribosomes. Although they are highly amplified, rRNA genes represent only 1%–2% of total nucleolar DNA. In HeLa cells, approximately 540 rRNA genes are present in each nucleus. However, only a maximum of 150 of these genes are simultaneously active in each cell, where they are localised within one to several nucleoli.

Very recently, the first proteomic analysis of the human nucleolus identified 271 proteins, among which over 30% are uncharacterised, 22% are nucleotide and nucleic-acid binding proteins, 14% are ribosomal proteins and 11% are RNA modifying enzymes (Andersen *et al.*, 2002). The main proteins engaged in the rRNA transcription and processing are: RNA polymerase I (RPI), Upstream Binding Factor (UBF), DNA topoisomerase I, nucleolin or C23 protein, fibrillarin, and numatrin or B23 protein.

- *RPI* is a heteromeric enzyme with a relative molecular mass, M_r, of 500–600 kDa, which is made up of 12 sub-units with 3 associated factors. The two largest sub-units have a M_r of 190–127 kDa. It is known that approximately 15 000 RPI molecules are simultaneously engaged (i.e. are active in rRNA synthesis) within one cell. It takes less than 5 min for each RPI molecule to transcribe one rRNA gene. RPI is mainly localised within the fibrillar centres. The largest RPI sub-unit is one of the AgNOR proteins.

- *UBF* is a very abundant (50 000 copies per cell) transacting factor, with a M_r of 97–94 kDa, which contains numerous High Mobility Group (HMG) boxes. It has the ability to bind to the rDNA promoter (not exclusively) and to modify rRNA genes in order to recruit other activating factors. UBF is localised within both fibrillar centres and the dense fibrillar component, in contact with the latter. UBF is an AgNOR protein. Interestingly, the protein of the retinoblastoma gene (pRb) can bind to UBF in an inactive complex. In proliferating cells, UBF is released by the phosphorylation of pRb and can activate rRNA gene transcription.

- *DNA topoisomerase I* is a 91 kDa protein which makes transient single-strand breaks within rRNA genes during their transcription by RPI. It has been immunolocalised within the nucleolar components in which rRNA synthesis takes place, i.e. fibrillar centres and the dense fibrillar component, in contact with the latter. Interestingly, it is the main target for the anticancer drug Camptothecin derivatives which are increasingly used in treatment.
- *Nucleolin* is a 100 kDa protein and is a major AgNOR protein (Figure 7.2). Nucleolin is the most abundant nucleolar phosphorylated protein within proliferating cells. Its N-terminal domain is highly phosphorylated. It contains several highly acidic regions which bind and displace H1 histone (inducing chromatin decondensation) and is responsible for the silver-stainability of nucleolin. Nucleolin has four RNA-binding domains in its central part, thus suggesting binding to rRNAs. The C-terminal domain contains numerous glycine- and arginine-rich segments and can interact with RNA. Nucleolin is absent from fibrillar centres but is present within both the dense fibrillar and granular components.
- *Fibrillarin* is a 36 kDa phosphoprotein which is active in pre-rRNA processing and methylation. Its N-terminal part comprises a glycine- and arginine-rich domain. Its central region is similar to an RNA-binding domain. Fibrillarin is associated with U3, U8 and U13 small RNA which are involved in pre-rRNA processing. Fibrillarin is specifically localised in the dense fibrillar component.
- *Numatrin* is a 37–39 kDa protein which has ribonuclease activity and binds to rRNA molecules during the late stage of their processing. It can also shuttle between nucleus and cytoplasm and acts as a molecular chaperone by transporting ribosomal and non-ribosomal proteins from cytoplasm to nucleolus. In the nucleolus, numatrin is localised in both the dense fibrillar and granular components; it is also a major AgNOR protein (Figure 7.2).

In order precisely to identify both molecules and functions relative to the nucleolar components, many studies using cytochemical, immunocytochemical, *in situ* hybridisation and high resolution autoradiography have been performed. Their main conclusions were as follows.

1. FCs contain DNA in an extended configuration. Part of this DNA is composed of rDNA. rRNA has seldom been reported to be located in FCs. All the proteins of the transcriptional machinery were also found within FCs.
2. The DFC contains rDNA in scattered foci which are in close contact with the FCs. They also contain rRNA, UBF, fibrillarin, nucleolin and nucleophosmin.
3. GCs contain rRNA and proteins such as nucleolin and nucleophosmin.

A general consensus has been achieved regarding the functions of the nucleolar components. The peripheral portion of the FCs is considered to represent the site of rRNA transcription. Recently synthesised rRNA molecules accumulate

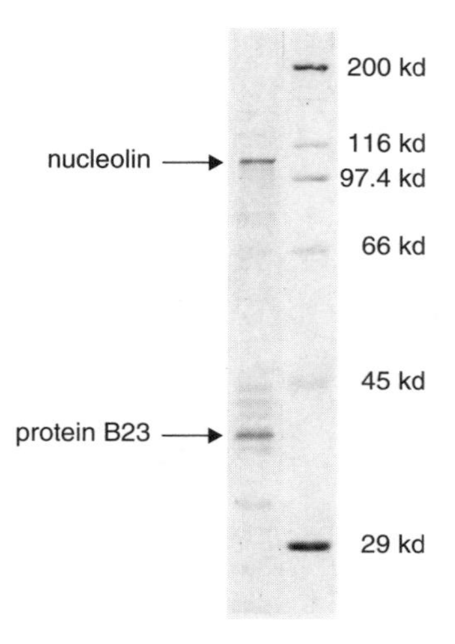

Figure 7.2 Silver-staining of nucleolar proteins extracted from a human cancer cell line, separated by SDS-polyacrylamide gel electrophoresis and transferred on nitrocellulose membranes. Two bands are clearly visible, with molecular weights of 105 and 39 kDa, corresponding to nucleolin and numatrin, respectively

in the DFC, where their processing initiates. Further rRNA processing occurs in the GC.

Structure and Function of Interphase AgNORs

In the 1980s a series of ultrastructural cytochemical studies demonstrated that, during interphase, the NORs are localised in the FCs and in the closely associated DFC. These components represent, therefore, the interphase counterpart of metaphase NORs. As a consequence of the presence of the AgNOR proteins in FCs and closely associated DFC, these nucleolar sub-structures are deeply stained with silver (Figure 7.3a) and can, therefore, be clearly visualised also by light microscopy, where they appear as well-defined black dots (Figure 7.3b). Confocal microscopy and three-dimensional reconstruction indicate that these dots give rise to several necklaces within the nucleolus (Ploton *et al.*, 1994). The silver-stained dots are defined as 'interphase AgNORs' in order to distinguish them from 'metaphase AgNORs'. Curiously, in their first applications in tumour pathology the interphase AgNORs were more simply indicated as AgNORs and, from then on, in pathology the term 'AgNOR' indicates interphase AgNORs.

As already mentioned, all the components necessary for rRNA transcription are located in the AgNORs and, indeed, it is within these structures that rRNA is synthesised. The AgNOR number is related to the level of ribosomal biogenesis. It ranges from one or two in resting lymphocytes (characterised by a very low rRNA transcriptional activity), to many dozens in proliferating neoplastic cells (characterised by a very high rRNA transcriptional activity).

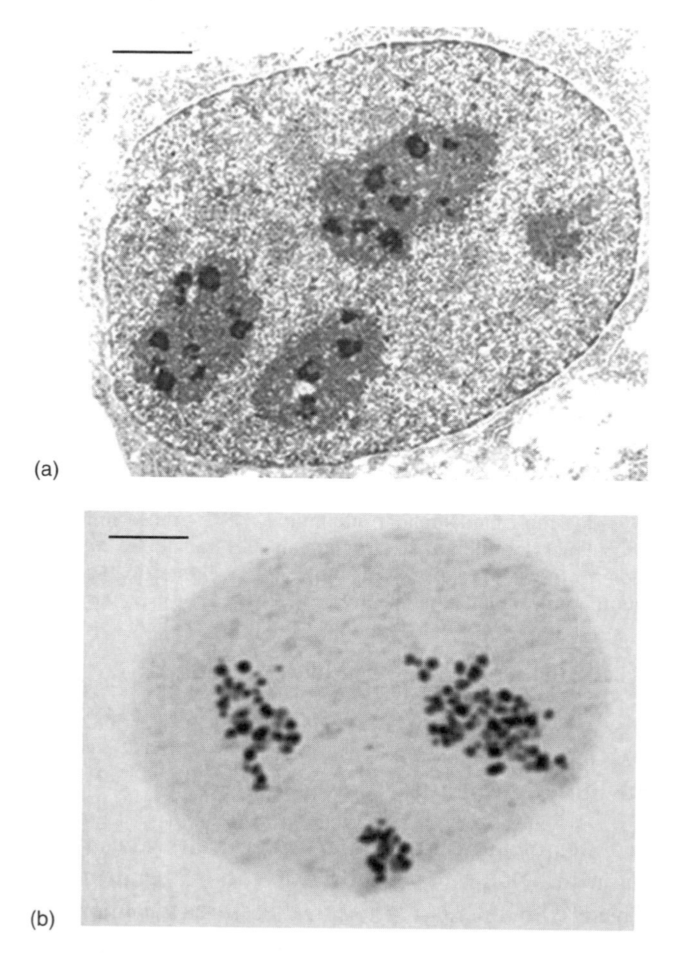

Figure 7.3 Silver-stained TG nuclei, visualised at electron (a) and light (b) microscopy. (a) In the nucleoli the silver deposits are localised in the fibrillar components. (b) These structures appear as darkly stained dots clustered in the nucleoli. Bar, 3 μm

The AgNOR distribution can, therefore, be considered to represent the morphological expression of the level of ribosomal biogenesis in the cell.

7.3 NOR silver-staining

NOR silver-staining was originally used to stain metaphase NORs on cytogenetic preparations and then applied by Ploton *et al.* (1986) in a simplified form for the visualisation of interphase NORs at light microscopy. Because of its simplicity, reliability and specificity, the silver-staining method introduced by Ploton *et al.* (1986) has become the most frequently used technique for AgNOR visualisation in routinely processed cyto-histological samples.

NOR silver-stainability is greatly influenced by several factors, including the fixatives used for tissue processing, as well as the temperature and duration of the silver-staining reaction. Concerning the fixatives, it is well known that alcohol-based fixatives give more intense and specific AgNOR protein visualisation than do formalin-containing fixatives. Moreover, in formalin-fixed specimens the intensity of the staining reaction is also greatly affected by the time of fixation and, consequently, by the penetration of the fixative within the tissue. The poor quality of silver-staining obtained in formalin-fixed histological samples results from the 'masking' effect of the fixative on cellular proteins. In order to eliminate this effect and improve NOR staining quality, Öfner *et al.* (1994) have introduced a 'retrieval' method, corresponding to the 'antigen retrieval' applied by pathologists to enhance immunohistochemical staining on routine sections. This technique involves the exposure of the sections in citrate buffer solution (10 mM, pH 6.0) at high temperatures (120°C) for 20 min Section heating can be obtained alternatively by wet autoclaving, pressure cooking or microwaving. Exposure of routine sections to high temperatures in appropriate aqueous salt solutions dramatically improves NOR silver-stainability; thus, its use in obtaining reproducible AgNOR visualisation in routinely processed histological samples is recommended.

In addition to the fixative used, the temperature and the time taken for the staining reaction also greatly affect NOR silver-stainability. The two variables are inversely related to one other: the higher the temperature, the shorter the time required for selective NOR visualisation. When the staining reaction is prolonged beyond the optimal time for selective NOR visualisation, other nucleolar components are progressively stained until single NORs are no longer detectable and the whole nucleolus appears homogeneously stained by silver.

Over recent years, NOR silver-staining has been applied to tissue samples processed by different fixatives and exposed to various temperatures (from a rather vague 'room temperature' to 50°C) and over different time spans (from 10 to 60 min). Consequently, the term 'AgNOR' has been indiscriminately attributed to either a single silver-stained NOR, or to a cluster of NORs that appear as a discrete silver-stained dot or to the whole nucleolus which has been over-stained with silver. This situation has caused wide disagreement between single studies on the same tumour types and has produced widespread scepticism regarding the effectiveness of the AgNOR method in tumour diagnosis and prognosis. In order to achieve a definitive standardisation of NOR silver-staining, in 1993 the International Committee on AgNOR Quantitation was established and, two years later, the Guidelines for AgNOR Quantification were defined. The protocol proposed by the Committee includes staining of the sections in the dark, at a constant temperature of 37°C, by using pre-warmed solutions and glassware; staining times vary depending on different preparations. The details of these guidelines are given in Table 7.1. It should also be mentioned that, in slides mounted on synthetic mountant media, the intensity

Table 7.1 The silver-staining protocol as proposed by the International Committee for AgNOR Quantitation

1. Cytological samples or chromosomes smears

- Smear the cells and air dry. Fix with Merckofix or 95% ethanol. Post-fix for 30 min in Carnoy's solution (absolute ethanol: glacial acetic acid 3 : 1 vol/vol) and hydrate through graded alcohols to ultrapure water
- Prepare a solution consisting of a 0.66% gelatin solution dissolved in ultrapure water, to which formic acid is added to make a final 0.33% solution. Pre-warm to 37°C the solution and the glassware where the staining reaction will be performed
- Dissolve silver-nitrate in the gelatin–formic acid solution to make a final 33.33% solution, and immediately immerse the slides in the solution obtained. Stain in the dark at a constant temperature of 37°C for 12 min
- Pour off the solution and wash the slides in several baths of ultrapure water
- Dehydrate and mount routinely (avoid synthetic slide-mounting media)

2. Frozen sections

- After drying, fix the sections in 95% ethanol or methacarn (methanol:chloroform:glacial acetic acid 6 : 3 : 1 vol/vol/vol), after which the samples can be stained with silver or stored at 20°C
- Before silver-staining, post-fix for 30 min in Carnoy's solution (absolute ethanol: glacial acetic acid 3 : 1 vol/vol) and hydrate through graded alcohols to ultrapure water
- Prepare a solution consisting of a 0.66% gelatin solution dissolved in ultrapure water, to which formic acid is then added to make a final 0.33% solution. Pre-warm to 37°C the solution and the glassware where the staining reaction will be performed
- Dissolve silver-nitrate in the gelatin–formic acid solution to make a final 33.33% solution, and immerse the slides immediately in the solution obtained. Stain in the dark at a constant temperature of 37°C for 14 min
- Pour off the solution and wash the slides in several baths of ultrapure water
- Dehydrate and mount routinely (avoid synthetic slide-mounting media)

3. Histological samples fixed in 95% ethanol or in other alcohol-based fixatives, and routinely paraffin-embedded

- Dewax the sections in xylene
- Post-fix for 30 min in Carnoy's solution (absolute ethanol:glacial acetic acid 3 : 1 vol/vol) and hydrate through graded alcohols to ultrapure water
- Prepare a solution consisting of a 0.66% gelatin solution dissolved in ultrapure water, to which formic acid is then added to make a final 0.33% solution. Pre-warm to 37°C the solution and the glassware where the staining reaction will be performed
- Dissolve silver-nitrate in the gelatin–formic acid solution to make a final 33% solution, and immerse the slides immediately in the solution obtained. Stain in the dark at a constant temperature of 37°C for 12 min
- Pour off the solution and wash the slides in several baths of ultrapure water
- Dehydrate and mount routinely (avoid synthetic slide-mounting media)

4. Histological samples fixed in buffered formalin and routinely paraffin-embedded

- Dewax the sections in xylene. Hydrate the sections through graded alcohols to ultrapure water avoiding any additional post-fixation in alcohol-based fixatives
- Immerse sections in sodium citrate buffer (10 mM sodium-citrate monohydrate, pH 6.0) in plastic Coplin jars and boil at 120°C for 20 min in wet autoclave or pressure cooker. Cool to room temperature and wash with ultrapure water
- Prepare a solution consisting of a 0.66% gelatin solution dissolved in ultrapure water, to which formic acid is then added to make a final 0.33% solution. Pre-warm to 37°C the solution and the glassware where the staining reaction will be performed

Table 7.1 (*continued*)

- Dissolve silver-nitrate in the gelatin–formic acid solution to make a final 33% solution, and immerse the slides immediately in the solution obtained. Stain in the dark at a constant temperature of 37°C for 13 min
- Pour off the solution and wash the slides in several baths of ultrapure water
- Dehydrate and mount routinely (avoid synthetic slide-mounting media)

of the staining reaction progressively decreases until AgNORs in the sample finally disappear. On the contrary, the use of natural mountant media, such as Canada Balsam, allows silver-stained slides to be perfectly preserved for a very long time.

7.4 Quantitative AgNOR analysis

Quantitative AgNOR analysis can be achieved according to either the *counting method* or the *morphometric method*. The counting method is the most commonly used technique for AgNOR quantitation in tumour pathology. It consists of the direct count by microscope of each silver-stained dot per cell, by carefully focusing through the section thickness at very high (100×) magnification. In spite of its very simple and inexpensive nature, the counting method has many drawbacks. It is time-consuming and subjective, particularly when single silver-stained dots cluster together or partially overlap. Using the counting method, several authors have reported significant inter-observer variations in the mean AgNOR numbers obtained in the same histological sections. It should also be noted that, since each aggregate that cannot be resolved in individual NORs is counted as 'one AgNOR', the counting method does not take into consideration the dimension of each silver-stained dot which, particularly in malignant cells, may be greatly variable.

In order to overcome these drawbacks and to obtain a more objective and reproducible AgNOR quantification, the morphometric method was introduced. This technique consists of the automatic or semiautomatic measurement of the silver-stained *area* within the nuclear profile by using image cytometry. In comparison with the counting method, the morphometric method is faster, more accurate, and – last but not least – more objective. Its only practical limitation is the need for adequate instrumentation (consisting of a digital camera mounted on a light microscope and linked to a PC equipped with basic morphometric software) and some familiarity with computers and statistics. Several authors have demonstrated that the morphometric method is more reproducible than the counting method and that the two quantitative methods do not yield comparable results. The AgNOR values obtained by both methods are affected by the fixative used for tissue processing as well as by the temperature and time of the staining reaction. However, in contrast to the counting method, the morphometric method may by applied to compensate for the staining variability produced by the use of different fixatives or silver-staining protocols. In reality, it is possible to obtain comparable data among tumour samples fixed with different fixatives or stained according to different staining procedures by relating the mean AgNOR area of cancer cells to that of control cells (lymphocytes or stromal cells) evaluated by image analysis. Taking these considerations into account, the **International Committee on AgNOR Quantitation** has indicated image analysis as the method of choice for AgNOR evaluation in routine cyto-histopathology.

7.5 AgNORs as a parameter of the level of cell proliferation

Just after the first application of the interphase AgNOR parameter in tumour pathology, a correlation between the quantity of AgNORs and the proliferative activity of neoplastic cells was demonstrated, independent of the tumour type. Several studies carried out on experimental models – such as human circulating lymphocytes stimulated to proliferate by phytohemagglutinin and rat hepatocytes stimulated to proliferate by partial hepatectomy – have demonstrated clearly that, in proliferating cells, AgNORs progressively increase from G1 to the S and G2 phases of the cell cycle. This increase is associated with a corresponding increase in ribosome biogenesis. Further studies have indicated that the relationship between AgNOR values and cell proliferation was more complex but, at the same time, much more interesting than was initially thought. In fact, a series of data obtained using *in vitro* cultured human cancer cell lines demonstrated that the AgNOR quantity is significantly related to the level of cell proliferation. Studies carried out on human carcinoma xenografts growing subcutaneously in nude (athymic) mice showed that the quantity of AgNORs was independent of the number of proliferating cells but, indeed, related only to the tumour mass doubling time. The close correlation between AgNOR values and rapidity of cell proliferation resulted from the different activity of ribosomal biogenesis in cells characterised by different doubling times. Proliferating cells have, in fact, to synthesise many proteins that are necessary for cell cycle progression and to produce an adequate ribosomal complement for the daughter cells. Therefore, the shorter the cell cycle time, the greater must be the ribosomal biogenesis per unit time. To this end, the cells increase the number of those nucleolar structures (AgNORs) that are involved in rRNA synthesis. Thus, the quantity of AgNORs is a parameter that indicates, in cycling cells, the rapidity of cell proliferation. The AgNOR parameter, therefore, is a unique tool for the evaluation of the level of cell proliferation in routine histological sections of tumour tissue at diagnosis (Devenzini, 2000).

7.6 Application of the AgNOR technique to tumour pathology

No Utility for the Diagnosis of Malignancy

AgNOR quantification was introduced in tumour pathology as a parameter to distinguish malignant cells from benign or normal cells. Quantitative AgNOR analysis, carried out in a wide series of human tumours (all the main human tumours were analysed in the three-year span following the first publication by Ploton *et al.*, 1986) showed that, generally speaking, cancer cells have a greater quantity of AgNORs than the corresponding normal or hyperplastic cells. Therefore, frequently, a malignant lesion can be distinguished from a benign lesion of the same tissue by evaluating the AgNOR quantity. However, the reliability of a diagnostic parameter in tumour pathology depends on its capacity to define clearly the nature of a particular tumour. From this point of view, AgNOR must be considered to be of little experience and therefore in most tumour types the AgNOR values of malignant lesions were found to overlap those of the corresponding benign lesions, thus indicating that the AgNOR quantification cannot be considered as an absolute parameter for the cyto-histological diagnosis of malignancy (Figure 7.4). The only types of tumours in which AgNOR quantification demonstrated a reliable use were melanotic lesions of the skin and the pleural neoplastic effusions. No overlapping of AgNOR values was found between nevocellular nevi and malignant melanocarcinomas nor between neoplastic (both metastatic carcinoma and mesothelioma) cells and reactive cells in human pleural effusions. The low diagnostic power of AgNORs depends on the fact that, as previously stated, AgNORs are the morphological expression of ribosomal rRNA transcriptional activity. Even if the synthesis of rRNA is frequently greater in malignant than in benign tumours or corresponding normal cells, it may be that: (a) some normal resting cells are characterised by a very high ribosomal biogenesis (see neurons, for example) and, therefore, by more AgNORs than in neoplastic cells, and (b) malignant tissues with a very low proliferative activity – lower than that of the corresponding benign lesions – show a lower rRNA transcriptional activity and, therefore, fewer AgNORs, thus making it impossible to distinguish the two forms of neoplasia on the basis of the AgNOR quantity alone.

Relevance in Terms of Tumour Prognosis

Great caution in the use of the AgNOR method for diagnostic purposes was recommended just after its introduction in tumour pathology. However, the concept of AgNORs as a diagnostic parameter spread so widely that, frequently, AgNORs have been considered exclusively as such. The scanty success of AgNORs as a diagnostic parameter, therefore, has induced many pathologists to reject AgNOR assessment for prognostic purposes also.

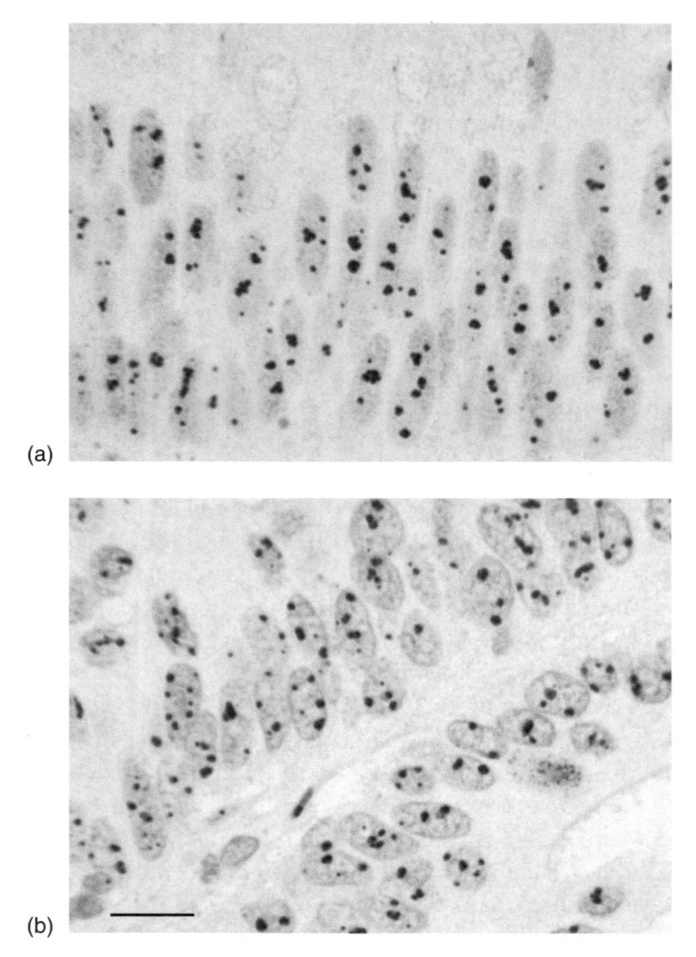

Figure 7.4 AgNOR distribution in formalin-fixed and paraffin-embedded samples of human adenomatous polyp (a) and adenocarcinoma (b) of the colon. No difference in the AgNOR amount is detectable between the two tumour lesions. Bar, 10 μm

Indeed, since the AgNOR score is a parameter of cell proliferation rate, its main use in tumour pathology lies in prognosis. Up to now, more than 400 studies have been published on the relationship between AgNORs and tumour prognosis. In most of these studies the AgNOR parameter proved to be a relevant prognostic factor for many types of neoplastic lesions: the greater the quantity of AgNORs, the worse the prognosis of the disease. In many of these studies, the predictive relevance of AgNORs has been compared with that of other well-established parameters in a multivariate analysis of survival, showing its independent prognostic value (Table 7.2). In relation to this, for the

Table 7.2 Human tumors in which multivariate analysis of survival demonstrated the AgNOR protein parameter to be an independent prognostic variable

Tumour type	Reference	Number of cases evaluated
Bladder carcinomas	Korneyev IA *et al. Mol. Pathol.* 2000; **53**: 129–32	62
	Masuda M *et al. J. Cancer Res. Clin. Oncol.* 1997; **123**: 1–5	90
	Lipponen PK *et al. Br. J. Cancer* 1991; **64**: 1139–44	229
Breast carcinomas (female)	Ceccarelli *et al. Micron* 2000; **31**: 143–9	217
	Biesterfeld S *et al. Virchows Arch.* 2001; **438**: 478–84	89
	Öfner D *et al. Breast Cancer Res. Treat.* 1996; **39**: 165–76	115
	Nakayama K and Abe R *J. Surg. Oncol.* 1995; **60**: 160–7	131
	Aubele M *et al. Pathol. Res. Pract.* 1994; **190**: 129–37	137
Breast carcinomas (male)	Pich A *et al. Am. J. Pathol.* 1994; **145**: 481–9	27
	Pich A *et al. Human Pathol.* 1996; **27**: 676–82	34
Choroidal melanomas	Tuccari G *et al. Anal. Cell Pathol.* 1999; **19**: 163–8	34
Colorectal carcinomas	Öfner D *et al. J. Pathol.* 1995; **175**: 441–8	92
	Joyce WP *et al. Ann. R. Coll. Surg. Engl.* 1992; **74**: 172–6	164
	Rüschoff J *et al. Pathol. Res. Pract.* 1990; **186**: 85–91	70
	Moran K *et al.* Br. J. Surg. 1989; **76**: 1152–5	51
Endometrial carcinomas	Giuffrè G *et al. Anal. Quant. Cytol. Histol.* 2001; **23**: 31–9	64
	Trerè D *et al. Gynecol. Oncol.* 1994; **211**: 282–5	45
Esophageal carcinomas	Morita M *et al. Cancer Res.* 1991; **1**: 5339–41	98
Gall-bladder carcinomas	Nishizawa-Takano JE *et al. Dig. Dis. Sci.* 1996; **41**: 840–7	76
Gastric carcinomas	Giuffrè G *et al. Virchows Arch.* 1998; **433**: 261–6	78
	Kakeji Y *et al. Cancer Res.* 1991; **51**: 3503–6	91
Glottic squamous cell carcinomas	Xie X *et al. Head Neck* 1997; **19**: 20–6	93
Hepatocellular carcinoma (without portal vein invasion)	Shimizu K *et al. Hepatology* 1995; **21**: 393–7	89
Laryngeal squamous cell carcinomas	Krecicki T *et al. Acta Otorhinolaryngol. Belg.* 1998; **52**: 215–21	154
Leukemia	Metze K *et al. Int. J. Cancer* 2000; **89**: 440–3	57
	Pich A *et al. J. Clin. Oncol.* 1998; **16**: 1512–18	40
	Trerè D *et al. Br. J. Cancer* 1994; **70**: 1198–1202	119
Lung carcinomas	Bernardi FD *et al. Mod. Pathol.* 1997; **10**: 992–1000	52
	Antonangelo L *et al. Chest* 1997; **111**: 110–14	81

(*continued overleaf*)

Table 7.2 (*continued*)

Tumour type	Reference	Number of cases evaluated
	Lee YC *et al. Thorac. Cardiovasc. Surg.* 1996; **44**: 204–7	73
	Oyama T *et al. Surg. Oncol.* 1993; **2**: 341–7	102
Malignant melanoma	Barzilai A *et al. Am. J. Dermatopathol.* 1998; **20**: 473–7	30
Multiple myeloma	Pich A *et al. Am. J. Surg. Pathol.* 1997; **21**: 339–47	116
Non-Hodgkin lymphomas	Korkolopoulou P *et al. Leuk. Lymphoma* 1998; **30**: 625–36	91
	Jakic-Razumovic J *et al. J. Clin. Pathol.* 1993; **46**: 943–7	61
Oral squamous cell carcinomas	Teixeira G *Am. J. Surg.* 1996; **172**: 684–8	4351 80
	Xie X *et al. Cancer* 1997; **79**: 2200–8	
	Piffko J *et al. J. Pathol.* 1997; **182**: 450–6	
Pharyngeal carcinomas	Pich A *et al. Br. J. Cancer* 1991; **64**: 327–32	61
Prostate carcinomas	Conytractor H *et al. Urol. Int.* 1991; **46**: 9–14	63
	Chiusa L *et al. Cancer* 1997; **15**: 1956–63	65
Renal cell carcinomas	Yasunaga Y *et al. J. Surg. Oncol.* 1998; **68**: 11–18	96
	Pich *et al. Oncol. Rep.* 1997; **4**: 749–51	21
	Shimazui T *et al. J. Urol.* 1995; **154**: 1522–6	59
	Delahunt B *et al. Cancer* 1995; **75**: 2714–19	206
Soft tissue sarcomas	Nakanishi H *et al. Oncology* 1997; **54**: 238–44	70
	Kuratsu S *et al. Oncology* 1995; **52**: 363–70	151
	Tomita T *et al. Int. J. Cancer* 1993; **54**: 194–9	194
Thymomas	Pich A *et al. Cancer* 1994; **1**: 1568–74	90

large number of patients evaluated, the studies carried out in transitional cell carcinomas of the bladder, colorectal cancers, soft tissue sarcomas, leukaemia, renal carcinomas, breast cancers and myelomas are very important. Figure 7.5 shows the distribution of AgNORs in two cases of breast cancer characterised by a good and a fatal prognosis, respectively, included in the study by Ceccarelli *et al.* (see Table 7.2). In this study the AgNOR parameter demonstrated an independent prognostic value, together with: Ki67 labelling index, lymph-node status, and tumour size. The lack of prognostic value of AgNORs that has been reported in a few studies, in most cases is the consequence of methodological errors in either the staining or the quantification of AgNORs. The most frequent mistake has been to consider whole nucleoli as single AgNORs: therefore, by evaluating the AgNOR parameter using the counting method, no information on the actual entity of ribosomal biogenesis and, consequently, on the level of cell proliferation of cancer cells could be obtained.

The observation that AgNOR quantity is an independent prognostic parameter indicates the great power of cell kinetics as a predictive factor. Indeed, the growth rate of a tumour mass inside the host represents one of the most important

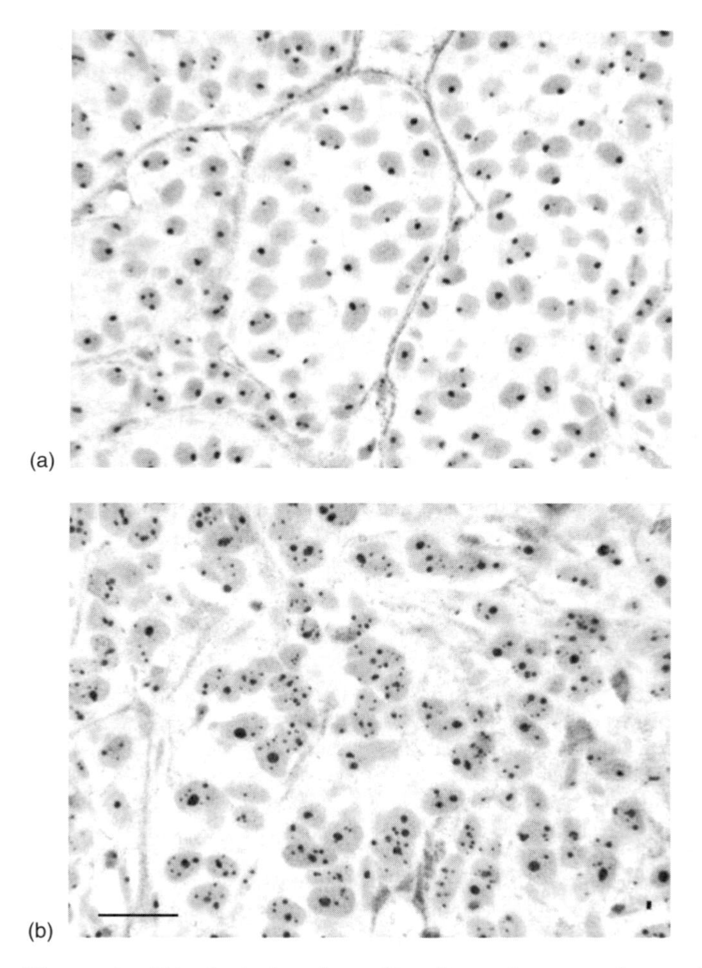

Figure 7.5 Silver-stained histological sections of two human breast carcinomas. The patient in (a) was alive and well 52 months after the surgical resection, while the patient in (b) died from the disease 30 months after diagnosis. Note the low AgNOR quantity of the case in (a), as compared to the greater quantity and more irregular AgNOR distribution in (b). Bar, 30 μm

prognostic factors in oncology. Tumour growth rate is mainly related to the numbers of cycling cells and to the level of cell proliferation. Many methods are available for the definition of the first parameter and, among them, the Ki67/MIB1-labelling index is certainly the most reliable and frequently applied procedure in routine diagnostic histopathology.

7.7 What future for AgNORs in tumour pathology?

Up to now, AgNOR staining and quantification have been standardised and the biological meaning of AgNOR variations in neoplastic cells have been elucidated. The future of AgNOR assessment in tumour pathology appears to be dependent on the relevance that oncologists attribute to cell kinetics to define the prognosis of cancer patients. The possibility of distinguishing groups of patients with different clinical outcomes by means of the use of the cell kinetic parameters should be very useful also for the identification of different therapeutic approaches based on the cell kinetic characteristics of tumours.

The success of future applications of AgNORs may also be dependent on the development of new methods for their more rapid and easy assessment. In relation to this, very recently the use of flow cytometry has been suggested as a means of quantification of AgNOR dots within cancerous cells (Jacquet *et al.*, 2001). These authors found a large decrease in the forward scattered light and correlated it with the area of AgNORs. This method, which is less time-consuming than image analysis, allows the quantification of AgNORs in large populations of cells. The use of flow cytometry for the quantification of silver-stained NORs or NOR-associated proteins could have great potential for the routine characterisation of cell proliferation rate within tumours.

7.8 References

Andersen JS, Lyon CE, Fox AH, Leung AK, Lam YW, Steen H, Mann M and Lamond AI (2002) Directed proteomic analysis of the human nucleolus. *Curr. Biol.* **12**: 1–11.

Anonymous (1987) NORs: a new method for the pathologist. *Lancet* **1**: 1413.

Crocker J and Skilbeck N (1987) Nucleolar organiser region associated proteins in cutaneous melanotic lesions: a quantitative study. *J. Clin. Pathol.* **40**: 885–9.

Derenzini M (2000) The AgNORs. *Micron.* **31**: 117–20.

Derenzini M, Thiry M and Goessens G (1990) Ultrastructural cytochemistry of the mammalian cell nucleolus. *J. Histochem. Cytochem.* **38**: 1237–56.

Jacquet B, Canet V, Giroud F, Montmasson MP and Brugal G (2001) Quantification of AgNORs by flow versus image cytometry. *J. Histochem. Cytochem.* **49**: 433–7.

Öfner D, Bankfalvi A, Riehemann K, Bier B, Bocker W and Schmid KW (1994) Wet autoclave pretreatment improves the visualization of silver-stained nucleolar organizer-region-associated proteins in routinely formalin-fixed and paraffin-embedded tissues. *Mod. Pathol.* **7**: 946–50.

Ploton D, Gilbert N, Menager M, Kaplan H and Adnet JJ (1994) Three-dimensional co-localization of nucleolar argyrophilic components and DNA in cell nuclei by confocal microcopy. *J. Histochem. Cytochem.* **42**: 137–48.

Ploton D, Menager P, Jeaesson P, Himber G, Pigeon F and Adnet JJ (1986) Improvement in the staining and in the visualization of the argyrophilic proteins of the nucleolar organizer region at the optical level. *Histochem. J.* **18**: 5–14.

8

Apoptosis and Cell Senescence

Lee B. Jordan and **David J. Harrison**

8.1 Introduction

Apoptosis is an essential, evolutionary conserved and ubiquitous process required for the function and development of multicellular organisms. Cell senescence is an integral and increasingly recognised aspect of cellular existence. Ageing and death are part of life! In this chapter, we aim to provide a brief overview of these processes, their features and, more importantly, discuss the mechanisms and difficulties of identifying, detecting and quantifying them.

8.2 Apoptosis

The word 'apoptosis' is derived from classical Greek referring to the falling of leaves from a tree in autumn or the loss of petals from a flower. Kerr *et al.* (1972) proposed its use after the morphological identification of unique cellular events, now synonymous with this process. Historically, apoptosis is contrasted with necrosis, as an active vs. passive form of death. A revised terminology has been proposed (Majno and Joris, 1995; Trump *et al.*, 1997) claiming that necrosis is an all encompassing ancient term to describe post mortem changes in any cell, regardless of the pre-lethal changes or events that lead to death. The term 'oncosis', derived from the Greek word for swelling, has been resurrected from the original use by Von Recklinghausen (1910) (Trump *et al.*, 1997; Von Recklinghausen, 1910). Consequently, the pre-lethal morphological changes of cell death have been divided into apoptosis and oncosis, with necrosis possible sequelae to both. However tempting this revision is, it adds to a field dominated by confused and conflicting terminology and as this has not been widely accepted, we will refrain from its use.

Molecular Biology in Cellular Pathology. Edited by John Crocker and Paul G. Murray
 ISBN: 0-470-84475-2

The next misinterpretation must be addressed; the concept of accidental vs. programmed death. Apoptosis gradually has become interchangeable with programmed cell death due to the increasing association with the physiological timed deletion of cells in processes such as development. Not all death under such circumstances is apoptotic and it is crucial to realise that every cell has the 'programme' for death under the appropriate stimulus (Trump *et al.*, 1997). However, the activation of that programme may produce varied morphology, although the biomechanics involved may be similar. Indeed, Sperandio *et al.* (2000) have described a non-apoptotic programmed cell death characterised by cytoplasmic vacuolation without nuclear fragmentation which, like apoptosis, is associated with active transcription and protein synthesis. They termed this process, 'parapoptosis'.

Apoptotic Stimuli

Apoptosis can be initiated by numerous stimuli in a wide variety of settings. For simplification, physiological and pathological division can be used (Jordan and Harrison, 2000).

Physiological stimuli include those encountered during development, where apoptosis is the most common pre-lethal morphology associated with programmed deletion on a schedule set by external and internal environment factors, for example hormones, body mass and nutrition (Columbano, 1995; Leist and Nicotera, 1997; Trump *et al.*, 1997; Webb *et al.*, 1997). This is also applicable to adult life, with regression of hyperplastic tissues no longer required or deletion of excess cells produced by mitosis. A further example would be the deletion of self-reactive T-cell clones within the thymus.

Pathological stimuli are those that are injurious to the cell outwith physiological expectation, such as chemical, microbiological, radiological or genetic factors, inducing an apoptotic response often with a self-protective motive. Such a response may be appropriate, preventing the spread of a genetic lesion that may lead to neoplasia. Alternatively, these may be inappropriate, for example after an occlusion of blood supply to the myocardium where both apoptosis and necrosis are initiated. In this example, the programme for death is activated by the ischaemic injury, provided there is sufficient time. If the blood supply is not reinstated, then these apoptotic cells die from necrosis as the energy to continue an active process is unavailable. If the blood supply returns rapidly, then many cells may be lost by completing their suicide programmes as part of reperfusion injury. This can be viewed as inappropriate as the injurious agent has fled and such losses may be detrimental to tissue, organ and organism function. This is the subject of intense research for the medical implications alone (Jordan and Harrison, 2000). Obviously, anti-apoptotic pathways exist to curb some of this loss, but it appears the that overall dangers of not responding to an apoptotic stimulus far outweigh any inappropriate loss of cells.

Features of Apoptosis

Overview

The apoptotic process can be subdivided into four distinct stages: initiation or priming, commitment or decision, execution and clearance (van Heerde *et al.*, 2000; Webb *et al.*, 1997). Initiation is the initial stage where an injurious stimulus such as DNA damage, growth factor absence or hypoxia/anoxia begins the cascade to cell suicide (Evans and Littlewood, 1998). This is the stage that can be initiated by Fas ligand and tumour necrosis factor (TNF) alpha [12]. Proteases, often referred to as upstream or decision caspases, may allow continuation to the next stage (Thornberry and Lazebnik, 1998). Commitment or decision is the stage where the cell can still escape death. It is often characterised by mitochondrial involvement (Green and Reed, 1998; Susin *et al.*, 1997), by release of procaspase 2, 3 and 9, cytochrome C and apoptosis-inducing-factor (Lutgens *et al.*, 1999; Susin *et al.*, 1999b). Execution is the point of no return and the main subject of this section, typified by downstream or effector caspases (Thornberry and Lazebnik, 1998) that are responsible for the majority of the typical morphological changes of apoptosis. Beyond the morphology, specific events can be identified including DNA cleavage, cell membrane changes and caspase cleavage product formation; all are detailed below and are the main bases of apoptotic recognition. Clearance is the final stage, when cellular remnants are removed.

In depth: execution of apoptosis

Kerr *et al.* (1972) were the first to describe the morphological changes synonymous with the execution phase. The changes are co-ordinated structural and biochemical events experienced throughout the cell. A classic morphological comparison of necrosis and apoptosis has been described throughout the literature; we provide a brief chronological overview in Table 8.1 (Freude *et al.*, 1998; Kerr *et al.*, 1972; Trump *et al.*, 1997; Webb *et al.*, 1997; Wyllie, 1980).

The easiest apoptotic events to visualise by microscopy are the nuclear changes (Figure 8.1). Chromatin condenses and aggregates in a crescent distribution within the nucleus. Examination of this chromatin reveals cleaved DNA fragments in multiples of 180–200 base pairs (Wyllie, 1980), cleavage being performed by a cation-dependent endonuclease initially into 30–50 kilobase pairs. This is not essential and does not occur in some cell types, as its inhibition does not halt the process or ultimate outcome (Jacobson *et al.*, 1994; Schulze-Osthoff *et al.*, 1994). Simultaneously with the condensation, nuclear integrity is altered. The nuclear lamina (intermediate fibre skeleton) responsible for membrane and nuclear pore stability is cleaved, allowing fragmentation of nuclear contents into membrane bound vesicles (Kaufmann, 1989; Lazebnik *et al.*, 1993; Reipert *et al.*, 1996; Ucker *et al.*, 1992). Remaining

Table 8.1 Morphological comparison of necrosis and apoptosis in chronological sequence

Apoptosis	Necrosis
Cell and cytosol shrinkage • Functional alteration of ion flux and build up of intracellular calcium • Mitochondrial permeability transition, release of AIF, cytochrome C, procaspases • Effector enzyme activation, e.g. caspase	*Plasma membrane alterations* • Blebbing, blunting, distortion of microvilli, failed integrity causing cellular oedema, loss of intercellular attachments. (Functional – loss of ion flux control, Ca^{2+} build-up) • Cytoplasmic protein denaturation and clumping
Nuclear changes • Chromatin condensation and clumping • Modified transcription/translation • Site-specific cleavage of DNA by a cation-dependent endonuclease, aided by numerous factors • Nuclear fragmentation (karryohexis)	*Variable mitochondrial changes* • Swelling, rarefaction, formation of phospholipid-rich amorphous densities. (Functional loss of ATP production, impairing Na^{+} K^{+} ATPase pump contributing to cellular swelling)
Cytoplasmic blebbing • Consequence of modified cytoskeleton produced by caspase/tTG action	*Dilatation of endoplasmic reticulum* • Detachment of ribosomes.
Variable endoplasmic reticulum changes	*Nuclear changes* • Disaggregation of nuclear skeleton and chromatin clumping • Karyolysis or karryohexis • Random dissociation and cleavage of DNA (uncontrolled enzymic release including those of lysosomal origin)
Condensation or no alteration of mitochondria.	*Cellular dissociation*
Cytoplasmic budding • 'Packaging': production of vesicles or apoptotic bodies containing chromatin/organelles. (Function of cytoskeletal modification)	*Clearance* • Probable activation of inflammatory pathways
Clearance. • Phagocytosis by surrounding cells without inducing an inflammatory response (ideal situation) • Relies on expression of external marker moieties, e.g. phosphatidyl–serine	
Energy required.	No energy requirement.

intracellular organelles remain structurally intact, although mitochondrial dysfunction occurs, typified by altered transmembrane potential and uncoupling of the electron transport pathway from adenosine triphosphate (ATP) generation with increased production-reactive oxygen species (Kroemer *et al.*, 1995). This is accompanied by the release of procaspase 2, 3 and 9, cytochrome C and apoptosis inducing factor (AIF) (Lutgens *et al.*, 1999; Susin *et al.*, 1999b),

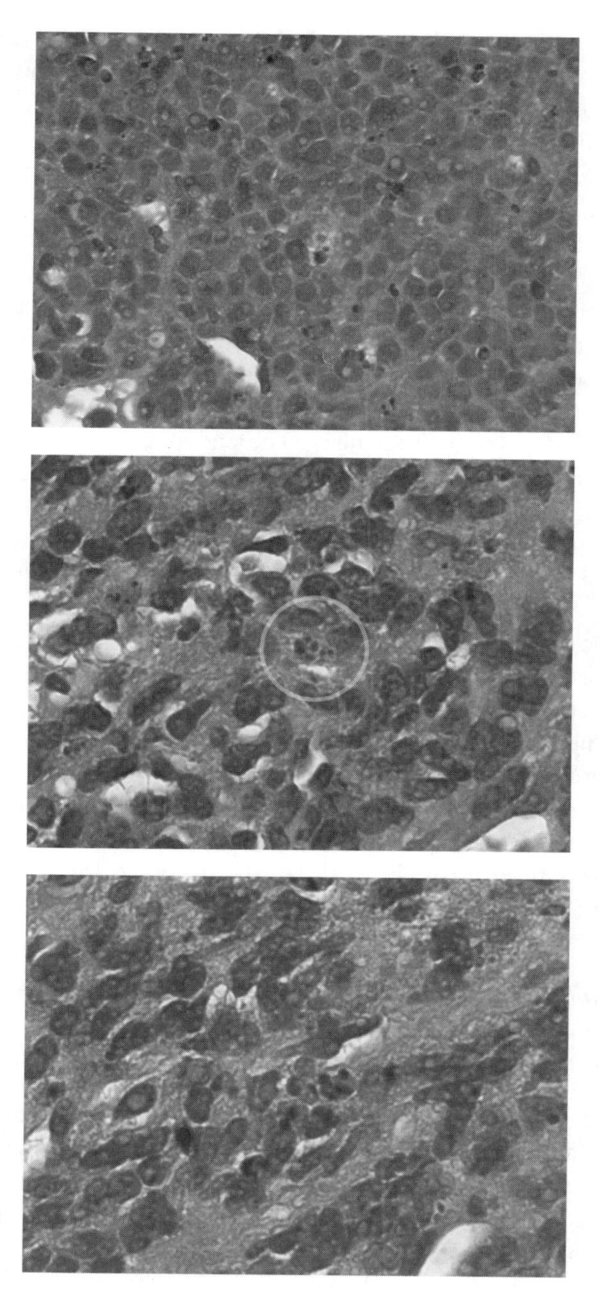

Figure 8.1 A series of three photomicrographs depicting a highly apoptotic diffuse large B-cell lymphoma. Upper – Low power [×200, haematoxylin and eosin (H&E)] image showing scattered apoptotic cells. Centre – High power (×400, H&E) showing the nuclear fragmentation and formation of apoptotic bodies (circled). Lower – High power (×400, H&E) showing a central group of apoptotic cells. A colour version of this figure appears in the colour plate section

which begins the terminal decline of cellular integrity. This step is regulated by the Bcl-2 protein family, which controls the opening of the mitochondrial permeability transition pore (Budihardjo *et al.*, 1999; Crompton *et al.*, 1999; Green and Reed, 1998; Kluck *et al.*, 1997; Susin *et al.*, 1996; Zamzami *et al.*, 1996). Apaf-1 and cytochrome C complex together activate caspases including caspase-3. Apaf-1 normally is inhibited by the Inhibitors of Apoptosis (IAP) family (Budihardjo *et al.*, 1999; Du *et al.*, 2000; Zou *et al.*, 1997). The caspase cascade begins the selective cleavage of DNA repair enzymes such as poly (ADP-ribose) polymerase (PARP) and others responsible for cellular integrity and homeostasis. Within the cytoplasm, tissue transglutaminase (tTg) cross-links various proteins (Fesus *et al.*, 1987) and cytoskeletal filaments aggregate in parallel arrays, partly driven by active caspases, which eventually contribute to the classical morphological changes. The endoplasmic reticulum may dilate and fuse with the plasma membrane; this combined with the loss of membrane phospholipid asymmetry, microvilli and cell–cell contacts destabilises the plasma membrane. The net result is the dissociation from neighbouring cells, continued shrinking and budding of apoptotic bodies into the surrounding tissue.

The time to perform the execution phase, once initiated, is rapid and conserved, possibly indicating a final common pathway (Bursch *et al.*, 1990). However, the time from exposure to the noxious stimulus to resolution of apoptosis is highly variable and appears to depend on the type of stimulus, cell type (phenotype, genotype and gene expression) and the internal/external environmental milieu. This appears to be related to the variability of reception and transduction of the apoptotic stimulus, followed by initiation of the response. Perhaps unsurprisingly, many of the pathways involved play a role in carcinogenesis (Webb *et al.*, 1997).

In depth: resolution of apoptosis

Perhaps one of the most striking differences between apoptosis and necrosis is, in the vast majority of cases, the lack of an inflammatory response to apoptotic cell death. Indeed, phagocytosis of apoptotic fragments can suppress immune responses and facilitate tolerance. The final stage of apoptosis is termed 'clearance', which is achieved by phagocytosis, both by the professional phagocyte, such as the macrophage, and the semi-professional phagocyte (Savill, 1996). Early studies implicated interactions of endogenous macrophage lectins with specific *N*-acetyl sugar moieties displayed on the surface of apoptotic cells with initiation of phagocytosis within resident macrophage tissue populations (Savill, 1997; Webb *et al.*, 1997). The monocyte derived macrophages and neutrophils, which are able rapidly to take up large numbers of apoptotic cells, employ a different recognition system. This revolves around $\alpha_v\beta_3$ vitronectin receptor and CD36. Co-operation from thrombospondin (TSP)-1, a soluble adhesive glycoprotein produced by many cells, enables a bridge from the macrophage to markers expressed upon the apoptotic cell/body (Duvall *et al.*, 1985; Savill, 1996, 1997; Savill *et al.*, 1990, 1992). Such markers include phosphatidylserine, a phospholipid normally localised to the inner aspect of the plasma membrane that is expressed externally on apoptosis (Fadok *et al.*, 1992; Martin *et al.*, 1995); Vermes *et al.*, 1995). The mammalian ATP-binding cassette transporter (Flora and Gregory, 1994), scavenger receptors and specific phosphatidylserine receptors (Savill *et al.*, 1990) have been associated with the initiation of the phagocytic pathway, ending with engulfment of the apoptotic body. It has been hypothesised that in some circumstances inflammation may be an actively pursued course to bring in more phagocytes to a region of high apoptosis (Savill, 1997).

Therefore, the uptake of apoptotic bodies is dependent upon the appropriate expression of surface markers on those bodies that shout 'eat me'; failure to do so or failure of recognition allows the contents of these bodies to leak as they undergo secondary necrosis, initiating an inflammatory response. This may be important in the understanding of the pathogenesis of inflammatory and immune disease (Savill, 1996). An excellent example is the widespread hepatic apoptosis induced in murine models after exposure to anti-Fas antibody, the local phagocyte reserves are exceeded and severe liver damage occurs through inflammatory processes (Ogasawara *et al.*, 1993). Furthermore, the lack of capable phagocytes or the poor phagocytic ability of the local cells within a tissue may lead to a low grade inflammatory response that recruits phagocytes, which in turn initiates repair damaging tissue architecture, e.g. the myocardial fibrosis seen in ischaemic heart disease (Jordan and Harrison, 2000).

Identification of Apoptosis

Originally, apoptosis was only identifiable by morphological criteria; subsequent developments have led to an expanding variety of detection methods, each with particular drawbacks that often resort to morphological criteria for confirmation or corroboration of those results. Morphology remains the gold standard at present, albeit accepting that not all apoptotic cells will be detected by this approach.

Features of apoptotic cells vary significantly depending on the nature of the apoptotic stimulus, cell type, stage of apoptosis (Darzynkiewicz *et al.*, 1998), and possible artefacts induced by attempts to detect the process. One must remember that any measure of apoptosis is a snapshot in time, and it is difficult to assess the rate of apoptosis within a given system (Darzynkiewicz *et al.*, 1998), although real-time imaging may be a solution to this problem.

Visual methods

Standard microscopic techniques

Microscopy is the original method of detection and is still useful. Light microscopy of cells is rapid and quantitative, enabling visualisation of nuclear changes and measurement of apoptotic frequency across the field (Figure 8.1). Problems are inevitable with light microscopy and a consensus between investigators as to the criteria for an apoptotic cell must be agreed as individual bias may influence results. Early stages are not visible and only during the execution stage do the classical features become apparent. Late stages of apoptosis may be difficult to distinguish from necrosis impairing specificity (see the comparative assessment in Table 8.1) and morphology is not unambiguous evidence that cells are apoptotic (Darzynkiewicz *et al.*, 1998; Loo and Rillema, 1998). On top of these problems is one of low sensitivity due to rapid removal of apoptotic debris from tissues, perhaps within 30–60 min of onset, depending on the tissue involved. Consequently, light microscopy should not be used alone, but combined with other methods the technique is more powerful. Furthermore, the use of real-time image capture can view the entire apoptotic process, thereby aiding evaluation. An example of an optical, real-time image capture is shown in a series of photomicrographs (Figure 8.2).

Electron microscopy provides the most definitive proof of apoptosis by visualising in exquisite detail the morphology of a cell at that precise time. This is the gold standard for morphology with high specificity and sensitivity. However, electron microscopy is not readily quantifiable and is tedious and difficult to apply for experimental and comparative use (Loo and Rillema, 1998).

Vital dyes

Vital dyes are another means of visual identification of cell death. These are based on the loss of membrane integrity of a cell on death, allowing access to dye that cannot penetrate a living cell. Examples are trypan blue (visible on light microscopy) and propidium iodide (visible on fluorescent microscopy) (Loo and Rillema, 1998). This must be used with other techniques such as morphology

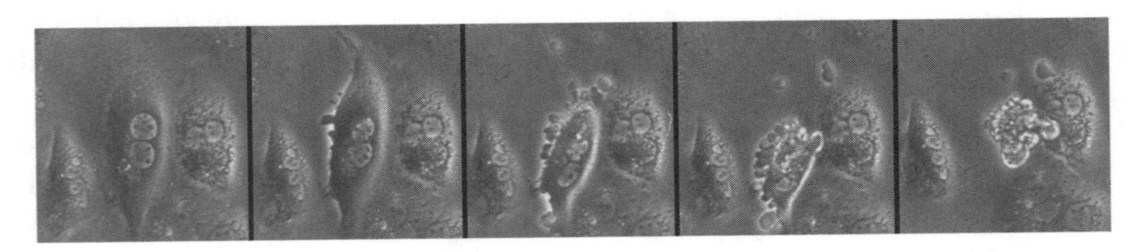

Figure 8.2 An optical real-time image capture of a murine hepatocyte culture recently exposed to apoptotic stimuli. The sequence from left to right are five time lapsed stills showing the characteristic morphological changes of apoptosis, from cell shrinkage to cytoplasmic budding of apoptotic bodies. (Courtesy of Dr Sandrine Prost)

or flow cytometry using a cell viability marker as it is the absence of staining that may indicate apoptosis. However, these dyes are non-specific and reliance on membrane permeability means that late apoptotic cells will stain making an underestimation of apoptosis in that tissue at that time.

Nuclear stains

Nuclear stains can be used to locate condensed or segregated nuclear chromatin; the dyes available are fluorescent and include acridine orange, bisbenzimide (Hoechst 33258 and 33342) and propidium iodide. The latter requires fixation and permeation of the cell prior to staining (Loo and Rillema, 1998). These dyes are rapid, easy to perform on many samples and can be used to quantify apoptotic death (Loo and Rillema, 1998) and are often used with flow cytometry.

DNA labelling methods

Further methods utilise chemical labelling and visualisation of internucleosomal DNA cleavage fragments. 'Terminal deoxynucleotidyl transferase-mediated dUTP-biotin nick end labelling' of DNA fragments, or TUNEL, incorporates biotinylated dUTP onto the 3′ ends of DNA fragments by utilising the activity of terminal deoxynucleotidyl transferase, and was originally developed for detection of apoptotic cells in tissue sections (Gavrieli *et al.*, 1992). Subsequent modifications have enabled its use in flow cytometry (Hotz *et al.*, 1994) (see the section 'Indirect Methods' below) and incorporation of an additional strand breakage induced by photolysis, which is proportional to DNA replication, allows both cell proliferation and apoptosis to be measured in a single step procedure (Li *et al.*, 1994, 1995; Moore *et al.*, 1998). The alternative method is '*in situ* end labelling' or 'ISEL' (occasionally referred to as '*in situ* nick transition/translation') and, like TUNEL, adds biotinylated dUTP to 3′ ends of DNA using DNA polymerase I (Klenow fragment). ISEL has followed the same evolution as TUNEL (Ansari *et al.*, 1993; Laio *et al.*, 1997; Loo and Rillema, 1998). After labelling, appropriate visualisation is achieved by the addition of fluorescent dyes or the peroxidase method (Loo and Rillema, 1998). Both methods are sensitive for detecting early DNA fragmentation and enable quantification of apoptosis. They may be combined with other cell staining techniques such as cell-specific markers to apoptosis to certain cell subsets (Loo and Rillema, 1998). The main drawback is that not all cells that are apoptotic undergo DNA cleavage: necrotic cells experience random cleavage and some forms of apoptosis in certain cells progress without cleavage. Many difficulties with ISEL and TUNEL have been described; an overview of factors influencing DNA end-labelling is given in Table 8.2. More recently, a DNA fragmentation ELISA assay has also become commercially available (Naruse *et al.*, 1994). Anti-single-stranded DNA antibodies have also been used but found to be non-specific for apoptosis (Aigner, 2002). Hairpin oligos can be used to detect

Table 8.2 Factors reported to influence DNA end labelling

Tissue

Tissue drying – increases the false positive rate (Grasi-Kraup *et al.*, 1995; Migheli *et al.*, 1994; Petito and Roberts, 1995)

Fixative

Fixative used – ethanol diminishes staining intensity and increases background versus 10% formalin (Mundle *et al.*, 1995)

Long formalin fixation times (>3 weeks) – increases the false-negative rate of detection when compared with morphology (Davidson *et al.*, 1995)

Delay in fixation – increases the false-positive rate (Soini *et al.*, 1998)

Inadequate fixation – increases the false-positive rate (Kockx *et al.*, 1998; Lutgens *et al.*, 1999)

Concentration of fixative – variable effect (Migheli *et al.*, 1994; Negoescu *et al.*, 1996; Petito and Roberts, 1995; Sasano, 1995)

Sections

Proximity to the section edge – staining errors (Jerome *et al.*, 2000)

DNA shearing during the sectioning process – false positivity as confirmed by confocal microscopy (Jerome *et al.*, 2000)

Pre-treatments

Pre-treatment used affects results – microwaving increases number of labelled cells by two-fold compared with proteinase-K (Negoescu *et al.*, 1996)

Duration with pepsin or proteinase-K – false negatives with too little incubation and false positives with over-exposure (Ansari *et al.*, 1993; Migheli *et al.*, 1994)

Technique

Duration and concentration of polymerase (ISEL) or transferase treatment – influences the number of nuclei stained (Jerome *et al.*, 2000; Kockx *et al.*, 1998; Lutgens *et al.*, 1999; Soini *et al.*, 1998)

DNA strand breaks – not solely associated with apoptosis (fragmentation may be present in necrosis or caused by DNA damaging agents) (Boe *et al.*, 1991; Cohen *et al.*, 1992; Collins *et al.*, 1992; Kockx *et al.*, 1998; Lutgens *et al.*, 1999; Oberhammer *et al.*, 1991, 1992; Soini *et al.*, 1998; Zakeri *et al.*, 1993)

Artefact of non-specific staining of Ca^{2+} filled vesicles and RNA splicing fragments – removed by an improved protocol (Kockx *et al.*, 1998; Lutgens *et al.*, 1999)

double strand breaks (Didenko *et al.*, 1994) but limitations like those pertaining to TUNEL exist. These difficulties reinforce the view that DNA labelling alone is insufficient.

Membrane alterations

The clearance of apoptotic bodies depends on expression of membrane factors to highlight the dying cell for phagocytosis; one example of such is phosphatidyl–serine (Fadok *et al.*, 1992; Martin *et al.*, 1995b; Vermes *et al.*, 1995). During apoptosis, the normal membrane asymmetry is lost and phosphatidyl–serine is expressed on the external cell surface. Annexin V has a high-affinity binding to phosphatidyl-serine, regulated by calcium ion presence (Andree *et al.*, 1990). This makes it a highly specific marker for apoptotic cells prior to and following membrane breakdown (Moore *et al.*, 1998). Labelling of annexin V enables immunofluorescent and/or immunohistochemical visualisation, and such labelling has been widely used to investigate and quantify apoptosis in various systems (Bennett *et al.*, 1995; Boersma *et al.*, 1996; Moore *et al.*, 1998; Pitti *et al.*, 1996; van Heerde *et al.*, 2000). Annexin V–biotin assays have been stated to be more flexible and more specific than other assays like TUNEL and have been used *in vivo* to identify apoptotic cells at sites of injury or development with subsequent tissue or animal harvesting for histological analysis (van Heerde *et al.*, 2000).

Synaptotagmin I, a syntaxin anchoring protein related to synaptophysin, binds to anionic phospholipids in cell membranes that become exposed during apoptosis. This can be detected by immunological means using Synaptotagmin I in a similar fashion to annexin V and has been successfully employed in flow cytometry (see the section 'Indirect Methods' below) and confocal microscopy (Zhao *et al.*, 2001).

Immunohistochemistry/immunocytochemistry

Immunohistological techniques involving the direct binding of antibodies to specific proteins to enable subsequent visualisation by fluorescence or peroxidase reactions are probably the most rapidly increasing method used for apoptotic marker detection. This is probably due to rapidity and ease of application to both frozen and, increasingly, fixed tissues; this enables an appreciation of the histological context to be retained. This area overlaps with some mentioned previously. However, this section will focus on the use of specific antibodies raised against both the enzymic effectors of apoptosis, e.g. enzymes that disable other enzymes, such as the caspases and their cleavage products, often referred to as neo-epitopes. Examples of neo-epitopes include PARP, which is involved in DNA repair (Sallmann *et al.*, 1997), Bcl-XL, a potent inhibitor of programmed cell death (Clem *et al.*, 1998) and many structural proteins such as actin (Yang *et al.*, 1998) and cytokeratin (CK) 18 (Leers *et al.*, 1999), the modification of which is responsible for the classic morphology observed in apoptotic cells. As such, identification of these specific proteins provides alternative apoptotic markers.

Detection of effectors. Caspases or cysteine proteases are the enzymes most frequently associated with apoptosis. Both are implicated in the initiation of apoptosis (upstream/decision caspases), commitment and subsequent execution of apoptosis (downstream/effector caspases) (Lutgens *et al.*, 1999; Susin *et al.*, 1999b; Thornberry and Lazebnik, 1998). At the time of writing there are commercial antibodies against 14 caspases, some for both the pro- and active forms, the most potentially useful of which is active caspase-3. Very little of the literature refers to problems revolving around caspase detection and staining, but a physiological role of caspase beyond that of apoptosis has been described (Chang and Yang, 2000; Coffey *et al.*, 2001; Fadeel *et al.*, 2000; Grutter, 2000).

Pochampally *et al.* (1998) have detailed this evidence, suggesting this is a consequence of non-apoptotic caspase processing; MDM2, an oncogenic protein, is a regulator of p53 and is cleaved by caspase-3 during apoptosis, yielding a 60 kDa fragment. An identical 60 kDa fragment has been detected in non-apoptotic tumour cells, whilst apoptosis-specific caspase-3 targets like PARP remain intact. This has implications for the specificity of active caspase detection. The confusing nomenclature and possible functions of caspases are detailed in Table 8.3.

Another effector is tissue transglutaminase (tTG), a member of a family of cross-linking proteins that catalyse Ca^{2+}-dependent reactions leading to post-translational protein modification (Marzari *et al.*, 2001; Melino *et al.*, 2000; Piacentini *et al.*, 2000; Szegezdi *et al.*, 2000). Physiological functions include haemostasis, semen coagulation and keratinocyte cornified envelope formation (Aeschlimann and Thomazy, 2000). Up-regulation of tTG occurs in various cells undergoing apoptosis (Fesus *et al.*, 1987; Marzari *et al.*, 2001; Melino *et al.*, 2000; Piacentini *et al.*, 2000; Szegezdi *et al.*, 2000) and inhibition of tTG can suppress apoptosis (Bernassola *et al.*, Oliverio *et al.*, 1999). Various reports have alleged that the use of tTG immunohistochemistry is more accurate than DNA end-labelling for the detection of apoptosis (Aschoff *et al.*, 2000; Rittmaster *et al.*, 1999). However, evidence from osteocyte and chondrocyte apoptosis has shown poor co-localisation of tTG staining with morphological apoptosis (Stevens *et al.*, 2000) and tTG has been identified in necrotic cells (Iwaki *et al.*, 1994). In addition to this, tTG has been found to be a substrate of caspase-3 and the cleavage fragment, a neo-epitope, has been said to be a valuable marker of caspase-3 activation and a marker of the late execution phase of apoptosis (Fabbi *et al.*, 1999).

Calpain, an intracellular Ca^{2+}-dependent protease that cleaves cytoskeletal and sub-membranous proteins during cell death is not as specific as caspase activation for the identification of apoptosis and its presence can also be demonstrated in necrotic cells (Wang, 2000).

Detection of products – neo-epitopes. Neo-epitopes generated from the action of caspases and other enzymic actions during apoptosis are potential targets for antibody detection. Typical cited examples include PARP, Actin/Fractin and CK 18. The number of newly discovered fragments grows virtually by the day; a list can be seen in Table 8.4. Currently, no substrate cleavage fragments generated by non-caspase enzymes are routinely detectable, reliable or specific.

Detection of others, including the Fas pathway. There are many other proteins, detectable by antibodies, which are claimed to be specific indicators of the apoptotic cascade. However, many may not be specific for eventual cell death. We present an overview in what follows.

Table 8.3 The caspase family

Protease	Alternative designations	Further information
Caspase-1	Interleukin-1β converting enzyme (ICE) Homology with Ced-3 (*C. elegans*)	Apoptosis and processing of pro-inflammatory cytokines (Taylor *et al.*, 2000; Wang and Leonardo, 2000)
Caspase-2	Nedd-2 (mouse), ICH-1	Apoptosis. Putative role in brain development (Eldadah and Faden, 2000) Three active subunits (p18, p14, p12) [91, 93]. Two alternate mRNAs yielding caspase-2L (pro-apoptotic) and caspase-2S (can antagonise cell death (Ito *et al.*, 2000)
Caspase-3	Yama, apopain, CPP32	Apoptosis – cleaves PARP (Casciola-Rosen *et al.*, 1996; Duan *et al.*, 1996; Fernandes-Alnemri *et al.*, 1996)
Caspase-4	ICH-2, ICErelII, TX, caspase-11 (mouse)	Apoptosis
Caspase-5	TY, ICErelIII, caspase-12 (mouse)	Apoptosis – cleaves own precursor, requires the cysteine 245 residue (Faucheu *et al.*, 1996; Munday *et al.*, 1995; Van de Craen *et al.*, 1997). Mutation associated with microsatellite instability associated tumours of colon, endometrium and stomach (Schwartz *et al.*, 1999)
Caspase-6	Mch2	Apoptosis – cleaves nuclear lamins (Takahashi *et al.*, 1996)
Caspase-7	Mch3, CMH-1, ICE-LAP3	Apoptosis – cleaves PARP (Casciola-Rosen *et al.*, 1996; Duan *et al.*, 1996)
Caspase-8	MACHa1, FLICE, Mch5	Apoptosis – binds to FADD (which is bound to active Fas and via TRADD to TNF-R) (Boldin *et al.*, 1996; Fernandes-Alnemri *et al.*, Muzio *et al.*, 1996), cleaves PARP and possibly caspase-1 (Boldin *et al.*, 1996)
Caspase-9	Mch6, ICE-LAP6	Apoptosis – cleaves PARP (Casciola-Rosen *et al.*, 1996; Duan *et al.*, 1996), regulated by Akt and p21-Ras via direct phosphorylation (Cardone *et al.*, 1998)

(*continued overleaf*)

Table 8.3 (*continued*)

Protease	Alternative designations	Further information
Caspase-10	Mch4	Apoptosis – cleaves caspase-3 and caspase-7 (Fernandes-Alnemri *et al.*, 1996)
Caspase-11		Apoptosis – role in oligodendrocyte death in various encephalitides (Hisahara *et al.*, 2001), further functions uncertain
Caspase-12		Apoptosis
Caspase-13		Apoptosis
Caspase-14	MICE (Mini-ICE)	Apoptosis – may be restricted to embryonic tissues, processed *in vitro* by caspase-8, caspase-10 (Ahmad *et al.*, 1998; Hu *et al.*, 1998; Van de Crean *et al.*, 1998)

Table 8.4 Detectable targets of apoptotic enzyme effectors (caspases)

Substrate/neo-epitope	Function and further information
PARP (Poly {ADP-ribose} polymerase)	Nuclear enzyme (116 kDa) activated by DNA strand breaks, indirect role in DNA repair (Lindahl *et al.*, 1995; Pieper *et al.*, 1999; Sallmann *et al.*, 1997; Saraste and Pulkki, 2000). Cleaved by caspases (caspase-3). Signature fragment is 85 kDa (Duriez and Shah, 1997; Kaufmann *et al.*, 1993; Lazebnik *et al.*, 1994)
Bcl-XL	A potent inhibitor of apoptosis, cleaved by caspases (Clem *et al.*, 1998)
MDM2	Oncogenic protein, regulator of p53, cleaved by caspase-3 yielding a 60 kDa fragment. Raises doubt over specificity of caspase detection representing purely apoptosis (Pochampally *et al.*, 1998)
Actin	Microfilament forming protein with multiple localisations and functions, e.g. regulation of cell shape in the cortical cytoskeleton (Brown *et al.*, 1997; Kayalar *et al.*, 1996; Mashima *et al.*, 1997)
Fractin	A caspase-3 product, derived from Actin that may be involved in blebbing (Yang *et al.*, 1998)
CK (cytokeratin)18 (M30, CytoDeath)	Intermediate filament protein in keratinocytes, endometrium, hepatobiliary and other epithelial cells cleaved by caspase action (Caulin *et al.*, Leers *et al.*, 1999; Morsi *et al.*, 2000; Saraste and Pulkki, 2000). Has been used to quantitate apoptosis by immunocytochemical means (Leers *et al.*, 1999)

Table 8.4 (*continued*)

Substrate/neo-epitope	Function and further information
CK (cytokeratin)19	Intermediate filament protein in keratinocytes and other epithelial cells cleaved by caspases (Caulin *et al.*, 1997; Saraste and Pulkki, 2000)
Spectrin/fodrin	Actin cross-linking protein in cortical cytoskeleton cleaved by caspases (Martin *et al.*, 1995a; Vanags *et al.*, 1996; Wang *et al.*, 1998)
Beta-catenin	Intracellular attachment protein in cell-to-cell junction sites cleaved by caspases (Brancolini *et al.*, 1997)
Gelsolin	Microfilament fragmenting protein that complexes with actin and fibronectin (Kothakota *et al.*, 1997; Lind and Janmey, 1984). Cleaved by caspase-3 to an active product that may sever actin filaments in a Ca^{2+}-dependent manner (Kothakota *et al.*, 1997)
Gas2	Microfilament organising protein cleaved by caspases (Brancolini *et al.*, 1995)
PAK2	Protein kinase involved in regulation of cytoskeleton, cleaved by caspases (Rudel and Bokoch, 1997)
MEKK-1	Regulate cell survival and morphology at cell–matrix and cell–cell contacts sites, cleaved by caspases (Cardone *et al.*, 1997)
FAK	Regulate cell adhesion at cell–matrix and cell–cell contacts sites, cleaved by caspases (Wen *et al.*, 1997)
Rabaptin 5	Membrane protein that regulates intracellular vesicle traffic, cleaved by caspases (Cosulich *et al.*, 1997)
Lamin A and B	Intermediate filament that forms the nuclear lamina, responsible for nuclear envelope integrity, cleaved by caspases (Moir *et al.*, 1995; Orth *et al.*, 1996)
DFF (DNA fragmentation factor)	Cleaved by caspase-3 into a 40 kDa fragment (DFF-40, CAD, CPAN) that induces DNA fragmentation and a 45 kDa fragment (DFF-45, ICAD) that inhibits the former (Liu *et al.*, 1997; Sakahira *et al.*, 1998)
Seryl-tRNA synthetase	Cleaved by caspase-3 (Casas *et al.*, 2001)
NuMa	Mediator of nuclear chromatin–matrix protein interactions, cleaved by caspases (Casiano *et al.*, 1996)

Prostate Apoptosis Response 4 (PAR4) protein has a putative nuclear localisation signal and is expressed by cells entering apoptosis; originally identified in prostatic epithelium, it is not inducible by growth factor stimulation, oxidative stress, necrosis or growth arrest (Sells *et al.*, 1994). Second Mitochondria-derived Activator of Caspase (SMAC), often referred to as DIABLO (Direct IAP [Inhibitor of Apoptosis Protein] Binding Protein with Low PI) on release into the cytosol, promotes caspase activation in the cytochrome-c/Apaf-1/caspase-9 pathway by binding IAPs and removing their inhibitory activity (Deveraux *et al.*, 1998; Du *et al.*, 2000; Fesus *et al.*, 1987; Verhagen *et al.*, 2000).

Apoptosis Inducing Factor (AIF) is a mitochondrial protein that translocates to the nucleus during apoptosis, inducing DNA fragmentation, chromatin condensation and enhancing cytochrome-c and caspase-9 release from mitochondria. Bcl-2, a mitochondrial permeability regulator, inhibits AIF release (Green and Reed, 1998; Kluck *et al.*, 1997; Susin *et al.*, 1996, 1999; Zamzami *et al.*, 1996).

Antibodies against cytochrome C conformational changes have been suggested as possible identifiers of apoptosis (Jemmerson *et al.*, 1999; Varkey *et al.*, 1999). Apaf-1, which forms the cytosol-based 'apoptosome' complex (cytochrome c, ATP and procaspase-9) for caspase-9 activation, can also be detected by similar means.

Cellular Apoptosis Susceptibility (CAS) protein is highly expressed in actively dividing cells and decreases during growth arrest. The role it plays is uncertain but it appears important in toxin and TNF-mediated cell death (Brinkmann *et al.*, 1995).

For completeness, the following can be detected by similar methods and should be mentioned. Again, these are associated with apoptosis but should be interpreted in context as members of pathways rather than specific indicators of death.

The best characterised apoptotic pathway, that which revolves around Fas, has numerous markers that can be used for the assessment of apoptosis. These include Fas, members of the TNF super-family, proteins with similar homology and their ligands, e.g. the Fas ligand, the TNF-related apoptosis-inducing ligand (TRAIL) (Wiley *et al.*, 1995) and TWEAK (Chicheportiche *et al.*, 1997). These pathways rely on a protein component termed the death domain (DD). The death domain is a cytoplasmic domain of approximately 80 amino acids required for the signal transduction of pro-apoptotic stimuli such as those induced by Fas and TNF-R ligation (Tartaglia *et al.*, 1993). Various members of the death domain family and signalling intermediates can be detected, including: FADD/MORT1 (binds to Fas) (Chinnaiyan *et al.*, 1995), TRADD (binds to TNF-R1) (Smith *et al.*, 1994), RIP (serine/threonine kinase that binds TRADD involved in TNF mediated apoptosis) (Stanger *et al.*, 1995), RAIDD/CRADD (binds RIP), Siva (binds a TNF-R named CD27 and is crucial to CD27-mediated apoptosis) (Prasad *et al.*, 1997), RICK/RIP2/CARDIAK (involved in NFκB activation and TNF-R apoptosis) (Inohara *et al.*, 1998) and ICEBERG (inhibits the anti-apoptotic role of interleukin-1β in inflammation) (Humke *et al.*, 2000). FAF1 (Fas-associated protein factor 1) enhances the efficiency of FAS-mediated apoptosis, especially when over-expressed. In contrast to TRADD, FADD and RIP, FAF1 lacks a DD and cannot induce apoptosis independently of FAS activation (Chu *et al.*, 1995). The net result of this cascade is the activation of FLICE or caspase-8. A further related protein is FLASH (FLICE associated huge protein), an additional component of the Fas–FADD–caspase–8 (DISC) complex and required for Fas-mediated apoptosis. FLASH shares homology with the *C. elegans* CED-4 protein and the mammalian Apaf-1 protein (Imai *et al.*, 1999).

Inhibitors include FLIP/Casper/I-FLICE/CLARP/FLAME-1/MRIT (FLICE/caspase-8 inhibitory protein), which has two isoforms; the short version contains two DDs and has homology to FADD, the long version also contains a caspase-like domain without catalytic activity with homology to FLICE/caspase-8. Overall FLIP is inhibitory to apoptosis (Deiss *et al.*, 1995; Hulkko *et al.*, 2000; Irmler *et al.*, 1997; Levy-Stumpf *et al.*, 1997; Yang *et al.*, 1997). Toso (identified as a cell surface protein that is expressed in lymphoid cells) blocks apoptosis mediated by members of the TNF family, including Fas, and has been shown to inhibit TCR-induced T-cell self-killing (Hitoshi *et al.*, 1998). SODD (silencer of death domains) binds to the DD of TNF-R1, thus inhibiting the recruitment of TRADD or TRAF2 to TNF-R1 resulting in the inhibition of TNF-mediated apoptosis (Jiang *et al.*, 1999).

Also important are the Death Associated Proteins (DAP1–DAP5); these do not contain DDs but have varying roles, including TNF-R1 initiation of apoptosis (DAP1) (Shu *et al.*, 1997), glucocorticoid receptor interactions (DAP3) (Hu *et al.*, 1997) and general suppression of translation (DAP5) (Han *et al.*, 1997). Further proteins associated with the Fas-mediated pathway are also identifiable and include Daxx, which enhances apoptosis by activating the Jun N-terminal Kinase (JNK), one of the end stages of the Fas pathway. Daxx is cooperative with, but independent of, the Fas–FADD–FLICE/caspase-8 pathway and is inhibited by Bcl-2 (Thome *et al.*, 1997).

The members of the Bcl-2 family are often seen as *the* regulators of the cellular life and death transition. An in-depth review of their function and interaction with the wealth of cellular protein cascades has been provided by Cory and Adams (2002). This family includes pro-apoptotic factors [Bax, Bcl-x_S, Bad, Bak, NBK (Bik), BID, Hrk. Bok, Bim, Noxa and Diva] and anti-apoptotic factors [Bcl-2. Bcl-x_L, Bcl-xγ, Bcl-xβ, Bcl-w, Mcl-1, Bag-1, A1 (Bfl-1), BAR and Bl-1 (TEGT)]. Bcl-10, also designated CIPER, c-CARMEN and mE10, is pro-apoptotic, was first identified in MALT B-cell lymphomas and has homology to equine herpesvirus-2 E10 gene. Bcl-10 is associated with a caspase recruitment domain (CARD), and is essential for NFκB activation in a NIK-dependent pathway and caspase-9 activation. Over-expression of CARD induces apoptosis (Koseki *et al.*, 1999; Willis *et al.*, 1999; Yan *et al.*, 1999).

RNA binding proteins TIA-1 and TIAR are purported to have a role in apoptosis. TIA-1 is located within cytotoxic granules of lymphocytes and TIAR is located within the nucleus, both are phosphorylated by Fas-Activated Serine/Threonine (FAST) kinase. Their modification immediately precedes DNA fragmentation (Anderson, 1995; Dember *et al.*, 1996; Taupin *et al.*, 1995; Tian *et al.*, 1995).

The p53 protein and its relatives, including p63 and p73, the GADD family, other tumour suppressor proteins (e.g. Rb, DICE1, Fhit, TID-1 and Beclin1) and other proteins involved in the induction or inhibition of apoptosis (e.g. PEA-15, PD-1, IEX-1 and the anti-apoptotic Nip protein family) are all integral to apoptosis, but detection, as with most of the above, depends upon the interpretation of surrounding events.

Combinations of apoptotic markers

The specificity of these visual methods is constantly being brought into doubt with their increasing use; this is why 'good' studies often combine the advanced techniques with the standard morphology, as so often described within the literature. Other researchers have used novel combinations of techniques such as purely immunofluorescence methods; ISEL for the detection of DNA fragmentation combined with a nuclear stain (cyanine dye, e.g. YOYO-1) to visualise the apoptotic chromatin condensations. This combination is claimed to give unequivocal identification of an apoptotic nucleus (Tatton *et al.*, 2001).

Indirect methods

Flow cytometry

The principal indirect method used is flow cytometry, which can measure the mechanism and extent of cell death by measuring the light-scattering properties together with cellular ability to take up and retain dyes or labels. This allows discrimination between cells with differing DNA contents; differing cell kinetics (via 5′-bromo-2′-deoxyuridine incorporation); those undergoing mitosis [detection of Histone (H3) phosphorylation]; those that are proliferating from those that are quiescent (G1/G0 vs. S/M phase); those in differing parts of the cell cycle (by specific cyclin staining vs. DNA content); and those undergoing apoptosis. Flow cytometric assessment can discriminate between apoptosis and necrosis by: (a) utilising morphology including the effect of nuclear condensation and loss of DNA fragments rendering apoptotic cells smaller with less DNA than a G_0/G_1 diploid cell; (b) the presence of phosphatidylserine on the membrane surface (e.g. Annexin V staining); (c) the collapse of mitochondrial cell membrane potential; (d) DNA fragmentation; and (e) caspase activation (Afanasyev *et al.*, 1993; Darzynkiewicz *et al.*, 2001; Koester *et al.*, 1997; Nicoletti *et al.*, 1991; Telford *et al.*, 1994). Obviously, further combinations of staining are possible and this indirect technique is often combined with one or more visual methods described above. Various overviews have been published recently on the use of flow cytometry in cell population dynamics (Darzynkiewicz *et al.*, 2001; Deckers *et al.*, 1993; Dive *et al.*, 1992; Ormerod *et al.*, 1993).

The problems of flow cytometry include the loss of histological context. Furthermore, particles that contain fractional DNA content are assumed to be apoptotic. However, these may be cellular fragments from artefact or necrotic lysis. This overestimates the apoptotic index [46]. Adding compensation filters to only measure certain fragment sizes can underestimate the index (Darzynkiewicz *et al.*, 1998). Not only these, but phagocytes that have ingested apoptotic bodies now contain altered DNA, abnormal plasma membranes and other characteristic apoptotic features. Thus, if flow cytometry is combined with a method that utilises these features, then phagocytic cells will be falsely labelled as apoptotic. The apoptotic index generated by flow cytometry is only the percentage of cells within a population that are apoptotic and not a measure of the rate of apoptosis or a quantitative measure of cell death, it is not cumulative (Darzynkiewicz *et al.*, 1998). Beyond this, the combinations of flow cytometry with other labelling methods inherit the difficulties surrounding those labelling methods, in addition to those of flow cytometry.

Biochemical and physiological methods

DNA cleavage fragments and DNA associated proteins Wyllie (1980) described the biochemical hallmark of apoptosis, namely the cleavage of genomic DNA into multiples of 180–200 base pair oligonucleotides. Various methods are available for isolating these oligonucleotides, and subsequently displaying the results on an ethidium bromide gel or autoradiographic detection after ^{32}P-labelling of the DNA fragments. The appearance seen has become known as DNA laddering. The identification of such is proof of cation-dependent endonuclease activity, which is energy requiring and specific to apoptosis and therefore an excellent method of identifying apoptosis *per se*. The process is semi-quantitative in that relative amounts of low molecular weight DNA may be determined using computerised means (Loo and Rillema, 1998). However, large numbers are cells are required and specific populations of cells cannot be examined (Darzynkiewicz *et al.*, 1998; Loo and Rillema, 1998). Although PCR techniques can overcome some sensitivity problems, there is still loss of the histological context. Furthermore, DNA laddering is a late feature and DNA fragmentation is not associated with all forms of apoptosis (Boe *et al.*, 1991; Cohen *et al.*, 1992; Collins *et al.*, 1992; Darzynkiewicz *et al.*, 1998; Oberhammer *et al.*, 1991, 1992; Zakeri *et al.*, 1993).

Enzyme-linked immunoassays have been commercially produced using monoclonal antibodies against DNA and histones. These enable measurement of apoptosis by quantitating histone-associated DNA fragments released from cultured cells; these systems are quick but have the same disadvantages as the DNA laddering method (Loo and Rillema, 1998).

Cytoplasmic and membrane modifications. Further *in vitro* tests can be performed to establish cytoplasmic changes; these include assessment of caspase activity from lysed cell populations. This is enabled by various commercially available fluorogenic and chromogenic substrates; most are non-permeable and require tissue homogenisation (Gurtu *et al.*, 1997), therefore direct single-cell assessment is not possible, unlike fixed tissue immunohistochemistry. The neoepitopes generated by caspase action and detected by immunohistochemistry (see the section 'Immunohistochemistry/Immunocytochemistry' above) may give an index of activity (Leers *et al.*, 1999; Sallmann *et al.*, 1997; Yang *et al.*, 1998).

Calcium flux measurements by indicators (e.g. fura-2) can be used, but these are neither specific nor indeed sensitive for apoptosis, as apoptosis without Ca^{2+} rise has been observed (Berridge *et al.*, 1999; McConkey, 1996; Whyte *et al.*, 1993).

Mitochondrial dysfunction can be shown by collapse of membrane potential (the permeability transition) as demonstrated by the diminished ability of fluorochromes to build up within mitochondria (Metivier *et al.*, 1998). This process is shown in Figure 8.3. In addition, the escape of procaspase 2, 3, 9, cytochrome C and AIF can be detected in the cytosol and assayed. Cytochemical detection of cytochrome C within the cytosol is possible, although the amount required for apoptotic death is not yet quantified (Willingham, 1999). These 'mitochondrial' methods only have a low sensitivity as pathways may exist that circumvent these processes (Bossy-Wetzel *et al.*, 1998; Kluck *et al.*, 1997; Krajewski *et al.*, 1999; Mancini *et al.*, 1998; Susin *et al.*, 1999a).

Membrane changes as detailed in the section 'Visual Methods' above, can also be detected indirectly by measurement of membrane extracts by two-dimensional thin layer chromatography to detect phosphatidylserine in combination with fluorescarmine (Fadok *et al.*, 1992). Membrane structural changes can be identified using merocyanine 540 (Fadok *et al.*, 1992). These methods are only used for *in vitro* analyses.

Others. An old method of detection of cell death relied on lactate dehydrogenase release on loss of plasma membrane integrity, which is common both to necrosis and late apoptotic processes. It is rapid, simple and quantitative (Loo and Rillema, 1998; Wroblewski and LaDue, 1955).

The MTT (3-(4,5-dimethylthiazol-2-yl)2,5-diphenyltetrazolium bromide)/XTT assay is an indirect measure of cell growth/cell death that relies on the conversion of MTT/XTT into a coloured formazan by mitochondria, allowing spectrophotometric measurement (Loo and Rillema, 1998). Dead or dying cells cannot achieve this conversion. Like measurement of lactate dehydrogenase, the assay is not specific and mitochondria must remain functional until late in apoptosis (Wyllie, 1980). Artefacts arise in proliferating cultures as population levels are not static, so percentages of cell death may become inaccurate (Loo and Rillema, 1998).

Extension to in vivo imaging

So far all the techniques of apoptotic detection have focused on identification of snapshots of the *in vivo* or *in vitro* circumstance. But what about direct real-time

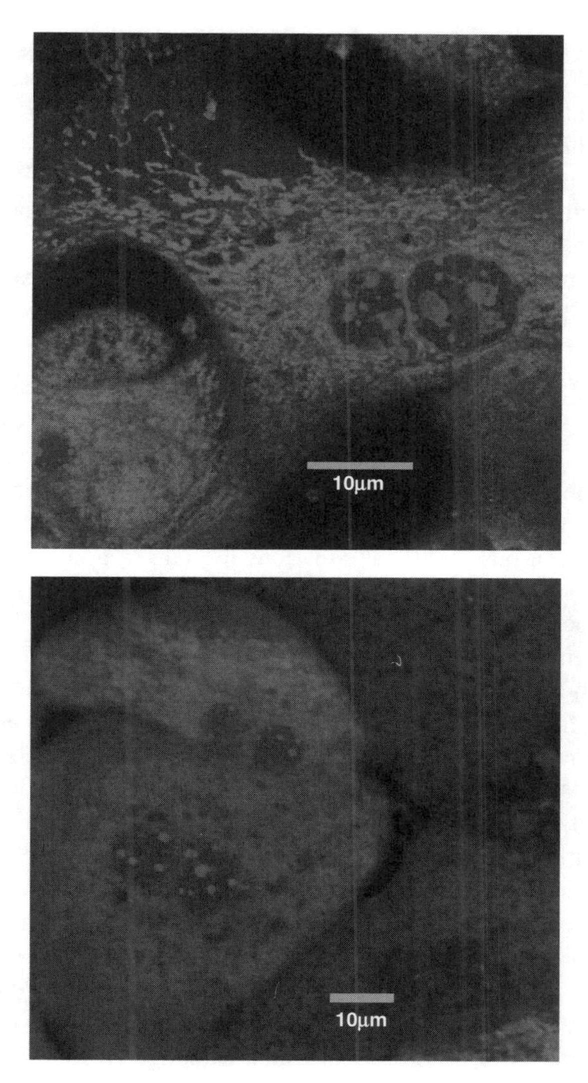

Figure 8.3 Confocal microscope images of cultured murine hepatocytes, nuclei are stained with ToPro 3 (a blue dye) and mitochondria stained with MitoTracker Red (a red dye). The uppermost image (A) shows the punctate pattern of healthy, polarised mitochondria. The lowermost (B) image displays cells harbouring mitochondria with a depolarised membrane. Staining shifts from concentrated to diffuse with the loss of membrane potential. (Courtesy of Dr B Tura). A colour version of this figure appears in the colour plate section

imaging of apoptosis? Recently, various research groups have begun exploring the possibility of real-time imaging of pathological processes, a natural extension of which is the identification of apoptosis. Zhao *et al.* (2001), utilising the C_2 domain of Synaptotagmin I labelled with superparamagnetic iron oxide nano-particles, were able to detect etopiside-induced apoptosis in the living examples of the murine lymphoma (EL4) tumour model via magnetic resonance imaging (MRI). In a similar study, Narula *et al.* (2001) used scintigraphic detection of Technetium-99m-labelled annexin V to identify phosphatidyl–serine expressed on the surface of apoptotic cardiomyocytes as a follow-up tool to identify rejection in human recipients of cardiac allografts. Problems with these innovations include accurate detection in the face of rapid clearance vs. overwhelming apoptosis and the differing phagocytic clearance potentials of tissues that make both these *in vivo* and *in situ* cytochemical methods extremely difficult (Willingham, 1999).

Counting: the significance of numbers

In many *in vitro* settings the frequency of apoptosis is high, so that counting individual cells is easy. Furthermore, in this situation it is relatively easy to demonstrate other biochemical properties that can be seen in apoptotic cells, such as nucleosomal ladders, caspase activation and cleavage of substrates such as PARP. However, in many clinical situations the rate of apoptosis may be much lower and perhaps be interpreted as evidence that apoptosis is not occurring (because no biochemical assay is positive) or that the amount of apoptosis is so small as to be insignificant. A powerful illustration of how this can lead to an underestimation of the importance of the phenomenon is provided from the field of immunology. Mice injected with a single dose of anti-CD4 antibody have a severe depletion of CD4 positive T lymphocytes that occurs within 24 h. In fact, after a single dose the number of cells is reduced by 50% at 48 h. Death appears to be the result of apoptosis, but despite this the maximal rate of apoptosis counted using morphological criteria on tissue sections is 1.3% (Howie *et al.*, 1994). The explanation for this apparent discrepancy is the extremely rapid time course of apoptosis and phagocytosis *in vivo*. Conversely, in a situation where TUNEL appears to suggest a rate of apoptosis exceeding 10% or 20% in clinical material one is wise to exercise scepticism. Is apoptosis likely to be that frequent in a disease where the evolution of injury and destruction of cells occurs over a period measured in months or years, such as congestive cardiac failure?

8.3 Cell senescence

What is Cell Senescence?

The term derives from experiments performed in the 1960s by Hayflick and others; they demonstrated that normal, non-transformed cells have only a finite capacity for cell division until they enter a viable but 'senescent' state (Campisi, 1996; Hayflick, 1965; Stanulis-Praeger, 1987). These cells will not proliferate further despite optimal conditions and consequently this is distinct from the 'quiescent' state, where cells can re-enter the cell cycle at any time. The number of divisions a cell can perform prior to senescence is referred to as the

Hayflick number (or limit) (Coates, 2002). Some researchers restrict this 'proof' of senescence to the term 'replicative senescence' (Campisi, 2001).

Mechanisms of cell senescence

This is a hotly debated topic that essentially revolves around several observations; the first is replicative cell senescence first proposed by Hayflick (1965) that has subsequently been attributed to the telomerase/telomere interaction described in many publications (Campisi, 2001; Chiu and Harvey, 1997; Lanza *et al.*, 2000; Sheils *et al.*, 1999; Whikehart *et al.*, 2000). Telomeres are DNA repeats [TTAGGG in vertebrates (Campisi, 2001)] present as end-caps on linear chromosomes. During the genomic replication, which occurs in preparation for cell division, these end-caps are progressively eroded owing to the biochemistry of DNA replication. Telomerase expressed in germ cells, and tumour cells enables replenishment of the telomere complement; this does not occur in most somatic cells. Reintroduction of telomerase in some somatic cells prevents senescence (Bodnar *et al.*, 1998; Lansdorp, 2000; Vaziri and Benchimol, 1998). Therefore, this provides the biological count for the (self) assessment of the replicative history by the cell and is the mechanism behind replicative senescence. In mice the telomere system fails to explain senescence as telomerase is ubiquitously expressed, but tissue culture of wild-type murine cells mimic the human fibroblast senescence after numerous divisions. The senescence signalling responsible for this appears to converge on p53 (Kamijo *et al.*, 1997; Serrano *et al.*, 1996). Further evidence that the telomere/telomerase theory does *not* explain all cell senescence is reported in at least five further mechanisms:

1. Certain types of DNA damage that cause a senescent arrest, including but not exclusively oxidative damage (Chen *et al.*, 1995; Di Leonardo *et al.*, 1994; Robles and Adami, 1998).
2. Expression of certain oncogenes (e.g. activated RAS and RAF), tumour suppressor genes or supraphysiological mitogenic signals (e.g. with E2F1) induce senescence and not transformation or apoptosis (Dimri *et al.*, 2000; Serrano *et al.*, 1997; Zhu *et al.*, 1998).
3. The level of DNA methylation, both 5-methylcysteine and CpG island methylation, is purported to decline with age (Singhal *et al.*, 1987; Wilson and Jones, 1983), this is reversed on transformation of cells (Slack *et al.*, 1999).
4. Elevated levels of extra- and intracellular ceramide, a product of sphingomyelin hydrolysis, have been observed to force cell senescence (Venable *et al.*, 1999).
5. The manipulation of chromatin density by pharmacological or genetic means induces senescence in human and murine cell lines (Jacobs *et al.*, 1999; Ogryzyko *et al.*, 1996).

Consequently, telomere/telomerase theory that explains replicative senescence is but a small part of the picture and the further observations provide a more complex view of cellular senescence.

Why is cell senescence important?

The importance of cell senescence falls into two main areas: ageing and neoplasia. The ageing theory assumes that the renewal reserve of any tissue is eroded as more constituent cells become senescent over time. This modifies the tissue phenotype and is purported to be responsible for disease processes seen in ageing (Campisi, 2001; Coates, 2002). The relation of telomere/telomerase theory with ageing (and neoplasia) on an organismal level has been a difficult one to prove and only reliably pertains to individual cell senescence (Campisi, 2001). The 'replicative mosaicism' expected in an organism has been proposed to explain the deficiency in the telomere theory of ageing by suggesting that exhausted proliferative potential in a tissue sub-population may be sufficient to induce age-dependent disease (Mikhelson, 2001). Problems include cells such as mature neurons and muscle that are post-mitotic without evidence of continual division; pure replicative senescence cannot apply to these tissues but they do show ageing-related changes in phenotype/function. A further theory of ageing revolving around free radical/oxidative damage, first proposed by Harman in 1956 (Terman, 2001), goes further to address this phenomenon. We expect biological systems to continuously renew structures, so how does oxidative damage accumulate over the cellular lifetime? Terman (2001) and others propose interference by the build-up of oxidatively damaged, indigestible material (e.g. lipofuscin) and that dilution of this material during cellular reproduction rejuvenates the daughter cells and inhibition of oxidation slows the accumulation. Once mature, a tissue with little turnover builds up these 'toxins', eventually resulting in a senescent population with altered phenotype leading to disease. Others suggest that the accumulation of somatic mutations both in the nucleus and mitochondria, the ability to synthesise PARP and to establish effective DNA repair are crucial to ageing (von Zglinicki *et al.*, 2001). Regardless of the exact mechanism, the rate of accumulation of injury varies depending on the tissues' production of, tolerance of, or ability to compensate for, free radicals. This form of cellular injury necessarily overlaps with the second area in which cell senescence has a clearer and experimentally affirmed role, the area of neoplasia.

There is evidence that a strong selection occurs during early tumorigenesis that enables replication beyond the natural Hayflick limit. This has been observed in tumour cell lines, cells transformed by viral and cellular oncogenes and those cells harbouring mutations in p53 or pRb. Indeed, mouse models with loss of tumour suppressor genes have cells with demonstrably extended Hayflick limits and a propensity toward neoplasia (Campisi, 1996, 1999, 2000, 2001; Reddel,

2000; Sager, 1991; Williams, 1957; Yeager *et al.*, 1998). Therefore, in one view, cell senescence appears to protect against neoplasia. This provides a potential pathway for cancer treatments by reinstating the Hayflick limit (Coates, 2002). More significantly, the connotation arising from the role of cell senescence in neoplasia is supportive of the 'antagonistic pleiotrophy' theory (Campise, 2001; Williams, 1957), which states that genes with good effects in early life (e.g. improved reproductive fitness) would be favoured by selection even if that advantage is small and long-term effects deleterious to the organism. So ageing may be the price paid by multicellular organisms for relative freedom from cancer in youth to be followed by a treacherous decline in function in later life with the ironic twist of an increasing risk of neoplasia, survival permitting. This leads us to the relation with cell death.

The relation between cell death and cell senescence

This is an extremely complex relation. Some authors have reported that replicative senescence induces a complex phenotype in cells that is more than just cessation of division, as there appears to be an increased resistance to apoptosis in some cell types and changes in differentiated function (Campisi, 2001). Apoptotic resistance is suggested to increase senescent cell populations in tissues, enhancing the aged phenotype; the disturbed secretory function of these senescent cells drives the division of local non-senescent pre-neoplastic cells (those that may harbour hidden somatic mutation), which explains the late increase in tumorigenesis observed in aged organisms (Campisi, 1997). Others suggest that the dysregulation of apoptosis contributes directly to the ageing process by enhancing the susceptibility of intact cells to apoptosis once they become injured. However, the resistance only occurs within certain damaged and possibly pre-neoplastic cells (Higami and Shimokawa, 2000), thus advocating the traditional view of apoptosis as a protection against tumorigenesis. In support of this latter concept, studies of mitochondrial changes in mice have demonstrated increased mitochondrial DNA mutation, less efficient oxidative phosphorylation and increased susceptibility to Ca^{2+}-induced permeability transition with increasing age. Therefore, older cells undergo apoptosis more readily on exposure to oxidative stress (Mather and Rottenberg, 2000). The perceived overlap of senescence and apoptotic regulation extends beyond the theoretical into the many described common pathways that can lead to neoplasia, senescence or death. The most striking experimental example of such overlap are knockout mice lacking p53 and Rb; their cells exhibit impaired senescence typified by marked telomere erosion by unrestricted division that can lead to massive cell death, termed 'crisis', that also coexists with increased chromosomal instability favouring recombination and neoplasia (Counter *et al.*, 1992; Hara *et al.*, 1991; Shay *et al.*, 1991). The relation between such pathways is well covered in a review by Bringold and Serrano (2000); they focus specifically on the overlap of the $p16^{INK4a}$/Rb, $p19^{ARF}$/p53/$p21^{Cip1}$ and PTEN/$p27^{Kip1}$

senescence-inducing pathways that link the accumulation of cell doublings to cellular senescence, tumour suppression and apoptosis. This overlap is unsurprising given the guardianship attributed to these pathways already in cell dynamics.

Markers of cell senescence

Taking the areas of cell senescence that we have covered, what markers can the researcher use to reliably identify senescent cells? The expression of the active subunit of telomerase (hTERT) has been used as a marker of telomerase activity within cells and to identify tumour cell populations (Keith *et al.*, 2001). Yet this is not specific and in many cases tumour cells do not express telomerase or, if they do, there is no evidence to separate a possible late neoplastic phenomenon to overcome impending senescence from constitutive expression in the cell of origin (stem cells have been shown to constitutively express telomerase) (Coates, 2002; Keith *et al.*, 2001). A more direct measure is to detect the telomere length; this can be achieved by *in situ* hybridisation using a complex methodology that is highly dependent on interpretation (Coates, 2002). Linking from that above, specific protein expressions associated with the Rb, p53 and PTEN pathways may be used but are non-specific due to reasons mentioned previously.

Dimri *et al.* (1995) identified an enzyme with an optimal pH of 6.0 that was said to be specific for cell senescence in culture and *in vivo*; it was named senescence-associated β-galactosidase (SA-β-Gal). This was detected by enzyme histochemistry specifically in senescent human fibroblast culture and in fibroblasts taken from skin of old people; no expression was identified in non-senescent controls. However, recently, Going *et al.* (2002) whilst examining the expression of SA-β-Gal in the gastrointestinal tract found strong expression in proliferating dysplastic epithelia. Indeed, there is evidence that SA-β-Gal is not a unique enzyme and nothing more than residual lysosomal acid β-galactosidase measured at a suboptimal pH (the optimal pH is 4.0) present in cells with copious enzyme expression, such as those containing increased lysosomal compartments such as lipofuscin replete senescent cells (Coates, 2002; Katz *et al.*, 1984; Kurz *et al.*, 2000; Sohal and Brunk, 1989). Furthermore, SA-β-Gal has been induced *in vitro* during reversible quiescence and differentiation (Krishna *et al.*, 1999; Mouton and Venable, 2000; Severino *et al.*, 2000; Yegorov *et al.*, 1998). Consequently, this is not a specific marker for cell senescence.

Unfortunately, unlike apoptosis, there are no reliable markers or even specific morphological changes that can assist in the identification of cell senescence. The current inability to detect and quantify these senescent populations and further elucidate their role in tissue dynamics is frustrating. Nevertheless, whilst utilising new senescence markers, we must bear in mind the lessons learned from work in the field of cell death, as problems such as the significance of numbers may be equally applicable to cell senescence.

8.4 Summary

Apoptosis has become recognised as a very important basic biological process, regulating tissue size and cell number. Increasingly, the overlap with cell senescence and their mutual coexistence is becoming more interrelated both in health and disease. In general terms there is a stereotypical set of morphological changes that are part of the apoptotic mechanism but no such correlate exists for senescence. Both processes may have associated and detectable biochemical changes, with apoptotic markers being by far the most developed. It is important to realise that within different tissues experiencing differing types of injury a selection occurs, compounded by the heterogeneity of cellular age. This influences the expected response of a tissue and the processes which the researcher may detect within the residual cell population. As a consequence, one must precisely define, in any given setting, what is being investigated and to appreciate that there is no 'gold standard' that defines apoptosis in every situation in which it might occur. Recall that apoptosis is a quiet killer; a low apparent frequency in a tissue section may hide the fact that dramatic cell loss and tissue remodelling is occurring. Cell senescence appears quieter still with an insidious nature, no reliable method of assessment and little understanding of its true role in disease processes. The challenge of understanding apoptosis and more so cell senescence is still in its infancy.

8.5 References

Aeschlimann D and Thomazy V (2000) Protein crosslinking in assembly and remodelling of extracellular matrices: the role of transglutaminase. *Connect. Tissue Res.* **41**: 1–27.

Afanasyev VN, Korol AB, Matylevich NP, Pechatnikov VA and Umansky SR (1993) The use of flow cytometry for the investigation of cell death. *Cytometry* **14**: 603–9.

Ahmad M, Srinivasula SM, Hegde R, Mukattash R, Fernandes-Alnemri T and Alnemri ES (1998) Identification and characterization of murine caspase-14, a new member of the caspase family. *Cancer Res.* **58**: 5201–5.

Aigner T (2002) Apoptosis, necrosis or whatever: how to find out what really happens? *J. Pathol.* **198**: 1–4.

Anderson P (1995) TIA-1: structural and functional studies on a new class of cytologic effector molecule. *Curr. Topics Microbiol. Immunol.* **198**: 131–43.

Andree HA, Reutelingsperger CPM, Hauptmann R, Hemker HC, Hermens WT and Willems GM (1990) Binding of vascular anticoagulant alpha (VAC alpha) to planar phospholipid bilayers. *J. Biol. Chem.* **265**: 4923–8.

Ansari B, Coates PJ, Greenstein BD and Hall PA (1993) *In situ* end-labelling detects DNA strand breaks in apoptosis and other physiological and pathological states. *J. Pathol.* **170**: 1–8.

Aschoff AP, Gunther E and Jirikowski GF (2000) Tissue transglutaminase in the small intestine of the mouse as a marker for apoptotic cells. Colocalization with DNA fragmentation. *Histochem. Cell. Biol.* **113**: 313–7.

Ashkenazi A and Dixit VM (1998) Death receptors: signalling and modulation. *Science* **281**: 1312–16.

Bennet MR, Gibson DF, Schwartz SM and Tait JF (1995) Binding and phagocytosis of apoptotic vascular smooth muscle is mediated in part by exposure of phosphatidylserine. *Circul. Res.* **77**: 1136–42.

Bernassola F, Rossi A and Melino G (1999) Regulation of transglutaminases by nitric oxide. *Ann. NY Acad. Sci.* **887**: 83–91.

Berridge M, Lipp P and Bootman M (1999) Calcium signalling. *Curr. Biol.* **9**: R157–59.

Bodnar AG, Ouellette M, Frolkis M, Holt SE, Chiu CP, Morin GB, Harley CB, Shay JW, Lichtsteiner S and Wright WE (1998) Extension of life span by introduction of telomerase into normal human cells. *Science* **279**: 349–52.

Boe R, Gjersten BT, Vintermyr OK, Houge G, Lanotte M and Doskeland SO (1991) The protein phosphatase inhibitor okadaic acid induces morphological changes of apoptosis in mammalian cells. *Exp. Cell. Res.* **195**: 237–46.

Boersma AWM, Nooter K, Oostrum RG and Stoter G (1996) Quantification of apoptotic cells with fluorescein isothiocyanate-labelled annexin V in chinese hamster ovary cell cultures treated with cisplatin. *Cytometry* **24**: 123–130.

Boldin MP, Goncharov TM, Goltsev YV and Wallach D (1996) Involvement of MACH, a novel MORT1/FADD-interacting protease, in Fas/APO-1 and TNF receptor-induced cell death. *Cell* **85**: 803–15.

Bossy-Wetzel E, Newmeyer DD and Green DR (1998) Mitochondrial cytochrome *c* release in apoptosis occurs upstream of DEVD-specific activation and independently of mitochondrial transmembrane depolarization. *EMBO J.* **17**: 37–49.

Brancolini C, Benedetti M and Schneider C (1995) Microfilament reorganization during apoptosis: the role of Gas2, a possible substrate for ICE-like proteases. *EMBO J.* **14**: 5179–90.

Brancolini C, Lazarevic D, Rodriquez J and Schneider C (1997) Dismantling cell–cell contacts during apoptosis is coupled to caspase-dependent proteolytic cleavage of beta-catenin. *J. Cell. Biol.* **139**: 759–71.

Bringold F and Serrano M (2000) Tumor suppressors and oncogenes in cellular senescence. *Exp. Gerontol.* **35**: 317–29.

Brinkmann U, Brinkmann E, Gallo M and Pastan I (1995) Cloning and characterisation of a cellular apoptosis susceptibility gene, the human homologue to the yeast chromosome segregation gene CSE1. *Proc. Natl. Acad. Sci. USA* **92**: 10427–31.

Brown SB, Bailey K and Savill J (1997) Actin is cleaved during constitutive apoptosis. *Biochem. J.* **323**: 233–7.

Budihardjo I, Oliver H, Lutter M, Luo X and Wang X (1999) Biochemical pathways of caspase activation during apoptosis. *Annu. Rev. Cell. Dev. Biol.* **15**: 269–90.

Bursch W, Paffe S, Putz B, Barthel G and Schulte-Hermann R (1990) Determination of the length of the histological stages of apoptosis in normal liver and in altered hepatic foci of rats. *Carcinogenesis* **11**: 847–53.

Butt A, Harvey N, Parasivam G and Kumar S (1998) Dimerization and Autoprocessing of the Nedd2 (caspase-2) Precursor Requires both the Prodomain and the Carboxyl-Terminal Regions. *J. Biol. Chem.* **12**: 6763–8.

Campisi J (1996) Replicative senescence: an old lives tale. *Cell* **84**: 497–500.

Campisi J (1997) Aging and cancer: the double-edged sword of replicative senescence. *J. Am. Geriatric Soc.* **45**: 1–6.

Campisi J (1999) Replicative senescence and immortalization. In: Stein G, Baserga R, Giordano A, Denhaldt D (eds). *The molecular basis of cell cycle and growth control.* Wiley, New York: 348–73.

Campisi J (2000) Cancer, aging and cellular senescence. *In Vivo* **14**: 183–8.

Campisi J (2001) From cells to organisms: can we learn about ageing from cells in culture? *Exp. Gerontol.* **36**: 607–18.

Cardone MH, Roy N, Stennicke HR, Salvesen GS, Franke TF, Stanbridge E, Frisch S and Reed JC (1998) Regulation of cell death protease caspase-9 by phosphorylation. *Science* **282**: 1318–21.

Cardone MH, Salvesen GS, Widmann C, Johnson G and Frisch SM (1997) The regulation of anoikis: MEKK-1 activation requires cleavage by caspases. *Cell* **90**: 315–23.
Casas C, Ribera J and Esquerda JE (2001) Antibodies against c-Jun N-terminal peptide cross-react with neo-epitopes emerging after caspase-mediated proteolysis during apoptosis. *J. Neurochem.* **77**: 904–15.
Casciola-Rosen L, Nicholson DW, Chong T, Rowan KR, Thornberry NA, Miller DK and Rosen A (1996) Apopain/CPP32 cleaves proteins that are essential for cellular repair: a fundamental principle of apoptotic death. *J. Exp. Med.* **183**: 1957–64.
Casiano CA, Martin SJ, Green DR and Tan EM (1996) Selective cleavage of nuclear autoantigens during CD95 (Fas/APO-1)-mediated T cell apoptosis. *J. Exp. Med.* **184**: 765–70.
Caulin C, Salvesen GS and Oshima RG (1997) Caspase cleavage of keratin 18 and reorganisation of intermediate filaments during epithelial cell apoptosis. *J. Cell. Biol.* **138**: 1379–94.
Chang HY and Yang X (2000) Proteases for Cell Suicide: Functions and Regulation of Caspases. *Microbiol. Mol. Biol. Rev.* **4**: 821–46.
Chen Q, Fischer A, Reagan JD, Yan LJ and Ames BN (1995) Oxidative DNA damage and senescence of human diploid fibroblast cells. *Proc. Natl. Acad. Sci. USA* **92**: 4337–41.
Chicheportiche Y, Bourdon PR, Xu H, Hsu YM, Scott H, Hession C, Garcia I and Browning JL (1997) TWEAK, a new secreted ligand in the tumor necrosis factor family that weakly induces apoptosis. *J. Biol. Chem.* **272**: 32401–10.
Chinnaiyan AM, O'Rourke K, Tewari M and Dixit VM (1995) FADD, a novel death domain-containing protein, interacts with the death domain of Fas and initiates apoptosis. *Cell* **81**: 505–12.
Chiu CP and Harley CB (1997) Replicative senescence and cell immortality: the role of telomeres and telomerase. *Proc. Soc. Exp. Biol. Med.* **214**: 99–106.
Chu K, Niu XH and Williams LT (1995) A Fas-associated protein factor, FAF1, potentiates Fas-mediated apoptosis. *Proc. Natl. Acad. Sci. USA* **92**: 11894–8.
Clem RJ, Cheng EH, Karp CL, Kirsch DG, Ueno K, Takahashi A, Kastan MB, Griffin DG, Earnshaw WC, Veliuona MA and Hardwick JM (1998) Modulation of cell death by Bcl-XL through caspase interaction. *Proc. Natl. Acad. Sci. USA* **95**: 554–9.
Coates PJ (2002) Markers of senescence? *J. Pathol.* **196**: 371–3.
Coffey RN, Watson RW and Fitzpatrick JM (2001) Signaling for the Caspases: Their role in Prostate Cell Apoptosis. *J. Urol.* **1**: 5–14.
Cohen GM, Sun X-M, Snowden RT, Dinsdale D and Skilleter DN (1992) Key morphological features of apoptosis may occur in the absence of internucleosomal DNA fragmentation. *Biochem. J.* **286**: 331–4.
Collins RJ, Harmon BV, Gobe GC and Kerr JFR (1992) Internucleosomal DNA cleavage should not be the only criterion for identifying apoptosis. *Int. J. Radiat. Biol.* **61**: 451–3.
Columbano A (1995) Cell death: current difficulties in discriminating apoptosis from necrosis in the context of pathological processes *in vivo*. *J. Cell. Biochem.* **58**: 181–90.
Cory S and Adams JM (2002) The Bcl-2 family: regulators of the cellular life-or-death switch. *Nat. Rev.* **2**: 647–56.
Cosulich SC, Horiuchi H, Zerial M, Clarke PR and Woodman PG (1997) Cleavage of rabaptin-5 blocks endosome fusion during apoptosis. *EMBO J.* **16**: 6182–92.
Counter CM, Avillon AA, LeFeuvre CE, Stewart NG, Greider CW, Harley CB and Bacchetti S (1992) Telomere shortening associated with chromosome instability is arrested in immortal cells which express telomerase activity. *EMBO J.* **11**: 1921–29.
Crompton M, Virji S, Doyle V, Johnson N and Ward JM (1999) The mitochondrial permeability transition pore. *Biochem. Soc. Symp.* **66**: 167–79.
Darzynkiewicz Z, Bedner E and Smolewski P (2001) Flow cytometry in analysis of cell cycle and apoptosis. *Sem. Hematol.* **38**: 179–93.
Darzynkiewicz Z, Bedner E, Traganos F and Murakami T (1998) Critical aspects in the analysis of apoptosis and necrosis. *Hum. Cell* **11**: 3–12.

Davidson FD, Groves M and Scaravilli F (1995) The effects of formalin fixation on the detection of apoptosis in human brain by *in situ* end-labeling of DNA. *Histochem. J.* **27**: 983–8.

Deckers CLP, Lyons AB, Samuels K, Sanderson A and Maddy AH (1993) Alternative pathways of apoptosis induced by methylprednisolone and valinomycin analyzed by flow cytometry. *Exp. Cell. Res.* **208**: 362–70.

Deiss LP, Feinstein E, Berissi H, Cohen O and Kimichi A (1995) Identification of a novel serine/threonine kinase and a novel 15-kD protein as a potential mediator of the gamma interferon-induced cell death. *Genes Dev.* **9**: 15–30.

Dember LM, Kim ND, Liu KQ and Anderson P (1996) Individual RNA recognition motifs of TIA-1 and TIAR have different RNA binding specificities. *J. Biol. Chem.* **271**: 2783–8.

Deveraux QL, Roy N, Stennicky HR, van Arsdale T, Zhou Q, Srinivasula SM, Alnemri ES, Salvesen GS and Reed JC (1998) IAPs block apoptosis events induced by caspase-8 and cytochrome c by direct inhibition of distinct caspases. *EMBO J.* **17**: 2215–23.

Di Leonardo A, Linke SP, Clarkin K and Wahl GM (1994) DNA damage triggers a prolonged p53-dependent G1 arrest and long-term induction of Cip 1 in normal human fibroblasts. *Genes Dev.* **8**: 2540–51.

Didenko VV, Tunstead JR and Hornsby PJ (1994) Biotin-labeled hairpin oligonucleotides: probes to detect double-strand breaks in DNA in apoptotic cells. *Am. J. Pathol.* **152**: 897–902.

Dimri GP, Itahana K, Acosta M and Campisi J (2000) Regulation of senescence checkpoint response by the E2F1 transcription factor and p14ARF tumor suppressor. *Mol. Cell. Biol.* **20**: 273–85.

Dimri GP, Lee X, Basile G, Acosta M, Scott G, Roskelly C, Medrano EE, Linskens M, Rubelj I, Pereira-Smith OM, Peacocke M and Campisi J (1995) A novel biomarker identifies senescent human cells in culture and in aging skin *in vivo*. *Proc. Natl. Acad. Sci. USA* **92**: 9363–7.

Dive C, Gregory CD, Phipps DJ, Evans DL, Milner AE and Wyllie AH (1992) Analysis and discrimination of necrosis and apoptosis (programmed cell death) by multiparameter flow cytometry. *Biochim. Biophys. Acta* **1133**: 275–85.

Du Ch, Fang M, Li Y, Li L and Wang X (2000) Smac, a Mitochondrial Protein that Promotes Cytochrome c-Dependent Caspase Activation by Eliminating IAP Inhibition. *Cell* **102**: 33–42.

Duan H, Orth K, Chinnaiyan AM, Poirier GG, Froelich CJ, He WW and Dixit VM (1996) ICE-LAP6, a novel member of the ICE/Ced-3 gene family, is activated by the cytotoxic T cell protease granzyme B. *J. Biol. Chem.* **271**: 16720–4.

Duriez PJ and Shah GM (1997) Cleavage of poly (ADP-ribose) polymerase: a sensitive parameter to study cell death. *Biochem. Cell. Biol.* **75**: 337–49.

Duvall E, Wyllie AH and Morris RG (1985) Macrophage recognition of cells undergoing programmed cell death (apoptosis). *Immunol.* **56**: 351–8.

Eldadah BA and Faden AI (2000) Caspase Pathways, Neuronal Apoptosis, and CNS Injury. *J. Neurotrauma* **10**: 811–29.

Evan G and Littlewood T (1998) A matter of life and cell death. *Science* **281**: 1317–22.

Fabbi M, Marimpietri D, Martini S, Brancolini C, Amoresano A, Scaloni A, Bargellesi A and Cosulich E (1999) Tissue transglutaminase is a caspase substrate during apoptosis. Cleavage causes loss of transaminating function and is a biochemical marker of caspase 3 activation. *Cell Death Differ* **6**: 992–1001.

Fadeel B, Orrenius S and Zhivotovsky B (2000) The Most Unkindest Cut of All: on the Multiple Roles of Mammalian Caspases. *Leukaemia* **8**: 1695–703.

Fadok VA, Voelker DR, Campbell PA, Cohen JJ, Bratton DL and Henson PM (1992) Exposure of phosphatidylserine on the surface of apoptotic lymphocytes triggers recognition and removal by macrophages. *J. Immunol.* **148**: 2207–16.

Fadok VA, Voelker DR, Campbell PA, Cohen JJ, Bratton DL and Henson PM (1992) Exposure of phosphatidylserine on the surface of apoptotic lymphocytes triggers specific recognition and removal by macrophages. *J. Immunol.* **148**: 2207–16.

Faucheu C, Blanchet AM, Collard-Dutilleul V, Lalanne JL and Diu-Hercend A (1996) Identification of a cysteine protease closely related to interleukin-1 beta-converting enzyme. *Eur. J. Biochem.* **236**: 207–13.

Fernandes-Alnemri TF, Armstrong RC, Krebs J, Srinivasula SM, Wang L, Bullrich F, Fritz LC, Trapani JA, Tomaselli KJ, Litwack G and Alnemri ES (1996) *In vitro* activation of CPP32 and Mch3 by Mch4, a novel human apoptotic cysteine protease containing two FADD-like domains. *Proc. Natl. Acad. Sci. USA* **93**: 7464–69.

Fesus L, Thomazy V and Falus A (1987) Induction and activation of tissue transglutaminase during programmed cell death. *FEBS Lett.* **224**: 104–8.

Flora PK and Gregory CD (1994) Recognition of apoptotic cells by human macrophage: inhibition of a monocyte/macrophage-specific monoclonal antibody. *Eur. J. Immunol.* **24**: 2625–32.

Freude B, Masters TN, Kostin S, Robicsek F and Schaper J (1998) Cardiomyocyte apoptosis in acute and chronic conditions. *Basic Res. Cardiol.* **93**: 85–9.

Gavrieli Y, Sherman Y and Ben-Sasson SA (1992) Identification of programmed cell death *in situ* via specific labelling of nuclear DNA fragmentation. *J. Cell. Biol.* **119**: 493–501.

Going JJ, Stuart RC, Downie M, Fletcher-Monaghan AJ and Keith WN (2002) 'Senescence-associated' β-galactosidase in the upper gastrointestinal tract. *J. Pathol.* **196**: 394–400.

Grasi-Kraup B, Ruttkay-Nedecky B, Koudelka H, Bukowska K, Bursch W and Schulte-Hermann R (1995) *In situ* detection of fragmented DNA (TUNEL) assay fails to discriminate among apoptosis, necrosis and autolytic cell death: a cautionary note. *Hepatology* **21**: 1465–8.

Green DR and Reed JC (1998) Mitochondria and apoptosis. *Science* **281**: 1309–12.

Grutter MG (2000) Caspases: Key Players in Programmed Cell Death. *Curr. Opin. Struct. Biol.* **6**: 649–55.

Gurtu V, Kain SR and Zhang G (1997) Fluorometric and colorimetric detection of caspase activity associated with apoptosis. *Anal. Biochem.* **251**: 98–102.

Han DKM, Chaudray PM, Wright ME, Friedmann C, Trask BJ, Riedel RT, Baskin DG, Schwartz SM and Hood L (1997) MRIT, a novel death-effector domain-containing protein, interacts with caspases and BclXL and initiates cell death. *Proc. Natl. Acad. Sci. USA* **94**: 11333–8.

Hara E, Tsurui H, Shinozaki A, Nakada S and Oda K (1991) Cooperative effect of antisense-Rb and antisense-p53 oligomers on the extension of life span in human diploid fibroblasts, TIG-1. *Biochem. Biophys. Res. Comm.* **179**: 528–34.

Hayflick L (1965) The limited *in vitro* lifetime of human diploid cell strains. *Exp. Cell. Res.* **37**: 614–36.

Higami Y and Shimokawa I (2000) Apoptosis in the aging process. *Cell Tissue Res.* **301**: 125–32.

Hisahara S, Yuan J, Momoi T, Okano H and Miura M (2001) Caspase-11 Mediates Oligodendrocyte CEll Death and Pathogenesis of Autoimmune-Mediated Demyelination. *J. Exp. Med.* **193**: 111–22.

Hitoshi Y, Lorens J, Kitada SI, Fisher J, LaBarge M, Ring HZ, Francke U, Reed JC, Kinoshita S and Nolan GP (1998) Toso, a cell surface, specific regulator of FAS induced apoptosis in T cells. *Immunity* **8**: 461–71.

Hotz MA, Gong J, Traganos F and Darzynkiewicz Z (1994) Flow cytometric detection of apoptosis: comparison of the assays of *in situ* DNA degradation and chromatin changes. *Cytometry* **15**: 237–44.

Howie SEM, Sommerfield AJ, Gray E and Harrison DJ (1994) Peripheral T lymphocyte depletion by apoptosis after CD4 ligation *in vivo*: selective loss of CD44- and 'activating' memory T cells. *Clin. Exp. Immunol.* **95**: 195–200.

Hu S, Snipas SJ, Vincenz C, Salvesen G and Dixit VM (1998) Caspase-14 is a novel developmentally regulated protease. *J. Biol. Chem.* **273**: 29648–53.

Hu S, Vincenz C, Ni J, Gentz R and Dixit VM (1997) I-FLICE, a novel inhibitor of tumor necrosis factor receptor-1 and CD-95-induced apoptosis. *J. Biol. Chem.* **272**: 17255–7.

Hulkko SM, Wakui H and Zilliacus J (2000) The pro-apoptotic protein death-associated protein 3 (DAP3) interacts with the glucocorticoid receptor and affects the receptor function. *Biochem. J.* **349**: 885–93.

Humke EW, Shriver SK, Starovasnik MA, Fairbrother WJ and Dixit VM (2000) ICEBERG: a novel inhibitor of interleukin-1 beta generation. *Cell* **103**: 99–111.

Imai Y, Kimura T, Murakami A, Yajima N, Sakamaki K and Yonehara S (1999) The CED-4-homologous protein FLASH is involved in Fas-mediated activation of caspase-8 during apoptosis. *Nature* **398**: 777–85.

Inohara N, del Peso L, Koseki T, Chen S and Nunez G (1998) RICK, a novel protein kinase containing a caspase recruitment domain, interacts with CLARP and regulates CD95-mediated apoptosis. *J. Biol. Chem.* **273**: 12296–300.

Irmler M, Thome M, Hahne M, Schneider P, Hofmann K, Steiner V, Bodmer J-L, Schröter M, Burns K, Mattmann C, Rimoldi D, French LE and Tschopp J (1997) Inhibition of death receptor signals by cellular FLIP. *Nature* **388**: 190–5.

Ito A, Uehara T and Nomura Y (2000) Isolation of Ich-1S (Caspase-2S)-Binding Protein that partially inhibits Caspase Activity. *FEBS Lett.* **3**: 360–4.

Iwaki T, Miyazono M, Hitosumatsu T and Tateishi J (1994) An immunohistochemical study of tissue transglutaminase in gliomas with reference to their cell dying processes. *Am. J. Pathol.* **145**: 776–81.

Jacobs JJ, Kieboom K, Marino S, DePinho RA and van Lohuizen M (1999) The oncogene and polycomb-group gene bmi-1 regulates cell proliferation and senescence through the *ink4a* locus. *Nature* **397**: 164–8.

Jacobson MD, Burne JF and Raff MC (1994) Programmed cell death and Bcl-2 protection in the absence of a nucleus. *EMBO J.* **13**: 1899–910.

Jemmerson R, Liu J, Hausauer D, Kong-Peng L, Mondino A and Nelson RD (1999) A conformational change in cytochrome c of apoptotic and necrotic cells is detected by monoclonal antibody binding and mimicked by association of the native antigen with synthetic phospholipid vesicles. *Biochemistry* **38**: 3599–609.

Jerome KR, Vallan C and Taggi R (2000) The TUNEL assay in the diagnosis of graft-versus-host disease: caveats for interpretation. *Pathology* **32**: 186–90.

Jiang Y, Woronicz JD, Liu W and Goeddel DV (1999) Prevention of constitutive TNF receptor 1 signaling by silencer of death domains. *Science* **283**: 543–6. Erratum: *Science* **283**: 1852.

Jordan LB and Harrison DJ (2000) *Apoptosis: a distinctive form of cell death*. In: Schunkert H and Riegger GAJ (eds). In: *Apoptosis in cardiac biology*. Kluwer Academic Publishers, Massachusetts, USA: 123–35.

Kamijo T, Zindy F, Roussel MF, Quelle DE, Downing JR, Ashmun RA, Grosveld G and Sherr CJ (1997) Tumor suppression at the mouse INK4a locus mediated by the alternative reading frame product p19ARF. *Cell* **91**: 649–59.

Katz ML, Robson WG, Hermann RK, Gosner AB and Bieri JG (1984) Lipofuscin accumulation resulting from senescence and vitamin E deficiency: spectral properties and tissue distribution. *Mech. Ageing Dev.* **25**: 149–59.

Kaufmann SH (1989) Induction of endonuclease DNA cleavage in human acute myelogenous leukaemia by etopiside, camptothecin, and other cytotoxic anticancer drugs: a cautionary note. *Cancer Res.* **49**: 5870–8.

Kaufmann SH, Desnoyers S, Ottaviano Y, Davidson NE and Poirer GG (1993) Specific proteolytic cleavage of poly(ADP-ribose) polymerase: An early marker of chemotherapy-induced apoptosis. *Cancer Res.* **53**: 3976–85.

Kayalar C, Ord T, Testa MP, Zhong LT and Bredesen DE (1996) Cleavage of actin by interleukin 1 beta-converting enzyme to reverse DNase I inhibition. *Proc. Natl. Acad. Sci.* **93**: 2234–38.

Keith WN, Evans TRJ and Glasspool RM (2001) Telomerase and cancer: time to move from a promising target to clinical reality. *J. Pathol.* **195**: 404–14.

Kerr JFR, Wyllie AH and Currie AR (1972) Apoptosis: a basic biological phenomenon with wide-ranging implications in tissue kinetics. *Br. J. Cancer* **26**: 239–57.

Kluck RM, Bossy Wetzel E, Green DR and Newmeyer DD (1997) The release of cytochrome *c* from mitochondria: a primary site for Bcl-2 regulation of apoptosis. *Science* **275**: 1132–6.

Kockx MM, Muhring J, Knaapen MWM and de Meyer GR (1998) RNA synthesis and splicing interferes with DNA *in situ* end labeling techniques used to detect apoptosis. *Am. J. Pathol.* **152**: 885–8.

Koester SK, Roth P, Mikulka WR, Schlossman SF, Zhang C and Bolton WE (1997) Monitoring early cellular responses in apoptosis is aided by the mitochondrial membrane protein-specific monoclonal antibody APO2.7. *Cytometry* **29**: 306–12.

Koseki T, Inohara N, Chen S, Carrio R, Merino J, Hottiger MO, Nabel GJ and Nunez G (1999) CIPER, a novel NF kappaB-activating protein containing a caspase recruitment domain with homology to herpesvirus-2 protein E10. *J. Biol. Chem.* **274**: 9955–61.

Kothakota S, Azuma T, Reinhard C, Klippel A, Tang J, Chu K, McGarry TJ, Kirschner MW, Koths K, Kwiatkowski DJ and Williams LT (1997) Caspase-3-generated fragment of gelsolin: effector of morphological change in apoptosis. *Science* **278**: 294–8.

Krajewski S, Krajewska M, Ellerby LM, Welsh K, Xie Z, Deveraux QL, Salvesen GS, Bredesen DE, Rosenthal RE, Fiskum G and Reed JC (1999) Release of caspase-9 from mitochondria during neuronal apoptosis and cerebral ischemia. *Proc. Natl. Acad. Sci. USA* **96**: 5752–7.

Krishna DR, Sperker B, Fritz P and Klotz U (1999) Does pH 6 beta-galactosidase activity indicate cell senescence? *Mech. Ageing Dev.* **109**: 113–23.

Kroemer G, Petit P, Naofal Z, Vayssiere J-L and Mignotte B (1995) The biochemistry of programmed cell death. *FASEB J.* **9**: 1277–87.

Kurz DJ, Decary S, Hong Y and Erusalimsky JD (2000) Senescence-associated (beta) galactosidase reflects an increase in lysosomal mass during replicative ageing of human endothelial cells. *J. Cell. Sci.* **113**: 3613–22.

Laio Y, Tang Z-Y, Liu K-D, Ye S-L and Huang Z (1997) Apoptosis of human BEL-7402 hepatocellular carcinoma cells released by antisense H-RAS DNA – *in vitro* and *in vivo* studies. *J. Cancer Res. Clin. Oncol.* **123**: 25–33.

Lansdorp PM (2000) Repair of telomeric DNA prior to replicative senescence. *Mech. Ageing Dev.* **118**: 23–34.

Lanza RP, Cibelli JB, Blackwell C, Cristofalo VJ, Francis MK, Baerlocher GM, Mak J, Schertzer M, Chavez EA, Sawyer N, Lansdorp PM and West MD (2000) Extension of life span and telomere length in animals cloned from senescent somatic cells. *Science* **288**: 665–9.

Lazebnik YA, Cole S, Cooke CA, Nelson WG and Earnshaw WC (1993) Nuclear events of apoptosis *in vitro* in cell-free mitotic extracts: A model system for analysis of the active phase of apoptosis. *J. Cell. Biol.* **123**: 7–22.

Lazebnik YA, Kaufmann SH, Desnoyers S, Poirier GG and Earnshaw WC (1994) Cleavage of poly (ADPribose) polymerase by a proteinase with properties like ICE. *Nature* **371**: 346–7.

Leers MP, Kolgen W, Bjorklund V, Bergman T, Tribbick G, Persson B, Bjorklund P, Ramaekers FC, Bjorklund B, Nap M, Jornvall H and Schutte B (1999) Immunocytochemical detection and mapping of a cytokeratin 18 neo-epitope exposed during early apoptosis. *J. Pathol.* **187**: 567–72.

Leist M and Nicotera P (1997) The shape of cell death. *Biochem. Biophys. Res. Comm.* **236**: 1–9.

Levy-Stumpf N, Deiss LP, Berissi H and Kimichi A (1997) DAP-5, a novel homolog of eukaryotic translation initiation factor 4G isolated as a putative modulator of gamma interferon induced programmed cell death. *Mol. Cell. Biol.* **17**: 1615–25.

Li H, Bergeron L, Cryns V, Pasternack M, Zhu H, Shi L, Greenberg A and Yuan J (1997) Activation of Caspase-2 in Apoptosis. *J. Biol. Chem.* **34**: 21010–7.

Li X, Traganos F, Melamed MR and Darzynkiewicz Z (1994) Simultaneous analysis of DNA replication and apoptosis during treatment of HL-60 cells with camptothecin and hyperthermia and mitogen stimulation of human lymphocytes. *Cancer Res.* **54**: 4289–93.

Li X, Traganos F, Melamed MR and Darynkiewicz Z (1995) Single-step procedure for labelling DNA strand breaks with fluorescein – or BODIPY – conjugated deoxynucleotides detection of apoptosis and bromodeoxyuridine incorporation. *Cytometry* **20**: 172–80.

Lind SE and Janmey PA (1984) Human plasma gelsolin binds to fibronectin. *J. Biol. Chem.* **259**: 13262–6.

Lindahl T, Satoh MS, Poirier GG and Klungland A (1995) Post-translational modification of poly (ADP-ribose) polymerase induced by DNA strand breaks. *Trends Biochem. Sci.* **20**: 405–11.

Liu X, Zou H, Slaughter C and Wang X (1997) DFF, a heterodimeric protein that functions downstream of caspase-3 to trigger DNA fragmentation during apoptosis. *Cell* **89**: 175–84.

Loo DT and Rillema JR (1998) Measurement of cell death. *Methods Cell Biol.* **57**: 251–64.

Lutgens E, Daemen MJAP, Kockx M, Doevendans P, Hofker M, Havekes L, Wellens H and de Muinick ED (1999) Atherosclerosis in APOE*3-Leiden transgenic mice: from proliferative to atheromatous stage. *Circulation* **99**: 276–83.

Majno G and Joris I (1995) Apoptosis, oncosis and necrosis. An overview of cell death. *Am. J. Pathol.* **146**: 3–15.

Mancini M, Nicholson DW, Roy S, Thornberry NA, Peterson EP, Casciola-Rosen LA and Rosen A (1998) The caspase-3 precursor has a cytosolic and mitochondrial distribution: implications for apoptotic signaling. *J. Cell. Biol.* **140**: 1485–95.

Martin SJ, OBrien GA, Nishioka WK, McGahon AJ, Saido TC and Green DR (1995a) Proteolysis of fodrin (non-erythroid spectrin) during apoptosis. *J. Biol. Chem.* **270**: 6425–8.

Martin SJ, Reutelingsperger CPM, McGahon AJ, Rader JA, Vanschie ECAA, Laface DM and Green DR (1995b) Early redistribution of plasma membrane phosphatidylserine is a general feature of apoptosis regardless of the initiating stimulus: inhibition by overexpression of Bcl-2 and Abl. *J. Exp. Med.* **182**: 1545–6.

Marzari R, Sblattero D, Florian F, Tongiorgi E, Not T, Tommasini A, Ventura A and Bradbury A (2001) Molecular dissection of the tissue transglutaminase autoantibody response in celiac disease. *J. Immunol.* **166**: 4170–6.

Mashima T, Naito M, Noguchi K, Miller DK, Nicholson DW and Tsuruo T (1997) Actin cleavage by CPP-32/apopain during the development of apoptosis. *Oncogene* **14**: 1007–12.

Mather M and Rottenberg H (2000) Aging enhances the activation of the permeability transition pore in mitochondria. *Biochem. Biophys. Res. Comm.* **273**: 603–8.

McConkey DJ (1996) Calcium flux measurement in cell death. In: Cotter TG, Martin SJ, editors, *Techniques in apoptosis: A user's guide*. London, Portland Press: 133–48.

Melino G, Candi E and Steinert PM (2000) Assays for transglutaminase in cell death. *Methods Enzymol.* **322**: 433–72.

Metivier D, Dallaporta B, Zamzami N, Larochette N, Susin SA, Marzo I and Kroemer G (1998) Cytofluorometric detection of mitochondrial alterations in early CD95/Fas/APO-1-triggered apoptosis of Jurkat T lymphoma cells. Comparison of seven mitochondrion-specific fluorochromes. *Immunol. Lett.* **61**: 157–63.

Migheli A, Cavalla P, Marino S and Schiffer D (1994) A study of apoptosis in normal and pathologic nervous tissue after *in situ* end-labeling of DNA strand breaks. *J. Neuropathol. Exp. Neurol.* **53**: 606–16.

Mikhelson VM (2001) Replicative mosaicism might explain the seeming contradictions in the telomere theory of aging. *Mech. Ageing Dev.* **122**: 1361–5.

Moir RD, Spann TP and Goldman RD (1995) The dynamic properties and possible functions of nuclear lamins. *Int. Rev. Cytol.* **162B**: 141–82.

Moore AM, Donahue CJ, Bauer KD and Mather JP (1998) Simultaneous measurement of cell cycle and apoptotic cell death. *Methods Cell Biol.* **57**: 265–78.

Morsi HM, Leers MP, Radespiel-Troger M, Bjorklund V, Kabarity HE, Nap M and Jager W (2000) Apoptosis, bcl-2 expression and proliferation in benign and malignant endometrial epithelium: An approach using multiparameter flow cytometry. *Gynecol. Oncol.* **77**: 11–7.

Mouton RE and Venable ME (2000) Ceramide induces expression of the senescence histochemical marker, beta-galactosidase, in human fibroblasts. *Mech. Ageing Dev.* **113**: 169–81.

Munday NA, Vaillancourt JP, Ali A, Casano FJ, Miller DK, Molineaux SM, Yamin TT, Yu VL and Nicholson DW (1995) Molecular cloning and pro-apoptotic activity of ICEreIII and ICEreIIII, members of the ICE/CED-3 family of cysteine proteases. *J. Biol. Chem.* **270**: 15870–6.

Mundle SD, Gao XZ, Khan S, Gregory SA, Preisler HD and Raza A (1995) Two *in situ* labeling techniques reveal different patterns of DNA fragmentation during spontaneous apoptosis *in vivo*, and induced apoptosis *in vitro*. *Anticancer Res.* **15**: 1895–904.

Muzio M, Chinnaiyan AM, Kischkel FC, O'Rourke K, Shevchenko A, Ni J, Scaffidi C, Bretz JD, Zhang M, Gentz R, Mann M, Krammer PH, Peter ME and Dixit VM (1996) FLICE, a novel FADD-homologous ICE/CED-3-like protease, is recruited to the CD95 (Fas/APO-1) death-inducing signaling complex. *Cell* **85**: 817–27.

Narula J, Acio ER, Narula N, Samuels LE, Fyfe B, Wood D, Fitzpatrick JM, Raghunath PN, Tomaszewski JE, Kelly C, Steinmetz N, Green A, Tait JF, Leppo J, Blankenberg FG, Jain D and Strauss HW (2001) Annexin-V imaging for non-invasive detection of cardiac allograft rejection. *Nature Med.* **7**: 1347–52.

Naruse I, Keino H and Kawarada Y (1994) Antibody against single stranded DNA detects both programmed cell death and drug induced apoptosis. *Histochem.* **101**: 73–8.

Negoescu A, Lorimer P, Labat-Moleur F, Drouet C, Robert C, Guillermet C, Brambilla C and Brambilla E (1996) *In situ* apoptotic cell labeling by the TUNEL method: an improvement and evaluation on cell preparations. *J. Histochem. Cytochem.* **44**: 959–68.

Nicoletti I, Migiorati G, Pagliacci MC, Grignani F and Riccardi C (1991) A rapid and simple method for measuring thymocyte apoptosis by propidium iodide staining and flow cytometry. *J. Immunol. Meth.* **139**: 271–9.

Oberhammer FA, Bursch W, Parzefall W, Breit P, Erber E, Stadler M and Schulte HR (1991) Effect of transforming growth factor beta on cell death of cultured rat hepatocytes. *Cancer Res.* **51**: 2478–85.

Oberhammer FA, Pavelka M, Sharma S, Tiefenbacher R, Purchio AF, Bursch W and Schulte-Hermann R (1992) Induction of apoptosis in cultured hepatocytes and in regressing liver by growth factor beta 1. *Proc. Natl. Acad. Sci. USA* **89**: 5408–12.

Ogasawara J, Waanabe-Fukunga R, Adachi M, Matsuzawa A, Kasgul T, Kitamura Y, Itoh N, Suda T and Nagata S (1993) Lethal effect of the anti-Fas antibody in mice. *Nature* **364**: 806–9.

Ogryzyko VV, Harai TH, Russanova VR, Barbie DA and Howard BH (1996) Human fibroblast commitment to a senescence-like state in response to histone deacetylase inhibitors is cell cycle dependent. *Mol. Cell. Biol.* **16**: 5210–18.

Oliverio S, Amendola A, Rodolfo C, Spinedi A and Piacentini M (1999) Inhibition of "tissue" transglutaminase increases cell survival by preventing apoptosis. *J. Biol. Chem.* **274**: 34123–8.

Ormerod MG, Sun X-M, Brown D, Snowden RT and Cohen GM (1993) Quantification of apoptosis and necrosis by flow cytometry. *Acta Oncol.* **32**: 417–24.

Orth K, Chinnayan AM, Garg M, Froelich CJ and Dixit VM (1996) The CED-3/ICE-like protease Mch2 is activated during apoptosis and cleaves the death substrate lamin A. *J. Biol. Chem.* **271**: 16443–6.

Petito CK and Roberts B (1995) Effect of post mortem interval on *in situ* end-labeling of DNA oligonucleosomes. *J. Neuropathol. Exp. Neurol.* **54**: 761–5.

Piacentini M, Rodolfo C, Farrace MG and Autuori F (2000) "Tissue" transglutaminase in animal development. *Int. J. Dev. Biol.* **44**: 655–62.

Pieper AA, Verma A, Zhang J and Synder SH (1999) Poly (ADP-ribose) polymerase, nitric oxide and cell death. *Trends Pharmacol. Sci.* **20**: 171–81.

Pitti RM, Marsters SA, Ruppert S, Donahue CJ, Moore A and Ashkenazi A (1996) Induction of apoptosis by APO-2 ligand, a new member of the tumor necrosis factor cytokine family. *J. Biol. Chem.* **271**: 12687–90.

Pochampally R, Fodera B, Chen L, Shao W, Levine EA and Chen J (1998) A 60kd MDM2 isoform is produced by caspase cleavage in non-apoptotic tumor cells. *Oncogene* **17**: 2629–36.

Prasad KV, Ao Z, Yoon Y, Wu MX, Rizk M, Jacquot S and Schlossman SF (1997) CD27, a member of the tumor necrosis factor receptor family, induces apoptosis and binds to Siva, a proapoptotic protein. *Proc. Natl. Acad. Sci. USA* **94**: 6346–51.

Reddel RR (2000) The role of immortalization in carcinogenesis. *Carcinogenesis* **21**: 477–84.

Reipert S, Reipert BM, Hickman JA and Allen TD (1996) Nuclear pore clustering is a consistent feature of apoptosis *in vitro*. *Cell Death Differ* **3**: 131–9.

Rittmaster RS, Thomas LN, Wright AS, Murray SK, Carlson K, Douglas RC, Yung J, Messieh M, Bell D and Lazier CB (1999) The utility of tissue transglutaminase as a marker of apoptosis during treatment and progression of prostate cancer. *J. Urol.* **162**: 2165–9.

Robles SJ and Adami GR (1998) Agents that cause DNA double strand breaks lead to p16INK4a enrichment and the premature senescence of normal fibroblasts. *Oncogene* **16**: 1113–23.

Rudel T and Bokoch GM (1997) Membrane and morphological changes in apoptotic cells regulated by caspase-mediated activation of PAK2. *Science* **276**: 1571–4.

Sager R (1991) Senescence as a mode of tumor suppression. *Environ. Health Prospect* **93**: 59–62.

Sakahira H, Enari M and Nagata S (1998) Cleavage of CAD inhibitor in CAD activation and DNA degradation during apoptosis. *Nature* **391**: 96–9.

Sallmann FR, Bourassa S, Saint-Cyr J and Poirier GG (1997) Characterization of antibodies specific for the caspase cleavage site on poly(ADP-ribose) polymerase: specific detection of apoptotic fragments and mapping of the necrotic fragments of poly(ADP-ribose) polymerase. *Biochem. Cell. Biol.* **75**: 451–6.

Sallmann FR, Bourassa S, Saint-Cyr J and Poirier GG (1997) Characterization of antibodies specific for the caspase cleavage site on poly(ADP-ribose) polymerase: specific detection of apoptotic fragments and mapping of the necrotic fragments of poly(ADP-ribose) polymerase. *Biochem. Cell. Biol.* **75**: 451–6.

Saraste A and Pulkki K (2000) Morphologic and biochemical hallmarks of apoptosis. *Cardiovasc Res.* **45**: 528–37.

Sasano H (1995) *In situ* end labeling, and its applications to the study of endocrine disease: How can we study programmed cell death in surgical pathology materials? *Endocrinol. Pathol.* **6**: 87–9.

Savill J (1996) Phagocyte recognition of apoptotic cells. *Biochem. Soc. Trans.* **24**: 1065–9.

Savill J (1997) Apoptosis in resolution of inflammation. *J. Leukocyte Biol.* **61**: 375–80.

Savill JS, Dransfield I, Hogg N and Haslett C (1990) Vitronectin receptor-mediated phagocytosis of cells undergoing apoptosis. *Nature* **343**: 170–3.

Savill JS, Hogg N, Ren Y and Haslett C (1992) Thrombospondin co-operates with CD36 and the vitronectin receptor in macrophage recognition of neutrophils undergoing apoptosis. *J. Clin. Invest.* **90**: 1513–22.

Schulze-Osthoff K, Walczak H, Droge W and Kramer PH (1994) Cell nucleus and DNA fragmentation are not required for apoptosis. *J. Cell. Biol.* **127**: 15–20.

Schwartz S Jr., Yamamoto H, Navarro M, Maestro M, Reventos J and Perucho M (1999) Frameshift mutations at mononucleotide repeats in caspase-5 and other target genes in endometrial and gastrointestinal cancer of the microsatellite mutator phenotype. *Cancer Res.* **59**: 2995–3002.

Sells SF, Wood DP Jr, Crist SA, Humphreys S and Rangnekar VM (1994) Commonality of gene programs induced by effectors of apoptosis in androgen-dependent and -independent prostate cells. *Cell Growth Differ* **5**: 457–66.

Serrano M, Lee H, Chin L, Cordon-Cardo C, Beach D and DePinho RA (1996) Role of the INK4a locus in tumor suppression and cell mortality. *Cell* **85**: 27–37.

Serrano M, Lin AW, McCurrach ME, Beach D and Lowe SW (1997) Oncogenic Ras provokes premature cell senescence associated with accumulation of p53 and p16INK4a. *Cell* **88**; 593–602.

Severino J, Allen RG, Balin S, Balin A and Cristofalo VJ (2000) Is beta-galactosidase staining a marker of senescence *in vitro* and *in vivo*? *Exp. Cell. Res.* **257**: 162–71.

Shay JW, Pereira-Smith OM and Wright WE (1991) A role for both Rb and p53 in the regulation of human cellular senescence. *Exp. Cell. Res.* **196**: 33–9.

Sheils PG, Kind AJ, Campbell KH, Waddington D, Wilmut I, Coleman A and Schnieke AE (1999) Analysis of telomere lengths in cloned sheep. *Nature* **399**; 316–7.

Shu HB, Halpin DR and Goeddel DV (1997) Casper is a FADD- and caspase-related inducer of apoptosis. *Immunity* **6**: 751–63.

Singhal RP, Mays-Hoopes LL and Eichhorn GL (1987) DNA methylation in aging of mice. *Mech. Ageing Dev.* **41**: 199–210.

Slack A, Cervoni N, Pinard M and Szyf M (1999) DNA methyltransferase is a downstream effector of cellular transformation triggered by simian virus 40 large T antigen. *J. Biol. Chem.* **274**: 10105–12.

Smith CA, Farrah T and Goodwin RG (1994) The TNF receptor superfamily of cellular and viral proteins: activation. Costimulation and death. *Cell* **76**: 959–62.

Sohal RS and Brunk UT (1989) Lipofuscin as an indicator of oxidative stress and aging. *Adv. Exp. Med. Biol.* **266**: 17–26.

Soini Y, Pääkkö P and Lehto V-P (1998) Histopathological evaluation of apoptosis in cancer. *Am. J. Pathol.* **153**: 1041–53.

Seperandio S, de Belle I and Bredesen DE (2000) An alternative, non-apoptotic form of programmed cell death. *Proc. Natl. Acad. Sci. USA* **97**: 14376–81.

Stanger BZ, Leder P, Lee TH, Kim E and Seed B (1995) RIP: a novel protein containing a death domain that interacts with Fas/APO-1 (CD95) in yeast and causes cell death. *Cell* **81**: 513–23.

Stanulis-Praeger B (1987) Cellular senescence revisited: a review. *Mech. Ageing Dev.* **38**: 1–48.

Stevens H, Reeve J and Noble BS (2000) Bcl-2, tissue transglutaminase and p53 protein expression in the apoptotic cascade in ribs of premature infants. *J. Anat.* **196**: 181–91.

Susin SA, Lorenzo HK, Zamzami N, Marzo I, Brenner C, Larochette N, Prevost MC, Alzari PM and Kroemer G (1999a) Mitochondrial release of caspase-2 and -9 during the apoptotic process. *J. Exp. Med.* **189**: 381–94.

Susin SA, Lorenzo HK, Zamzami N, Marzo I, Snow BE, Brothers GM, Mangion J, Jacotot E, Costantini P, Loeffler M, Larochette N, Goodlett DR, Aebersold R, Siderovski DP, Penninger JM and Kroemer G (1999b) Molecular characterization of mitochondrial apoptosis-inducing factor. *Nature* **397**: 441–6.

Susin SA, Zamzami N, Castedo M, Daugas E, Wang HG, Geley S, Fassy F, Reed JC and Kroemer G (1997) The central executioner of apoptosis: multiple connections between protease activation and mitochondrian Fas/APO-1/CD95-and ceramide-induced apoptosis. *J. Exp. Med.* **186**: 25–37.

Susin SA, Zamzami N, Castedo M, Hirsch T, Marchetti P, Macho A, Daugas E, Geuskens M and Kroemer G (1996) Bcl-2 inhibits the mitochondrial release of an apoptogenic protease. *J. Exp. Med.* **184**: 1331–41.

Szegezdi E, Szondy Z, Nagy L, Nemes Z, Friis RR, Davies PJ and Fesus L (2000) Apoptosis-linked *in vivo* regulation of tissue transglutaminase gene promoter. *Cell Death Differ* **7**: 1225–33.

Takahashi A, Alnemri ES, Lazebnik YA, Fernandes-Alnemri T, Litwack G, Moir RD, Goldman RD, Poirier GG, Kaufmann SH and Earnshaw WC (1996) Cleavage of lamin A by Mch2 alpha but not CPP32: multiple interleukin 1 beta-converting enzyme-related proteases with distinct substrate recognition properties are active in apoptosis. *Proc. Natl. Acad. Sci. USA* **93**: 8395–400.

Tartaglia LA, Ayres TM, Wong GHW and Goeddel DV (1993) A novel domain within the 55kd TNF receptor signals cell death. *Cell* **74**: 845–53.

Tatton NA, Tezel G, Insolia SA, Nandor SA, Edward PD and Wax MB (2001) *In situ* detection of apoptosis in normal pressure glaucoma. A preliminary examination. *Surv. Opthalmol.* **45**: S268–72.

Taupin JL, Tian Q, Kedersha N, Robertson M and Anderson P (1995) The RNA-binding protein TIAR is translocated from the nucleus to the cytoplasm during Fas-mediated apoptotic cell death. *Proc. Natl. Acad. Sci. USA* **92**: 1629–33.

Taylor S, Hanlon L, McGillivray C, Gault EA, Argyle DJ, Onions DE and Nicolson L (2000) Cloning and Sequencing of Feline and Canine Ice-Related cDNAs Encoding Hybrid Caspase-1/Caspase-13-like Propeptides. *DNA Seq.* **10**: 387–94.

Telford WG, King LE and Fraker PJ (1992) Comparative evaluation of several DNA binding dyes in the detection of apoptosis-associated chromatin degradation by flow cytometry. *Cytometry* **13**: 137–43.

Telford WG, King LE and Fraker PJ (1994) Rapid quantification of apoptosis in pure and heterogeneous cell populations using flow cytometry. *J. Immunol. Meth.* **172**: 1–16.

Terman A (2001) Garbage catastrophe theory of aging: imperfect removal of oxidative damage? *Redox Report* **6**: 15–26.

Thome M, Schneider P, Hoffman K, Fickenscher H, Meinl E, Neipel F, Mattmann C, Burns K, Bodmer JL, Schröter M, Scaffadi C, Krammer PH, Peter ME and Tschopp J (1997) Viral FLICE-inhibitory proteins (FLIPs) prevent apoptosis induced by death receptors. *Nature* **386**: 517–21.

Thornberry NA and Lazebnik Y (1998) Caspases: enemies within. *Science* **281**: 1312–16.

Tian Q, Taupin J, Elledge S, Robertson M and Anderson P (1995) Fas-activated serine/threonine kinase (FAST) phosphorylates TIA-1 during Fas-mediated apoptosis. *J. Exp. Med.* **182**: 865–74.

Trump BF, Berezesky IK, Chang SH and Phelps PC (1997) The pathways of cell death: oncosis, apoptosis, and necrosis. *Toxicol. Pathol.* **25**: 82–8.

Ucker DS, Meyers J and Obermiller PS (1992) Activation-driven T cell death. II. Quantitative differences alone distinguish stimuli triggering non-transformed T cell proliferation or death. *J. Immunol.* **149**: 1583–92.

Van de Craen M, Van Loo G, Pype S, Van Criekinge W, Van den Brande I, Molemans F, Fiers W, Declercq W and Vandenabeele P (1998) Identification of a new caspase homologue: caspase-14. *Cell Death Differ* **5**: 838–46.

Van de Craen M, Vandenabeele P, Declercq W, Van den Brande I, Van Loo G, Molemans F, Schotte P, Van Criekinge W, Beyaert R and Fiers W (1997) Characterization of seven murine caspase family members. *FEBS Letts.* **403**: 61–9.

van Heerde WL, Robert-Offerman S, Dumont E, Hofstra L, Doevendans PA, Smits JFM, Daemen MJAP and Reutelingsperger CPM (2000) Markers of apoptosis in cardiovascular tissues: focus on Annexin V. *Cardiovasc Res.* **45**: 549–559.

Vanags DM, Porn-Ares MI, Coppola S, Burgess DH and Orrenius S (1996) Protease involvement in fodrin cleavage and phosphatidylserine exposure in apoptosis. *J. Biol. Chem.* **271**: 31075–85.

Varkey J, Chen P, Jemmerson R and Abrams JM (1999) Altered cytochrome c display precedes apoptotic cell death in Drosophila. *J. Cell. Biol.* **144**: 701–10.

Vaziri H and Benchimol S (1998) Reconstitution of telomerase activity in normal human cells leads to elongation of telomeres and extended replicative life span. *Curr. Biol.* **8**: 279–82.

Venable ME, Lee J, Smyth MJ, Bielawska A and Obeid LM (1995) Role of ceramide in cellular senescence. *J. Biol. Chem.* **270**: 30701–8.

Verhagen AM, Ekert PG, Pakusch M, Silke J, Connolly LM, Reid GE, Moritz RL, Simpson RJ and Vaux DL (2000) Identification of DIABLO, a Mammalian that promotes apoptosis by binding to and antagonizing IAP proteins. *Cell* **102**: 43–53.

Vermes I, Haanen C, Steffens-Nakken H and Reutelingsperger C (1995) A novel assay for apoptosis. Flow cytometric detection of phosphatidylserine expression on early apoptotic cells using fluorescein labelled annexin V. *J. Immunol. Methods* **184**: 39–51.

Von Recklinghausen F (1910) *Untersuchungen uber Rachitis und Osteomalacie*. Jena: Verlag Gustav Fischer.

von Zglinicki T, Bürkle A and Kirkwood TBL (2001) Stress, DNA damage and ageing – an integrative approach. *Exp. Gerontol.* **36**: 1049–62.

Wang J and Lenardo MJ (2000) Role of Caspases in Apoptosis, Development, and Cytokine Maturation Revealed by Homozygous Gene Deficiencies. *J. Cell. Sci.* **113**: 753–57.

Wang KK (2000) Calpain and caspase: can you tell the difference? *Trends Neurosci.* **23**: 20–6.

Wang KK, Posmantur R, Nath R, McGinnis K, Whitton M, Talanian RV, Glantz DB and Morrow JS (1998) Simultaneous degradation of alphaII- and betaII-spectrin by caspase 3 (CPP32) in apoptotic cells. *J. Biol. Chem.* **273**: 22490–7.

Webb SJ, Harrison DJ and Wyllie AH (1997) In: Kaufmann SH, ed. *Apoptosis: pharmacological implications and therapeutic opportunities*. Academic Press, New York, NY, USA: 1–34.

Wen LP, Fahrni JA, Troie S, Guan JL, Orth K and Rosen GD (1997) Cleavage of focal adhesion kinase by caspases during apoptosis. *J. Biol. Chem.* **272**: 26056–61.

Whikehart DR, Register SJ, Chang Q and Montgomery B (2000) Relationship of telomeres and p53 in aging bovine corneal endothelial cell cultures. *Invest. Opthalmol. Vis. Sci.* **41**: 1070–5.

Whyte MK, Hardwick SJ, Meagher LC, Savill JS and Haslett C (1993) Transient elevations of cytosolic free calcium retard subsequent apoptosis in neutrophils *in vitro*. *J. Clin. Invest.* **92**: 446–53.

Wiley SR, Schooley K, Smolak PJ, Din WS, Huang C-P, Nicholl JK, Sutherland GR, Smith TD, Rauch C, Smith CA and Goodwin RG (1995) Identification and characterization of a new member of the TNF family that induces apoptosis. *Immunity* **3**: 673–82.

Williams GC (1957) Pleiotrophy, natural selection and the evolution of senescence. *Evolution* **11**: 384–411.

Willingham MC (1999) Cytochemical Methods for the Detection of Apoptosis. *Journal of Histochem. Cytochem.* **47**: 1101–9.

Willis TG, Jadayel DM, Du MQ, Peng H, Perry AR, Abdul-Rauf M, Price H, Karran L, Majekodunmi O, Wlodarska I, Pan L, Crook T, Hamoudi R, Isaacson PG and Dyer MJ (1999) Bcl10 is involved in t(1;14)(p22;q32) of MALT B cell lymphoma and mutated in multiple tumor types. *Cell* **96**: 35–45.

Wilson VL and Jones PA (1983) DNA methylation decreases in aging but not in immortal cells. *Science* **220**: 1055–7.

Wroblewski F and LaDue JS (1955) Lactate dehydrogenase activity in blood. *Proc. Soc. Exp. Biol. Med.* **90**: 210–3.

Wyllie AH (1980) Glucocorticoid-induced thymocyte apoptosis is associated with endogenous endonuclease activation. *Nature* **284**: 555–556.

Yan M, Lee J, Schilbach S, Goddard A and Dixit V (1999) mE10, a novel caspase recruitment domain-containing proapoptotic molecule. *J. Biol. Chem.* **97**: 683–6.

Yang F, Sun X, Beech W, Teter B, Wu S, Sigel J, Vinters HV, Frautschy SA and Cole GM (1998) Antibody to caspase-cleaved actin detects apoptosis in differentiated neuroblastoma and plaque-associated neurons and microglia in Alzheimer's disease. *Am. J. Pathol.* **152**: 379–89.

Yang X, Khosravi-Far R, Chang HY and Baltimore D (1997) Daxx, a novel Fas-binding protein that activates JNK and apoptosis. *Cell* **89**: 1067–76.

Yeager TR, DeVries S, Jarad DF, Kao C, Nakada SY, Moon TD, Bruskewitz R, Stadler WM, Meisner LF, Goldchrist KW, Newton MA, Waldman FM and Reznikoff C (1998) Overcoming cellular senescence in human cancer pathogenesis. *Genes Dev.* **12**: 163–74.

Yegorov YE, Akimov SS, Hass R, Zelenin AV and Prudovsky IA (1998) Endogenous β-galactosidase activity in continuously non-proliferating cells. *Exp. Cell. Res.* **243**: 207–11.

Zakeri ZF, Quaglino D, Latham T and Locksin RA (1993) Delayed internucleosomal DNA fragmentation in programmed cell death. *FASEB J.* **7**: 470–8.

Zamzami N, Susin SA, Marchetti P, Hirsch T, Gomez-Monterrey I, Castedo M and Kroemer G (1996) Mitochondrial control of nuclear apoptosis. *J. Exp. Med.* **183**: 1533–44.

Zhao M, Beauregard DA, Loizou L, Davletov B and Brindle KM (2001) Non-invasive detection of apoptosis using magnetic resonance imaging and a targeted contrast agent. *Nat. Med.* **11**: 1241–4.

Zhu J, Woods D, McMahon and Bishop JM (1998) Senescence of human fibroblasts induced by oncogenic Raf. *Genes Dev.* **12**: 2997–3007.

Zou H, Henzel WJ, Liu XS, Lutschg A and Wang XD (1997) Apaf-1, a human protein homologous to *C. elegans* Ced-4, participates in cytochrome c-dependent activation of caspase-3. *Cell* **90**: 405–13.

9

The Polymerase Chain Reaction

Timothy Diss

9.1 Introduction

The polymerase chain reaction (PCR) has proved itself to be the most important innovation in molecular biology over the last 20 years. It has greatly simplified and accelerated gene analysis. Furthermore, in histopathology, it has made gene analysis possible on a routine basis, as normally fixed and processed samples, including minute biopsy specimens, can be used as a source of DNA or RNA.

The technique was first described in 1985 (Saiki *et al.*, 1985) and has been used as a tool for molecular genetic analysis of histological specimens since 1987 (Impraim *et al.*, 1987). Since then new applications have been constantly emerging.

The PCR is a simple method for amplification of specific segments of DNA or cDNA reverse transcribed from RNA. Although this does not sound dramatic, it greatly simplifies genetic analysis and permits the study of virtually all types of clinical samples including archival formalin-fixed paraffin-wax-embedded tissue. Amplification of the gene segment of interest creates sufficient DNA in a rapid, single tube reaction for simple study of product size and sequence, without the complicating presence of the remainder of the genome and without time-consuming cloning or hybridization procedures.

The first published application was the study of mutations in the β-globin gene for prenatal diagnosis in sickle cell anaemia (Saiki *et al.*, 1985). Since then the PCR has been used to generate an enormous amount of important data concerning the molecular genetics of disease, such as the study of the p53 tumour suppressor gene in many malignancies (Kikuchi *et al.*, 1994). The first publication with a direct application to histopathological diagnosis described amplification of t(14;18) in follicular lymphoma (Lee *et al.*, 1987). Since then studies and applications have been increasingly published to the present. More

Molecular Biology in Cellular Pathology. Edited by John Crocker and Paul G. Murray
 ISBN: 0-470-84475-2

widely used protocols that have become an essential part of the diagnostic process have been developed for lymphomas (Diss and Pan, 1997) and sarcomas (Ladanyi and Bridge, 2000).

The histopathology laboratory is a vital area for molecular genetic research, where the observational skills of the histopathologist can be combined with molecular genetic data that are readily gained by the use of the PCR. The vast archives in which tissue samples are matched with clinical and histopathological information can be exploited to reveal inherited and somatic gene defects, presence of infectious agents and much more nucleic-acid-related information. Pathologists, who spend many hours correlating immunophenotype with morphological and cytological features to assist in the diagnosis and understanding of disease processes, can now add molecular genetic data to their knowledge base. This information can then be used to correlate genetic aberrations to disease onset, progression and response to therapy.

Many developments of the basic PCR method have been reported, such as quantitative PCR, *in situ* PCR and high throughput strategies (see Chapters 11 and 12), but this chapter aims to outline the principles of the basic PCR methods and to introduce the applications of the technique in modern histopathology. The available methods and applications will be outlined and their future potential discussed. The contribution of molecular genetics to the diagnosis of the lymphomas and soft tissue sarcomas will be used to illustrate the application of PCR and reverse transcriptase (RT)-PCR to diagnostic cellular pathology.

9.2 Principles

The Basic Method

The PCR is an *in vitro* method for amplification of specific short fragments of DNA. RNA sequences can be amplified after conversion into cDNA by reverse transcriptase (RT-PCR). Typical reasons for amplifying particular gene sequences are to determine the presence or absence of the sequence, for example DNA of infectious agents, such as mycobacteria (Perosio and Frank, 1993) or chromosome translocation (Lee *et al.*, 1987), to determine whether a mutational hotspot contains a disease-related mutation [for example c-kit (Sotlar *et al.*, 2000)] or to study clonality of rearranged antigen receptor genes in lymphocyte populations (Diss and Pan, 1997).

Oligonucleotide 'primers' which bind to the flanking regions of the target sequence are used to initiate specific copying of DNA strands by a DNA polymerase (Figure 9.1).

The requirements for the reaction are:

1. The template DNA to be studied, such as a genomic DNA or cDNA sample from a patient.

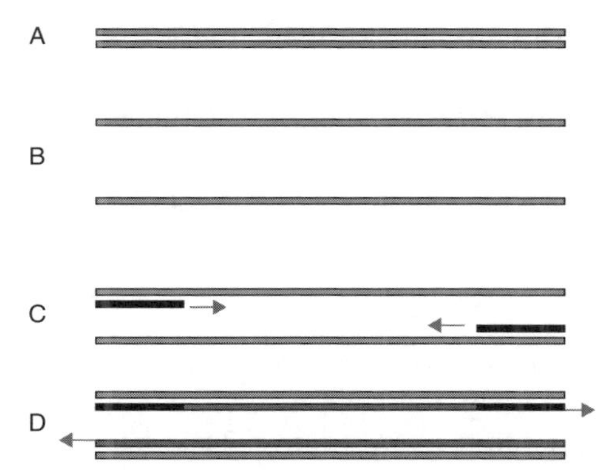

Figure 9.1 A single cycle of PCR. The double-stranded target DNA molecule (green; A) is denatured by heating to 93–95°C (B). Single-stranded oligonucleotide primers (black) bind by complementary base pairing to the flanking regions of each target strand at 50–60°C (C). DNA polymerase catalyses the formation of a new complementary strand (red; D) at 72°C thus doubling the number of target sequences. Repetition of cycles of PCR results in exponential amplification of the target sequence

2. Short single-stranded DNA 'primers', complementary to opposite strands of the flanking regions of the fragment of interest.
3. The four nucleotide triphosphates ('dNTPs').
4. A thermostable DNA polymerase.
5. An appropriate buffer solution.

The reaction mixtures are subjected to cycles of heating and cooling, which permit template denaturation, primer annealing and specific fragment DNA copying ('primer extension'; Figure 9.1). Each cycle theoretically doubles target fragment copy number, therefore leading to an exponential increase in the gene fragment of interest. After a typical protocol involving 35 cycles of the PCR, a many-thousand-fold increase in target copy number may be achieved, even allowing for reaction inefficiencies and limitations of amplification at high product concentrations. This provides sufficient specific DNA for simple analysis such as electrophoresis to determine product size and direct sequencing reactions to determine primary structure and permit mutation detection.

The machines that automate the temperature cycling during the PCR ('thermal cyclers' or 'PCR machines') are simply heating blocks that can be programmed according to the requirements of a given experiment. A typical protocol involves 35–40 cycles of 30 s at 93°C, 30 s at 55–60°C and 1 min at 72°C. Over the years the design of these machines has improved in terms of reliability and speed and amplification can usually be completed within 2 h. Thermal cyclers are available in different formats to suit different applications, including 0.5 ml and 0.2 ml microtubes, 96-well plates and glass slides for *in situ* PCR (Chapter 11).

The PCR process is essentially the same for all targets, although minor adjustments of the magnesium chloride and cycling parameters, especially the primer annealing temperature, are required for each set of primers to optimize the reactions.

The PCR is extremely sensitive and 5–10 template sequences can typically be amplified with the standard PCR. Using highly sensitive nested methods (see below) single target molecule sensitivity can be achieved (Marafioti *et al.*, 2000).

DNA Polymerase Enzymes

The original PCR experiments were performed using non-heat-stable DNA polymerases that were added to reaction mixes after each cycle, as they were destroyed at the denaturation temperature. These have been replaced by thermostable enzymes which survive high-temperature cycling and thus need to be introduced only once at the start of amplification. Reaction conditions and thermostable polymerases have been developed for high fidelity PCR (Eckert and Kunkel, 1991) and amplification of long fragments (Richie *et al.*, 1999).

Primer Design

Appropriate primer design is essential to the success of PCR experiments. Primers are generally in the region of 20 bases long, span the sequence of interest and are exactly complementary to their target. However, degenerate primers can be designed to target variable or partially known sequences. Primers should not have regions that can bind internally or to their partner primer by complementary base pairing, and they should not have multiple binding sites within the target genome. Computer programs to assist in primer design are now widely available, as are databases containing the required target sequences, for example the human genome from the Human Genome Project (see Chapter 15).

Nested PCR

Two-stage PCR protocols have been used widely and involve an initial amplification with outer primers followed by re-amplification of the products using an inner set of primers (Figure 9.1). This approach can greatly improve the sensitivity and specificity of the reaction, although care must be taken not to contaminate subsequent reactions with first round products.

Multiplex PCR

Multiple primers can be mixed in individual PCR tubes to allow simultaneous amplification of more than one sequence. Different sequences can be recognized by their characteristic size or by different fluorochromes in Gene Scanning experiments. Careful design is necessary, so that all primers have similar optimal binding temperatures to ensure equal amplification efficiencies. Multiplex PCR is useful in the analysis of rearranging genes such as the immunoglobulin genes in which the exact gene fragment usage is not known and several possibilities need to be included in the reaction mix (see the section 'Lymphomas' within Section 9.10 below). A mixture of primers that amplify a range of product sizes is useful as a control reaction for DNA quality, as poor DNA will not support amplification of relatively long fragments.

Paraffin-wax-embedded Material

The PCR can be applied successfully to paraffin-wax-embedded material, but there are limitations on the sizes of fragment that can be amplified. Primers need to be designed to amplify fragments of less than 250 bp and preferably of around 100 bp, as the DNA will be highly degraded. It is often necessary to increase the number of cycles, for example from 35 to 40, in order to achieve sufficient yield of product. Samples fixed or processed inappropriately may fail to amplify and PCR of very poor quality extracts may result in the introduction of false 'mutations' by PCR error. This can be a problem if sequence analysis of products is to be conducted. For optimum DNA preservation, fixation should not exceed 24 h. Suitable DNA rarely can be extracted from tissue samples kept for long periods in formalin pots (often a problem with post-mortem material); blocks promptly processed to wax are generally suitable.

9.3 Analysis of products

Basic Analysis

The simplest methods of product analysis are agarose gel electrophoresis or polyacrylamide gel electrophoresis (Figure 9.2), followed by ethidium bromide staining and viewing under UV illumination. The choice depends on the product

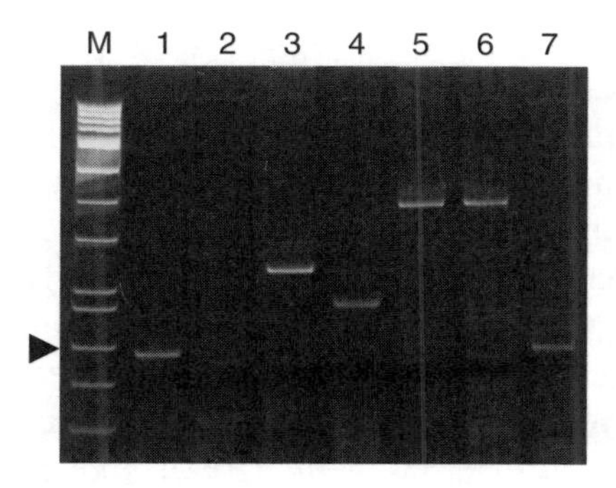

Figure 9.2 Detection of a chromosome translocation. Polyacrylamide gel of t(14;18) PCR products. Lanes were loaded as follows: M – molecular weight markers (the 118 bp fragment is indicated); 1 – positive control follicular lymphoma carrying t(14;18); 2 – negative control (no DNA); 3–7 – follicular lymphomas showing amplification of different sized fragments in different cases. The products in lanes 5 and 6 were amplified from different tissue samples from a single patient

size and resolution required. For larger products (300 bp and larger) agarose gels are more convenient and use less toxic reagents. If smaller fragments must be accurately sized, then polyacrylamide gels are required.

A basic analysis is performed in order to:

- determine whether the fragment is present or absent, for example to identify an infectious agent such as Epstein–Barr Virus (EBV) or *Helicobacter pylori* or to identify a chromosome translocation,
- determine the size of the fragment to confirm specificity or to identify insertion or deletion, and
- provide a suitable template for sequencing reactions

Gene scanning is a method that uses fluorescently-labelled primers and product analysis on an automated sequencing type gel. Results appear as plots of fluorescence against time that can be used to size products in comparison with control products of known size. This is performed in order to size products accurately and to quantify product yield (see Chapter 12). This method is well suited to a high throughput approach.

Single-stranded Conformational Polymorphism Analysis

The basic analysis of PCR products relies on the fact that double-stranded DNA molecules migrate through the gels on the basis of their size; larger products migrate more slowly than smaller ones. If the products are run whilst single-stranded under non-denaturing conditions, however, then their rate of migration is determined by their primary sequence as a result of sequence-specific self-folding. This is determined by the amount and type of complementary sequence binding within the DNA strand. A single base change results in altered mobility of products in the majority of cases. Products can be denatured and encouraged to remain single-stranded using buffer systems high in concentration of formamide and by running them on non-denaturing polyacrylamide gels at low temperatures (for example, 5°C). This method is known as single-stranded conformational polymorphism (SSCP) analysis. It is a very powerful method for screening for mutations in gene fragments of interest, for example mutational hotspots in oncogenes (Yamashita *et al.*, 2001). By amplifying small fragments, advantage can be taken of archival paraffin-wax-embedded material to permit the study of large numbers of cases of diseases of interest, for example BAX mutations in lymphomas (Martini *et al.*, 2000) and p53 mutations in breast carcinoma (Lukas *et al.*, 2000). A pitfall of this technique is that PCR error may contribute falsely mutated sequences to the product pool. This possibility is enhanced when copy number and DNA quality are low. Thus, DNA quality must be assessed and all novel sequences identified must be verified by duplication.

Restriction Fragment Length Polymorphism

Restriction digestion of products may be carried out to identify specific mutations that alter restriction sites, for example in haemochromatosis (Lynas, 1997). This analysis can only be applied to products containing restriction sites, but may have advantages over SSCP techniques that miss some mutations.

Heteroduplex Analysis

This involves heat denaturation of heterogeneous PCR products followed by rapid cooling on ice, which encourages the formation of mismatched products rather than perfectly matched ones. Any mismatched 'heteroduplexes' can be identified by altered electrophoretic mobility. Heteroduplex analysis can be used to search for point mutations (Xerri *et al.*, 1995) or to resolve genuine monoclonal products from polyclonal products derived from rearranged antigen receptor genes (Bottaro *et al.*, 1994).

Sequence Analysis

Sequencing is now a rapid, inexpensive and reliable method, using commercially-available user-friendly kits. The most demanding phase of the process is the generation of a high-quality template. This is relatively easy to achieve, using the PCR, especially with the development of polymerase enzymes and optimized cycling parameters for amplifying long fragments (up to several kilobases). The size of fragment that can be amplified from degraded DNA extracts from paraffin-wax-embedded samples is limited but fragments of up to 300 bp have been sequenced directly without cloning of PCR products (Goodrow *et al.*, 1992).

9.4 RT-PCR

RT-PCR is important because it permits analysis of mRNA and therefore the study of gene transcription and easier analysis of multiple exons. RNA is extracted and mRNA reverse transcribed using reverse transcriptase. mRNA can be non-specifically transcribed into cDNA using a polyT primer or random hexamers, or specifically transcribed using gene-specific primers. The latter, conveniently the antisense primer used for the PCR, have been found to be most efficient for fixed tissue samples. The standard PCR is carried out after the generation of cDNA to amplify the desired region. Applications include translocation detection (Liu *et al.*, 2002), the study of RNA viruses such as Hepatitis C or Measles (Komminoth *et al.*, 1994), or to quantify cytokine expression (El-Sherif *et al.*, 2001). Crucially, methods have been designed that permit analysis of paraffin samples although there are limitations on the fragment sizes that can be amplified.

9.5 Quantitative PCR

The generation of PCR products can be monitored during the reaction process using fluorescently labelled probes, a process known as real-time PCR (Lehmann and Kreipe, 2001). Accurate quantification of DNA and RNA and high-throughput can be achieved using TaqMan, LightCycler and iCycler, technology. This is a rapidly expanding field that is currently expensive but will become more affordable as competition increases. Real-time PCR is particularly important in minimal residual disease studies (see Chapter 12).

9.6 DNA and RNA extraction

Suitable DNA or RNA for PCR analysis can be obtained from most tissue sources, including fresh or frozen unfixed tissue, cytology smears, stained sections and blood films. In histopathology, a major breakthrough was the discovery that DNA suitable for molecular genetic analysis could be extracted from formalin-fixed paraffin-wax-embedded tissue. Although the DNA obtained is highly degraded when compared with that from fresh or frozen tissue samples and methods such as Southern blot analysis are impossible, PCR amplification of small fragments is successful in most cases. The average DNA size also differs in extracts from different specimens; this is thought to result mainly from differences in fixatives or duration of fixation (Greer *et al.*, 1991). As a result some samples cannot be PCR amplified but most will succeed, given appropriate handling.

Some fixatives such as Bouin's, must be avoided but the most widely used solutions, such as buffered formalin, are suitable if the fixation time is not prolonged. The tissue can remain in the wax block for many years without impairing DNA quality, but if it remains in formalin for several days or more then the PCR is unlikely to be successful. Much frustration has occurred for this reason in studies of non-urgent material such as *post-mortem* cases, where fixation is usually prolonged.

DNA extraction from paraffin-wax-embedded samples is a relatively quick and simple procedure and the complicated methodology normally associated with tissue DNA extraction using unpleasant organic extraction media such as phenol and chloroform is not required. Because of the sensitivity of the PCR, only small amounts of tissue are needed, typically a single 5 μm section will suffice. Removal of wax is followed by digestion with a protease enzyme. The simplest procedures use heat inactivation of protease and direct use of the supernatant without further purification. In some cases the extract contains sufficiently high concentrations of PCR inhibitors for the reaction to fail, necessitating further purification. This is particularly likely in some tissue types such as skin and liver. Commercially available spin column methods for digestion and purification of extracts are simple, reliable and safe to use. The extraction process can be easily performed in a single day and therefore permits rapid throughput of samples. However, overnight digestion may be necessary to improve yields in some tissues.

RNA extraction from formalin-fixed paraffin-wax-embedded tissue samples can be performed using commercially available kits. As with DNA, the RNA may be relatively degraded and amplification of small fragments is necessary. It must be anticipated that some samples will not provide any usable RNA.

Although nucleic acids generally can be extracted from paraffin-wax sections, in some cases and in some applications it will be necessary to use fresh or frozen unfixed material for optimal results.

9.7 Correlation of the PCR with morphology

It is sometimes important to localize specific nucleic acid sequences to their tissue component or cell of origin. Other approaches, such as fluorescence *in situ* hybridization (FISH) may not be possible. Cells or tissue components of interest can be microdissected prior to DNA or RNA extraction and the PCR. This is a powerful tool with which to study the genetics of disease and the distribution of infectious agents.

Microdissection can be performed at several levels. For example, relatively large areas of tissue sections, such as lymphoid infiltrates, may be identified by haematoxylin and eosin staining and separated from the remainder of the section using a scalpel blade with low power microscopy or by eye. Smaller regions of tissue or groups of cells can be dissected under medium power microscopy using fine tools (Pan *et al.*, 1994). Finally, single cells can be analysed using dedicated equipment, such as laser capture microdissection. This is explored further in Chapter 10.

Applications of microdissection include localization of infectious agents, correlation of abnormal cytology with genetic aberrations in tumours and improvement of the sensitivity of clonality analysis and therefore tumour detection. Single cell PCR can be used to correlate genetic features with selected cells, such as immunoglobulin genes in lymphocytes at particular stages of differentiation and Reed–Sternberg cells in Hodgkin's lymphoma (Marafioti *et al.*, 2000).

In situ PCR is a controversial and demanding technique in which specific nucleic acid sequences are amplified within tissue sections and localized to their cell of origin. These techniques are discussed in detail in Chapter 11.

9.8 Problems

Poor DNA or RNA quality is the most common reason for PCR failure, this is best minimized by appropriate tissue preparation, especially fixation and processing. Because PCR is sensitive to a few target copies, false positive amplification may result from cross-contamination of tissue samples. This may occur before, during or after the extraction process. It is important to confirm PCR results by analysing patient material from more than one source, such as from additional biopsies or repeat extractions. The major danger of false positivity, however, arises from contamination of samples or reaction mixes with previously-amplified product. Strict precautions must be taken to minimize this risk. Meticulous care must be exercised in all aspects of the process and separation of PCR set-up and product analysis is essential. The use of pipette tips with filter barriers and molecular biology grade reagents and plastic-ware is advisable.

9.9 Applications

Typical applications of the PCR, which provide important information regarding the molecular genetics of disease but which do not, at present, have routine diagnostic or prognostic use are:

1. Mutation screening of apoptosis-related or tumour suppressor genes such as FAS and p53 to aid understanding of tumour biology and the role of specific genes in malignant transformation (Lukas *et al.*, 2000).
2. X-linked clonality analysis of androgen receptor (Tamura *et al.*, 1998).
3. Loss of heterozygosity (LOH) studies of pre-cancerous states (Lakhani *et al.*, 1995).

Further applications of the PCR are listed in Table 9.1. The PCR is still an under-used resource in histopathology; many diseases are still not well understood and the archives can provide a regular source of material for the study and disease correlation of new genes as they are discovered.

Table 9.1 Applications of PCR in cellular pathology

Antigen receptor gene rearrangement studies in lymphoma
Translocations in sarcomas, lymphomas and other tumours
Point mutations in tumour suppressor genes and oncogenes
Gene amplification
Infectious agents
Identity analysis (microsatellites)
Probe generation for *in situ* hybridization

9.10 Diagnostic applications

In diagnostic histopathology, important areas are those in which diagnosis is difficult by the use of conventional morphological and immunophenotypic methods and in which molecular aberrations that can readily be detected are strongly associated with diagnostic categories of disease. These criteria are currently met in two areas, namely: lymphomas and sarcomas, which will be considered in more detail below.

Lymphomas

The use of the PCR has been vital in the study and development of diagnostic techniques in lymphomas and leukaemias. First, it has enabled the cumbersome gene rearrangement techniques of Southern blot analysis to be replaced by rapid, inexpensive strategies for clonality analysis and the detection of chromosome aberrations that can be applied to paraffin-wax-embedded samples (Table 9.2). Secondly, it has simplified amplification and sequence analysis of immunoglobulin and T-cell receptor genes and this has greatly improved our understanding of the biology of the lymphomas and provided clone-specific markers for the study of dissemination, progression and response to therapy.

Chromosome aberrations

The earliest application of the PCR to diagnosis of lymphomas was the amplification of the t(14;18) translocation (Lee *et al.*, 1987). This translocation brings the BCL2 gene on chromosome 18 into the immunoglobulin heavy chain locus on chromosome 14, usually in the J region. This causes up-regulation of expression of BCL2 protein and protects the transformed cell from apoptosis. The translocation is a useful marker of follicular lymphoma, since it rarely is found in other lymphoma types other than diffuse large B-cell lymphomas that have transformed from follicular lymphomas. Its detection can be used to confirm the

Table 9.2 Gene targets for PCR identification of lymphomas

Gene target	Lymphoma	Purpose
Immunoglobulin heavy chain	B-cell	Benign vs. malignant
Immunoglobulin light chains	B-cell	Benign vs. malignant
T-cell receptor beta chain	T-cell	Benign vs. malignant
T-cell receptor gamma chain	T-cell	Benign vs. malignant
t(14;18)	Follicular	Classification
t(11;14)	Mantle cell	Classification
t(11;18)	MALT	Classification
t(8;14)	Burkitt's	Classification
t(2;5)	ALCL	Classification

diagnosis or to follow up the disease after therapy. Primers have been designed that span the breakpoint region. However, not all breakpoints are clustered in this region, necessitating the design of further primers targeting the BCL2 region. Even so, a proportion of rearrangements cannot be amplified using a simple PCR strategy, because of widespread breakpoints, so there is a significant false negative rate.

Other important translocations that have been targeted include t(11;14) in mantle cell lymphomas, which is a useful marker for diagnosis and follow-up and t(11;18) in MALT lymphomas, which identifies cases which do not respond to anti-*Helicobacter pylori* treatment (Liu *et al.*, 2002). t(11;14) often can be amplified from DNA, as many breakpoints are clustered within a close region. However, the same problem of false negativity occurs, as additional breakpoints are outside the major translocation cluster. t(11;18) is best amplified using RT-PCR because the breakpoints are widely spread within introns. Nevertheless, reliable procedures for RT-PCR in paraffin-wax-embedded material have been developed. t(2;5) is a characteristic feature of anaplastic large cell lymphomas, which are mainly of T-cell origin. Identification of this translocation by the PCR and RT-PCR has played a major role in the understanding of this disease (Kadin and Morris, 1998), although immunohistochemistry and FISH are the preferred diagnostic procedures.

Clonality analysis

Clonality analysis of lymphoid populations is more straightforward than that of other cell types because lymphocytes carry clone-specific antigen receptor gene rearrangements (Figure 9.3). B-cells rearrange their immunoglobulin heavy and

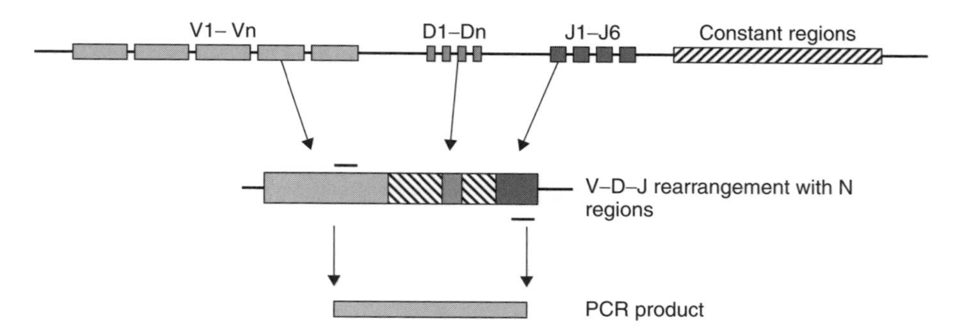

Figure 9.3 Immunoglobulin gene rearrangement and PCR clonality analysis. The germline immunoglobulin heavy chain gene with separate clusters of V, D and J regions (top) rearranges to form the coding sequence for the variable region (centre) containing single V, D and J regions with junctional N region addition (shaded). The PCR primers (narrow bars; centre) are designed to amplify across the highly variable junctional regions to yield products (bottom) of variable size depending on the specific rearrangement. A colour version of this figure appears in the colour plate section

light (κ and/or λ), chain genes, whereas T-cells rearrange their T-cell receptor genes (γ and δ and/or α and β). These rearrangements (Figure 9.3) are specific to a given lymphocyte or clone of lymphocytes. Identification of a predominance of a single rearrangement indicates a monoclonal or malignant proliferation of lymphocytes. The PCR can be used to amplify the variable regions of immunoglobulin or T-cell receptor genes and gel electrophoresis used to separate the fragments. Monoclonal populations are characterized by one or two dominant bands, representing a single or bi-allelic rearrangement, whereas polyclonal populations appear as a broad smear of products (Figures 9.4 and 9.5). This simple technique has proved to be of great help in the diagnosis of difficult cases of B- and T-cell lymphoma, in which morphological and immunohistochemical means have not provided a definite answer.

Certain types of lymphoproliferation cause particular problems at diagnosis, where it is unclear whether the lymphocyte proliferation is reactive or malignant. These include B-cell proliferations of the mucosal sites such as stomach and salivary gland (MALT lymphomas) and T-cell proliferations of skin and intestine. The PCR has provided a useful means of confirming a monoclonal nature in lymphomas of these types. However, the current methods have a relatively high false negative rate (in the region of 20%), which means that a negative result is unhelpful. Attempts are currently being made by a European Consortium (Biomed-2) to optimize and standardize methodologies in order to improve monoclonality detection rates and to ensure that all laboratories use appropriate, standardized protocols.

Clone-specific markers

The variable regions of rearranged antigen receptor genes are ideal clone-specific markers that can be amplified readily using the PCR. It can be deduced whether

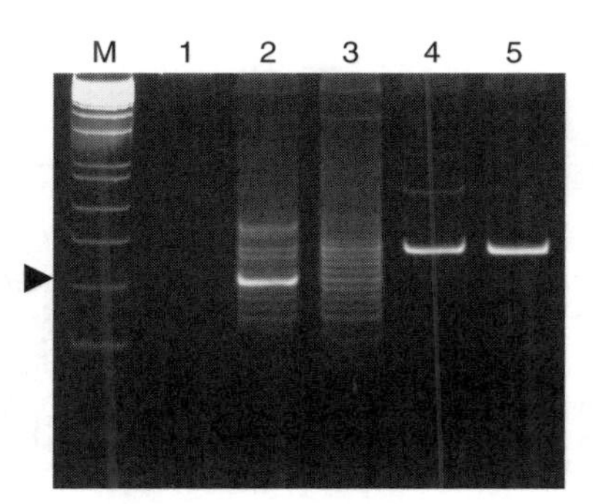

Figure 9.4 B-cell clonality analysis. Polyacrylamide gel of PCR products showing amplification of immunoglobulin heavy chain gene. Lanes were loaded as follows: M – molecular weight markers (the 82 bp fragment is indicated); 1 – negative control (no DNA); 2 – monoclonal control (B-cell lymphoma) showing a dominant band with a background smear; 3 – polyclonal control (reactive lymph node) showing a broad smear; 4 and 5 – duplicated test sample showing a reproducible dominant band indicating monoclonal B-cell expansion

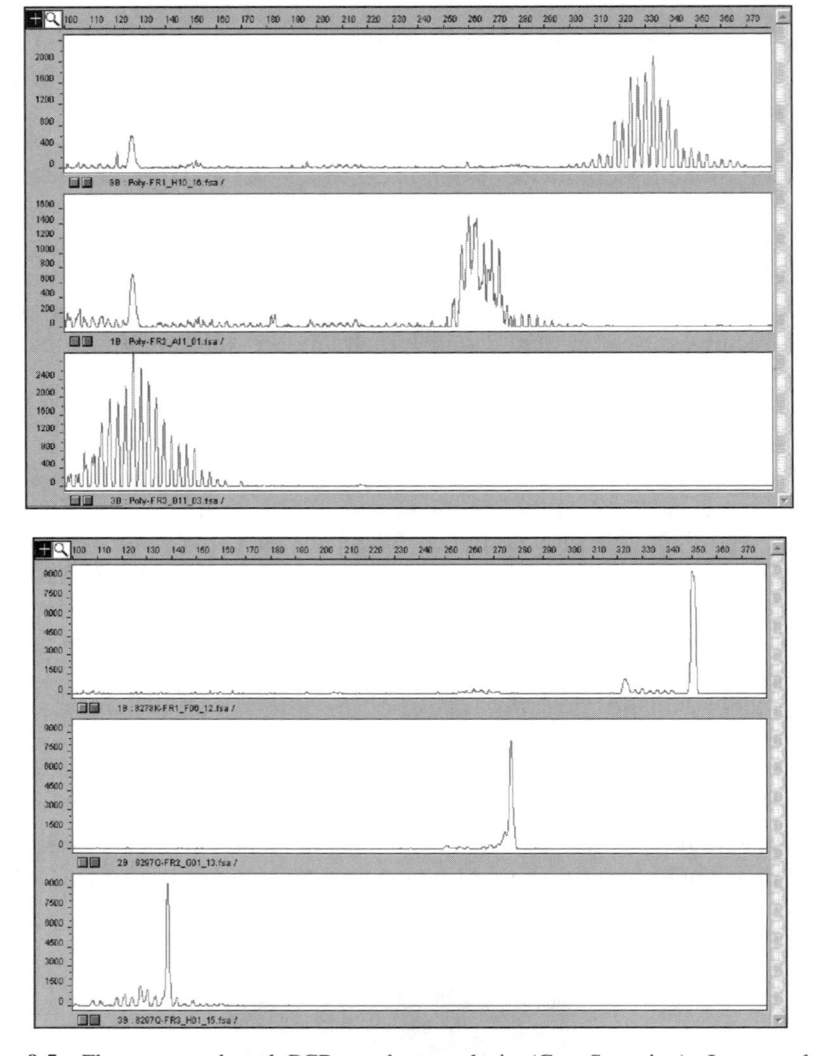

Figure 9.5 Fluorescence-based PCR product analysis (GeneScanning). Immunoglobulin heavy chain gene products from polyclonal (reactive lymph node; top) and monoclonal (B-cell lymphoma; bottom) B-cell populations. Three different primer sets were used amplifying from V region framework 1 (top panels), V region framework 2 (centre panels) and V region framework 3 (bottom panels) to J regions. The polyclonal patterns have a characteristic Gaussian distribution (top) whilst the monoclonal patterns show single dominant peaks (bottom). The image was kindly provided by Dr Helen White

B- or T-cell lymphomas arising at different sites or time points during the course of a disease are derived from a common clone. Clone-specific primers can be designed that will amplify only the tumour clone, greatly increasing the sensitivity of detection. This is useful in studies of minimal residual disease (Du *et al.*, 2000).

Sequence analysis of antigen receptor genes

Analysis of the sequence of the variable regions of the immunoglobulin genes in B-cell lymphomas allows deduction of the differentiation stage of the tumour clone and can help to classify the lymphoma. Naïve B-cells have no somatic mutations in their variable regions. This is a characteristic of small lymphocytic lymphoma and most mantle cell lymphomas. B-cells that have undergone the follicle centre reaction carry somatic mutations in their Ig variable regions; this also applies to plasmacytomas and follicular lymphomas. The latter can be distinguished by the presence of ongoing mutations; that is, different cells within the tumour clone are mutated to different degrees (Bahler and Levy, 1992). PCR amplification of Ig variable regions followed by sequence analysis therefore provides considerable information concerning the natural history of the disease, though this is not currently routinely applicable to diagnosis.

Sarcomas

Sarcomas are a group of aggressive tumours in which the diagnosis on histological grounds alone may be very problematic. Certain types, presenting in unusual sites, may be difficult to classify without the help of molecular genetic data. Many sarcomas carry recurrent chromosome translocations that can be used as diagnostic markers (Table 9.3, Figure 9.6). These translocations usually result in chimaeric fusion transcripts with a DNA- or RNA-binding domain combined with a transcription factor, for example EWS/FLI1, although there are examples of fusion proteins that act as tyrosine kinases [t(12;15)] or growth factors [t(17;22); Ladanyi *et al.*, 2000]. The translocations can be identified by conventional metaphase cytogenetics, Southern blot analysis, RT-PCR (Figure 9.7), immunohistochemistry or FISH.

The RT-PCR approach is the most practical and widely applicable, and analysis of these translocations has now become a routine part of the diagnosis

Table 9.3 Diagnostically important translocations in the sarcomas (Graadt van Roggen *et al.*, 1999)

Tumour	Translocation	Genes involved
Ewing's sarcoma/PNET	t(11;22)(q24;q12)	EWS/FLI1
	t(21;22)(q22;q12)	EWS/ERG
	t(7;22)(p22;q12)	EWS/ETV1
Clear cell sarcoma	t(12;22)(q13;q12)	EWS/ATF-1
Synovial sarcoma	t(X;18)(p11.2;q11.2)	SYT/SSX1;SSX2
Myxoid chondrosarcoma	t(9;22)(q22;q11-12)	EWS/CHN
Alveolar rhabdomyosarcoma	t(2;13)(q35;q14)	PAX3/FKHR
	t(1;13)(p36;q14)	PAX7/FKHR
Myxoid liposarcoma	t(12;16)(q13;p11)	CHOP/TLS

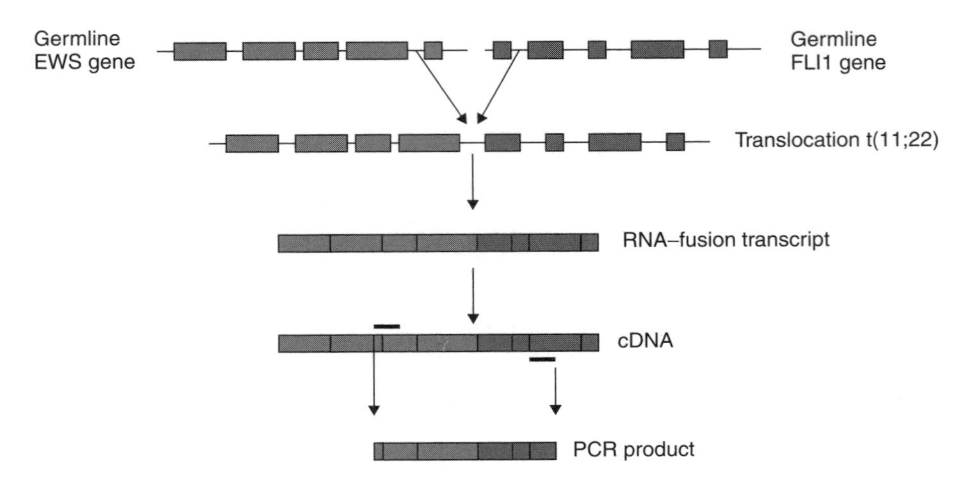

Figure 9.6 RT-PCR of chromosome translocation t(11;22) in Ewing's sarcoma, showing (top to bottom) the germline genes, fusion gene, fusion transcript, cDNA and PCR product. Chromosome 11 derived sequences are shown in red, chromosome 22 sequences in green. PCR primers are shown as black bars. (Re-drawn from Graadt van Roggen *et al.*, 1999.) A colour version of this figure appears in the colour plate section

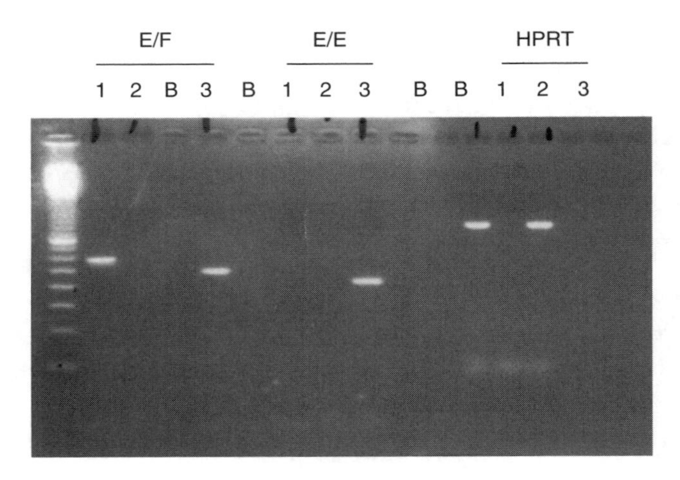

Figure 9.7 Agarose gel of RT-PCR products. 1 – Ewing's sarcoma tumour sample; 2 – water control; 3 – positive controls; B – blank lanes. E/F – amplification of t(11;22) showing positivity in the tumour sample and control. E/E – amplification of t(21;22) showing positivity in the control only. HPRT – control RT-PCR showing positivity in the tumour sample and positive control, but not in the negative control. The image was kindly provided by Professor PCW Hogendoorn

of the sarcomas in an increasing number of laboratories (Letson and Muro-Cacho, 2001).

RT-PCR of mRNA is required rather than PCR of genomic DNA, as the breakpoints are scattered within introns. Fresh or frozen tissue samples yield the most reliable mRNA, but routinely formalin-fixed, paraffin-wax-embedded samples usually can be used. An efficient RNA extraction process must be used, usually involving extended protease digestion and small fragments must be amplified (<150 bp) owing to degradation of RNA. Fixation must be of short duration and a proportion of archival blocks will yield no useful mRNA.

The stages in the technique are as follows:

- extraction of mRNA;
- reverse transcription/PCR amplification of the fusion transcript; and
- detection of specific product by gel electrophoresis or real-time, quantitative PCR.

The use of real-time PCR is becoming more popular, as this gives quantitative data and is suitable for a highly reproducible, high throughput strategy. This approach is also more sensitive and therefore suited to assessment of minimal residual disease. Further elucidation of the molecular genetic aberrations that contribute to the malignant transformation of the sarcomas will lead to further markers helpful in the diagnosis and prognosis of this important group of tumours. For example, activating c-KIT mutations have been found in gastrointestinal stromal tumours (Allander *et al.*, 2001). Highly sensitive and specific PCR and RT-PCR strategies will become of primary importance in the assessment of these diseases.

9.11 Infectious diseases

Modern diagnostic microbiology laboratories have embraced the PCR as a powerful and reliable means of complementing traditional methods for the identification of infectious agents. This trend has not been followed in many histopathology departments in a routine diagnostic sense but many PCR-based studies have been undertaken to investigate infectious agents in disease processes using paraffin-wax-embedded tissue samples. There is great potential to extend this area of study. Methods for demonstration of an enormous number of organisms including bacteria, viruses, parasites and fungi have been reported. Useful targets have included (a) *H. pylori* to enable correlation of infection with development of gastric ulcer and cancer, (b) *Mycobacterium tuberculosis* to aid diagnosis of difficult cases (Mehrotra *et al.*, 2002) and (c) HPV to study the role of the virus in cervical cancer (Wang-Johanning *et al.*, 2002).

9.12 Identity

Modern forensic science uses PCR analysis of highly polymorphic regions of the genome, which permits identification of individuals. The most informative and easily analysed regions are the microsatellites, or short tandem repeats, usually di- or tri-nucleotide repeats. PCR amplification across these regions can determine the numbers of repeats that can be related to individuals. Amplification of several such regions gives an individual 'fingerprint'. Finding a common pattern in different individuals other than identical twins is extremely unlikely. Therefore this method can be used to identify mixed-up histological tissue samples and identify cross-contaminated blocks (Abeln *et al.*, 1995).

9.13 The future

PCR and RT-PCR are now well-established methods for the study of nucleic acid sequences in histological material. The challenge is to put the technology to full use by identification of the genetic changes that are important in human disease. Knowledge of these changes will permit the establishment of rapid, cost-effective tests for diagnosis and monitoring of disease as well as providing a basic understanding of their underlying causes. These developments will be enhanced by our unravelling of the complete human genome and by improvements in techniques for gene amplification using histological samples.

Optimization and standardization of techniques by multi-centre collaboration are required to ensure that the PCR and related methods are used to their full potential in patient management.

9.14 References

Abeln EC, van Kemenade FD, van Krieken JH *et al.* (1995) Rapid identification of mixed up bladder biopsy specimens using polymorphic microsatellite markers. *Diagn. Mol. Pathol.* **4**: 286–91.

Allander SV, Nupponen NN, Ringner M *et al.* (2001) Gastrointestinal stromal tumors with KIT mutations exhibit a remarkably homogeneous gene expression profile. *Cancer Res.* **61**: 8624–8.

Bahler DW and Levy R (1992) Clonal evolution of a follicular lymphoma: evidence for antigen selection. *Proc. Natl Acad. Sci. USA* **89**: 6770–4.

Bottaro M, Berti E, Biondi A *et al.* (1994) Heteroduplex analysis of T-cell receptor gamma gene rearrangements for diagnosis and monitoring of cutaneous T-cell lymphomas. *Blood* **83**: 3271–8.

Diss TC and Pan L (1997) Polymerase chain reaction in the assessment of lymphomas. *Cancer Surv.* **30**: 21–44.

Du MQ, Diss TC, Dogan A *et al.* (2000) Clone-specific PCR reveals wide dissemination of gastric MALT lymphoma to the gastric mucosa. *J. Pathol.* **192**: 488–93.

Eckert KA and Kunkel TA (1991) DNA polymerase fidelity and the polymerase chain reaction. *PCR Methods Appl.* **1**: 17–24.

El-Sherif AM, Seth R, Tighe PJ *et al.* (2001) Quantitative analysis of IL-10 and IFN-gamma mRNA levels in normal cervix and human papillomavirus type 16 associated cervical precancer. *J. Pathol.* **195**: 179–85.

Goodrow TL, Prahalada SR, Storer RD *et al.* (1992) Polymerase chain reaction/sequencing analysis of ras mutations in paraffin-embedded tissues as compared with 3T3 transfection and polymerase chain reaction/sequencing of frozen tumor deoxyribonucleic acids. *Lab. Invest.* **66**: 504–11.

Graadt van Roggen JF, Bovee JV, Morreau J and Hogendoorn PC (1999) Diagnostic and prognostic implications of the unfolding molecular biology of bone and soft tissue tumours. *J. Clin. Pathol.* **52**: 481–9.

Greer CE, Peterson SL, Kiviat NB *et al.* (1991) PCR amplification from paraffin-embedded tissues. Effects of fixative and fixation time. *Am. J. Clin. Pathol.* **95**: 117–24.

Impraim CC, Saiki RK, Erlich HA and Teplitz RL (1987) Analysis of DNA extracted from formalin-fixed, paraffin-embedded tissues by enzymatic amplification and hybridization with sequence-specific oligonucleotides. *Biochem. Biophys. Res. Commun.* **142**: 710–6.

Kadin ME and Morris SW (1998) The t(2;5) in human lymphomas. *Leuk.* Lymphoma **29**: 249–56.

Kikuchi Y, Kishi T, Suzuki M, Furusato M and Aizawa S (1994) Polymerase chain reaction–single strand conformation polymorphism analysis of the p53 gene in paraffin-embedded surgical material from human renal cell carcinomas. *Virchows Arch.* **424**: 229–33.

Komminoth P, Adams V, Long AA *et al.* (1994) Evaluation of methods for hepatitis C virus detection in archival liver biopsies. Comparison of histology, immunohistochemistry, in-situ hybridization, reverse transcriptase polymerase chain reaction (RT-PCR) and *in-situ* RT-PCR. *Pathol. Res. Pract.* **190**: 1017–25.

Ladanyi M and Bridge JA (2000) Contribution of molecular genetic data to the classification of sarcomas. *Hum. Pathol.* **31**: 532–8.

Lakhani SR, Collins N, Stratton MR and Sloane JP (1995) Atypical ductal hyperplasia of the breast: clonal proliferation with loss of heterozygosity on chromosomes 16q and 17p. *J. Clin. Pathol.* **48**: 611–5.

Lee MS, Chang KS, Cabanillas F *et al.* (1987) Detection of minimal residual cells carrying the t(14;18) by DNA sequence amplification. *Science* **237**: 175–8.

Lehmann U and Kreipe H (2001) Real-time PCR analysis of DNA and RNA extracted from formalin-fixed and paraffin-embedded biopsies. *Methods* **25**: 409–18.

Letson GD and Muro-Cacho CA (2001) Genetic and molecular abnormalities in tumors of the bone and soft tissues. *Cancer Control* **8**: 239–51.

Liu H, Ye H, Ruskone-Fourmestraux A *et al.* (2002) t(11;18) is a marker for all stage gastric MALT lymphomas that will not respond to *H. pylori* eradication. *Gastroenterology* **122**: 1286–94.

Lukas J, Niu N and Press MF (2000) p53 mutations and expression in breast carcinoma *in situ*. *Am. J. Pathol.* **156**: 183–91.

Lynas C (1997) A cheaper and more rapid polymerase chain reaction-restriction fragment length polymorphism method for the detection of the HLA-H gene mutations occurring in hereditary hemochromatosis. *Blood* **90**: 4235–6.

Marafioti T, Hummel M, Foss HD *et al.* (2000) Hodgkin and Reed–Sternberg cells represent an expansion of a single clone originating from a germinal center B-cell with functional immunoglobulin gene rearrangements but defective immunoglobulin transcription. *Blood* **95**: 1443–50.

Martini M, D'Alo F, Pierconti F *et al.* (2000) Bax mutations are an infrequent event in indolent lymphomas and in mantle cell lymphoma. *Haematologica* **85**: 1019–23.

Mehrotra R, Metz P, Kohlhepp S (2002) Comparison of in-house polymerase chain reaction method with the Roche Amplicortrade mark technique for detection of *Mycobacterium tuberculosis* in cytological specimens. *Diagn. Cytopathol.* **26**: 262–5.

Pan LX, Diss TC, Peng HZ *et al.* (1994) Clonality analysis of defined B-cell populations in archival tissue sections using microdissection and the polymerase chain reaction. *Histopathology* **24**: 323–7.

Perosio PM and Frank TS (1993) Detection and species identification of mycobacteria in paraffin sections of lung biopsy specimens by the polymerase chain reaction. *Am. J. Clin. Pathol.* **100**: 643–7.

Richie KL, Goldsborough MD, Darfler MM *et al.* (1999) Long PCR for VNTR analysis. *J. Forensic Sci.* **44**: 1176–85.

Saiki RK, Scharf S, Faloona F *et al.* (1985) Enzymatic amplification of beta-globin genomic sequences and restriction site analysis for diagnosis of sickle cell anemia. *Science* **230**: 1350–4.

Sotlar K, Marafioti T, Griesser H *et al.* (2000) Detection of c-kit mutation Asp 816 to Val in microdissected bone marrow infiltrates in a case of systemic mastocytosis associated with chronic myelomonocytic leukaemia. *Mol. Pathol.* **53**: 188–93.

Tamura M, Fukaya T, Murakami T *et al.* (1998) Analysis of clonality in human endometriotic cysts based on evaluation of X chromosome inactivation in archival formalin-fixed, paraffin-embedded tissue. *Lab. Invest.* **78**: 213–8.

Wang-Johanning F, Lu DW, Wang Y *et al.* (2002) Quantitation of human papillomavirus 16 E6 and E7 DNA and RNA in residual material from ThinPrep Papanicolaou tests using real-time polymerase chain reaction analysis. *Cancer* **94**: 2199–210.

Xerri L, Parc P, Bouabdallah R *et al.* (1995) PCR-mismatch analysis of p53 gene mutation in Hodgkin's disease. *J. Pathol.* **175**: 189–94.

Yamashita K, Yoshida T, Shinoda H and Okayasu I (2001) Novel method for simultaneous analysis of p53 and K-ras mutations and p53 protein expression in single histologic sections. *Arch. Pathol. Lab. Med.* **125**: 347–52.

9.15 Online information

Nucleotides

`http://www.ncbi.nlm.nih.gov/entrez/query.fcgi/`

Genome

`http://www.nature.com/genomics/`

`http://www.gdb.org/`

Antibody sequences

`http://www.mrc-cpe.cam.ac.uk/imt-doc/public/INTRO.html`

Primer design

`http://www-genome.wi.mit.edu/cgi-bin/primer/primer3.cgi/`

`http://bioinformatics.weizmann.ac.il/mb/bioguide/pcr/software.html`

10

Laser Capture Microdissection: Techniques and Applications in the Molecular Analysis of the Cancer Cell

Amanda Dutton, Victor Lopes and **Paul G. Murray**

10.1 Introduction

Over recent years dramatic advances in molecular technology have provided new tools to analyse the underlying processes that result in disease. However, the quality, reliability and interpretation of the data are highly dependent on the homogeneity of the cell type under investigation. Fundamentally, tissues are heterogeneous and complicated structures composed of admixtures of cell types.

Microdissection of tissue sections and cytological preparations has become an increasingly important technique for cell isolation, facilitating the *in vivo* examination of cells that are otherwise inaccessible by traditional methodology. Several methods of microdissection have been developed, ranging from ablation of unwanted cellular material to manual micromanipulation; however, these approaches frequently are time consuming, tedious, imprecise and are highly dependent on operator dexterity. Such factors limit their practical usefulness.

Laser capture microdissection (LCM) is a novel technique that permits a one-step procurement of pure cells from tissue and cell preparations. Developed at the National Cancer Institute of the National Institute of Health (NIH), Bethesda, by Michael Emmert-Buck and colleagues (Emmert-Buck *et al.*, 1996), LCM overcomes many of the drawbacks of traditional microdissection techniques. Commercialisation of this technology was achieved in 1997 as part of

Molecular Biology in Cellular Pathology. Edited by John Crocker and Paul G. Murray
 ISBN: 0-470-84475-2

a Collaboration Research and Development Agreement partnership between the NIH group and Arcturus Engineering Inc. The incorporation of LCM within the 'cancer gene anatomy project' (CGAP), which aims to identify genes specific to non-cancerous, pre-cancerous and malignant cells, has led to its accelerated development (Strauberg *et al.*, 1997).

This chapter describes the principle behind LCM, focusing on the important technical requirements for successful cell isolation and the advantages and drawbacks associated with the procedure. Finally, the growing applications of LCM are reviewed with particular emphasis on its use in cancer research.

10.2 The principle of LCM

LCM uses the precision of lasers to selectively adhere target cells to a thermoplastic transfer membrane (Figure 10.1). Initially, the cells of interest are identified within a dehydrated uncoverslipped tissue or cell preparation using an inverted microscope. Prior staining is often necessary to visualise cell and tissue morphology; a simple haematoxylin stain usually is sufficient, although other staining procedures such as immunohistochemistry may be employed (see below). A thermoplastic transparent membrane, made from ethylene vinyl acetate (EVA), mounted on an optically clear cap-like structure is then placed over the preparation using the LCM transfer film carrier arm mechanism. A low-energy infrared laser beam is focused on a cell or a group of cells using a joystick-controlled stage. The laser is then activated at the push of a button resulting in focal transient melting of the EVA onto the underlying cell(s) of interest. The melted polymer expands and fuses with the target, filling the interstices of the cell(s); within milliseconds the polymer solidifies and forms an integral part of the tissue composite. Stronger bonds are formed between the plastic film and the cell(s) than those that exist between the cell(s) and the glass slide. Consequently, when the transfer film is manually lifted from the preparation the chosen cellular material is also lifted. The EVA transfer film has a diameter of 6 mm, which facilitates the capture of approximately 3000 cells per film. These multiple functions are integrated into the PixCell II™ LCM System, which is commercially available from Arcturus Engineering Inc. (Mountain View, CA, web site `http://www.arctur.com/`). The LCM system has three adjustable laser beam sizes, 7.5, 15 and 30 μm, which enable the capture of single cells and small clusters of cells. Further adjustable laser parameters include impulse duration, frequency and beam power, all of which can be altered to optimise the capture conditions for a particular tissue type.

Following microdissection, the cells attached to the transfer film can be processed for the particular downstream application. The cap with the cells attached can be directly fitted onto a standard 0.5-mL microcentrifuge tube containing buffer solution. Although transferred tissue is firmly bound to the transfer film, it

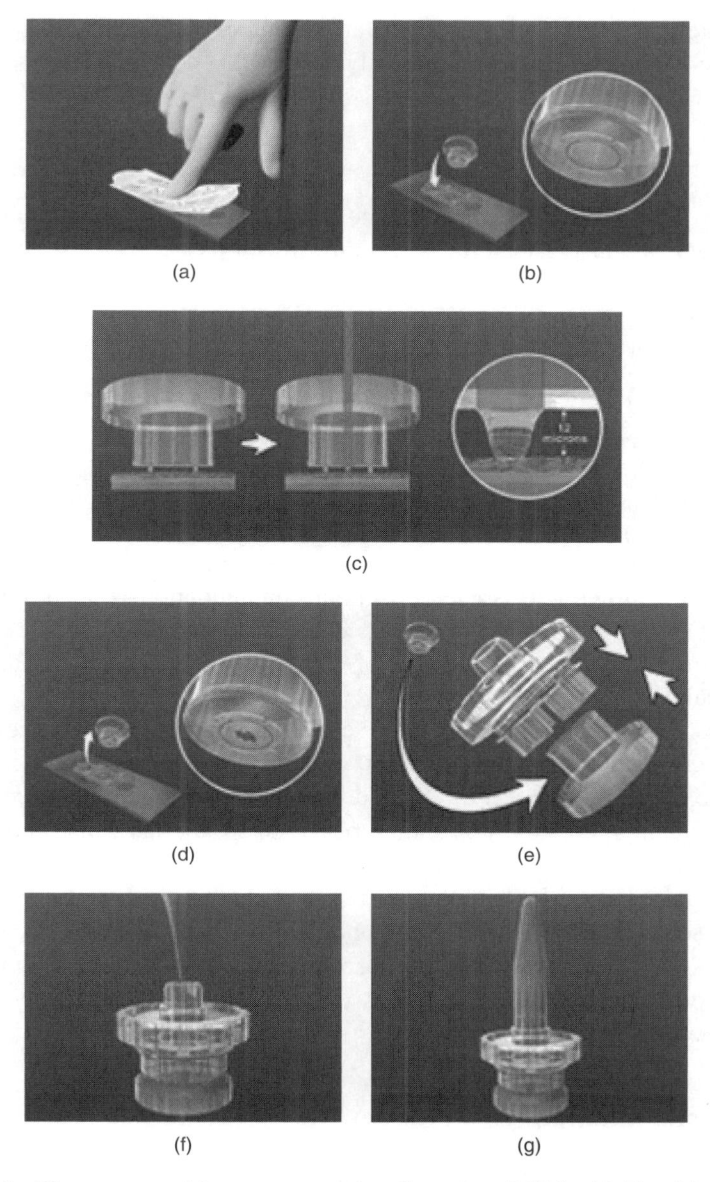

Figure 10.1 The process of laser capture microdissection (LCM). (a) The dried uncover-slipped tissue preparation is ready for laser capture. PrepStrips® are gently placed over the tissue; these remove loose debris that would otherwise adhere to the transfer film. (b) The cap containing transfer film is placed over the tissue. Shown here are the high sensitivity caps with circular inner rail that prevent direct contact between cap and tissue. (c) The laser is fired and melts the transfer film, which forms a bond between the cap and the chosen cell(s). (d) The cap is lifted together with adherent cells. (e) The ExtracSure® device is then placed over the cap to create a chamber into which (f) buffer can be pipetted. (g) A microcentrifuge tube is then placed over the ExtracSure® device. (Figure reproduced with permission from Arcturus Ltd). A colour version of this figure appears in the colour plate section

is completely accessible to aqueous solutions and extraction of cell components (DNA, RNA and protein) can be readily achieved.

10.3 Technical considerations

LCM is a powerful technical aid to the isolation of pure cell samples from a wide variety of starting material, including frozen tissue sections, conventional formalin-fixed, paraffin-wax-embedded tissue sections and cytological preparations such cytospin preparations or smears. However, there are several technical requirements for successful LCM.

Section Preparation

Tissue sections should be cut onto uncoated sterile (for optimum preservation of nucleic acids) glass slides. The optimal thickness is considered to be 7 μm; cells do not lift easily from thicker section and because thicker sections are >1 cell thick, single cell isolation without contamination is accordingly more difficult. On the other hand, captured cells from thinner sections might not contain a representative complement of cell constituents and important genetic or biochemical information may be lost.

LCM is compatible with the use of section adhesives; however, cell transfer can be more difficult if they are used. Adequate section adhesion is required so that the tissue remains on the slide during the necessary pre-treatment steps, including staining, but adhesion should not be so strong that cells cannot be captured. Section adhesives are just one aspect of section preparation that affects ease of capture, others include the tissue type, whether heat treatments are used to adhere sections, and the use of immunohistochemistry or *in situ* hybridisation prior to LCM, which will inevitably lead to a greater chance of section detachment. In practice, optimum conditions can be determined relatively easily by 'trial and error'.

Fixation and Staining

The success of a particular downstream application is highly dependent on cellular fixation and preservation prior to LCM. Typically, frozen sections or cytospin preparations are fixed in alcohol. Although aldehyde-based fixation can also be used, such cross-linking agents diminish the quality and quantity of nucleic acid and protein. Frozen sections or cytospins are usually required for RNA and protein work. While DNA can be successfully obtained from paraffin-wax-embedded material, the yield and quality of DNA is superior from frozen sections. For the successful preservation of RNA, special procedures are essential both during tissue preparation and in subsequent laser capture.

These include the use of RNAse-free reagents and equipment throughout. Ideally, LCM should be performed immediately after sectioning but if this is not possible then frozen sections can be stored at −70°C in airtight dessicators. Prior to use, sections are slowly brought to room temperature before the dessicator is opened; this prevents condensation onto the glass slide. One of the most efficient ways to prevent RNAse activity is to keep the section completely free of water. The use of RNAse-inhibitors is also recommended; Ambion Ltd (`http://www.ambion.com/catalog/`) supplies a diverse range of anti-RNAse reagents. Water content also has a considerable effect on the efficiency of cell transfer, evidently even minute quantities can prevent capture occurring. Consequently, immediately prior to capture, tissue sections must be completely dehydrated and washed in fresh xylene. A useful set of tissue preparation protocols is available at `http://www.arctur.com/technology/protocols.html`.

10.4 Advantages and disadvantages of LCM

Advantages

Prior to the development of LCM, the isolation of cells was generally achieved by relatively crude microdissection using manual tools, by so-called 'ablation' techniques, where all but the wanted cells were removed or destroyed, and by a number of micromanipulator-based approaches. LCM has major advantages compared with many of these methods. For example, its simple one step transfer procedure is markedly faster and easier to perform than most other methods. When preparation conditions are optimised, LCM facilitates the procurement of thousands of cells within minutes. Furthermore, there is no requirement for any manual micromanipulation, thus the need for a high standard of operator dexterity is reduced. Although single cell collection requires optimisation of the capture conditions, it is possible to reliably and accurately obtain single cells from both paraffin wax and frozen tissue sections (Figure 10.2). Procurement of cell clusters (Figure 10.3) can be rapidly achieved by repeatedly pulsing the laser over the required area. At the time of laser activation the tissue is exposed to peak temperatures of approximately 90°C for several milliseconds. Such brief thermal transients do not adversely affect preservation of DNA, RNA or proteins (Emmert-Buck *et al.*, 1996; Goldstein *et al.*, 1998).

One of the major benefits of LCM is that the morphology of the captured cells is maintained. Through the use of the LCM computer software, images of the microdissection process can be captured and stored. In particular, the transfer film can be visualised after capture allowing verification that the appropriate cells were microdissected and that unwanted cells or debris were not collected. Following microdissection, the surrounding unselected tissue remains intact on the slide and is therefore fully accessible for further LCM.

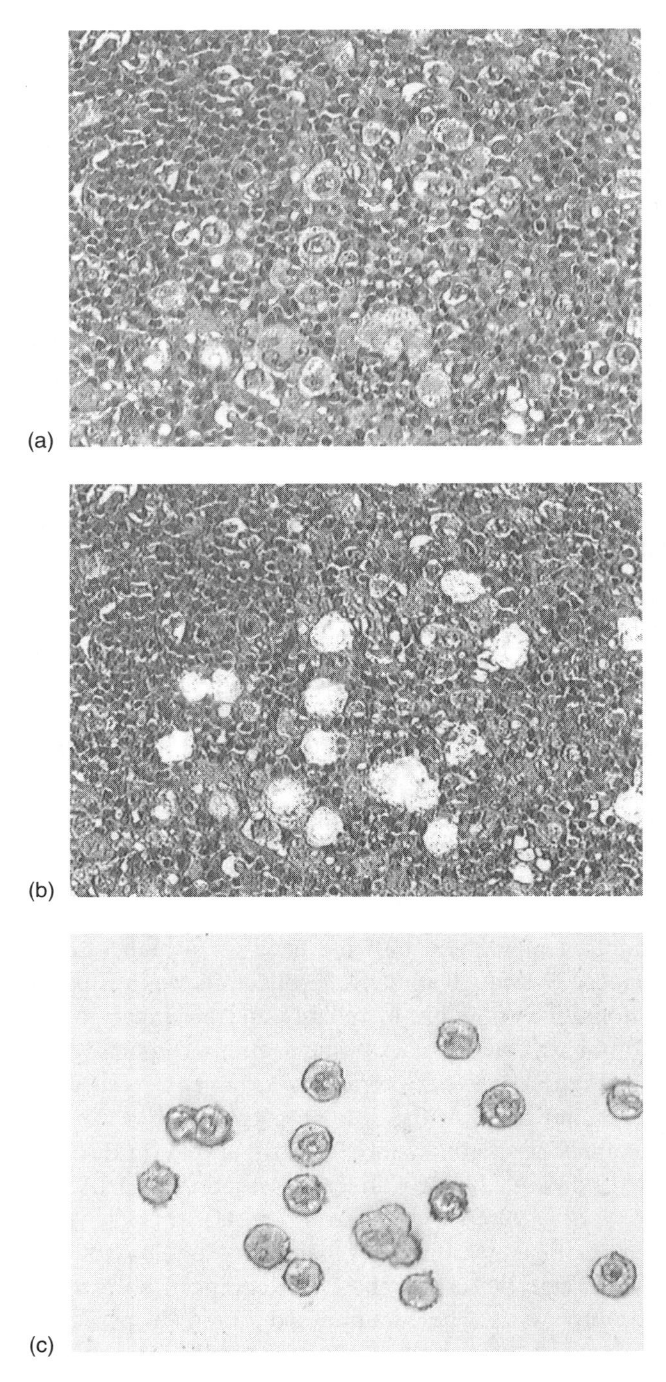

Figure 10.2 Microdissection of single Hodgkin/Reed–Sternberg (HRS) cells: (a) before capture; (b) after capture; (c) image on cap. A colour version of this figure appears in the colour plate section

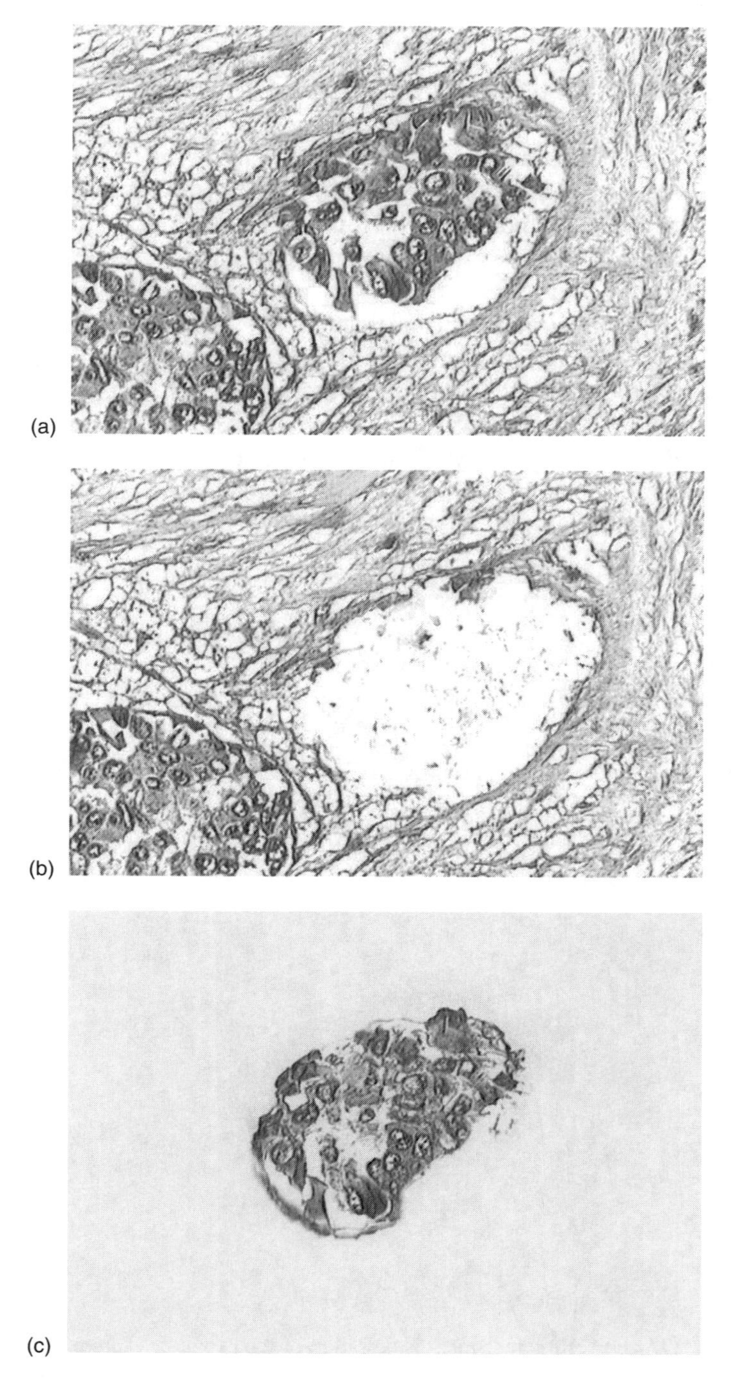

Figure 10.3 Microdissection of a cluster of breast cancer cells. (Figure courtesy of D. Lissauer, Dept Pathology, University of Birmingham). A colour version of this figure appears in the colour plate section

Disadvantages

The main limitation of this system is that optical resolution is reduced owing to the use of air-dried uncoverslipped tissue sections. Consequently, routine stains such as haematoxylin and eosin may not be sufficient for precise microdissection; this can be particularly troublesome in tissues with diffuse admixtures of cell types with indistinct cell morphology. However, such problems can be partially overcome with the application of immunohistochemical stains prior to LCM (Figure 10.4).

Occasionally, cells do not lift from section to transfer film, although usually this is resolved either by increasing the power of the laser or by adjusting the tissue preparation conditions (i.e. by reducing section adhesion). A drawback of increasing laser power is that the captured area increases and so can make single cell capture difficult. Conversely, if the section is too loosely bound to the glass slide, unwanted cellular material may adhere non-specifically to the transfer film. To help overcome this, 'high sensitivity' (HS) capture transfer films can

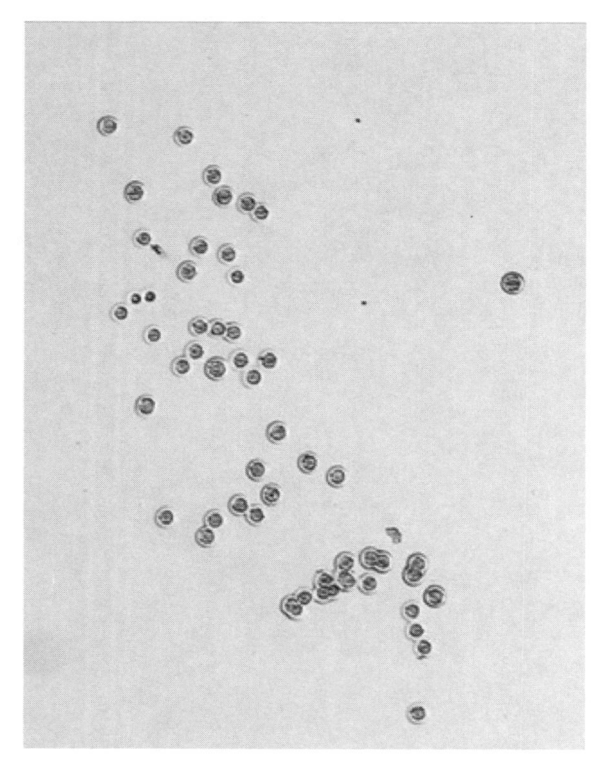

Figure 10.4 Low power microscopic visualisation of the cap after capture of CD30-stained HRS cells, showing multiple CD30 staining cells in the absence of any unwanted material. It is important to check the cap at this stage to ensure that no unwanted material is collected. (Figure courtesy of C. Lister, Dept Pathology, University of Birmingham). A colour version of this figure appears in the colour plate section

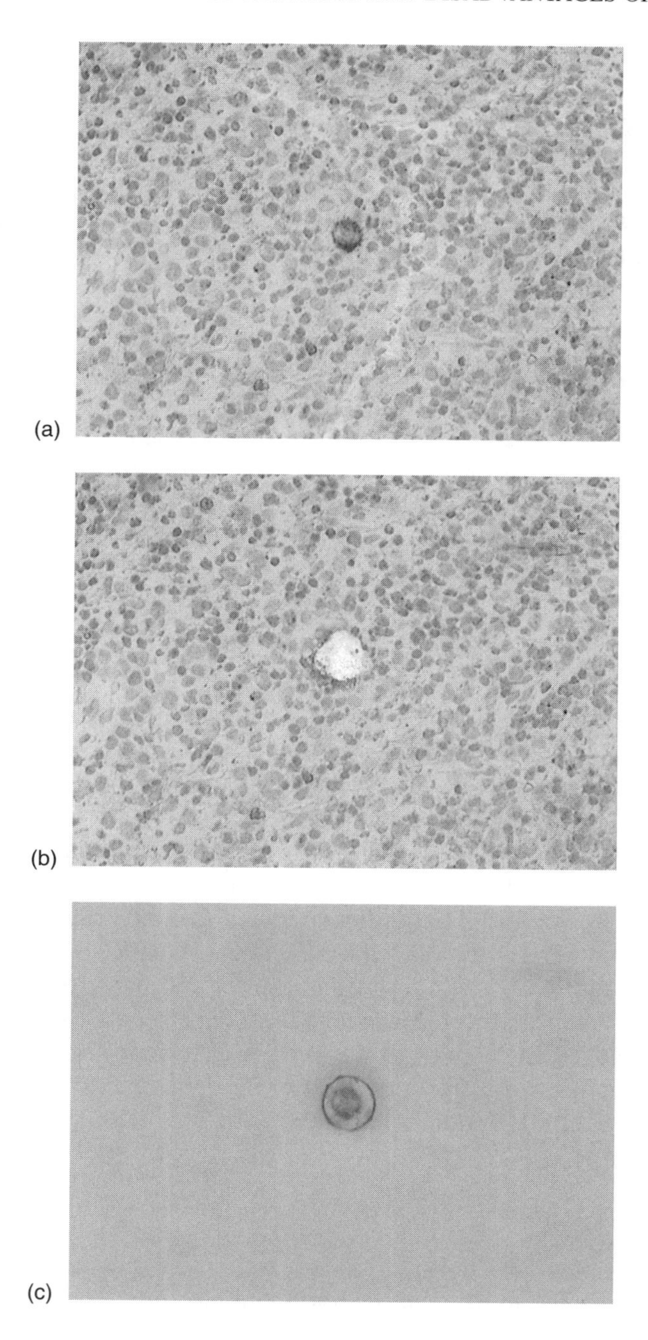

Figure 10.5 Immunohistochemistry can be used to direct the laser capture to defined cell types. Shown here is the microdissection of a single latent membrane protein-1 (LMP1) expressing cell within a tonsil from a patient with infectious mononucleosis. A colour version of this figure appears in the colour plate section

be used. These caps have a rail-like structure that ensures that the transfer film does not make direct contact with the tissue sections. After capture, ExtracSure® attachments can be fitted directly onto the HS caps. Once attached to the caps these form a sealed chamber that allows small volumes of buffer to be pipetted directly onto the cells and are ideally suited to the extraction of cell components from small cell numbers. Caps should always be visualised microscopically after capture and the cap discarded if any unwanted material is present. This is often best performed using one of the lower power objectives (Figure 10.5).

10.5 Applications of LCM

Since tumours are comprised of heterogeneous populations of cells there is a need for techniques that enable the analysis of pure tumour cell populations. Providing that the starting material is of sufficient quality, DNA, RNA and protein can be successfully extracted from tumour cells harvested through LCM, even at the single cell level.

DNA

DNA-based analyses of cells are easily facilitated with LCM. There usually is no need for long complicated DNA extraction and purification procedures because of the small numbers of cells involved. Cellular material is easily transferred from the film into solution by proteinase K digestion. Single-step DNA polymerase chain reaction (PCR) can then be applied. In fact, in their first paper, Emmert-Buck and colleagues described DNA analysis of LCM-procured cells (Emmert-Buck *et al.*, 1996). They reported the loss of heterozygosity (LOH) at several loci in several tumour types, including the BRCA1 gene in familial breast cancer, chromosome 8p in prostate cancer and the p16 gene in invasive oesophageal squamous cancer. Since then there have been a multitude of studies that have employed LCM prior to the genetic analysis of tumour cell populations (For examples, see Cohn *et al.*, 2001; Glasow *et al.*, 2001; Leonard *et al.*, 2001; Ling *et al.*, 2001; Liston *et al.*, 2001; Ortiz *et al.*, 2001; Takeshima *et al.*, 2001.)

Clonality studies

LCM is ideally suited to the analysis of tumour clonality. For example, Kremer *et al.* (2001) utilised LCM to analyse clonality in a case of Hodgkin's disease and a cutaneous T-cell lymphoma arising in the same patient. Sequencing of PCR products from pooled immunohistochemically defined Hodgkin/Reed–Sternberg (HRS) cells showed all HRS cells to have identical Ig heavy chain gene rearrangements. This confirmed that they were clonal B cells, unrelated to the neoplastic T cells. The same group performed similar studies on immunostained paraffin wax sections of rare cases of composite B-cell lymphomas that showed two morphologically and phenotypically distinct cell populations (Fend *et al.*, 1999a). Facilitated by LCM, DNA from each cell population was examined along with whole tissue DNA. Two distinct PCR products were derived from the two microdissected cell samples, indicating that they arose from unrelated B-cell clones. Amplification and detection of both cell populations was expected in the whole tissue analyses as it contained both cell populations. However, only a single band was identified, implying false monoclonality. Therefore, the preferential amplification of dominant cell populations that can occur in whole tissue PCR experiments is avoided if LCM is used to select the different cell populations within a tissue section.

LCM has also been used to examine the relatedness of different histological variants within the same tumour. In neuroblastoma, for example, two different tumour elements – neuroblastic and schwannnian stromal cells – can be observed, and although a common pluripotent cell of origin for both cell types has been proposed this has not been conclusively demonstrated. Mora *et al.* (2001) demonstrated by allelotype analysis at chromosomes 1p36, 11q23, 14q32 and 17q and by analysis of MYCN copy number that in virtually all cases these cell types had the same genetic composition.

The study of the non-malignant cells of tumours that in many cases comprise a significant fraction of the total tumour mass, has largely been ignored. However, LCM offers the opportunity to study these cells and their relationship to the malignant populations. A study by Kurose and colleagues (Kurose *et al.*, 2001) suggests a more important role for these cells in breast cancers than previously supposed. In their study, they microdissected neoplastic epithelium and stromal cells from sporadic invasive adenocarcinomas of the breast. The authors detected frequent LOH in both neoplastic and stromal compartments, and particular genetic changes could be seen to co-exist in both stroma and epithelium within the same tumour. Their results led them to propose a model of multi-step carcinogenesis involving both epithelium and stromal cells.

Comparative genomic hybridisation

Comparative genomic hybridisation (CGH) is a *in situ* hybridisation technique used to characterise chromosomal abnormalities where there is a net loss (deletion) or gain (duplication, insertion or amplification) of genetic material. It is based on the competitive hybridisation of labelled tumour DNA and normal DNA to normal metaphase chromosomes. CGH is not useful for the detection of balanced translocations or inversions where there is no overall gain or loss of material. The main advantage of CGH compared with conventional cytogenetics is that it is not necessary to obtain metaphase chromosome spreads from the tumour under investigation. Because of this, CGH is ideally suited to the analysis of samples where it is difficult to obtain good quality metaphase chromosomes. Unlike conventional cytogenetics, CGH is applicable to the study of archival tumour specimens (frozen and paraffin).

Recently, using a degenerate oligonucleotide primed-polymerase chain reaction (DOP-PCR) method it has been possible to amplify sufficient DNA for CGH even from minute amounts of starting material and in turn this has meant that CGH can now be regularly performed on microdissected samples. The combination of LCM and CGH has enabled the study of small lesions previously not accessible to global analysis of genetic changes. For example, it is now possible to microdissect small pre-malignant lesions and to begin to unravel the genetic changes that might characterise the early changes leading to malignancy. Umayahara and colleagues (Umayahara *et al.*, 2002) used CGH with microdissection to detect genetic alterations in normal epithelium, CIN and invasive carcinoma tissues colocalised in tumours from 18 patients with squamous cell carcinoma of the uterine cervix. Gains on chromosome 1 and on 3q and losses on 2q, 3p, 4, 6p, 11q and 17p were frequent alterations in CIN and invasive carcinoma lesions. The frequency and average number of genetic alterations corresponded directly to the extent to which the cervical carcinoma had progressed. Similarly, Marchio *et al.* (2001) examined hepatocellular carcinomas (HCCs) and nodules containing liver cell dysplasia and cirrhosis adjacent to the tumours. Seven cases of large cell changes (LCC) and three cases of small cell changes (SCC) were analysed. They found that a subset of chromosomal alterations present in HCCs was also present in the adjacent SCC, supporting the preneoplastic nature of SCC.

More recent developments in this technology are based on the use of array-CGH; instead of normal metaphase chromosomes, the targets for the labelled DNA in this method are genomic sequences (covering the entire genome or selected sequences), which are arranged on an array. Array-CGH has been applied to cancer cell populations isolated by LCM (Daigo *et al.*, 2001). In contrast to conventional CGH, array-CGH provides greater resolution and by the inclusion of target sequences spanning common translocation breakpoints, reciprocal translocations can also be detected. Since metaphase chromosomes are not required as targets for array-CGH, problems with user identification of metaphase chromosomes are eliminated.

Detection of viral sequences

LCM is ideally suited to the detection of viral DNA (or RNA) in tumours. This might be important when attempting to establish an aetiological link between a particular virus and a tumour. Where viral copy number in such cells is low, sensitive *in situ* hybridisation is required which is often challenging and requires isotopic-based methods. Whole section PCR generally is unsuitable since this will detect the presence of virus in both malignant and non-malignant cells, which may be a problem for ubiquitous viruses, such as the Epstein–Barr virus (EBV). Several groups have reported the successful demonstration of viral sequences in LCM-procured cells. One major advantage of this approach is that the nature of the cell type infected can be ascertained. This is well illustrated in the study by Torres-Munoz *et al.* (2001) who demonstrated the presence of HIV-1 DNA in neurons isolated from post-mortem brains by LCM. Studies of viral loads in tumours can be assessed by quantitative PCR. However, if viral copy number in tumour cells is to be measured, then non-malignant stromal cells must be eliminated from the analysis. This can be readily achieved by LCM. For example, Klussmann and co-workers (Klussmann *et al.*, 2001) used such an approach to measure HPV16 load in tumour cells laser captured from tonsillar carcinomas and their metastases. In our own studies, we have been able to reproducibly measure EBV loads in virus-associated tumour cells after LCM (Lissauer *et al.*, unpublished).

RNA

As mentioned earlier, RNA is extremely susceptible to degradation by RNAses and more stringent procedures during tissue preparation are required. Despite this, RNA is readily detectable from captured cells, even down to the single cell level.

Gene expression by RT-PCR

The identification of gene expression in tumour cells can provide important information on the nature and origin of cancers and also yield diagnostically or prognostically valuable data. Although immunohistochemistry and *in situ* hybridisation are extremely valuable tools to assess gene expression *in situ*, each has its own drawbacks. Immunohistochemistry is limited by the availability of antibody reagents. Currently, the identification and characterisation of new genes far exceeds the development of antibodies directed to their protein products. Although *in situ* hybridisation can detect mRNA species, frequently it is insufficiently sensitive to detect low abundance transcripts. Consequently, RT-PCR of captured cells provides an attractive alternative. RT-PCR analysis of single cells captured by LCM has been widely reported (Jin *et al.*, 1999). Furthermore, immunostaining prior to laser capture can be compatible with RNA preservation (Fend *et al.*, 1999b, 2000). We are able to amplify mRNAs from single cells or small numbers of cells following immunohistochemistry but only after a modified short (30 min) procedure incorporating RNA protection at all steps (Figure 10.6). Quantitative RT-PCR has been applied to microdissected cells (Suzuki *et al.*, 2001; Trogan *et al.*, 2002; Walch *et al.*, 2001). Recent developments in the application of RT-PCR analysis to paraffin-wax-embedded

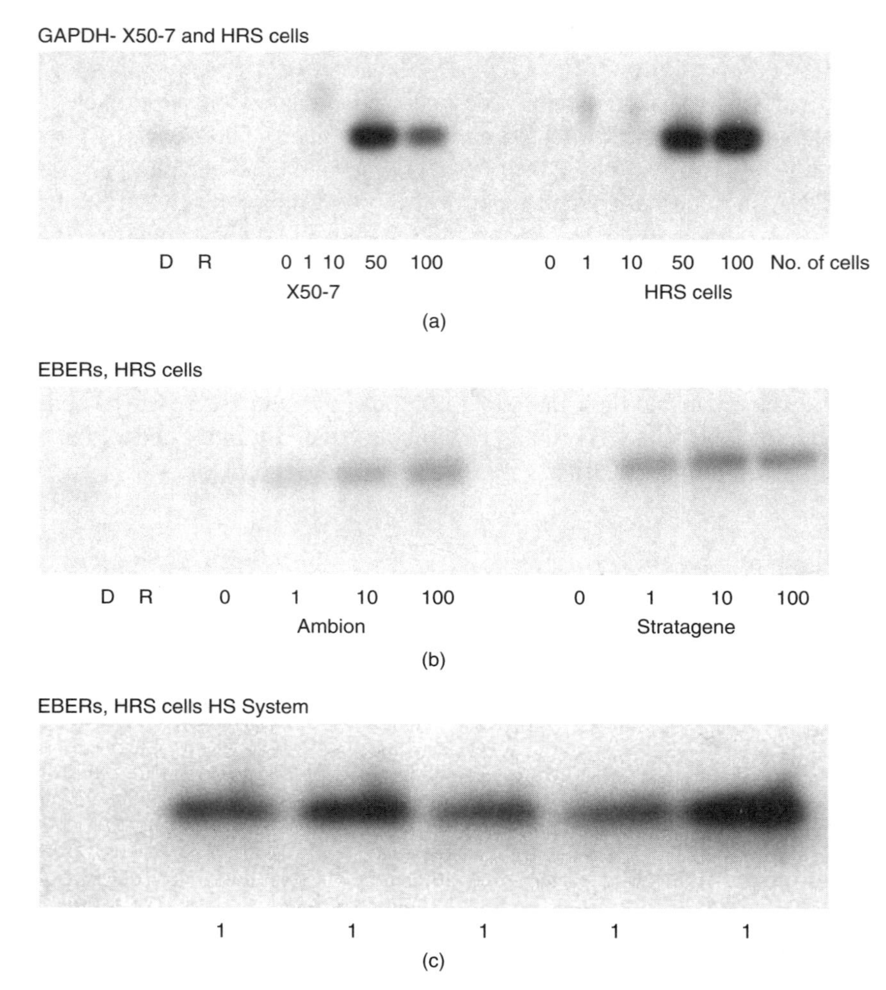

Figure 10.6 RT-PCR analysis of microdissected cells. (a) Detection of GAPDH transcripts in both X50-7 and HRS cells at 50 cells; (b) detection of EBER expression in HRS cells using two different kits; (c) HS system gives improved sensitivity at the one-cell level

tissues will greatly facilitate future gene expression studies of cells captured from archival tumour samples (Specht *et al.*, 2001).

Gene expression microarrays

Gene expression profiling using microarrays is a powerful technology that can provide information on the mRNA complement of tumour cells and reveal genes that characterise tumour types. The use of this approach in the molecular classification of cancers and its relevance to identifying clinically distinct tumour subgroups has been demonstrated in a number of important recent studies (Alizadeh *et al.*, 2000, 2001; Ramaswamy *et al.*, 2001; van't Veer *et al.*,

2002; West *et al.*, 2001). The application of microarray technology has not generally involved the study of gene expression from individual cell types residing in a given tissue/organ. Such studies would greatly facilitate an understanding of the complex interactions that exist *in vivo* between neighbouring cell types in normal and diseases states and would avoid the 'averaging out' of gene profiles that occurs when gene expression microarray analysis is applied to whole tissue samples. There have been relatively few such studies, mainly because the large number of cells required for gene expression microarrays can be prohibitive. However, the CLONTECH's PCR-based SMART technology (Switch Mechanism At the 5′end of RNA Templates) bypasses this problem by allowing accurate cDNA amplification from nanogram quantities of total RNA. This amount of RNA ensures that SMART amplification yields a pool of cDNA, which reflects the sample's original complexity and relative abundance of the original RNA sample (Herrler, 2000). Such approaches should in the future enable more widespread use of microarray-based gene expression profiling on microdissected cells.

Protein

Proteomics

Traditionally, proteomics uses 2-D PAGE and image subtractions to locate differences between experimental and control protein populations (see Chapter 17). Combining proteomic analysis with LCM provides a powerful means to study protein from pure populations of tumour cells. However, for the analysis of microdissected cells by 2-D-PAGE, approximately 50 000 cells are required (Banks *et al.*, 1999). The collection of such cell numbers by LCM is possible in cases where clusters of cells can be microdissected but is time consuming and would preclude the analysis of a large series of tumours.

An alternative approach is to use surface-enhanced laser desorption/ionisation (SELDI), which provides a rapid, high throughput and highly sensitive method for analysing patterns of protein expression in tissues. Proteins of interest are directly applied to a surface utilising a defined chemical chromatographic characteristic (i.e. hydrophobic, hydrophilic, cationic, anionic) or biochemical ligands such as proteins, receptors, antibodies or DNA oligonucleotides. The bound proteins are treated with wash buffers to remove non-specific binding partners and then analysed by time-of-flight mass spectrometry to produce a mass map (a so-called 'retentate' map). The protein 'fingerprint' generated is based on the combined precise molecular weight signatures of each individual protein bound and ionised off the specific bait surface employed. The success of the SELDI process is defined in part by the miniaturisation and integration of these multiple functions on a single surface (the ProteinChip Array). The Ciphergen ProteinChip system (Ciphergen Ltd, Palo Alto, CA) consists of the ProteinChip Arrays, the ProteinChip Reader and associated software.

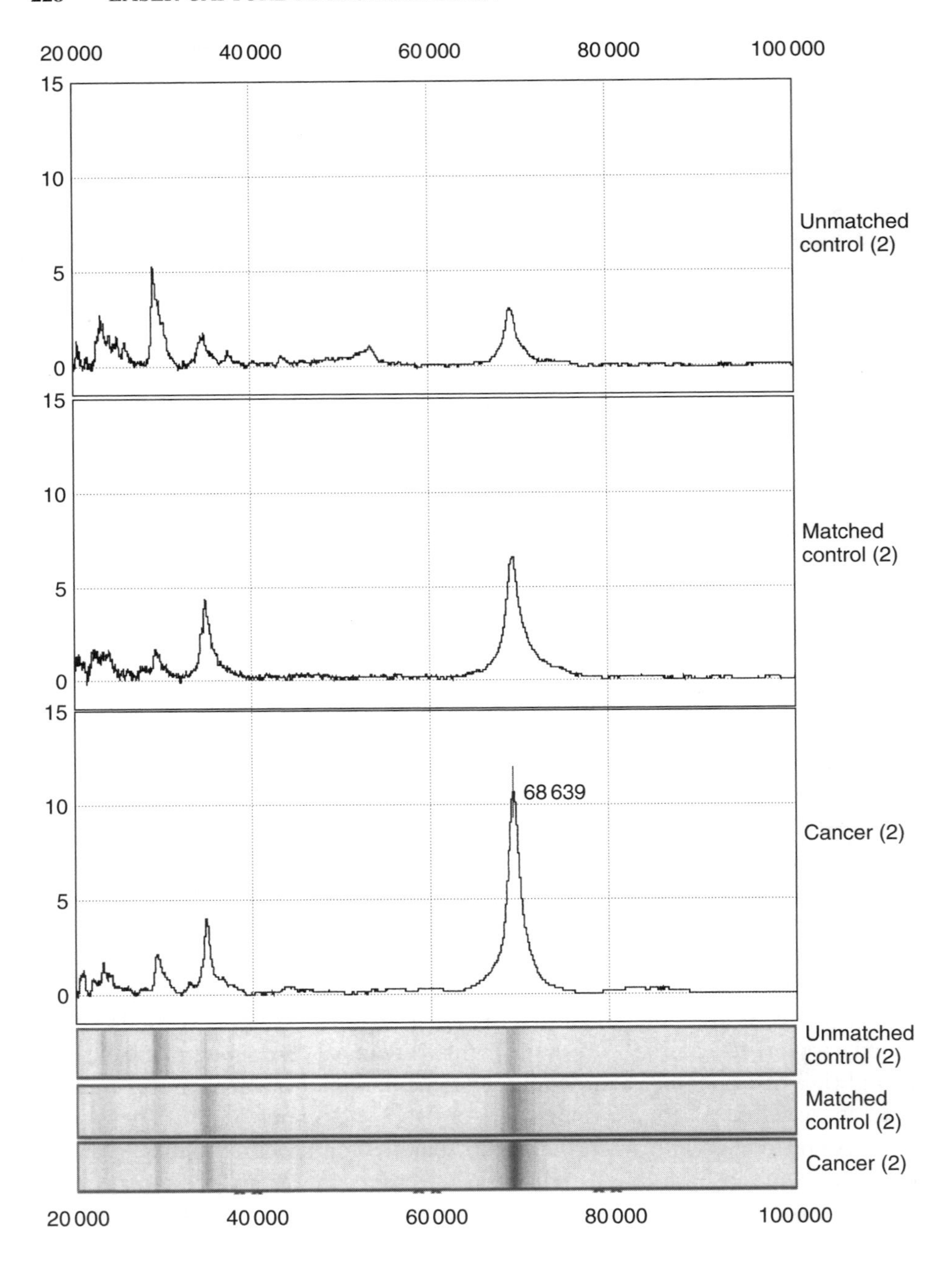

Figure 10.7 Proteomic analysis of microdissected oral cancer cells using the Ciphergen ProteinChip® System. Analysis of the higher molecular weight range shows the upregulation of a 68.6 KDa protein in the cancer when compared to the matched and unmatched normal tissues

The system can be used to profile changes in protein expression patterns that might define specific disease states or characterise disease progression. Fewer cells (approximately 2000–3000 depending on cell type) are required to generate a SELDI profile. Candidate biomarkers could be purified from LCM starting material, although these purification approaches are still in development and at the present time may require more cells than can be conveniently collected by LCM. A further disadvantage of SELDI is that only small proteins (<10 000 Da) can be easily analysed from complex mixtures. Nevertheless, successful application of SELDI profiling to microdissected samples has been reported by several groups (Paweletz *et al.*, 2001; von Eggeling *et al.*, 2000). We have been able to obtain reproducible SELDI profiles from microdissected oral cancer cells (Figure 10.7). For an excellent review of the application of LCM in proteomics research see Craven *et al.* (2001).

10.6 Future perspectives

There is little doubt that developments in microdissection will continue to make important contributions to the study of tumour biology. The analysis of pure tumour cell populations allied to the powerful emerging technologies of gene expression arrays and proteomics is already providing rapid advances in our understanding of the pathogenesis of cancer. There is evidence that such knowledge will not only influence the way in which cancer is diagnosed, but will also lead to more accurate prognostication and to improved treatments for cancer patients.

10.7 Acknowledgements

We are grateful to Steve Cleverley for assistance with the proteomic analysis and to Jane Starczynski for help in the preparation of frozen sections.

10.8 References

Alizadeh AA, Eisen MB, Davis RE, Ma C, Lossos IS, Rosenwald A, Boldrick JC, Sabet H, Tran T, Yu X, Powell JI, Yang L, Marti GE, Moore T, Hudson J Jr, Lu L, Lewis DB, Tibshirani R, Sherlock G, Chan WC, Greiner TC, Weisenburger DD, Armitage JO, Warnke R and Staudt LM *et al.* (2000) Distinct types of diffuse large B-cell lymphoma identified by gene expression profiling. *Nature* **403**: 503–11.

Alizadeh AA, Ross DT, Perou CM and van de Rijn M (2001) Towards a novel classification of human malignancies based on gene expression patterns. *J. Pathol.* **195**: 41–52.

Banks RE, Dunn MJ, Forbes MA, Stanley A, Pappin D, Naven T, Gough M, Harnden P and Selby PJ (1999) The potential use of laser capture microdissection to selectively obtain distinct populations of cells for proteomic analysis – preliminary findings. *Electrophoresis* **20**: 689–700.

Cohn DE, Mutch DG, Herzog TJ, Rader JS, Dintzis SM, Gersell DJ, Todd CR and Goodfellow PJ (2001) Genotypic and phenotypic progression in endometrial tumorigenesis: determining when defects in DNA mismatch repair and KRAS2 occur. *Genes Chromosomes Cancer* **32**: 295–301.

Craven RA and Banks RE (2001) Laser capture microdissection and proteomics: possibilities and limitation. *Proteomics* **1**: 1200–4.

Daigo Y, Chin SF, Gorringe KL, Bobrow LG, Ponder BA, Pharoah PD and Caldas C (2001) Degenerate oligonucleotide primed-polymerase chain reaction-based array comparative genomic hybridization for extensive amplicon profiling of breast cancers: a new approach for the molecular analysis of paraffin-embedded cancer tissue. *Am. J. Pathol.* **158**: 1623–31.

Emmert-Buck MR, Bonner RF, Smith PD, Chuaqui RF, Zhuang Z, Goldstein SR, Weiss RA and Liotta LA (1996) Laser capture microdissection. *Science* **274**: 998–1001.

Fend F, Quintanilla-Martinez L, Kumar S, Beaty MW, Blum L, Sorbara L, Jaffe ES and Raffeld M (1999a) Composite low grade B-cell lymphomas with two immunophenotypically distinct cell populations are true biclonal lymphomas. A molecular analysis using laser capture microdissection. *Am. J. Pathol.* **154**: 1857–66.

Fend F, Emmert-Buck MR, Chuaqui R, Cole K, Lee J, Liotta LA and Raffeld M (1999b) Immuno-LCM: laser capture microdissection of immunostained frozen sections for mRNA analysis. *Am. J. Pathol.* **154**: 61–6.

Fend F, Kremer M and Quintanilla-Martinez L (2000) Laser capture microdissection: methodical aspects and applications with emphasis on immuno-laser capture microdissection. *Pathobiology* **68**: 209–14.

Glasow A, Horn LC, Taymans SE, Stratakis CA, Kelly PA, Kohler U, Gillespie J, Vonderhaar BK and Bornstein SR (2001) Mutational analysis of the PRL receptor gene in human breast tumors with differential PRL receptor protein expression. *J. Clin. Endocrinol. Metab.* **86**: 3826–32.

Goldstein SR, McQueen PG and Bonner RF (1998) Thermal modeling of laser capture microdissection. *Appl. Optics* **37**: 7378–91.

Herrler M (2000) Use of SMART-generated cDNA for differential gene expression studies. *J. Mol. Med.* **78**: B23.

Jin L, Thompson CA, Qian X, Kuecker SJ, Kulig E and Lloyd RV (1999) Analysis of anterior pituitary hormone mRNA expression in immunophenotypically characterised single cells after laser capture microdissection. *Lab. Invest.* **79**: 511–12.

Klussmann JP, Weissenborn SJ, Wieland U, Dries V, Kolligs J, Jungehuelsing M, Eckel HE, Dienes HP, Pfister HJ and Fuchs PG (2001) Prevalence, distribution, and viral load of human papillomavirus 16 DNA in tonsillar carcinomas. *Cancer* **92**: 2875–84.

Kremer M, Sandherr M, Geist B, Cabras AD, Hofler H and Fend F (2001) Epstein–Barr virus-negative Hodgkin's lymphoma after mycosis fungoides: molecular evidence for distinct clonal origin. *Mod. Pathol.* **14**: 91–7.

Kurose K, Hoshaw-Woodard S, Adeyinka A, Lemeshow S, Watson PH and Eng C (2001) Genetic model of multi-step breast carcinogenesis involving the epithelium and stroma: clues to tumour–microenvironment interactions. *Hum. Mol. Genet.* **10**: 1907–13.

Leonard N, Chaggar R, Jones C, Takahashi M, Nikitopoulou A and Lakhani SR (2001) Loss of heterozygosity at cylindromatosis gene locus, CYLD, in sporadic skin adnexal tumours. *J. Clin. Pathol.* **54**: 689–92.

Ling G, Persson A, Berne B, Uhlen M, Lundeberg J and Ponten F (2001) Persistent p53 mutations in single cells from normal human skin. *Am. J. Pathol.* **159**: 1247–53.

Liston BW, Gupta A, Nines R, Carlton PS, Kresty LA, Harris GK and Stoner GD (2001) Incidence and effects of Ha-ras codon 12 G$\rightarrow$A transition mutations in preneoplastic lesions induced by N-nitrosomethylbenzylamine in the rat esophagus. *Mol. Carcinogen.* **32**: 1–8.

Marchio A, Terris B, Meddeb M, Pineau P, Duverger A, Tiollais P, Bernheim A and Dejean A (2001) Chromosomal abnormalities in liver cell dysplasia detected by comparative genomic hybridisation. *Mol. Pathol.* **54**: 270–4.

Mora J, Cheung NK, Juan G, Illei P, Cheung I, Akram M, Chi S, Ladanyi M, Cordon-Cardo C and Gerald WL (2001) Neuroblastic and Schwannian stromal cells of neuroblastoma are derived from a tumoral progenitor cell. *Cancer Res.* **61**: 6892–8.

Ortiz BH, Ailawadi M, Colitti C, Muto MG, Deavers M, Silva EG, Berkowitz RS, Mok SC and Gershenson DM (2001) Second primary or recurrence? Comparative patterns of p53 and K-ras mutations suggest that serous borderline ovarian tumors and subsequent serous carcinomas are unrelated tumors. *Cancer Res.* **61**: 7264–7.

Paweletz CP, Liotta LA and Petricoin EF 3rd (2001) New technologies for biomarker analysis of prostate cancer progression. Laser capture microdissection and tissue proteomics. *Urology* **57**: Suppl 1: 160–3.

Ramaswamy S, Tamayo P, Rifkin R, Mukherjee S, Yeang CH, Angelo M, Ladd C, Reich M, Latulippe E, Mesirov JP, Poggio T, Gerald W, Loda M, Lander ES and Golub TR (2001) Multiclass cancer diagnosis using tumor gene expression signatures. *Proc. Natl. Acad. Sci. USA* **98**: 15149–54.

Specht K, Richter T, Muller U, Walch A, Werner M and Hofler H (2001) Quantitative gene expression analysis in microdissected archival formalin-fixed and paraffin-embedded tumor tissue. *Am. J. Pathol.* **158**: 419–29.

Strauberg RL, Dahl CA and Klausher RD (1997) New opportunities for uncovering the molecular basis of cancer. *Nat. Genet.* **15**: 5415–16.

Suzuki K, Matsui H, Hasumi M, Ono Y, Nakazato H, Koike H, Ito K, Fukabori Y, Kurokawa K and Yamanaka H (2001) Gene expression profiles in human BPH: utilization of laser-capture microdissection and quantitative real-time PCR. *Anticancer Res.* **21**(6A): 3861–4.

Takeshima Y, Amatya VJ, Daimaru Y, Nakayori F, Nakano T and Inai K (2001) Heterogeneous genetic alterations in ovarian mucinous tumors: application and usefulness of laser capture microdissection. *Hum. Pathol.* **32**: 1203–8.

Torres-Munoz J, Stockton P, Tacoronte N, Roberts B, Maronpot RR and Petito CK (2001) Detection of HIV-1 gene sequences in hippocampal neurons isolated from postmortem AIDS brains by laser capture microdissection. *J. Neuropathol. Exp. Neurol.* **60**: 885–92.

Trogan E, Choudhury RP, Dansky HM, Rong JX, Breslow JL and Fisher EA (2002) Laser capture microdissection analysis of gene expression in macrophages from atherosclerotic lesions of apolipoprotein E-deficient mice. *Proc. Natl. Acad. Sci. USA* **99**: 2234–9.

Umayahara K, Numa F, Suehiro Y, Sakata A, Nawata S, Ogata H, Suminami Y, Sakamoto M, Sasaki K and Kato H (2002) Comparative genomic hybridization detects genetic alterations during early stages of cervical cancer progression. *Genes Chromosomes Cancer* **33**: 98–102.

van 't Veer LJ, Dai H, van de Vijver MJ, He YD, Hart AA, Mao M, Peterse HL, van der Kooy K, Marton MJ, Witteveen AT, Schreiber GJ, Kerkhoven RM, Roberts C, Linsley PS, Bernards R and Friend SH (2002) Gene expression profiling predicts clinical outcome of breast cancer. *Nature* **415**: 530–6.

von Eggeling F, Davies H, Lomas L, Fiedler W, Junker K, Claussen U and Ernst G (2000) Tissue-specific microdissection coupled with ProteinChip array technologies: applications in cancer research. *Biotechniques* **29**: 1066–70.

Walch A, Specht K, Bink K, Zitzelsberger H, Braselmann H, Bauer M, Aubele M, Stein H, Siewert JR, Hofler H and Werner M (2001) Her-2/neu gene amplification, elevated mRNA expression, and protein overexpression in the metaplasia–dysplasia–adenocarcinoma sequence of Barrett's esophagus. *Lab. Invest.* **81**: 791–801.

West M, Blanchette C, Dressman H, Huang E, Ishida S, Spang R, Zuzan H, Olson JA Jr, Marks JR and Nevins JR (2001) Predicting the clinical status of human breast cancer by using gene expression profiles. *Proc. Natl. Acad. Sci. USA* **98**: 11462–7.

11

The *In-situ* Polymerase Chain Reaction

John J. O'Leary, Cara Martin and **Orla Sheils**

11.1 Introduction

Solution phase polymerase chain reaction (PCR) has become one of the most revolutionary tools in molecular pathology. The technique, conventionally performed in a tube, permits the amplification of single copy mammalian/viral targets (either DNA or RNA) in fresh/frozen and archival paraffin-wax-embedded material. However, the inability to visualise and localise amplified product within cells and tissue specimens has been a major limitation, especially for pathologists attempting to correlate genetic events with pathological changes.

In-situ hybridisation (ISH) has partly addressed this problem by permitting localisation of specific nucleic acid sequences at the individual cell level, but most conventional non-isotopic *in-situ* detection systems do not detect single copy genes, except for those incorporating elaborate sandwich detection techniques, as described by Herrington *et al.* (1992).

Before amplification can be performed in solution phase PCR (SPPCR), nucleic acid extraction is first carried out, which necessitates cellular destruction. Subsequent correlation of results with histological features is consequently not possible. In the last few years, a number of studies have described 'hybrid' techniques coupling PCR with ISH (Bagasra *et al.*, 1992, 1993; Boshoff *et al.*, 1995; Haase *et al.*, 1990; Herrington *et al.*, 1992). The techniques initially were not accepted owing to technological problems encountered during amplification of the desired nucleic acid sequence.

In-situ amplification technologies now have been extended to include a number of modifications of the initially described technique of *in-situ* PCR, to

Molecular Biology in Cellular Pathology. Edited by John Crocker and Paul G. Murray
 ISBN: 0-470-84475-2

include primed *in-situ* labelling (PRINS), cycling PRINS, *in-situ* peptide nucleic acid PCR (PNA PCR), *in-situ* PNA PCR, *in-situ* immuno-PCR, IS-TaqMan® PCR and allele-specific amplification. Other in-cell amplification technologies such as nucleic acid base amplification (NASBA) can also be used. These employ T3 and T7 RNA technology and are isothermal replication methodologies for the amplification of RNA in cells and tissues sections.

11.2 Overview of the methodology

All the in-cell PCR techniques attempt to create double-stranded or single-stranded DNA/cDNA amplicons within the cell, which can either be detected directly or following an ISH step. For successful application of the techniques, a fine balance between adequate digestion of cells (allowing access of amplification reagents) and maintaining localisation of amplified product within the cellular compartment and preserving tissue/cell morphology is achieved.

To do this, the techniques require specific cyclical thermal changes to occur at the individual cell level, akin to what occurs in SPPCR. The first step involves denaturation of double-stranded DNA (dsDNA) to single-stranded DNA (ssDNA), if one is amplifying a DNA target. For RNA target-specific amplification, the RNA template is already single stranded and reverse transcription is carried out to create a cDNA template.

Secondly, (primers) are then annealed to the respective ends of the desired target sequence and a thermostable enzyme (Taq DNA polymerase or Klenow fragment) is used to extend or ligate [DNA ligase, in IS-LCR (see below)] the correctly positioned primers.

Specific Taq polymerases have now been designed for use in in-cell PCR analyses including IS-Taq polymerase, which has a higher concentration and greater unity polymerase activity.

Subsequent rounds of thermocycling increase the copy number of the desired target sequence, in a nucleic acid amplification reaction.

However, unlike SPPCR, an exponential increase in the amount of amplified product is never achieved, with linear amplification occurring in most situations. This is due to the relative inefficiency of the techniques, owing to problems of accessibility of amplification reagents to the desired nucleic acid sequence, because of the compactness of the nuclear compartment of the cell (containing dsDNA, ssDNA, pre-mRNA and histone proteins).

Once the amplicon is created, detection must be carried out. If a labelled primer or nucleotide is used, then the amplicon is labelled and can be demonstrated directly by immunocytochemical techniques. In general, however, direct labelled approaches using labelled nucleotides in the PCR mix or labelled primers have largely been abandoned owing to non-specific signal generation and non-specific label incorporation into the PCR amplicon.

Alternatively, and more acceptably, an ISH step (using a single-stranded oligo probe or a double-stranded genomic probe) is carried out post-amplification, adhering to the general rules of standard ISH kinetics and applying stringent post-hybridisation washing conditions.

11.3 In-cell PCR technologies

Several techniques are now described, including the following (Figure 11.1):

1. *DNA in-situ PCR (IS-PCR):* PCR amplification of cellular DNA sequences in tissue specimens using a labelled nucleotide (i.e. dUTP) within the PCR reaction mix. The labelled product is then detected using standard detection techniques as for conventional ISH or immunocytochemistry. This technology is not recommended for use.
2. *Labelled primer driven in-situ amplification (LPDISA):* amplification of DNA sequences using a labelled primer within the PCR reaction mix. The labelled product is then detected as for DNA IS-PCR. This technology is not recommended for use (Figure 11.2).
3. *PCR ISH:* PCR amplification of cellular DNA sequences in tissue specimens followed by ISH detection of the amplified product using a labelled internal or genomic probe. The labels used can either be isotopic (^{32}P, ^{35}S) or non-isotopic (e.g. biotin, digoxigenin, fluorescein) (Figure 11.2).
4. *Reverse transcriptase in-situ PCR (RT IS-PCR):* amplification of mRNA sequences in cells and tissue specimens by first creating a copy DNA template (cDNA) using reverse transcriptase (RT) and then amplifying the

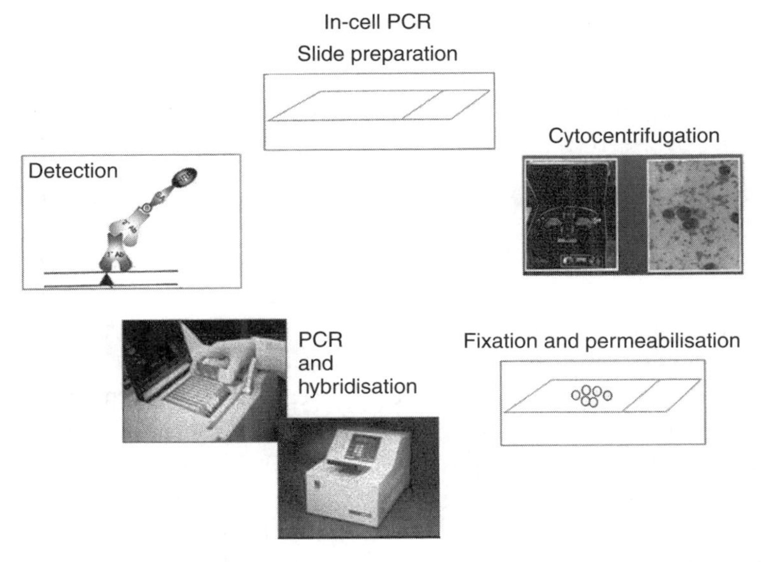

Figure 11.1 Schematic representation of in-cell PCR, indicating the critical steps involved in the process from cell/tissue adhesion to fixation permeabilisation to PCR amplification to detection of amplicon. A colour version of this figure appears in the colour plate section

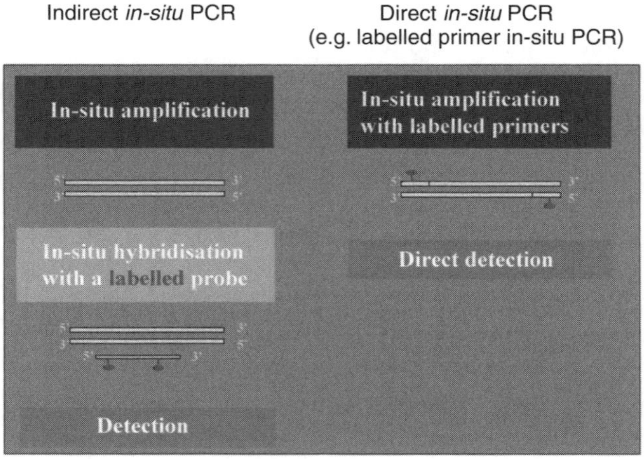

Figure 11.2 Schematic of PCR-ISH vs. labelled primer driven in-cell DNA PCR. A colour version of this figure appears in the colour plate section

newly created DNA template as for DNA IS-PCR. Again, this technique has the inherent problems with non-specific label incorporation, and should not be used for the analysis of RNA templates in cells and tissues sections.

5. *Reverse transcriptase PCR ISH:* amplification of RNA sequences in cells and tissues specimens by creating a cDNA template using RT. The newly created cDNA is then amplified, and the amplicon probed with an internal oligonucleotide as in PCR-ISH.

6. *PRINS and cycling PRINS:* amplification of specific genetic sequences in metaphase chromosome spreads or interphase nuclei, using one primer to generate single-stranded PCR product. If many rounds of amplification are utilised, then the technique is called cycling PRINS (Figures 11.3 and 11.4) (Gosden *et al.*, 1991).

7. *In-situ PNA PCR (IS-PNA-PCR) and PCR-PNA-ISH:* IS-PNA-PCR refers to amplification of DNA targets using a DNA mimic molecule, peptide nucleic acid (PNA). PNA is a simple molecule, made up of repeating N-(2-aminoethyl)-glycine units linked by amide bonds (Figure 11.5). Purines (A and G) and pyrimidine bases (C and T) are attached to the backbone by methylene carbonyl linkages. When used as a primer, PNA is not elongated by Taq DNA polymerase and therefore can be used in primer exclusion assays, which allows the discrimination of point mutations and direct individual cell haplotyping. The second reaction, which employs PNA, is PCR-PNA-ISH; here, a 15-20-mer PNA probe is used for the ISH step following amplification. PNAs have a higher T_m (melting temperature) than DNA oligo probes and single point mutations in a PNA-DNA duplex lower the T_m by approximately 15°C as compared with the corresponding DNA-DNA mismatch duplex (De-Mesmaeker *et al.*, 1995).

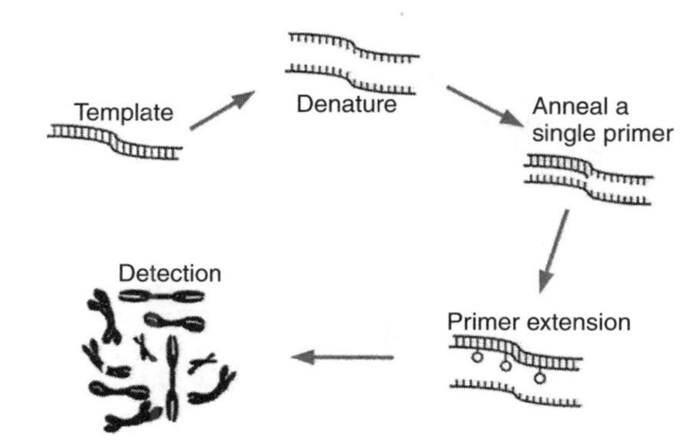

Figure 11.3 Schematic representation of PRINS. This technique utilises a single primer for DNA amplification

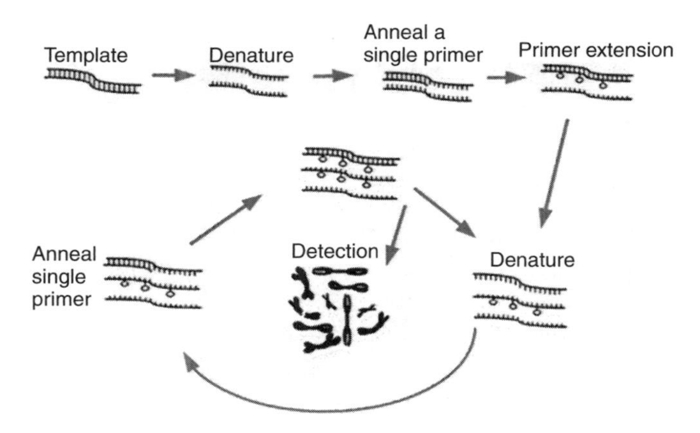

Figure 11.4 Schematic representation of cycling PRINS

8. *IS-TaqMan® PCR:* amplification of DNA sequences using a conventional primer pair as in standard PCR. However, an internal TaqMan® probe is added to the amplification mix. A fluorescent reporter molecule (FAM, HEX, etc.) is placed at the 5′ end of the probe. At the 3′ end, a quencher molecule (again fluorescent, usually TAMRA) is positioned. Once the probe is linearised and intact, the proximity of the quencher to the reporter molecule does not allow any fluorescence from the reporter molecule. Taq DNA polymerases possess two properties important for the reaction: (a) a Y-strand fork displacement (which allows Taq DNA polymerase to lift off a single strand in a Y configuration and (b) 5′–3′ exonucleolytic activity, which causes cleavage of the linker arm that attaches the reporter molecule to the 5′ end of the TaqMan® probe, thereby giving rise to fluorescence if and only if specific amplification has occurred (Holland *et al.*, 1991; Lawyer *et al.*, 1989).

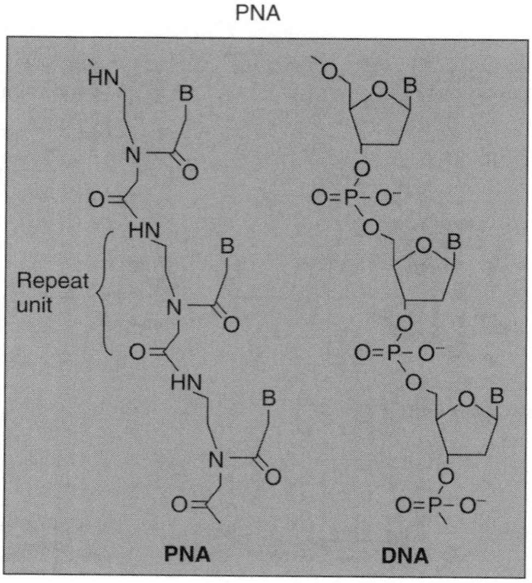

PNA

Figure 11.5 Structure of a PNA molecule

9. *In-situ allele specific amplification (IS-ASA):* this technique utilises amplification refractory mutation system (ARMS) PCR, which has the ability to detect point polymorphisms in human DNA sequences using artificially created base pair mismatches at the 3′ end of PCR primers. If the polymorphism matching that of the primer sequence is present, then amplification of that sequence will preferentially occur.

10. *In-situ immuno-PCR:* *in-situ* immunoassays do not allow the detection of the minute numbers of target molecules accessible with IS-PCR. *In-situ* immuno-PCR, in which the DNA marker is linked to target molecules through an antibody-biotin-avidin bridge, is subsequently amplified by IS-PCR. Amplified DNA sequences are then detected by ISH. The originators of this technique claim that the technique may be the only one available to detect minute quantities of biological macromolecules such as proteins, carbohydrates and lipids in intact cells or tissue sections (Cao *et al*., 2000).

11.4 In-cell amplification of DNA

Equipment

Several *in-situ* amplification instruments ranging from standard thermal cycles (using modifications), thermocycling ovens, and specifically dedicated thermocyclers (Applied Biosystems Gene Amp *in-situ* PCR system 1000, Hybaid Omnigene/Omnislide and MJB slide thermocycler) can be used for in-cell DNA and RNA amplification.

If one uses a standard thermal cycler, then an amplification chamber must be created for the slide. This can be made from aluminium (aluminium foil boat). This boat, which contains the slide, is placed on the thermal cycler, covered with mineral oil and then wrapped completely. However, optimisation of thermal conduction is never completely achieved. 'Thermal lag' (i.e. differences in temperature between the block face, the glass slide and the PCR reaction mix at each temperature step of the reaction cycle) is commonly encountered (O'Leary *et al.*, 1994a).

This problem is overcome by use of the specifically designed *in-situ* amplification machines, which offer an in-built slide temperature calibration curve with greater thermodynamic control.

Starting Material

Many techniques have been described for performing in-cell amplification. In 1990, Haase initially described ISPCR in intact fixed single cells, suspended in PCR reaction buffer. After amplification, cells were cytocentrifuged onto glass slides and the amplified product detected using ISH (Haase *et al.*, 1990). Initially, some investigators used pieces of glass slides (with cells from cytocentrifuge preparations) in standard eppendorf tubes, incubated directly in PCR reaction buffer (Spann *et al.*, 1991). More recently, techniques using tissues and cells attached to microscope slides have been used (Bobroski *et al.*, 1995; Cheng and Nuovo, 1994).

For IS-PCR, PCR-ISH, IS-LCR, IS-PNA PCR and IS-TaqMan® PCR, fixed cells and tissues including archival paraffin-wax-embedded material can be used for amplification. The best results are obtained with freshly fixed cells and tissues, although successful amplification with old archival material (up to 40 years) has been achieved.

Fixed metaphase chromosome spreads and interphase nuclei can be used for the detection of specific sub-chromosomal regions using PRINS and cycling PRINS (Gosden *et al.*, 1991).

For successful *in-situ* amplification to occur, a rigid cellular cytoskeleton must be created, providing a suitable microenvironment that allows access of amplification reagents with minimal leakage of amplified product. Satisfactory results are obtained with tissues fixed in 1%–4% paraformaldehyde, neutral buffered formaldehyde (NBF) and 10% formalin (12–24 h for biopsy/solid tissue; 10–30 min for cytological preparations) (Nuovo *et al.*, 1993; O'Leary *et al.*, 1994b). Less consistent results are obtained with ethanol and acetic acid fixed tissues (Nuovo *et al.*, 1993).

Fixation of cells with formaldehyde fixatives provides a number of drawbacks. Formaldehyde is not easily removed from tissues after tissue processing. Aldehyde groups react with DNA and histone proteins to form DNA–DNA and DNA–histone protein cross-links. Formaldehyde fixation also 'nicks' DNA template (random breaks in dsDNA), which may be blunt ended (i.e. have non-overlapping ends). These nicks may subsequently act as potential priming sites for Taq DNA polymerase leading to incorporation and elongation of labelled and unlabelled nucleotides (i.e. dATP, etc.) analogous to *in-situ* end labelling for apoptosis. This process occurs at room temperature leading to spurious results with DNA IS-PCR. This is the predominant reason why in-cell DNA and RNA PCR (i.e. without a hybridisation step) have been largely abandoned.

Because repeated cycles of heating and cooling are used during *in-situ* amplification, cells and tissues must be adequately attached to a solid support (usually glass), so that detachment does not occur. Glass slides are pre-treated with coating agents to ensure maximal section adhesion, the most commonly used being aminopropyltriethoxysilane (APES), Denhardt's solution and Elmer's glue.

Cell and Tissue Permeabilisation

Cells must be adequately digested and permeabilised to facilitate access of reagents. This can be achieved by protease treatment (e.g. proteinase K, pepsin or trypsin) and/or mild acid hydrolysis (0.01–0.1 N HCl). Maximal digestion times and protease concentrations have to optimised for the particular tissue/cytological preparation employed.

Long digestion times inevitably compromise cellular morphology. Short digestion times result in incomplete dissociation of histone protein–DNA cross-links, which ultimately can hinder the progression of Taq DNA polymerase along native DNA templates. Acid hydrolysis probably acts by driving such cross-links to complete dissociation. Alternatively, microwave irradiation of cells and tissue sections can be used to expose nucleic acid templates. Short pulses of microwave irradiation (with or without proteolysis) using a citrate buffer, analogous to antigen retrieval for immunocytochemistry, allow access of amplification reagents to the desired target sequence, and in addition facilitate post-amplification immunocytochemistry.

Following this (if a non-isotopic labelling method is used), a blocking step must be employed depending on the method used for post-amplification detection of product (e.g. peroxidase or alkaline phosphatase detection systems). In the former, endogenous peroxidase is quenched by incubation in a 3% H_2O_2 solution with sodium azide, while 20% ice-cold acetic acid blocks intestinal alkaline phosphatase.

DNA Amplification Protocols

Successful amplification is governed by

- careful optimisation of cycling parameters;
- appropriate design of primer pairs (taking into account their T_m, i.e. specific melting temperature, their ability to form primer–dimers and uniqueness); and

- optimisation of Mg^{2+} concentrations, needed to drive the amplification reaction.

Owing to reagent sequestration (see below), higher concentrations of amplification reagents are required during *in-situ* amplification than for conventional solution phase techniques, including primers, dNTP's and Mg^{2+}. Mg concentrations in particular have to be carefully optimised, with satisfactory amplification occurring for most applications at 2.5–5.5 mM. Amplification reagent volumes usually vary depending on the surface area of the cell preparation/tissue section used. This is important as 'patchy' amplification may occur over the surface of the slide owing to volume variations consequent to localised amplification failure. When using the Gene Amp *in-situ* PCR 1000 system, typically 25–50 μL are used.

Initial denaturation of DNA can be achieved before amplification either during permeabilisation or following the fixation process. Alternatively, it may be performed at the beginning of the amplification protocol itself. Denaturation can be achieved using heat, heat/formamide or alkaline denaturation (Bagasra *et al.*, 1994). In addition, most investigators advocate the use of 'hot start' PCR to reduce mispriming and primer oligomerisation, and Nuovo *et al.* (1991a) have suggested the addition of single strand binding (SSB) protein derived from *Escherichia coli*. The precise mode of action of this protein is unknown, but it functions in DNA replication and repair by preventing primer mispriming and oligomerisation.

Optimisation of cycling parameters has to be performed for each particular assay. Most protocols employ 25–30 rounds of amplification; exceptionally 50 cycles. Some investigators have performed two successive 30-cycle rounds with the addition of new reagents, including primers and DNA polymerase between each round. A modification of this is 'nested' PCR, where internal primers 'nested' within the amplicon produced during the first round are added. However, this is not recommended for routine use.

Primer selection has evolved around two basic strategies: single primer pairs or multiple primer pairs with or without complementary tails. Multiple primer pairs have been designed to generate longer/overlapping product with the obvious advantage of localisation of amplicons and minimal product diffusion. However, if 'hot-start' PCR is employed, then single primer pairs usually are sufficient to ensure successful amplification.

Post-amplification Stringency, Washing and Amplicon Fixation

Post-fixation with 4% paraformaldehyde and/or ethanol may be employed to maintain localisation of amplified product. If one is performing PCR-ISH, then an oligonucleotide or genomic probe is applied at this stage. Maximum specificity is achieved using probes hybridising to sequences internal to the amplified product only. Genomic probes are not restricted to these sequences and appear to provide comparable results.

Following *in-situ* amplification, most protocols include a post-amplification washing step [using sodium chloride sodium citrate (SSC), formamide and varying washing temperatures] to remove diffused extracellular product which may result in non-specific staining and generation of false-positive results. The 'stringency' of the wash is defined by the set of conditions employed (i.e. SSC concentration, percentage formamide used and the washing temperature). The investigator attempts to achieve a washing 'window' where the signal to background noise ratio is maximal; however, post-hybridisation washing stringencies are derived empirically.

11.5 Detection of amplicons

Non-isotopic labels (e.g. biotin, digoxigenin, fluorescein) are more widely used and, when used in conjunction with a 'sandwich' immunohistochemical detection technique, appear to provide similar degrees of detection sensitivity to isotopic labels. These may vary from one-step to five-step detection systems as for conventional immunocytochemistry. The product is finally visualised by either a colour reaction (e.g. NBT/BCIP or AEC chromagens) or fluorescence (Figures 11.6 and 11.7).

Nuovo (1992) has described the performance of post-amplification conventional immunocytochemistry, the obvious advantage being co-localisation of product with the cell of interest, e.g. endothelial or macrophage. However, attempts to reproduce this by various groups, including our own, have been disappointing and it is likely that most epitopes do not withstand repetitive thermal cycling.

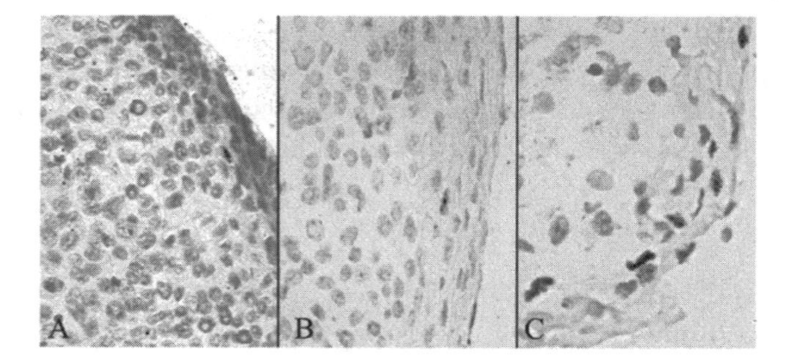

Figure 11.6 ISH and PCR-ISH detection of HPV in a cervical biopsy. Panel A: ISH, one-step immunocytochemical detection, sensitivity 20–30 genomes per cell. Panel N: ISH, three-step detection, sensitivity 10–20 genomes per cell. Panel C: PCR-ISH with three-step detection, sensitivity 1 genome per cell. A colour version of this figure appears in the colour plate section

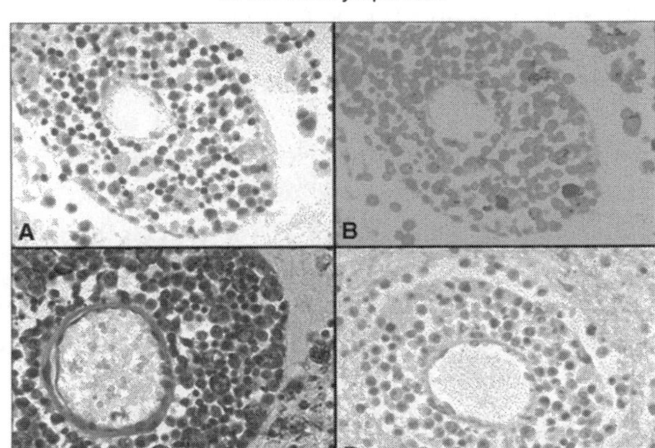

Figure 11.7 Detection of Human Herpes Virus 8 (HHV 8) in an AIDS CNS lymphoma. Panel A: PCR-ISH. Panel B: parallel image analysis. Panel C CD68 immunocytochemistry. Panel D: negative control. A colour version of this figure appears in the colour plate section

11.6 Reaction, tissue and detection controls for use with in-cell DNA PCR assays

These include parallel solution phase PCR, omission of primers and/or Taq DNA polymerase, irrelevant primers and/or probes, known negative controls and reference control genes. The number of controls performed depends on the amount of tissue/cells available for the reaction.

The following controls are required in PCR-ISH:

- Reference control gene PCR-ISH, e.g. beta globin or Pyruvate dehydrogenase (PDH)
- DNase digestion of target tissues/cells
- RNase digestion of target tissues/cells
- Target primers with irrelevant probe
- Irrelevant primers with target probe
- Irrelevant primers with irrelevant probe
- Reference control gene primers with the target probe
- Target primer one only (asymmetric PCRISH)
- Target primer two only
- No Taq polymerase
- No primers

- Omit the reverse transcriptase step in RT IS-PCR or RT PCR-ISH,
- ISH controls for PCR-ISH and RT PCR-ISH
- Detection controls for immunocytochemical detection systems.

Reference control genes, including the use of a single copy mammalian gene such as PDH, are important to assess the degree of amplification in the tissues section/cell preparation. When amplifying DNA targets, the addition of DNases should abolish the signal. If this does not occur, then the signal may have resulted from spurious amplification or, alternatively, may represent either RNA/cDNA. RNase pre-treatment is mandatory in the assessment of RNA targets (*in-situ* RT PCR) and should be included in the amplification of DNA templates to minimize false-positive signals originating from cellular RNA.

The use of a reference control gene primer pair in conjunction with target-specific probe assesses the degree of 'stickiness' of the target probe sequence and the creation of a false-positive result. The addition of only one primer in the amplification mix generates an 'asymmetric PCR' with a quantitative reduction in the amount of product synthesised. Irrelevant primers with irrelevant probe should not generate a signal. The specificity of the ISH component of PCR-ISH is assessed by employing an irrelevant probe with target-specific primers.

The role of primer–primer dimerisation and primer oligomerisation in the generation of false-positive signals is assessed by excluding Taq DNA polymerase, whereas the contribution of non-specific elongation of nicked DNA in tissue sections is examined by the exclusion of primers. The latter is an extremely important control for IS-PCR.

As in routine ISH, hybridisation controls and detection controls are essential to exclude false-positive/false-negative results due to failure of the ISH step or aberrant staining of tissues by the detection system.

11.7 In-cell RNA amplification

Cell and Tissue Preparation

The techniques specifically designed for amplification of RNA targets are RT IS-PCR and RT PCR-ISH. In general, RNA targets are easier to amplify than DNA targets, because of the increased number of starting copies of target (Figures 11.8 and 11.9). Once cells or tissues are removed from the body, RNA degradation begins almost instantaneously. RNases are ubiquitous in the environment, fingers, gloves and bench tops being among some of the many sources. Fixative solutions contain specific RNases that degrade RNA, and tissue processing

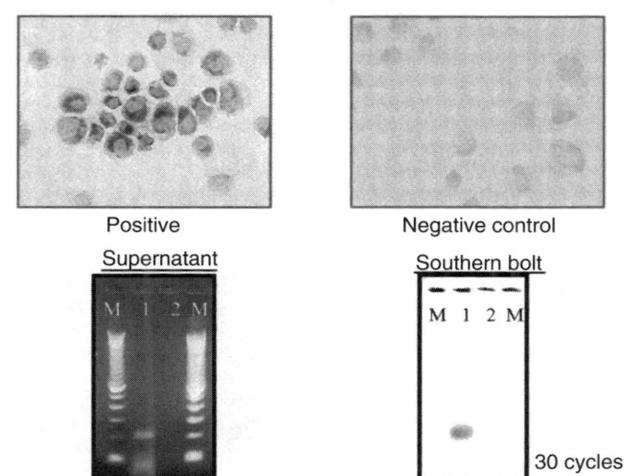

Figure 11.8 RT PCR-ISH detection of dopamine receptor 2 (DRD 2) in hamster ovarian epithelial cells (transfectants) and slab gel and southern blot analysis of created amplicon. A colour version of this figure appears in the colour plate section

In-cell allelotyping using in-cell PCR

Exon 7

DNase & RNase digestion

RT 112 carcinoma cells

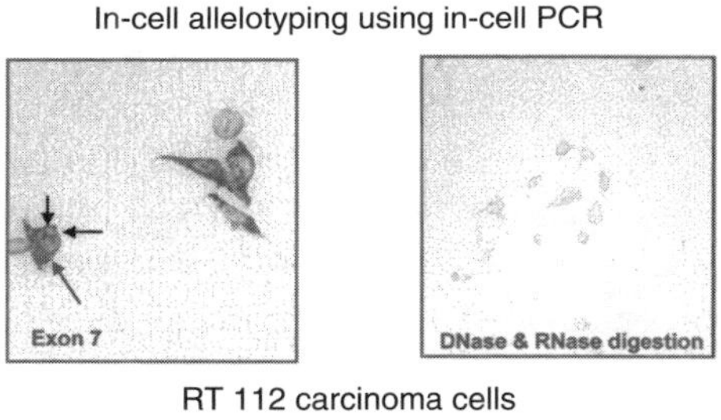

Figure 11.9 In-cell allelotyping for p53 exon 7 in a bladder carcinoma cell line. Note two copies of p53 in the nucleus of the cell left panel and transcript visible in the cytoplasm. DNase and RNase digestion abolished both signals. A colour version of this figure appears in the colour plate section

again contaminated by RNases minimises the amount of target RNA that can be amplified.

Initially, an RNase-free working environment should be created. For optimal preservation of RNA in tissue sections, immediate fixation in RNase-free solutions should be carried out. Alcohol, acetic acid–alcohol and neutral buffered formaldehyde fixatives made up in autoclaved di-ethylpyrocarbonate (DEPC) treated water should be used. Again, all protocols for unmasking nucleic acid should, where possible, employ RNase-free conditions.

Amplification Methodology and Chemistries

Amplification of RNA targets in cells and tissue is reasonably simple using currently available chemistries. A cDNA template is created initially using a reverse transcriptase enzyme, usually Moloney mouse leukaemia virus (MMLV) RT followed by amplification of the newly synthesised cDNA template. A post-amplification ISH step may be employed (RT PCR-ISH) as for DNA PCR-ISH. These techniques employ a two-step approach, i.e. reverse transcription and then amplification. Our group has described a single-step methodology using the rTth DNA polymerase enzyme that obviates the need for splitting the reaction. rTth polymerase possesses both reverse transcriptase and DNA polymerase activity.

Controls for RNA *in-situ* Amplification

The same controls are used as in DNA *in-situ* amplification. Omission of the reverse transcriptase step obviously will yield a faint or negative result. Of importance for RNA assays is the optimal digestion of DNA in cells and tissue sections, which in many cases can be difficult to remove. Our group has found that a combination of RNase A and T1 provides optimal results.

11.8 Problems encountered with in-cell PCR amplification

Many groups have encountered problems with *in-situ* PCR and as yet no universally applicable technique is available (O'Leary *et al.*, 1995, 1996; O'Leary and Herrington, 1996; Teo and Shaunak, 1995). Important factors include: the starting material, fixation conditions and target. DNA *in-situ* PCR is particularly fraught with difficulties, especially with paraffin-wax-embedded material where non-specific incorporation of nucleotide sequences may occur in the presence of Taq DNA polymerase, and it is our opinion that this technology should not be used.

PCR-ISH appears to be more specific, especially if a 'hot-start' modification is employed or alternatively if multiple primer pairs are used. PCR-ISH protocols, in general, are more sensitive but, in contrast to solution phase PCR, are less efficient with apparent linear amplification only. The degree of amplification is difficult to assess. Nuovo (1992, 1994) and Nuovo *et al.* (1991a, 1991b, 1993, 1995) have reported a 200–300-fold increase in product, in contrast to Embretson and colleagues (Embretson *et al.*, 1993) who estimate an increase of 10–30-fold only. Our experience would tend to support the latter figure.

The major limitation with DNA IS-PCR, as previously mentioned, is the non-specific incorporation of nucleotides into damaged DNA by Taq DNA polymerase. This is cycle and DNA polymerase dependent and may occur in the absence of primers and/or with a 'hot-start' modification. Therefore, the routine use of DNA *in-situ* PCR is as yet not feasible owing to the risk of generating false signals. Gosden *et al.* (1991) have previously reported the use of strand break joining in chromosomal work to eliminate spurious incorporation during DNA *in-situ* PCR. Pre-treatment with di-deoxy blockage has also been documented to eliminate non-specific incorporation, but this is not always successful. Our group has utilised strand 'super-denaturation', i.e. where dsDNA is denatured at high temperatures. The DNA is then maintained in a denatured state for an extended period of time (5–10 min), but again this has produced inconsistent results.

Reagent Sequestration

Increased concentrations of reagents are required for successful *in-situ* amplification (usually of the order of 2–5 times). This is thought to be due to reagent sequestration as a result of reagents adhering to slides or to the coating materials used. An additional possibility is that the reagents may intercalate with fixative residues left in tissues. However, it has been documented that pre-treating slides with 0.1%–1% bovine serum albumin (BSA) allows a reduction in reagent concentration, which may function by blocking this sequestration (O'Leary *et al.*, 1995).

11.9 Amplicon diffusion and back diffusion

An inevitable consequence of *in-situ* amplification is product diffusion from the site of synthesis, which may occur as a result of permeabilisation and/or cell truncation. One approach is to reduce the number of cycles. Another frequently employed strategy is to post-fix the slides in ethanol or paraformaldehyde which helps to maintain localisation of product. Alternative approaches include overlaying the tissue section with agarose and/or incorporation of biotin-substituted nucleotides (analogous to IS-PCR). This latter modification promotes the generation of bulkier products, which are less likely to diffuse.

Patchy Amplification/Incomplete Amplification

Patchy amplification is commonly encountered, with between 30% and 80% of cells containing the target sequence of interest staining at any one time. There are many reasons for this, including non-uniform digestion with variations in cell permeability, failure to completely disassociate DNA–histone protein cross-linkages and cell truncation. This latter factor is an inevitable consequence of microtome sectioning where cell 'semi-spheres' are created. As a result, the nuclear contents are truncated giving rise to two possibilities: (a) the desired target sequence may not be present and (b) the target sequence may be present but the product may have diffused out.

11.10 Future work with in-cell PCR-based assays

The recent developments of in-cell TaqMan® PCR (Figures 11.10 and 11.11) allow investigators to directly quantitate transcripts within cell and tissue specimens. The ability to use two-colour detection systems allows simultaneous detection of a housekeeping gene and a target gene.

In-cell TaqMan® PCR demonstrates classical TaqMan® probe kinetics, with fluorescent quenching visible after a certain number of cycles. This is due to the volume-limited space of the cytoplasm/nucleus and is brought about by the proximity of the quencher and reporter sequences in molar excess in the 'freestate' following hydrolysis of the probe.

The discovery of in-cell TaqMan® PCR theoretically should make it possible to introduce TaqMan® arrays for direct quantitative analysis. This would offer a major advantage over conventional cDNA and SNP hybridisation array platforms, and would for the first time allow gene quantitation of multiple genetic

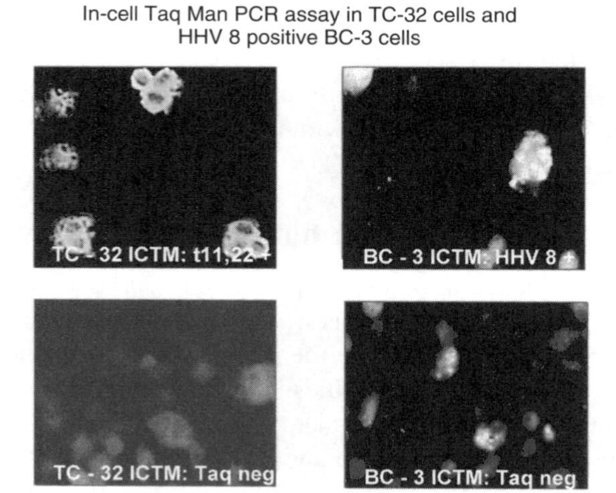

Figure 11.10 In-cell TaqMan® analysis of the t11;22 translocation, characteristic of PNET in the TC-32 cell line, derived from a PNET. A colour version of this figure appears in the colour plate section

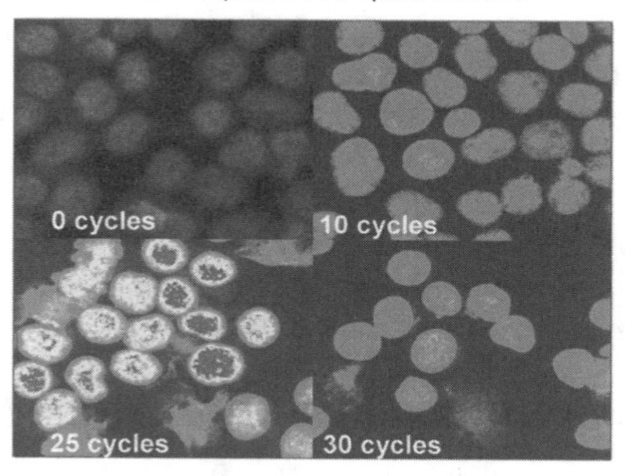

Figure 11.11 Quasi-real time TaqMan® PCR analysis of HHV 8 in an effusion lymphoma cell line BC-3. Note the diminution of fluorescence after 25 cycles due to the release of the quencher and reporter into the volume-limited space of the nucleus in the BC-3 cell line. A colour version of this figure appears in the colour plate section

Table 11.1 Uses of in-cell PCR technologies in cellular pathology

DNA and RNA viruses
HIV 1,2
HPV 6, 11, 16, 18, 31, 33 etc.
HBV, HCV
CMV
Measles
HHV 6, 7 and 8
HSV DNA
LGV (lymphogranuloma venereum)
Oncogenes, tumour suppressor genes and markers of malignancy
p53 mutations
Ras mutations (H-Ki, N-ras)
Gene rearrangements (t11;22, t11;14)
Chromosome mapping (PRINS and cycling PRINS)
T-cell receptor rearrangements
Metalloproteinases and their inhibitors
EGF receptor expression
Nitric oxide synthase

loci simultaneously. Table 11.1 lists the current use of in-cell PCR analyses in cellular pathology.

11.11 References

Bagasra O, Hauptman SP, Lischner HW, Sachs M and Pomerantz RJ (1992) Detection of human immunodeficiency virus type 1 provirus in mononuclear cells by *in situ* polymerase chain reaction. *N. Engl. J. Med.* **326**: 1385–91.

Bagasra O, Seshamma T, Hanson J, Bobroski L, Saikumari P, Pestaner JP and Pomerantz RJ (1994) Applications of *in-situ* PCR methods in molecular biology: I. Details of methodology for general use. *Cell Vision* **1**: 324–35.

Bagasra O, Seshamma T and Pomerantz RJ (1993) Polymerase chain reaction *in situ*: intracellular amplification and detection of HIV-1 proviral DNA and other gene sequences. *J. Immunol. Meth.* **158**: 131–45.

Bobroski L, Pestnaer JP, Seshamma T, Pomerantz RJ and Bagasra O (1995) Localisation of HIV-1 in cardiac tissue utilizing the *in-situ* polymerase chain reaction. *FASEB J.* **9**: PA1051.

Boshoff C, Schultz TF, Kennedy MM, Graham AK, Fisher C, Thomas A, McGee JO'D, Weiss RA and O'Leary JJ (1995) Kaposi's sarcoma associated herpes virus (KSHV) infects endothelial and spindle cells. *Nat. Med.* **1**: 1274–8.

Cao Y, Kopplow K and Liu GY (2000) *In-situ* immuno-PCR to detect antigens. *Lancet* **356**(9234): 1002–3.

Cheng JD and Nuovo GJ (1994) Utility of reverse transcriptase (RT) *in situ* polymerase chain reaction in the diagnosis of viral infections. *J. Histotechnol.* **17**: 247–51.

De-Mesmaeker A, Altmann KH, Waldner A and Wendeborn S (1995) Backbone modifications in oligonucleotides and peptide nucleic acid systems. *Curr. Opin. Struct. Biol.* **5**(3): 343–55.

Embretson J, Zupancic M, Beneke J, Till M, Wolinsky S, Ribas JL, Burke A and Haase AT (1993) Analysis of human immunodeficiency virus – infected tissues by amplification and *in situ* hybridisation reveals latent and permissive infection at single cell resolution. *Proc. Natl. Acad. Sci. USA* **90**: 357–61.

Gosden J, Hanratty D, Starling J, Fnates J, Mitchell A and Porteous D (1991) Oligonucleotide primed *in situ* DNA synthesis (PRINS): a method for chromosome mapping, banding and investigation of sequence organisation. *Cytogenet. Cell. Genet.* **57**: 100–4.

Haase AT, Retzel EF and Staskus KA (1990) Amplification and detection of lentiviral DNA inside cells. *Proc. Natl. Acad. Sci. USA* **87**: 4971–5.

Herrington CS, de Angelis M, Evans MF, Troncone G and McGee JO'D (1992) Detection of high risk human papilloma virus in routine cervical smears: strategy for screening. *J. Clin. Pathol.* **45**: 385–90.

Holland PM, Abramson RD, Watson R and Gelfand DH (1991) Detection of specific polymerase chain reaction product by utilizing the 5′ to 3′ exonuclease activity of *Thermus aquaticus* DNA polymerase. *Proc. Natl Acad. Sci. USA* **88**: 7276–80.

Lawyer FC, Stoffel S, Saiki RK, Myambo KB, Drummond R and Gelfand DH (1989) Isolation, characterization, and expression in *Escherichia coli* of the DNA polymerase gene from the extreme thermophile, *Thermus aquaticus*. *J. Biol. Chem.* **264**: 6427–37.

Nuovo GJ (1992) *PCR In-situ Hybridisation. Protocols and Applications*. Raven Press, New York.

Nuovo GJ (1994) *In situ* detection of PCR amplified DNA and cDNA; a review. *J. Histotechnol.* **17**: 235–46.

Nuovo GJ, Gallery F, Horn R, MacConnell and Bloch W (1993) Importance of different variables for enhancing *in situ* detection of PCR amplified DNA. *PCR Meth. Appl.* **2**: 305–12.

Nuovo GJ, Gallery F, MacConnell P, Becker P, and Bloch W (1991a) An improved technique for the detection of DNA by *in-situ* hybridisation after PCR amplification. *Am. J. Pathol.* **139**: 1239–44.

Nuovo GJ, MacConnell P, Forde A and Delvenne P (1991b) Detection of human papilloma virus DNA in formalin fixed tissues by *in-situ* hybridisation after amplification by PCR. *Am. J. Pathol.* **139**: 847–50.

Nuovo GJ, MacConnell PB, Simsir A, Valea F and French DL (1995) Correlation of the *in situ* detection of polymerase chain reaction-amplified metalloproteinase complementary DNAs and their inhibitors with prognosis in cervical carcinoma. *Cancer Res.* **55**: 267–75.

O'Leary JJ, Browne G, Bashir MS, Landers RJ, Crowley M, Healy I, Lewis FA and Doyle CT (1995) Non-isotopic detection of DNA in tissues, in: *Non-isotopic Methods in Molecular Biology – A Practical Approach* (eds ER Levy and CS Herrington). IRL Press, Oxford.

O'Leary JJ, Browne G, Johnson MI, Landers RJ, Crowley M, Healy IB, Street JT, Pollock AM, Lewis FA, Andrew A, Cullinane C, Mohamdee O, Kealy WF, Hogan J and Doyle CT (1994a) PCR *in-situ* hybridisation detection of HPV 16 in fixed CaSki and fixed SiHa cells – an experimental model system. *J. Clin. Pathol.* **47**: 933–8.

O'Leary JJ, Browne G, Landers RJ, Crowley M, Healy IB, Street JT, Pollock AM, Murphy J, Johnson MI, Lewis FA, Mohamdee O, Cullinane C and Doyle CT (1994b) The importance of fixation procedures on DNA template and its suitability for solution phase polymerase chain reaction and PCR *in-situ* hybridisation. *Histochem. J.* **26**: 337–46.

O'Leary JJ, Chetty R, Graham AK and McGee JO'D (1996) *In situ* PCR: pathologists dream or nightmare?. *J. Pathol.* **178**: 11–20.

O'Leary JJ and Herrington CS (eds) (1996) *PCR In-situ Amplification – A Practical Approach*. IRL Press, Oxford.

Spann W, Pachmann K, Zabnienska H, Pielmeier A and Emmerich B (1991) *In-situ* amplification of single copy gene segments in individual cells by the polymerase chain reaction. *Infection* **19**: 242–4.

Teo IA and Shaunak S (1995) Polymerase chain reaction *in situ*: an appraisal of an emerging technique. *Histochem. J.* **27**: 647–59.

12

TaqMan® Technology and Real-Time Polymerase Chain Reaction

John J. O'Leary, Orla Sheils, Cara Martin
and **Aoife Crowley**

12.1 Introduction

Real-time polymerase chain reaction (PCR) can be performed using a variety of technologies including the 5′ nuclease assay, hair-pin primers, Scorpion probes, intercalating dyes, e.g. SYBR green, and FRET (free-resonance energy transfer technology) and using a dual probe hybridisation systems (molecular beacons) (Figures 12.1 and 12.2).

In this chapter we discuss primarily the 5′ nuclease assay (also called TaqMan® PCR), examine the assay functionally and demonstrate specific examples of the technology for use in pathology laboratories.

The use of PCR in molecular diagnostics has increased to the point where it is now accepted as the norm for detecting nucleic acids from a number of origins. Reverse transcription PCR (RT-PCR) is the most sensitive method for detection of low abundance mRNA, often obtained from limited tissue samples. The introduction of fluorescent-based kinetic PCR adds a further dimension to the potential scope of this technology.

Real-time quantitative PCR is the reliable detection and measurement of products generated during each cycle of the PCR which are directly in proportion to the amount of template prior to the start of the PCR process.

The applications of real-time quantitative PCR are diverse and numerous, including such areas as DNA/cDNA copy number measurement in genomic or

Molecular Biology in Cellular Pathology. Edited by John Crocker and Paul G. Murray
 ISBN: 0-470-84475-2

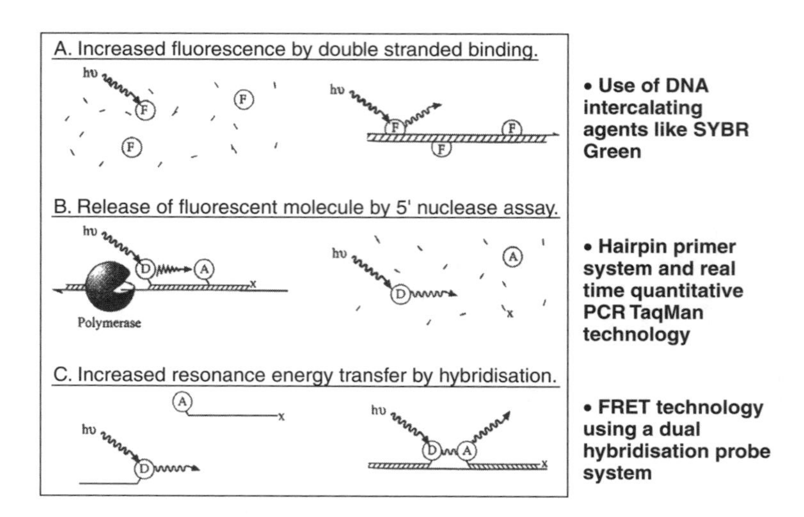

Figure 12.1 Fluorescent monitoring of DNA and RNA in solution by three different technologies: (a) intercalating dyes; (b) 5′ nuclease assay; (c) FRET technology

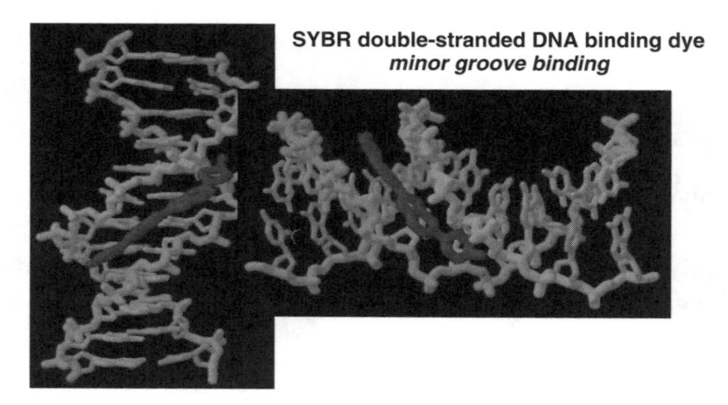

Figure 12.2 SYBR green: an intercalating dye for double-stranded DNA. A colour version of this figure appears in the colour plate section

viral DNA/RNA, functional genomics including mRNA expression analysis, allelic discrimination assays and confirmation of microarray data.

Upstream modifications to the technique, including the use of laser capture microdissection, add an extra dimension in facilitating the production of homogenous cell populations from complex tissue sections for more accurate quantitative analysis.

12.2 Probe technologies

DNA Binding Fluorophores

The basis of non-specific sequence detection methods is the use of intercalating dyes such as SYBR green or ethidium bromide (Figure 12.2). The unbound dye exhibits little fluorescence in solution, but during elongation increasing amounts of dye bind to the nascent double-stranded DNA. When monitored in real time, the increase in fluorescence can be observed during the polymerisation step. This method eliminates the need for target-specific fluorescent probes, but it is important to note that the primers used determine the overall specificity of the reaction. In addition, because the presence of any double-stranded DNA is capable of generating fluorescence, the assay is no more specific than conventional RT-PCR.

Hybridisation Probes; Molecular Beacons

In this system two hybridisation probes are employed. One probe carries a fluorescein donor molecule at its 3′ end whose emission spectrum overlaps with an acceptor probe on the 5′ end of a second probe. In solution, the two dyes are separate. However, after the denaturation step they hybridise with the target sequence during annealing, and adopt a head-to-tail configuration. This brings the two fluorescent molecules into close proximity, and the fluorescein moiety can transfer its energy to the second. The light energy emitted by the second molecule can then be monitored, and increased levels of fluorescence correspond with the amount of DNA synthesised during the PCR reaction. An additional advantage of this system is the fact that the probes are not hydrolysed, thus allowing the possibility of the generation of melting curves. These can be used to monitor amplification efficiency.

Hydrolysis Probes

The TaqMan® assay uses the 5′-3′ exonuclease activity of *Taq* or *rTth* DNA polymerase to hydrolyse a hybridisation probe bound to the target amplicon (Holland *et al.*, 1991; Lawyer *et al.*, 1989; Lyamichev *et al.*, 1993). The technique allows for direct detection of PCR product by the specific release of a fluorescent reporter molecule during the PCR reaction (Lee *et al.*, 1993). In this system, three oligonucleotides are required to bind to the target (Figure 12.3). Two template-specific primers define the endpoints of the amplicon and serve as a first level of specificity. A further level of specificity is achieved by the inclusion of a specific probe, which binds internally to the points defined by the primers. The characteristics of the TaqMan® probe include a reporter molecule at the 5′ end, whose emission is quenched by a second molecule at the 3′ end. The spatial proximity of the reporter to the quencher in an intact probe ensures that no net fluorescence is detected. The release of the fluorescence only occurs if target-specific amplification occurs, obviating the need to confirm the amplicon following amplification.

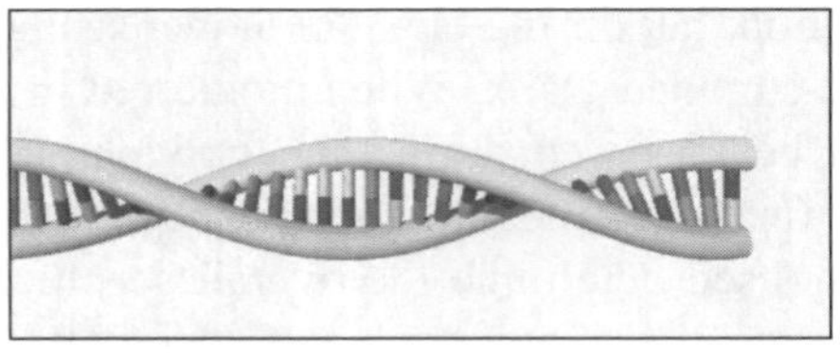

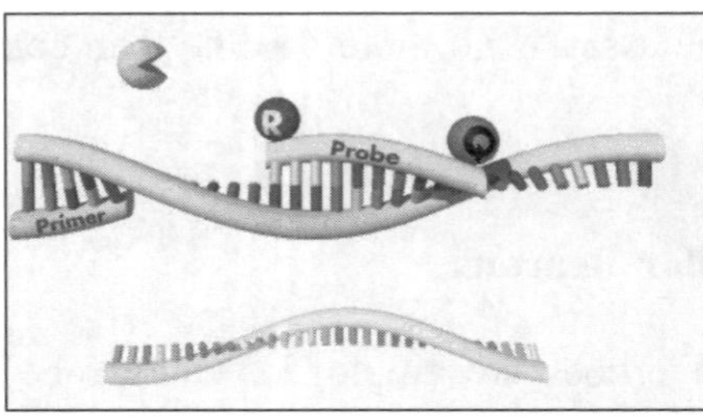

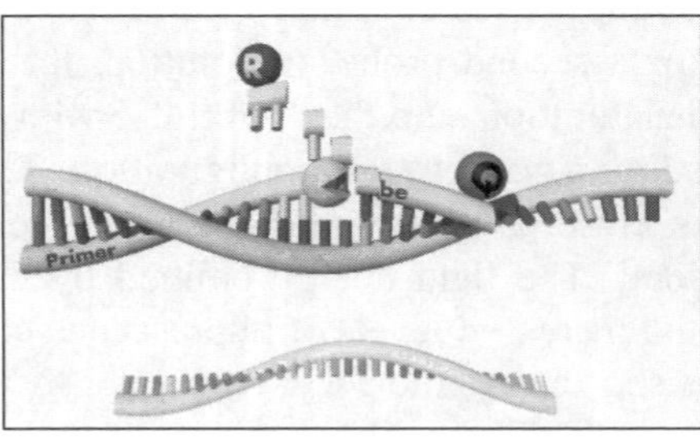

Figure 12.3 Schematic representation of TaqMan® PCR, using an internal probe labelled with a reporter (R) and quencher sequence (Q). Details of the reaction are given in the figure. A colour version of this figure appears in the colour plate section

12.3 TaqMan® probe and chemistry (first generation)

Three factors determine the specific performance of a dual-labelled fluorogenic probe in TaqMan® PCR:

- effective quenching of the reporter sequence in an intact probe;
- effective hybridisation efficiency of the probe; and
- efficient cleavage of TaqMan® by AmpliTaq DNA polymerase.

The original TaqMan® probe consists of an oligonucleotide, usually 20–30 bases in length, with a 5′ reporter dye, a 3′ quencher dye and a 3′ blocking phosphate.

Reporter Dyes

The fluorescence *reporter dye*, e.g. FAM (6-carboxy-fluoroscein), VIC or JOE, is covalently linked to the 5′ end of the oligonucleotide probe. Tetrachloro-6-carboxy-fluoroscein (TET) and hexachloro-6-carboxy-fluoroscein (HEX) can also be used as fluorescent reporter dyes in this system. Each of these reporters is quenched by TAMRA (6-carboxy-tetramethyl-rhodamine) (*the quencher dye*), which is attached by a linker-arm-modified nucleotide (LAN) to the 3′ end of the probe. The probe is chemically phosphorylated at its 3′ end, which prevents probe extension during PCR applications. When the probe is intact (linearised), the proximity of the reporter dye to the quencher dye results in direct suppression of the fluorescence from the reporter dye by Forster-type energy transfer (Forster, 1948; Lakowicz, 1983). During PCR the probe will specifically anneal between the forward primer (primer 1) and the reverse primer (primer 2). The 5′–3′ exonuclease activity of the AmpliTaq DNA polymerase will cleave the hybridised probe between the reporter and the quencher molecules. Once the reporter molecule is cleaved from its spatial bind with the quencher, this results in an increase in fluorescence that is proportional to the amount of product that has accumulated.

Placement of the Quencher Sequence

Fluorescence quenching depends on the spatial proximity of the reporter and quencher dyes and is dependent on Forster-type energy transfer efficiency (Forster, 1948; Lakowicz, 1983). This varies with the inverse sixth power of the distance between fluorophores.

Originally, it was thought that quencher dyes needed to be close to the 5′ end in order to achieve adequate quenching (Lee *et al.*, 1993). However, the quencher dye TAMRA can be attached to any nucleotide position in a probe and still effectively quench the fluorescence emission of a reporter dye attached to the 5′ end. The flexibility of a single-stranded probe apparently allows for conformational adaptations to be made between the reporter and quencher dyes in order to achieve close enough energy transfer. Probes with a reporter dye at the 5′ end and TAMRA attached to the 3′ end exhibit sufficient quenching for most fluorogenic 5′ nuclease assays.

The effect of quenching, however, may be variable and the placement of the quencher dye can have a major impact on the efficiency of cleavage of the probe by AmpliTaq DNA polymerase. If one is to observe an increase in reporter fluorescence intensity in this assay, the probe must be effectively cleaved between the reporter and quencher dyes. When dyes are on opposite ends of the probe (5′ end and 3′ end) any cleavage of the probe is therefore detected.

Therefore an oligonucleotide with a reporter dye on the 5′ end and TAMRA attached to the 3′ end directly provides the best chance of having a probe that reliably detects PCR amplification in TaqMan® PCR assays. Another advantage of placing TAMRA on the 3′ end is that maximum hybridisation efficiency is achieved, as steric interference is minimised. The quencher dye is attached to the probe by post-synthetic coupling of TAMRA-NHS-ester to a LAN. Linker arm nucleotide is a thymidine nucleotide with a 6-carbon linker arm attached to a 4′ position of the thymidine ring. The presence of the linker arm combination partially disrupts the base-pairing capabilities of thymidine.

An oligonucleotide that contains a LAN–TAMRA will have a reduced T_m (melting temperature) and therefore lower hybridisation efficiency. The effect on T_m is least severe when LAN–TAMRA is at the very end of the oligonucleotide. Specifically, the 3′ nucleotide of a fluorogenic probe need not be directly homologous to the target sequence and in general the best fluorogenic probe designs make 3′ end nucleotide a LAN–TAMRA (whether or not the selected probe sequence ends in T).

There is one exception to placing LAN–TAMRA at the 3′ end, and this is for allelic discrimination. Lee *et al.* (1993) have described allele-specific probes that have the quencher molecule located 7 nucleotides from the 5′ end of the probe sequence. The process allowed them to distinguish the normal human cystic fibrosis allele from Δ F508 mutant.

Quenching

Factors other than the dye placement can affect the degree of quenching of the fluorogenic probe. It has been found that positioning a G next to the reporter dye can cause direct quenching. This can persist even after cleavage of the fluorogenic probe. In general, it is best to avoid a G at the 5′ end position of the probe. When one places TAMRA at the 3′ end, quenching appears to be directly dependent on the flexibility of the single-stranded probe itself. Any structure that directly interferes with this flexibility can potentially reduce the degree of quenching observed in the intact (linearised) probe. Structures to be avoided therefore include hair-pins, self-hybridising sequences, hybridisation to either the reverse or the forward PCR primer and oligomer formation such as tetramers that can form when an oligonucleotide has four or more Gs in a row.

The purity of the probe can also have a marked effect on the observed quenching. Any molecule that has a reporter dye but does not have a quencher dye can be a source of contamination. The presence of unquenched reporter raises the initial intensity of the reporter fluorescence. Subsequently, it can be difficult to discern changes in the reporter signal. Because of this problem, the reporter dye is attached to the 5′ end using a phosphoramidite, so that incomplete sequences generated during the oligonucleotide synthesis are not labelled with the reporter dye. In general all fluorogenic probes should be HPLC-purified so as to minimise the amount of unquenched reporter present.

12.4 Second generation TaqMan® probes

Minor Groove Binding Probes

Certain naturally occurring antibiotics and synthetic molecules, called minor groove binders (MGBs), are able to fit into the minor groove of the helix formed by double-stranded DNA. It has been discovered that certain MGBs, are beneficial and broadly enabling for DNA analysis by stabilising DNA duplexes and increasing the mismatch discrimination character (that is the ability of an oligonucleotide probe to bind only to its complete complementary sequence) when bound to an oligonucleotide probe. The MGB can be incorporated in several reagents that are useful in different analyses, e.g. 5′ nuclease assays and capture detection arrays.

MGB Probes and the 5′ Nuclease Assay

MGB fluorogenic probes have a number of characteristics that make them superior to traditional fluorogenic probes. The MGB can be attached to the 3′ end, the 5′ end, or to an internal nucleotide of an oligonucleotide. MGB probes bind more tightly to their complement, which raises the T_m of probes and allows for more flexible assay design. The 5′-nuclease PCR assay requires probes to have a T_m close to the temperature of the thermal-stable polymerase extension step, i.e. 65–72°C. Studies have shown that the introduction of an MGB onto a 12-mer probe with a T_m of 20°C increases its T_m to 65°C. This T_m is equivalent to the T_m of a 27-mer probe without an MGB. It is recommended that probe T_m be >10°C higher than primer T_m, so the introduction of an MGB contributes significantly to assay design, allowing the use of a wider range of primers (Gibson *et al.*, 1996; Kutyavin *et al.*, 2000; Orlando *et al.*, 1998).

The use of an MGB allows shorter probes to be used. In 5′-nuclease assays, probe length can be as short as 12 bases while maintaining an appropriate T_m. Probes usually are designed to a T_m close to the extension temperature (65–72°C) of the thermal-stable polymerase extension step in PCR. MGB–oligonucleotides allow better mismatch discrimination. MGB–oligonucleotides demonstrate an increased difference between the T_m of match and single-base mismatch oligonucleotides, therefore increasing the discriminatory power of hybridisation assays.

Another remarkable feature of MGB probes is the low background fluorescence and, as a consequence, a greater signal-to-noise ratio. The background fluorescence of intact MGB probes increases slightly with probe length but is still several times lower than probes without an MGB. MGB probes stabilise A–T bonds more than G–C bonds in a DNA duplex. This reduces the influence of target sequence on T_m. It has been found also that mismatch discrimination is improved when the mismatch is placed under the MGB. Typically an oligonucleotide–MGB conjugate is designed such that the MGB resides at the 3′ end of an oligonucleotide containing an A–T rich region of about 6–7 bases. It was observed that mismatches positioned within the MGB region were better discriminated, showing an increased free energy difference between match and mismatches for the different mismatch pairs under the MGB.

MGB–oligonucleotides can also be used for sequence-specific clamping to inhibit, for example, thermal-stable polymerase extension of high copy number templates that are not of interest. Clamping refers to an oligonucleotide that binds tightly and inhibits the extension or amplification of a particular sequence if it exists within the region defined by the PCR primers. MGB–oligonucleotides so designed can make effective clamps due to their high T_m, short length and the particular attributes of the MGB. Studies have found that a DNA target present at a concentration of 0.3% compared with a closely related allele can be specifically detected.

12.5 Hybridisation

For effective cleavage of the fluorogenic probe, hybridisation to the target sequence in the specimen is required. The general rules for designing hybridisation probes also apply directly to fluorogenic probes. Avoidance of secondary structure, keeping the GC content as close to 50% and avoiding long runs of single nucleotides such as Gs or As are advised. Importantly, the probe must not hybridise with either primer. If such hybridisations occur, this will directly lower the effective concentration of the probe and the primer, thereby reducing PCR hybridisation efficiency. Probe–primer hybrids are also a source of primer–dimer contamination, which can form during the amplification reaction. Importantly, PCR primers and the fluorogenic probe must hybridise to the template strand during the annealing step of the PCR amplification reaction, otherwise reaction failure is encountered. Primer–template hybrids are stabilised when AmpliTaq DNA polymerase extends the primer in the polymerisation step. The fluorogenic probe is not extended, so its hybridisation is not stabilised. Therefore the probe–template hybrid must be more stable and have a higher T_m than the primer–template hybrids. This is important when one designs a TaqMan® probe. In designing a probe that forms a more stable hybrid than PCR primers, estimated T_m values must be used. Although it is not completely accurate, the 'nearest neighbour method' (Rhoads, 1990) is the best algorithm available for estimating T_m values of oligonucleotides.

In general, the estimated T_m of the probe should be at least 5–10°C higher than the matched T_m values of the PCR primers. Importantly, T_m calculations must use the same set of salt and oligonucleotide concentrations for both primers and TaqMan® probe. When one is calculating the T_m, then these concentration constraints should ideally match those of the suggested PCR. In general, the estimated T_m values do not accurately reflect the melting temperatures under PCR conditions, but usually are adequate for comparative purposes. Optimal annealing temperatures of primers are empirically determined by trial and error, based on PCR product yield.

Probes ideally are at least 20–30 bases in length in order to achieve adequate specificity. Longer probes tend to have a higher T_m thereby giving rise to more successful TaqMan® PCR conditions. Importantly, however, when the length of the probe increases beyond 30 nucleotides, several problems arise.

Longer probes reduce the yield of oligonucleotide synthesis and the longer the probe the greater the chance of forming inter- or intra-molecular structures. This can create purification problems during synthesis, impair hybridisation efficiency during probe annealing, and impair cleavage of the probe or can directly interfere with the PCR.

12.6 TaqMan® PCR conditions

Importantly, the fluorogenic probe is not stabilised by being extended during the PCR, and this has consequences for the extension step used in amplification. The success of the reaction is largely determined by three events:

- the primers hybridising and extending;
- probe hybridising; and
- probe cleavage and primer extension being completed.

Since the fluorogenic probe must be directly hybridised for cleavage to occur, extension must be carried out to ensure probe binding. In general, to keep the probes hybridised, extension at or below 72°C is carried out. Increasing magnesium chloride concentrations can be exploited in order to keep the probe hybridising and to allow extension to occur. By increasing the magnesium chloride concentration, the extension temperature can also be slightly increased. By prolonging the annealing step at a temperature below the probe's T_m, all three events required for a successful 5′ nuclease assay results.

A two-step protocol using a combined annealing /extension temperature below the T_m of the probe increases the hybridisation cleavage efficiency of the TaqMan® system. As noted previously, higher concentrations of magnesium chloride (3.5–6 mM) tend to stabilise probe binding and thus promote the use of higher temperatures during the combined anneal–extension step.

Higher temperatures are desirable so that the extension rate of AmpliTaq DNA polymerase is as high as possible. Optimal annealing temperatures for any system are largely determined empirically.

Starting DNA Template

The investigator should start with enough copies of the DNA template to obtain an amplification signal after 25–30 cycles. Lower concentrations of DNA require higher cycle numbers to produce a sufficient signal. Typically 50–100 ng of DNA are used per assay.

The efficiency of DNA amplification is dependant on important variables, such as magnesium concentration, primer and probe concentration. Ideally, a

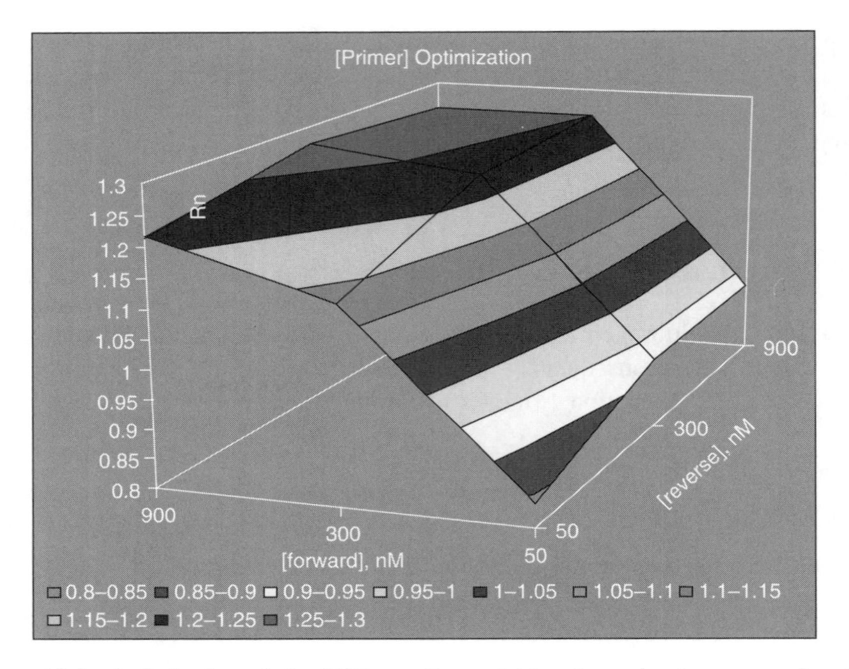

Figure 12.4 Optimisation of the PCR amplicon yield: primer titration curve. A colour version of this figure appears in the colour plate section

primer and probe matrix should first be performed to assess where the exponential and plateau phases of the PCR occur in any particular assay (Figure 12.4).

Starting RNA Template

Typically 50–100 ng of RNA is required for an RT TaqMan® PCR assay. The efficiency of RNA amplification is dependent on the presence of RNases in the environment, and the buffering capacity of the RT buffer used for RNA amplification. RNA amplification can be carried out using a one-step or two-step strategy. The one-step strategy employs an enzyme such as rTth polymerase, which possesses RT and DNA polymerase activity. The two-step strategy uses a separate RT (e.g. MuLV, Superscript) to generate cDNA followed by amplification with a DNA polymerase. In general, one-step strategies are favoured since they reduce the chances of contamination and minimise sample handling.

12.7 Standards for quantitative PCR

In order to assess the linear range of amplification, standards of known concentration and similar composition to the target can be employed. It is noteworthy that a DNA standard is not the optimal choice when dealing with an RNA assay system. With the increased interest in functional genomics, assays, which rely on the measurement of mRNA expression levels, are gaining in popularity. In such an assay, DNA would be inappropriate as a standard, as it would not take into account inter-sample variations due to the RT step. The standard of choice in RNA assays is the use of cRNA (copy RNA), or at least RNA of known concentrations. These standards can be generated by amplifying a sequence slightly larger than the TaqMan® PCR amplicon using a set of outer primers. The amplicon can then be cloned into a suitable vector and transcribed into cRNA.

12.8 Interpretation of results

An increase in fluorescence can be detected using a luminescence spectrophotometer such as the Applied Biosystems LS-50B, 5700, 7200 and 7700 thermal cyclers for end-point detection (i.e. at cycles 35, 40, etc.). This allows a relatively easy quantitative detection of amplified nucleic acid and is reliable and simple to perform. Detection is achieved using a 96-well plate reader and avoids

Real time TaqMan® PCR-7700 DNA sequencer technology

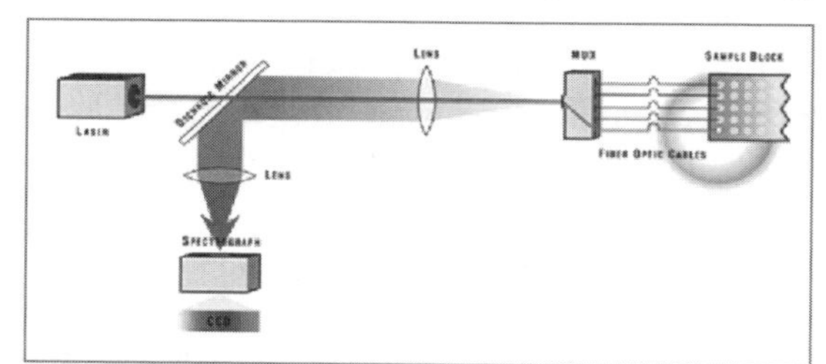

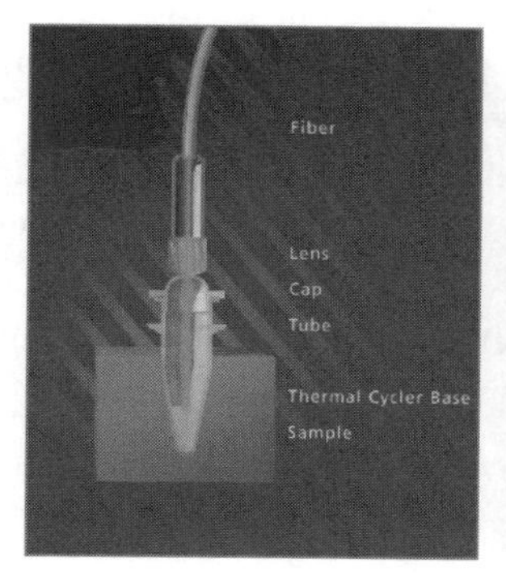

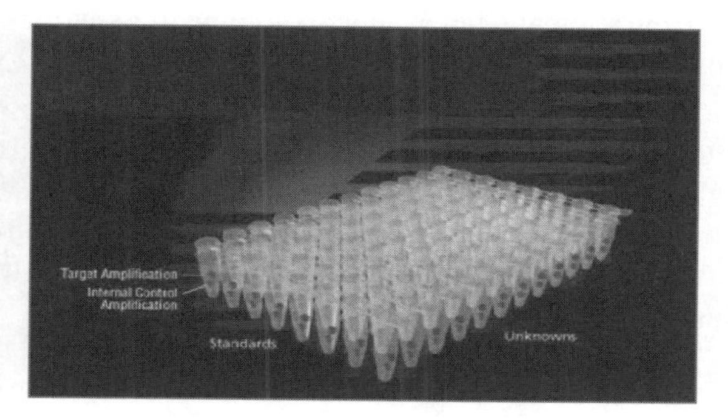

Figure 12.5 The ABI PRISM 7700 Sequencer Detector, offers a unique closed tube assay. The 7700 system collects the fluorescent emission between 500 and 660 nm in each of the 96 sample wells, once every few seconds. A laser directed to the sample vials through fibre optic cables excites the fluorescent dyes. These cables then carry the fluorescent emissions back to a CCD camera, where they are detected according to their individual wavelengths. The 7700 system uses specifically designed reaction tubes that remain closed during the PCR amplification and detection process, reducing the chances of contamination. A colour version of this figure appears in the colour plate section

the complexities of gel electrophoresis followed by ethidium bromide staining or autoradiography.

Real-time amplicon detection can be achieved using the Applied Biosystems 7700, 7000 or 7900 DNA sequencer detector, which utilises a laser scanning format to detect increases in fluorescence at defined time-points in the thermo-cycling protocol (Figure 12.5). In addition, the light cycler from Roche also facilitates the detection of released fluorescent reporter molecules in solution. All of these DNA sequence detectors can also be used for SYBR green, scorpion probe and molecular beacon technologies with little modification. In addition, non-PCR-based technologies, such as nucleic acid sequence based amplification (NASBA), which is used in our laboratory for the detection of HPV E6/E7 transcripts, can be used successfully with the above DNA sequence detection technology.

12.9 End-point detection

End-point detection is performed when the PCR run has been completed and the data are then collected. For end-point detection, a luminescence spectrophotometer is used (e.g. an ABI Prism 7200). The signals obtained during end-point detection and real-time detection assays result from changes in the fluorescence emission intensity of the reporter dye following cleavage of the probe. For end-point detection, interfering fluorescence fluctuations are normalised by applying two calculations.

The first calculation employs the quencher dye as a passive internal standard. This is accomplished by dividing the emission intensity of the reporter dye by the emission intensity of the quencher dye to give a ratio, defined as the *RQ* (reporter: quencher) for a particular reaction. Complicating factors such as concentration changes due to volume fluctuations are directly normalised by this ratio.

Other fluctuations that are not due to the PCR-related nuclease digestion are normalised by taking the *RQ* value for a tube that contains all the components including the target, which is defined as the $RQ+$, and subtracting this from the *RQ* value of the NTC (no template control), defined as the $RQ-$. The final value (ΔRQ) reliably indicates the magnitude of the signal, which is generated during the PCR.

This can be expressed mathematically as

$$\Delta RQ = (RQ+) - (RQ-)$$

where

$$RQ+ = \frac{\text{the emission intensity of reporter}}{\text{the emission intensity of quencher}} \quad \text{(PCR with target)}$$

$$RQ- = \frac{\text{the emission intensity of reporter}}{\text{the emission intensity of quencher}} \quad \text{(PCR without target)}$$

Determination of Threshold ΔRQ for End-point Detection

To ensure a statistically high confidence level in the results of the TaqMan® PCR, the protocol should include at least three NTCs per microtitre plate assay. Based on t-distribution values (Orlando *et al.*, 1998) any ΔRQ value above the threshold ΔRQ has a 99% confidence level of being a positive result.

For example, if the mean $RQ-$ for a system is 0.15 with a 0.01 standard deviation (SD), then the threshold ΔRQ is $6.965(n-2) \times 0.01 = 0.07$. Therefore, any ΔRQ greater than 0.07 is a positive result and indicates that the sample contains the target. Increasing the number of NTCs decreases the standard deviation. Values can be obtained from the standard t-table distribution.

Greater confidence can be obtained when reactions are carried out in duplicate or triplicate and the master mix is common to all reactions. The use of positive displacement pipettes also improves precision. To evaluate reproducibility, one can calculate the coefficient of variation (CV) on replicate samples. Inconsistent results, such as CVs exceeding 10%, can be caused by pipetting errors and incomplete mixing of DNA solutions. To achieve smaller CVs, preparing a mastermix for a replicate and if necessary dilution of the template with TE buffer or distilled water and may be employed.

12.10 Real-time detection

The 7700 real-time detector monitors the PCR at every cycle. Quantitative data are generated based on the PCR at early cycles, when PCR fidelity is highest. In addition, the 7700 has a linear dynamic range of at least five orders of magnitude, obviating the need for serial dilutions. In addition, the 7700 adopts a closed tube format, reducing the chance of contamination. The 7700 collects fluorescent emission between 500 and 600 nm in each of the 96 wells of the thermal cycler, once every few seconds. A laser directed through fibre optic cables excites the fluorescent dyes in the tube. These cables then carry the fluorescent emissions back to a charged couple device (CCD) camera, where they are detected according to their individual wavelengths. A value ΔR_n is then calculated for each sample by subtracting the reporter signal before the PCR from the normalised reporter signal. In addition, the in-built software calculates a value called the threshold cycle (C_t), a point where the amplification plot crosses a defined fluorescence threshold.

12.11 Relative quantitation

Relative quantitation is a common procedure used to assess differences in target sequences among different samples. It is particularly relevant in the assessment of functional variations in mRNA molecules. The objective of an RT PCR relative quantitative experiment is to determine the ratio of a target mRNA molecule to a different target molecule or to itself under different conditions.

The result is then normalised to an endogenous control (reference gene) and reported as a fold difference relative to a *calibrator sample*.

There are two methods for calculating the relative quantitation of target gene expression: the standard curve method, and the comparative C_t method (*User Bulletin 2: Relative Quantitation*. Applied Biosystems, USA, 1997).

For the *standard curve method*, standard curves are generated for both the target and the endogenous control. For each test sample the amount of target and control quantity is determined from the appropriate standard curve. Then the target amount is divided by the endogenous control quantity to obtain a normalised target value. To calculate the relative expression levels the normalised target values are divided by the calibrator normalised target values.

For the *comparative C_t method* the relative quantification values are calculated from the C_t values generated during the PCR. The C_t value is the cycle at which a statistically significant increase in PCR product is first detected. The comparative C_t method for relative quantitation calculates the relative gene expression using the following equation:

$$\text{relative quantity} = 2^{-\Delta\Delta C_t}$$

The ΔC_t is calculated by normalising the C_t of the target sample with the C_t of the endogenous control (C_t target − C_t endogenous control). The $\Delta\Delta C_t$ is then calculated by subtracting the average ΔC_t for the calibrator sample from the corresponding average C_t for the target sample. The relative levels of the target gene expression are then expressed as a fold change relative to the calibrator sample. For the comparative C_t method the endogenous control can be amplified in a separate tube or in a single multiplex reaction, which uses more than one primer pair/probe in the same tube.

12.12 Reference genes

A reference gene can be assayed separately or together with the unknown target and their final ratio calculated. The reference gene can be an endogenous or exogenous mRNA target. In both cases this strategy has the advantage of compensating for the high variability in the efficiency of the RT reaction. The use of an endogenous reference gene can also prevent errors that arise due to inaccurate estimation of total RNA concentration and quality in the original sample. The expression of a housekeeping gene (b-actin, GAPDH, b2-microglobin, ribosomal RNA, etc.) is commonly used as an endogenous reference mRNA target. However, caution must be exercised in the selection of a housekeeping control gene, as expression can be variable among different tissue types or even cell-cycle steps depending on the reference gene selected. In addition, expression of the housekeeping gene may be very high compared with that of the unknown target, and in such a situation the efficiency of amplification between the two targets can differ considerably. For accurate quantitation in reactions that amplify more than one target in the same tube it is important that the two reactions do not compete. This can be avoided by limiting the concentration of primers used in the PCR, which is dependant on the relative abundance of the different targets.

Alternatively, an exogenous target mRNA may be co-amplified with the target, with the advantage that the quantity of the reference gene can be adjusted to fixed concentrations within the range of the unknown target. A potential drawback to this rationale is that alterations in the quality of mRNA in the initial sample, which are inevitable in archival material, are not taken into consideration and may lead to erroneous results.

12.13 Specific TaqMan® PCR applications

TaqMan® Cytokine Card

The cytokine expression card (Applied Biosystems, Foster City, CA, USA) consists of a specially developed 96-well consumable divided into 24 sets of replicates, one for every cytokine assay. Each well contains lyophilized TaqMan® MGB probes (FAM labelled) and primers for one cytokine mRNA target. The total volume in each well including mastermix and template is 1 μL. The card measures the following 24 cytokines IL-1α, IL-1β, IL-2, IL-3, IL-4, IL-5, IL-6, IL-7, IL-8 IL-10, IL-12p35, IL-12p40, IL-13, IL-15, IL-17, IL-18, G-CSF, GM-CSF, M-CSF, IFN-gamma, LT-β, TGF-β, TNF-α and TNF-β. The 18S rRNA endogenous control TaqMan® MGB primers and probe (VIC labelled) are used in a multiplex reaction for relative quantitation. Relative RNA quantification assays on this cytokine card are performed in a two-step RT-PCR. In the first step, cDNA is reverse transcribed from extracted total RNA (50 ng–2 μg) using random hexamers and MultiScribe (Reverse Transcriptase). The second step includes amplification of the cDNA product using TaqMan® Universal Mastermix (AmpliTaq Gold DNA polymerase) (Figures 12.6 and 12.7).

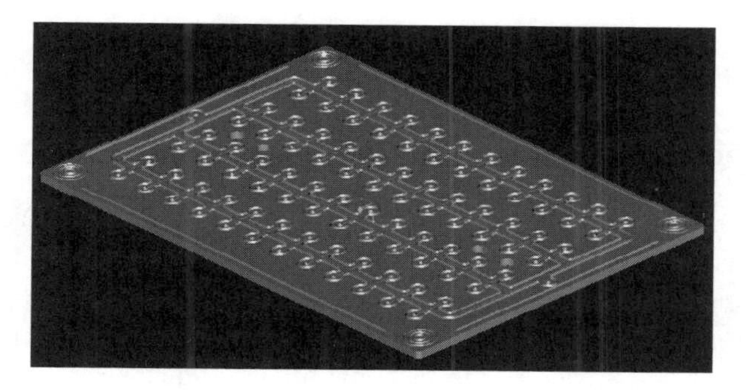

Figure 12.6 The cytokine card

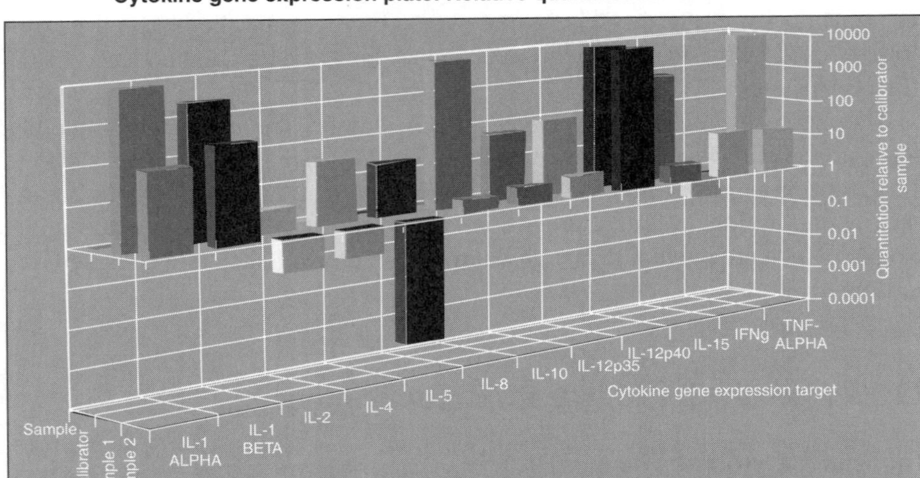

Figure 12.7 Relative expression of cytokines using the cytokine plate format. A colour version of this figure appears in the colour plate section

Allelic discrimination using TaqMan PCR

Strain A

ACTGACC

TGACTGG

Strain B

ACTTCCC

TGAAGGG

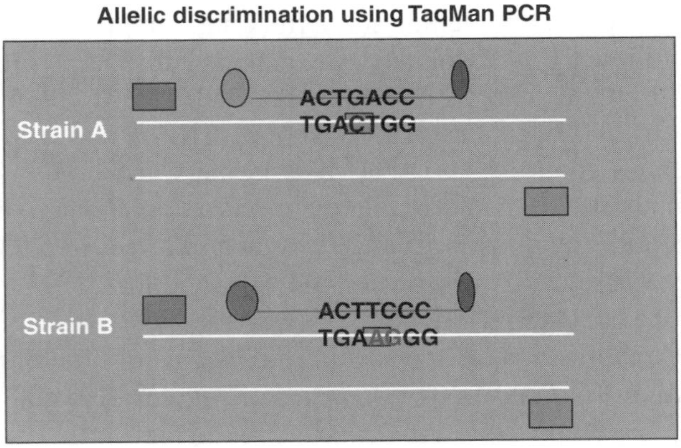

Figure 12.8 Allelic discrimination using TaqMan® PCR. A colour version of this figure appears in the colour plate section

Allelic discrimination and SNP genotyping TaqMan® assays

Quantitative SNP genotyping analyses are now possible using TaqMan® PCR technology and a dual probe approach (see Figures 12.8 and 12.9). Complex genotyping and haplotype analyses are possible using such an approach. The sensitivity of allelic discrimination assays is less than that of a comparative TaqMan® PCR. This is due to the fact that allelic TaqMan® PCRs are competitive PCR reactions (see Figure 12.8), with competition between the wild and

Allele Discrimination Data View

Normalized

Call: Allele 1

Graph

Legend: 1, 2, 1/2, No DNA, ?

Allele 2: -0.10, 0.10, 0.30, 0.50, 0.70, 0.90, 1.10

Allele 1: -0.10, 0.00, 0.10, 0.20, 0.30, 0.40, 0.50, 0.60, 0.70, 0.80, 0.90, 1.00, 1.10

Tray

	1	2	3	4	5	6	7	8	9	10	11	12
A	No Amp	No Amp	No Amp	No Amp	No Amp	No Amp	No Amp	No Amp	1	1	1	1
B	1	1	1	1	2	2	2	2	2	2	2	2
C	1 and 2	1 and 2	1 and 2	1 and 2	1 and 2	1 and 2	1 and 2	1 and 2	1 and 2	1 and 2	1 and 2	1 and 2
D	1 and 2	1 and 2	1 and 2	1 and 2	1 and 2	1 and 2	1 and 2	1 and 2	1 and 2	1 and 2	1 and 2	1 and 2
E	1 and 2	1 and 2	1 and 2	1 and 2	1 and 2	1 and 2	1 and 2	1 and 2	1 and 2	1 and 2	1 and 2	1 and 2
F	1 and 2	1 and 2	1 and 2	1 and 2	1 and 2	1 and 2	1 and 2	1 and 2	1 and 2	1 and 2	1 and 2	1 and 2
G	1 and 2	1 and 2	1 and 2	1 and 2	1 and 2	1 and 2	1 and 2	1 and 2	1 and 2	1 and 2	1 and 2	1 and 2
H	1 and 2	1 and 2	1 and 2	1 and 2	1 and 2	1 and 2	1 and 2	1 and 2	1 and 2	1 and 2	1 and 2	1 and 2

Figure 12.9 Allelic discrimination plot: typical output. A colour version of this figure appears in the colour plate section

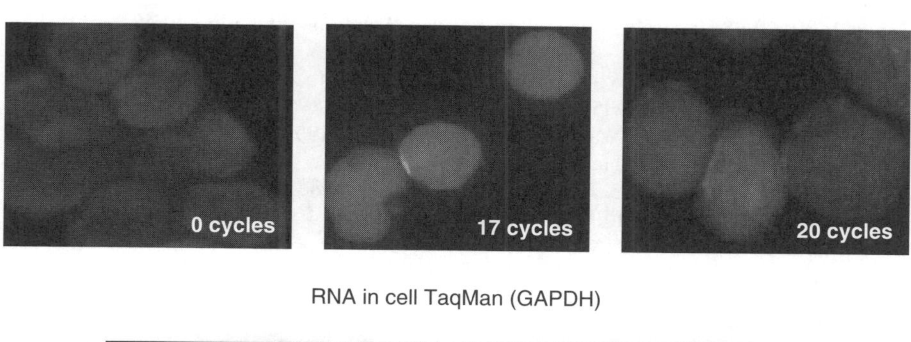

RNA in cell TaqMan (GAPDH)

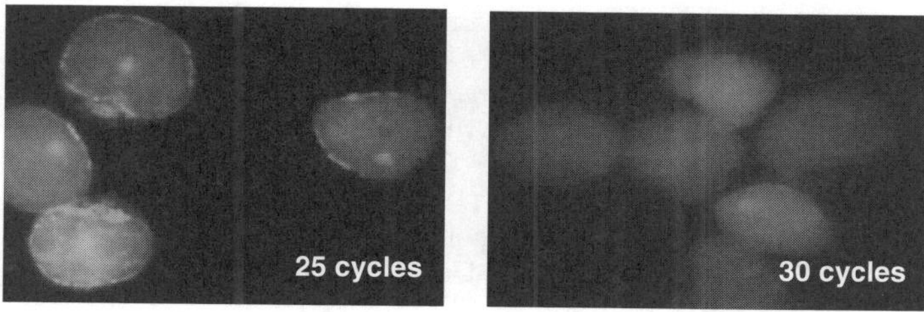

Figure 12.10 In-cell TaqMan® PCR detection of GAPDH in intact cells. Note quenching of the reporter molecule after 30 cycles. A colour version of this figure appears in the colour plate section

mutant specific probe occurring at each PCR cycle, but particularly so in the first rounds of PCR.

Assays on demand

Applied Biosystems have recently introduced a new service, which allows TaqMan® PCR users access to the recently published public and Celera Corporation genome databases. The assays on demand are available for DNA and RNA assays and consists of designed TaqMan® primers and probes sets for all known sequenced genes. Several hundred thousand TaqMan® sets are now available for use and can be obtained from the Applied Biosystems website.

In-cell TaqMan® PCR

It is now possible to perform in-cell TaqMan® PCR assays in cells and tissues sections (see also Chapter 11). DNA and RNA targets can be easily amplified, particularly in cell suspensions and whole cell isolates. Both end-point and quasi-real-time PCR can be performed using slight modifications of the solution phase PCR methodologies (Figure 12.10).

12.14 References

Forster VTH (1948) Zwischenmolekulare Energie-Wanderungund Fluoreszenz. *Ann. Phys. (Leipzig)* **2**: 55–75.

Gibson UEM, Heid CA and Williams MP (1996) A novel method for real time quantitative RT-PCR. *Genome Res.* **6**: 995–1001.

Holland PM, Abramson RD, Watson R and Gelfand DH (1991) Detection of specific polymerase chain reaction product by utilizing the 5′ to 3′ exonuclease activity of *Thermus aquaticus* DNA polymerase. *Proc. Natl. Acad. Sci. USA* **88**: 7276–80.

Kutyavin IV, Afonina IA, Mills A, Gorn VV, Lukhtanov EA, Belousov ES *et al.* (2000) 3′-Minor groove binder–DNA probes increase sequence specificity at PCR extension temperatures. *Nucleic Acids Res.* **28**(2): 655–61.

Lakowicz JR (1983) Energy transfer, in: *Principles of Fluorescent Spectroscopy*, Plenum Press, New York. pp. 303–39.

Lawyer FC, Stoffel S, Saiki RK, Myambo KB, Drummond R and Gelfand DH (1989) Isolation, characterization, and expression in *Escherichia coli* of the DNA polymerase gene from the extreme thermophile, *Thermus aquaticus*. *J. Biol. Chem.* **264**: 6427–37.

Lee LG, Connell CR and Bloch W (1993) Allelic discrimination by nick-translation PCR with fluorogenic probes. *Nucleic Acids Res.* **21**: 3761–6.

Lyamichev V, Brow MAD and Dahlberg JE (1993) Structure-specific endonucleolytic cleavage of nucleic acids by eubacterial DNA polymerases. *Science* **260**: 778–83.

Orlando C, Pinzani P and Pazzagli M (1998) Developments in quantitative PCR. *Clin. Chem. Lab. Med.* **36**(5): 255–69.

Rhoads RE (1990) Optimization of the annealing temperature for DNA amplification *in vitro*. *Nucleic Acids Res.* **18**: 6409–12.

13

Gene Expression Analysis Using Microarrays

Sophie E. Wildsmith and **Fiona J. Spence**

13.1 Introduction

Microarrays are of increasing interest to both industry and academia as tools for 'gene hunting' and also as quantitative methods for routine analysis of large numbers of genes. Techniques such as real-time polymerase chain reaction (RT-PCR) TaqMan™ and SybrMan™ are generally considered to be more accurate, robust, larger in dynamic range and less capital intensive, but for rapid, large-scale gene expression analysis using limited mRNA, microarrays and gene chips are preferred.

13.2 Microarray experiments

Platforms

Global gene expression platforms are now available in multiple formats, including cDNA arrays, oligonucleotides spotted onto slides or *in situ* synthesised oligonucleotide arrays manufactured using photolithography. Commercial sources for these include Stratagene (La Jolla, CA), Memorec (Köln, Germany) and BD BioSciences (Oxford, UK) for cDNA microarrays, Mergen Ltd (San Leandro, CA) for spotted oligomers and Affymetrix (Palo Alto, CA) for oligoarrays synthesised *in situ*. Purchasing from a supplier is more expensive than generating microarrays in-house, although the latter is beneficial in labour-intensive institutions or when proprietary gene information is utilised.

Molecular Biology in Cellular Pathology. Edited by John Crocker and Paul G. Murray
 ISBN: 0-470-84475-2

'Off-the-shelf' microarrays may also have the advantage of rigorous quality control and standardised protocols.

It is possible to produce oligonucleotide spotted arrays in-house, by designing oligonucleotide sequences that match genes of interest and then purchasing purified oligonucleotides to spot down on glass or other substrates. Alternatively there are new systems such as that available from CombiMatrix (Mukilteo, WA, USA) for computer-aided design and *in situ* synthesis of oligonucleotides. However, production of cDNA microarrays is currently the most affordable and popular method and is now well established. Numerous sources of information on cDNA microarray fabrication are available in the literature and on the internet (Bowtell, 1999; Cheung *et al.*, 1999; Wildsmith and Elcock, 2001 and `http://cmgm.stanford.edu/pbrown`). Thus, this chapter will focus on the implementation of experiments and analysis of data from cDNA microarrays. The experimental procedure differs slightly according to the number of fluorophores (or channels) and the type and manufacturer of the array. We have attempted to describe a generic process, indicating where possible the different options. Figure 13.1 demonstrates the procedure for a two-colour hybridisation.

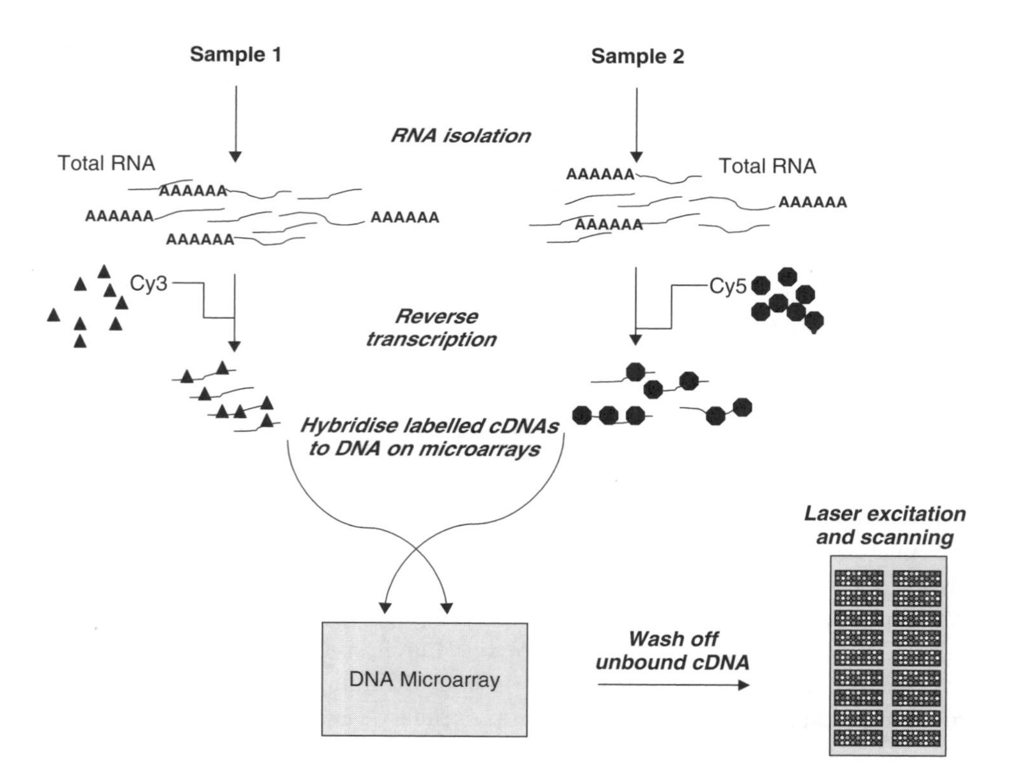

Figure 13.1 The microarray experimental process for two-colour hybridisations

RNA Extraction

First, RNA is extracted from the tissue or cells of interest. The quality of the RNA extracted is paramount to the overall success of the microarray experiment, as impurities in the sample can effect both the probe labelling efficiency and also stability of the fluorescent label (Hegde *et al.*, 2000). Snap-freezing of tissue in liquid nitrogen, immediately after harvesting, is used to preserve RNA integrity. Any further sectioning of the tissues should be carried out under RNAse-free conditions (Fernandez *et al.*, 1997). Total RNA can be extracted using kits such as TRIzol® (Invitrogen, Paisley, Scotland) and Rneasy (Qiagen, GmbH, Hilden, Germany). Some researchers perform a further extraction of mRNA; this results in a purer starting material but has the disadvantage of lower yields. Affymetrix recommend between 5 and 40 μg of total RNA is required for their GeneChips™ and 10 μg or less is the required amount of starting material for cDNA microarrays (Hegde *et al.*, 2000).

Sample Labelling

The mRNA is transcribed *in vitro*, with the concomitant inclusion of labelled nucleotides. The labels may be fluorescent or radioactive. In the case of dual channel/colour hybridisations, two samples will be labelled with dyes that fluoresce at different wavelengths, with different emission spectra. Example fluorophores, available coupled to nucleotides, are Cy3, Cy5, fluorescein and lissamine. Wildsmith *et al.* (2001) have demonstrated that AlexaFluor 546dUTP™ (Molecular Probes, Leiden, The Netherlands) gives a significantly higher signal than Cy3dCTP (Amersham Biosciences, Piscataway, NJ, USA). When performing two-colour hybridisations the control sample and 'test' sample are labelled with different fluorophores and the subsequent cDNA is then mixed together and hybridised simultaneously (Nuwaysir *et al.*, 1999). An advantage of simultaneously hybridising control and treated sample is that it obviates the need to control for differences in hybridisation conditions or between microarrays. A specific example of the huge impact this technique has had includes its use in the first published account of gene expression data of the entire genome of *Saccharomyces cerevisiae* (DeRisi *et al.*, 1997). In two-colour hybridisations, one would assume that the properties of the two fluorescent dyes being used are equivocal. In fact, for Cy5 and Cy3 this is not the case as Cy5 has been reported to give higher background fluorescence and also is more sensitive to photobleaching than Cy3 (Van Hal *et al.*, 2000). In addition, there is evidence from several independent sources that the combination of Cy3 and Cy5 dye labelling can affect data in certain genes. That is to say, when experiments are repeated and the dye combination for the two probes reversed, inconsistent results are obtained with certain genes (Taniguchi *et al.*, 2001). Despite these

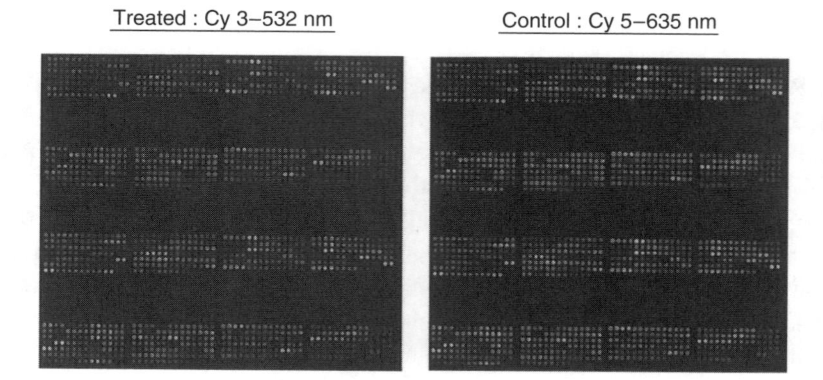

Figure 13.2 Two-colour fluorescent scan of human gene cDNA array. The probe mix consists of DNA from HepG2 control cells and cells treated with buthionine sulfoximine for 6 h. A colour version of this figure appears in the colour plate section

facts, two-colour hybridisations are widely accepted throughout the microarray community and an example of an image is shown in Figure 13.2.

Hybridisation and Processing

After labelling, the cDNA is purified (to remove unincorporated nucleotides), mixed with a hybridisation buffer and then applied to a cDNA microarray slide. The sample and the slide are heated prior to hybridisation in order to separate double-stranded DNA. A coverslip is applied (Shalon *et al.*, 1996), or preferably a hybridisation chamber is used to avoid evaporation and enable an even hybridisation. The hybridisation and subsequent wash steps are carried out at a buffer stringency and temperature that enables hybridisation of complementary strands of DNA but reduces non-specific binding.

Image Capture and Image Analysis

After hybridisation the microarray slides are scanned, using either a laser or a phosphorimager (depending on the type of label used). There are many different suppliers and models of fluorescence scanner, for example the ScanArray 5000 (Perkin Elmer Life Sciences, Zaventem, Belgium), GenePix 4000B (Axon GRI, Essex, UK) and the GeneArray® (Affymetrix, Santa Clara, CA). The choice of scanner is determined by sensitivity, resolution, flexible wavelength, file size generated, throughput and technical support available.

Images are analysed using software that measures the intensity of the signal from the hybridised spotted genes (spots), which provides a measurement of the amount of cDNA bound. Thus the initial concentration of messenger RNA is inferred. Early software packages 'drew' grids around the spots and integrated across the whole area of the grid. This overcame problems associated with accurate location of the spots, which is problematic, especially if the spots on the printed arrays are poorly aligned. More recent versions of software 'draw' circles around the spots themselves and perform measurements within and outside of this boundary. For example, the background may be calculated from a region outside the spot boundary. The intensity of the signal from the spot may be calculated using median, mode or mean values of the pixels within the spot. Researchers differ in their preferences regarding using median or mean values (Hegde *et al.*, 2000) and this is likely to depend upon the protocols and software used.

Image analysis software commonly is supplied with scanners or can be bought from the same supplier. This has the advantage of being optimised for that specific type of microarray and the benefit of upgrades and technical support. Software for microarray analysis is available from BioDiscovery (`http://www.biodiscovery.com`), Imaging Research (`http://imagingresearch.com`), GenePix Pro (Amersham Biosciences, Piscataway, NJ), arraySCOUT 2.0 (`http://www.lionbioscience.com`), NIH (`http://www.nhgri.nih.gov/DIR/LCG/15K/HTML/img_analysis.html`), Stanford University (`http://rana.Stanford.EDU/software`) Media Cybernetics (Silver Spring, MD, USA) and TIGR (`http//www.tigr.org/softlab`). Important criteria for image analysis software include speed, ease of use, automation and the ability to distinguish artefact from real signal (Wildsmith and Elcock, 2001).

As the technology has evolved and more experience gained, it has become more and more apparent that the most significant issues facing microarray users are the processing of the vast quantities of data generated and deciding exactly what tools are the most appropriate for data analysis. Because of the enormity of this, we have dedicated a complete section to describing the current status of this area.

13.3 Data analysis

It is important to be cognisant of the fact that the practical laboratory aspects of using microarrays are only part of gene expression analysis. Many researchers generate vast volumes of data, without a clear understanding of how to manage and interpret them. Furthermore, the variability in microarray data confers additional problems for analysis. In some cases the purpose of the experiment will be a gene-hunting exercise, in which case a cursory indication of potential gene biomarkers is sufficient analysis. In other instances, such as pathway mapping and screening studies, it is paramount that results are statistically meaningful and valid. The next few sections detail some relatively simple analysis methods and recommendations for the benefit of researchers with minimal statistical

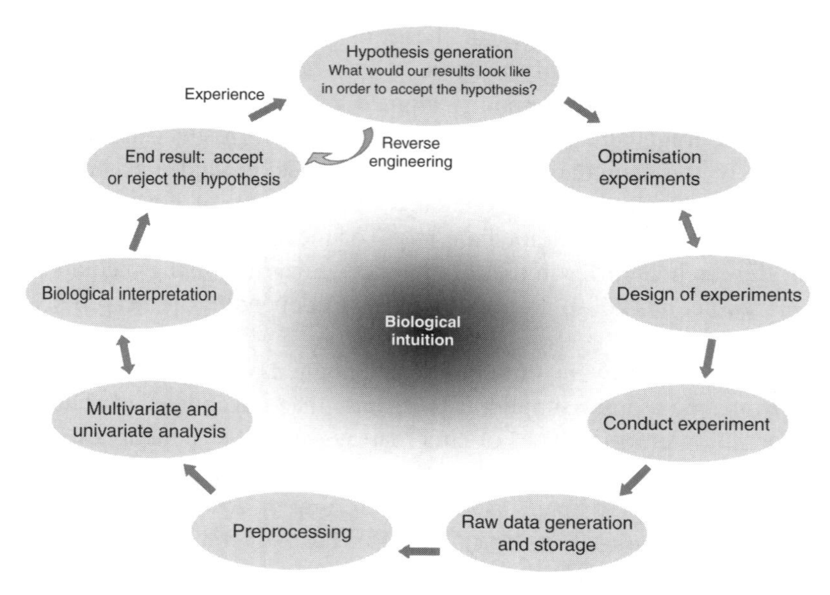

Figure 13.3 The ideal microarray experimental design and process

training. There are also suggestions for more advanced analysis for those who have the assistance of a statistician or specialist data analyst.

Most of the steps before, during and after performing a microarray experiment are optimally conducted with regard for statistics and data analysis. Careful planning before implementation facilitates the downstream analysis and interpretation of data. The following model summarises the entire microarray process with integration of the biological and data analysis components (Figure 13.3).

Hypothesis Generation

Any study is conceived for the purpose of investigating or obtaining supporting evidence for a biological hypothesis. Giving time at this early stage to consider downstream implications will pay dividends later. It is helpful if, rather than simply stating the aims of the experiment, the researcher asks the question 'What results do I expect?' or 'what answer will validate/invalidate my hypothesis?'. This 'reverse-engineering' proves useful in focusing the project, assessing the feasibility of the work, providing early preparation for data management and analysis and, importantly, in managing expectations with regard to outcomes.

A good example of careful experimental planning is demonstrated by Golub *et al.* (1999) in the classification of acute leukaemias in order to distinguish between acute lymphoblastic leukaemia (ALL) and acute myeloid leukaemia (AML). Distinguishing between ALL and AML using conventional techniques is known to be a difficult task. The researchers maximised their probability of success by choosing an easier, more defined model (normal kidney vs. renal cell carcinoma), on which to validate their analytical methods. In doing so they established that their techniques were suitable for classifying tissues according to disease and gained confidence in their approach before using the samples of real interest.

Optimisation Experiments

Although microarrays are becoming increasingly accessible to all, using these tools requires experience and it is unlikely that successful experiments will be conducted immediately. It is usual that some time is given to optimising a system for any specific application, for example for a given tissue or cell type. Additionally, the requirements for a given system may warrant some modifications. The standard approach for a scientist to take is to vary one parameter, whilst keeping all others constant. This is time-consuming and does not take into account the interactions between different factors. Well-designed, multifactorial experiments (Box *et al.*, 1978), provide a faster route for optimisation, with a statistical measure of confidence. An example of this technique is in the optimisation of microarray experimental conditions for preparation of fluorescent probes from rat liver tissue (Wildsmith *et al.*, 2001). When a major source of variation is revealed this can be investigated further with a view to minimising it or providing sufficient replicates to account for it.

Design of Experiments

Once confidence in the experimental procedure has been obtained the researcher is likely to have gained an insight into the reproducibility of the system. This assists in the design of the experiments, in particular in determining the minimal number of replicates necessary. Replication can be implemented at many stages – from biological samples through to microarray slides.

Owing to the enzyme-catalysed transcription reactions, a large amount of variation occurs during the probe-making stages in microarray experiments. Our work indicated that replicates should be made at this step and a minimum of six replicate probes are made for microarray experiments (Wildsmith *et al.*, 2001). These can be pooled or hybridised separately onto six microarray slides.

Lee *et al.* (2000) have examined the effect of the different location of cDNA spots on the glass slides and concluded that replicates are essential to provide meaningful data and to enable reliable inferences to be drawn.

With regard to commercially available gene chip systems, such as that available from Affymetrix (see Figure 13.4), the variation between chips, within a batch, is likely to be low due to stringent quality control and highly automated manufacture. The use of an automated wash station also reduces variability in intensities between chips. However, using a multi-step approach in the probe preparation and subsequent antibody binding steps may lead to variation between

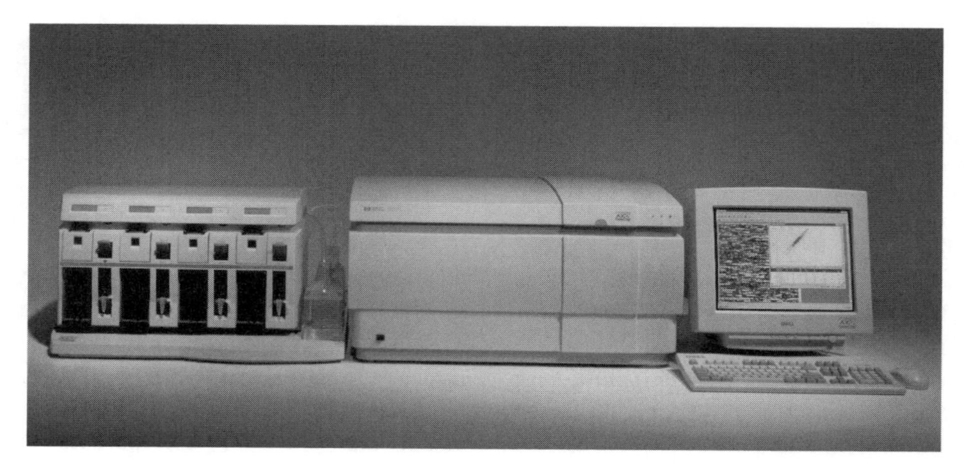

Figure 13.4 The GeneChip® Instrument System. From left to right, the hybridisation station, scanner and workstation. Image courtesy of Affymetrix. A colour version of this figure appears in the colour plate section

replicate samples prepared on different days. Pooling of reagents within an experiment, and analysing controls together with treated samples, will both reduce the variability within a given experiment.

Conduct of Experiment

At this stage some attention may be required for verifying and validating processes. For example, checking that the imaging instruments give consistent results across the slide, on repeat use and from day to day. If two imagers are used it is important to verify that the results from both machines are comparable. Some laboratories read fluorescence of one channel and then adjust the laser intensity of the second channel in order to obtain comparable readings. This is a method of normalising for the difference in intensities of fluorophores. It is important to be aware that this approach has a number of drawbacks. The arbitrary value of the second laser intensity setting will vary from experiment to experiment; thus comparisons of this channel cannot be made across experiments. Also the response of the fluorophore may not be linear across the laser intensity settings and this can lead to additional errors.

Another area for investigation prior to running the study itself is the image analysis component. Depending on the software used, the image analysis package may process the data to some extent, for example automatic background subtraction. Full understanding of the software is required so that it is clear at what point the data are 'raw', and the extent of inherent, inseparable manipulation. Effort may be required to determine the optimum settings for any software parameters.

As data are generated it is important to be aware of the data integrity – for example ensuring that all data are collected, so that there are no missing data that

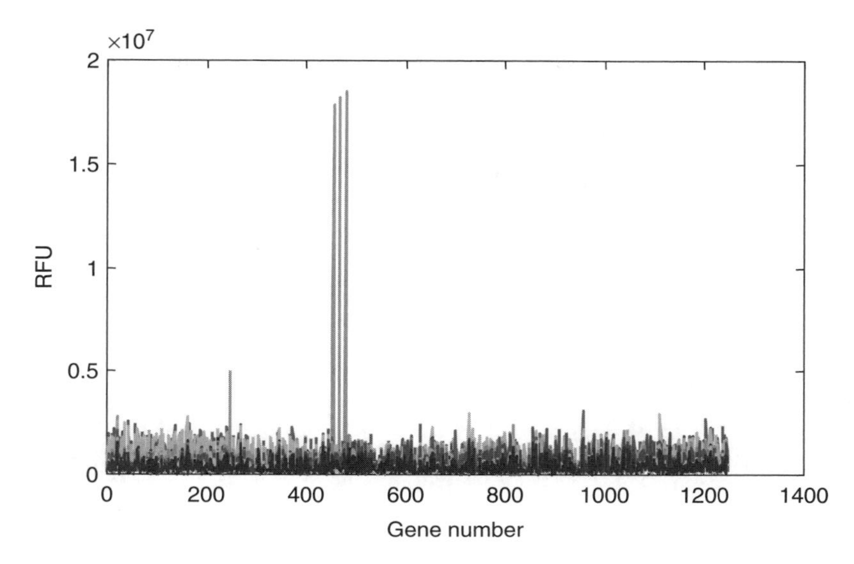

Figure 13.5 The relative fluorescence units (RFUs) of 1248 genes on seven microarray slides that were hybridised with cDNA made from the liver of a rat treated with acetaminophen. Note the gene outliers at approximately gene number 480

can complicate analysis later. The researcher may be intuitively aware of any spurious results and should be alert for anything extraordinary that could indicate problems, for example hybridisation intensities appearing inconsistent from sample to sample. Data analysis at this point can be a rapid indicator of dubious results. For example, Figure 13.5 shows a plot of the fluorescent intensities of 1248 genes that were hybridised with probe derived from acetaminophen-treated rat liver tissue. The data appear consistent, with the exception of peaks in intensity on one slide at around gene 4800. Further investigation of the microarray revealed a large artefact that had been missed by the image analysis process (Figure 13.6).

Raw Data Generation and Storage

One issue that arises when carrying out microarray analyses is how much data to store and in what form. For Good Laboratory Practice (GLP) purposes, often required in industry, storage of the raw data is necessary. This could be construed as the microarray image. Storing the image analysis results requires far less storage space and is easier to visualise, but it has the drawback that image analysis cannot be redone should superior software be available in the future. In reality, the methods used for microarrays are continually changing and the likelihood of revisiting old images on which the analysis has been performed, using outdated protocols is quite small.

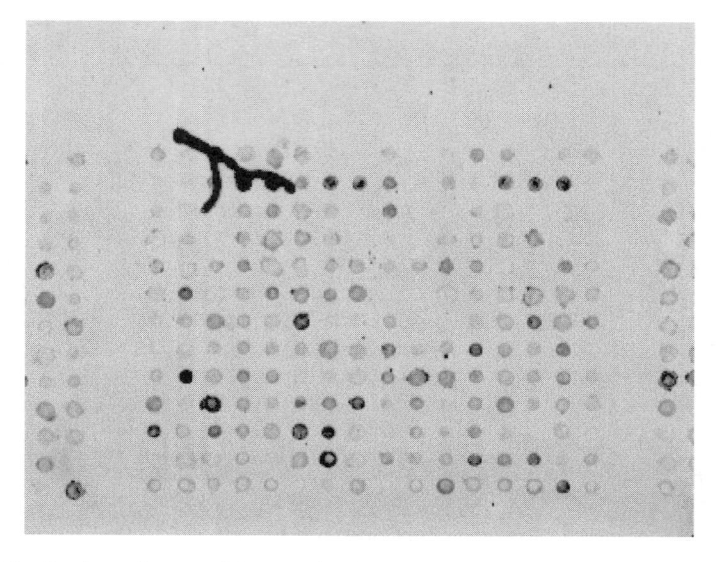

Figure 13.6 Portion of scanned image showing region where artefact occurred that caused very high signals, which were classed as outliers

Pre-processing

A number of pre-processing steps are often used in microarray analysis. These include filtering, log transformation, normalisation and background subtraction. Filtering may be used before or after transformation in order to extract data from preferred regions of interest, or in order to remove outliers (see the above example relating to image analysis artefact). One example of filtering is the removal of individual gene replicates that lie outside a given number (for example 5) of standard deviations from the mean. Alternatively, data points that lie in the top/bottom few percentiles (e.g. 0.1%) of the data can be removed. This method of removing outliers is also called 'trimming'. It is acceptable if there is a large volume of data where only a small proportion of data is removed and if the same method is applied consistently across all data. Care must be taken in the way in which this is carried out in order not to delete genuine data. For example, if one gene is consistently high or low in expression across replicates, then it is unlikely to be an outlier.

Another method of detecting outliers is to plot all genes (see the section 'Conduct of experiment' above) or to perform PCA analysis (see the section 'Multivariate analysis' below) to detect replicate outliers. The use of a PCA plot to detect outliers is shown in Figure 13.7.

Log transformation of data is accepted universally because the fluorescence data that are generated from microarrays tend to be skewed towards lower values. There are scientifically valid reasons why ratios of raw expression values should not be used (Nadon and Shoemaker, 2002). When using two-colour

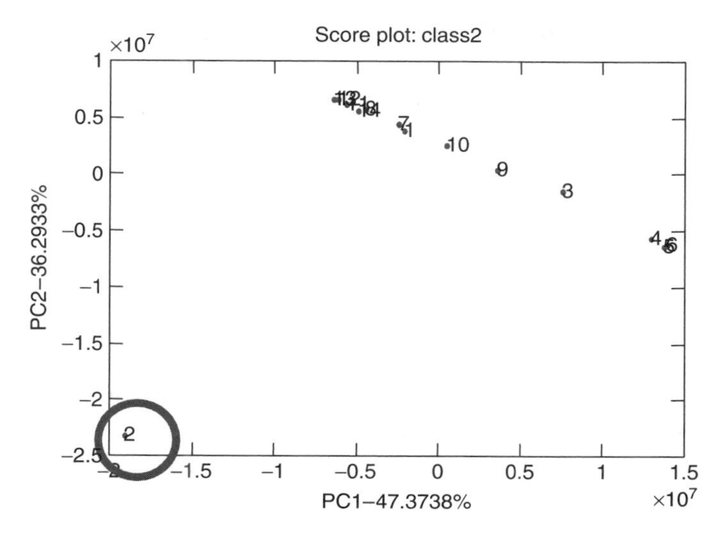

Figure 13.7 PCA plot of data used in Figure 13.4 showing one microarray (number 2) as an outlier

hybridisations it is common to express the ratio of treated to control as a logarithm in base 2 (Quackenbush, 2001). Thus genes up-regulated by a factor of 2 have a $\log_2$ (ratio) of 1, and genes down-regulated by a factor of 2 have a $\log_2$ (ratio) of -1.

Normalisation and background subtraction techniques are methods of data manipulation and their use is more subjective and often debated. The purpose of these techniques is to reduce the error (variability) that occurs between replicates and thus enable a comparison of data across samples.

The theory behind background subtraction is that during hybridisation there will be non-specific binding to the slide. This will effectively 'darken' the image and give falsely high readings of fluorescent intensity. Correcting for the non-specific hybridisation should reduce error due to background staining. Background subtraction often occurs automatically in microarray image analysis packages. The software may circle the spot of interest and use the region beyond the periphery as the measurement of background. In cases of uneven hybridisation this method enables locally high background to be subtracted on a regional basis. One criticism of this approach is that the slide surface beyond the periphery is not similar, in chemical terms, to that where the nucleic acid has been deposited, and therefore cannot act as a real control for non-specific binding. A more accurate measurement of non-specific binding can be gained from using a region where spotting chemicals have been deposited, but no target is present. This concept is the basis of a method using local 'blank spots' (Wu *et al.*, 2001).

A number of methods exist for normalisation of data. These include normalising to total signal or to a 'known' spot or gene, standardisation, or proprietary methods. Normalising to total signal is the simplest approach, whereby the gene intensity is expressed as a percentage or proportion of the signal intensity for the entire array. This method works best when the total intensities for the microarrays are similar and the number of changes is small compared with the number of genes. However, we often find, when using arrays of around 1000 genes, that pathological disease can up-regulate a large number of genes simultaneously. In this case, when total signal normalisation is applied, highly up-regulated genes will appear less up-regulated and genes that do not change from the control will appear down-regulated.

Normalisation to a control value is a more popular technique. A control value can be obtained from using a gene known to remain constant under the conditions of the experiment. DeRisi *et al.* (1997) used a panel of 90 housekeeping genes for normalisation, but found considerable variation in their gene expression. Unfortunately it is very difficult to know with certainty that a gene will not change and there is evidence to suggest that so-called 'housekeeping genes' are variable (Savonet *et al.*, 1997). Other control genes can be derived from an alternative species; these should not be expected to hybridise. We have used yeast and Arabidopsis genes as negative controls for hybridisations of rat tissues. No orthologs were known to the genes selected; however, in most cases non-specific binding occurred.

If two or more microarray replicates appear to be different, but they are expected to be the same, then they can be standardised. An example of this might occur if the total intensity of one microarray is greater than another, but the genes are proportionally equally up- or down-regulated. If the microarray sample spot data are assumed to be drawn from a normal distribution, then the 'z-transform' can be used. This requires that the mean and the standard deviation of the intensity values for each microarray are determined. The mean is subtracted from each individual gene value and the remainder is divided by the standard deviation. The intensity values from each microarray will then have the same mean and standard deviation. This has the advantage of facilitating comparison of microarrays with different dynamic ranges as well as total intensities. If the data are not normally distributed, then alternative non-parametric methods can be used, such as normalising to the median.

Univariate Analysis

Univariate methods of analysis involve examining one variable, or gene, at a time. This can be a very laborious task when examining a large volume of data, yet it is the preferred method of biologists. The simplest technique is to compare control and treated values and express the result as a 'fold-change' ratio. Typically, when examining small volumes of data, fold changes greater than 2 and less than 0.5 are considered meaningful (Quackenbush, 2001). This cut-off is essentially arbitrary and has the distinct drawback that microarray data are not homoscedastic; that is, there is more variation about the mean at low values than there is at high values (Draghici, 2002).

A second method for finding up- or down-regulated genes uses the standard deviation (SD) of the replicate gene data. Thus if changes greater than, say, 2 SD from the log mean ratio are considerably greater than changes associated with 'noise', then they are considered significant. This technique means that when looking at a large number of genes that are normally distributed there will be up- and down-regulated genes, regardless of whether there are (biological) changes (Draghici, 2002).

Rather than using arbitrary cut-off values it is far more meaningful to express the fold-change in terms of either confidence intervals, or a 'p-value' (that is, the probability of the value occurring by chance). Thus, a fold-change of 1.1 may be associated with a p-value of 0.001 and thus the probability that the gene is *not* up-regulated is 1 in 1000. Naturally, such small fold-changes may then be queried in terms of biological significance. One must then ask the question: Are large fold-changes more important (biologically) than small ones? Simple calculation of p-values for two data sets can be obtained using t-test functions in standard spreadsheet software. A number of replicates are necessary for this approach, and the data must be normally distributed. We have recently developed a method for calculating p-values for fold-changes that is not influenced by the distribution of the data or outliers and applied it to TaqMan™ and microarray data. Other complex and computationally intensive methods for calculating p-values are described in Draghici (2002) and Nadon and Shoemaker (2002).

Multivariate Analysis

Multivariate analysis of gene expression data is becoming increasingly popular in the microarray community and in other biological domains where large volumes of data are generated. Multivariate analysis methods include principal component analysis (PCA), factor analysis, multivariate analysis of variance (MANOVA) and cluster analysis. Currently, cluster analysis is the most widely-used method in the microarray community but PCA is growing in popularity (Crescenzi and Giuliani, 2001; Konu *et al.*, 2001).

Quackenbush (2001) provides a good review of clustering tools that is rather unique in the regard that different clustering algorithms and linkage methods are presented. Clustering methods are unsupervised, and they are powerful tools for gaining insight into huge data sets. They enable the data to be partitioned in order to facilitate interpretation; however, they do suffer from subjectivity. This is because the user selects various parameters, such as the algorithm used, linkage type, distance metric and, sometimes, cluster size. Whatever the data, clusters will always be identified, thus there is also a tendency to over-interpret the data – trying to attach meaning to clusters that may have no biological relevance.

Software available for cluster analysis includes Cluster, the output of which is viewed in Treeview; both available from `http://rana.stanford.edu/software`. This tool is particularly useful for clustering genes to identify genes that are co-regulated.

The PCA is a visualisation tool that enables complex, high-dimensional data to be represented in two or three dimensions. It facilitates identification of groups of similar data, thus enabling inferences to be made about the samples.

An example is shown in Figure 13.8. The figure shows the gene data for one sample (control rat liver) that was hybridised to seven microarrays according to the method used in Wildsmith *et al.* (2001). Each microarray contained two replicate gene sets; thus there were 14 replicate gene sets in total. The gene sets comprised 1248 genes (with controls). All the data (14×1248 data points) was input into the analysis and the PCA plot displays the 14 replicates individually. The two axes are principal components 1 and 2. Principal component 1 (PC1) accounts for 65.5% of the variation in the data, whereas PC2 represents only 13%. This means that the model accounts for 78.5% of variation in the data.

The first principal component (PC1) accounts for as much as possible of the variation in the original data and subsequent components (e.g. PC2) are of decreasing importance. Thus, samples 8 and 9 are very different from samples 1 and 2. In terms of interpreting the PCA plot, it is immediately clear that there are three or four distinct clusters of data. These are marked by circles. Datapoints tend to cluster in pairs; for example replicates 1 and 2, 3 and 4, 5 and 6, etc. These are the duplicate gene sets on the same microarray. This indicates that the variation within the microarray is lower than the variation between

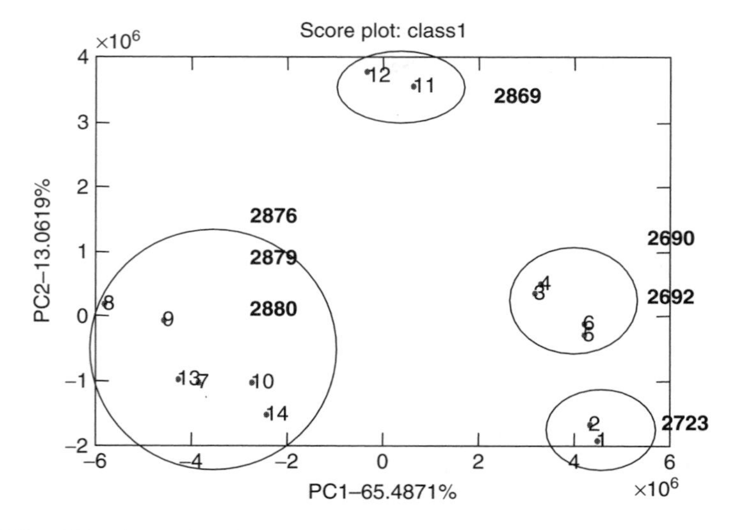

Figure 13.8 PCA plot of the microarray results from seven slides (2690, 2692, 2723, 2869, 2876, 2879, 2880), each with two replicate spot sets (labelled 1–14), after hybridisation with control rat liver. See text for explanation

replicate microarrays. However, given that the same sample is applied to all the microarrays, we must ask why we get further separation of the replicates. The answer lies in the associated data. The four-digit numbers associated with the clusters are the microarray slide numbers that indicate when they were printed. Numbers that are more similar seem to be more closely related, and thus we can hypothesise that there were some differences between the slides, such as differences in slide backgrounds or changes during the printing process or change-over of batches, between slides in the 2600–2700 region and the 2800s.

The PCA provides a clearer overview of the data than does cluster analysis. It is a rapid method for gaining an insight into the results, in particular where biological meaning can be attached to the components (Crescenzi *et al.*, 2001). There are a number of packages for multivariate data analysis, including SIMCA-P (Umetrics, AB, Umea, Sweden) and The Unscrambler (Camo ASA, Norway), both of which are useful for PCA.

Other tools for data visualisation include software packages such as Spotfire.net (Spotfire Inc., Cambridge, MA, USA) and GeneSpring (Silicon Graphics, San Carlos, CA, USA). Spotfire is particularly useful for visualisation of multidimensional data and for visualisation of temporal data. It is possible to use these tools to identify genes that are co-ordinately expressed over time.

Biological Interpretation

After developing a sound experimental strategy, ensuring that the results are statistically valid, and after analysis of the data, it is down to the biologist to assemble the pieces of information that have been obtained. This intertwined information may include unexpected results that are contradictory to intuition or to published literature. One way to untangle the data is to map the relevant genes onto existing pathways and known functions. The Kyoto Encyclopedia of Genes and Genomes (KEGG), available at `http://www.genome.ad.jp/kegg/`, is a useful source of information, especially where the gene products are enzymes. It enables visualisation of the position of up- or down-regulated genes in metabolic pathways.

The gene expression data obtained may differ from protein expression data, or information on gene product activity or location. When initiating a study it is useful to consider additional endpoints that can assist in the interpretation of the data. For *in vitro* studies, these might include cytotoxicity endpoints, metabolites, key signalling molecules or perhaps protein expression. Waring *et al.* (2001) used tetrazolium dye reduction (MTT) as a measure of hepatocyte cell viability for their studies of gene expression in response to hepatotoxic insult. For *in vivo* studies, expression information on the tissue of interest could be supported by pathology, histology and blood chemistry measurements. Gene expression results could be confirmed by *in situ* hybridisations or protein activity assays.

13.4 Recent examples of microarray applications

One area of rapid progress using microarray technology is the increased understanding of cancer. Molecular pathologists are subgrouping cancers of tissues such as blood, skin and breast, based on differential gene expression patterns. For example, within a small group of breast cancer tissue samples, Perou *et al.* (2000) distinguished two broad subgroups representing those expressing or alternatively lacking expression of the oestrogen receptor-α gene. The work was not conclusive, but never has progress in this field been so rapid when compared with the previous methods of gene identification.

Another example of the impact of this technology is in the identification of two biomarkers for prostate cancer, namely hepsin and PIM1 (Dhanasekaran *et al.*, 2001).

Microarray technology has also accelerated the understanding of the molecular events surrounding pulmonary fibrosis. Specifically, two distinct clusters of genes associated with inflammation and fibrosis have been identified in a disease where, for years, the pathogenesis and treatment have remained unknown (Katsuma *et al.*, 2001).

13.5 Conclusions

Important factors in gene expression experiments include sensitivity, precision and reproducibility in the measurement of specific mRNA sequences (Schmittgen *et al.*, 2000). These quality metrics can be maximised by using, or fabricating, high-quality microarrays, and by optimising each step of the microarray process. From conception to conclusion it is important to bear in mind the original hypothesis.

Having considered the complexity of the microarray experiment, the value obtained from a meticulously designed experiment should not be underestimated. As the number of high-quality gene expression studies increases, we hope that the literature will contain increasingly detailed information that will help interpret complex gene expression changes, and thus elucidate the mechanisms of disease.

13.6 Acknowledgements

The authors express their thanks to W. Wu for Figures 13.7 and 13.8, and to W. Wu, S. Roberts and P. Lane for useful comments and discussions.

13.7 References

Bowtell DL (1999) Options available – from start to finish – for obtaining expression data by microarray. *Nature Genet.* **21**: 25–32.

Box GEP, Hunter WG and Hunter JS (1978) Chapter 7, in: *Statistics for Experimenters: An Introduction to Design, Data Analysis and Model Building*. John Wiley and Sons, New York.

Cheung VG, Morley M, Aguilar F *et al.* (1999) Making and reading microarrays. *Nature Genet.* **21**: 15–19.

Crescenzi M and Giuliani A (2001) The main biological determinants of tumor line taxonomy elucidated by a principal component analysis of microarray data. *FEBS Lett.* **507**: 114–18.

DeRisi JL, Iyer VR and Brown PO (1997) Exploring the metabolic and genetic control of gene expression on a genomic scale. *Science* **278**: 680–6.

Dhanasekaran SM, Barrette TR, Ghosh D *et al.* (2001) Delineation of prognostic biomarkers in prostate cancer. *Nature* **412**: 822–4.

Draghici S (2002) Statistical intelligence: effective analysis of high-density microarray data. *Drug Discov. Today*, in press.

Fernandez PM, Pluta LJ, Fransson-Steen R *et al.* (1997) Reverse transcription-polymerase chain reaction-based methodology to quantify differential gene expression directly from microdissected regions of frozen tissue sections. *Mol. Carcinog.* **20**: 317–26.

Golub TR, Slonim DK, Tamayo P *et al.* (1999) Molecular classification of cancer: class discovery and class prediction by gene expression monitoring. *Science* **286**: 531–7.

Hegde P, Qi R, Abernathy K *et al.* (2000) A concise guide to cDNA microarray analysis. *Biotechniques* **29**(3): 548–62.

Katsuma S, Nishi K, Tanigawara K *et al.* (2001) Molecular monitoring of bleomycin-induced pulmonary fibrosis by cDNA microarray-based gene expression profiling. *Biochem. Biophys. Res. Commun.* **288**: 747–51.

Konu O, Kane JK, Barrett T *et al.* (2001) Region-specific transcriptional response to chronic nicotine in rat brain. *Brain Res.* **909**: 194–203.

Lee TML, Kuo FC, Whitmore GA *et al.* (2000) Importance of replication in microarray gene expression studies: statistical methods and evidence from repetitive cDNA hybridisations. *Proc. Natl. Acad. Sci. USA* **97**(18): 9834–9.

Nadon R and Shoemaker J (2002) Statistical issues with microarrays: processing and analysis. *Trends Genet.* **18**(5): 265–71.

Nuwaysir EF, Bittner M, Trent J *et al.* (1999) Microarrays and toxicology: the advent of toxicogenomics. *Mol. Carcinog.* **24**: 153–9.

Perou CM, Sorlie T, Eisen MB *et al.* (2000) Molecular portraits of human breast tumours. *Nature* **406**: 747–52.

Quackenbush J (2001) Computational analysis of microarray data. *Nature Rev. Genet.* **2**: 418–27.

Savonet V, Maenhaut C, Miot F *et al.* (1997) Pitfalls in the use of several 'housekeeping' genes as standards for quantitation of mRNA: the example of thyroid cells. *Anal. Biochem.* **247**: 165–7.

Schmittgen TD, Zakrajsek BA, Mills AG *et al.* (2000) Quantitative reverse transcription-polymerase chain reaction to study mRNA decay: comparison of endpoint and real-time methods. *Anal. Biochem.* **285**: 194–204.

Shalon D, Smith SJ and Brown PO (1996) A DNA microarray system for analysing complex DNA samples using two-colour fluorescent probe hybridisation. *Genome Res.* **6**: 639–45.

Taniguchi M, Miura K, Iwao H *et al.* (2001) Quantitative assessment of DNA microarrays – comparison with northern blot analyses. *Genomics* **71**: 34–9.

Van Hal NLW, Vorst O, Van Houwelingen MML *et al.* (2000) The application of DNA microarrays in gene expression analysis. *J. Biotech.* **78**: 271–80.

Waring JF, Ciurlionis R, Jolly RA *et al.* (2001) Microarray analysis of hepatotoxins *in vitro* reveals a correlation between gene expression profiles and mechanisms of toxicity. *Toxicol. Lett.* **120**: 359–68.

Wildsmith SE and Elcock FJ (2001) Microarrays under the microscope. *J. Mol. Pathol.* **54**: 8–16.

Wildsmith SE, Archer GEB, Winkley AJ *et al.* (2001) Maximisation of signal associated with cDNA microarrays. *BioTechniques* **30**: 202–8.
Wu W, Wildsmith SE, Winkley AJ *et al.* (2001) Chemometric strategies for normalisation of microarray data. *Anal. Chim. Acta.* **446**: 449–64.

13.8 Further reading

Chatfield C and Collins AJ (1980) *Introduction to Multivariate Analysis.* Chapman & Hall, London.
Heck DE, Roy A and Laskin JD (2001) *Nucleic Acid Microarray Technology for Toxicology: Promise and Practicalities. Biological Reactive Intermediates VI.* Kluwer Academic Press.
Schena M (ed.) (2001) *Microarray Biochip Technology.* Eaton Publishing.

13.9 Useful websites

Protocols:

`http://cmgm.stanford.edu/pbrown`
`http://cmgm.stanford.edu/pbrown/protocols.html`

Image analysis software:

`http://www.genome.ad.jp/kegg/`
`http://rsb.info.nih.gov/nih-image`
`http://rana.stanford.edu/software`
`http://www.nhgri.nih.gov/DIR/LCG/15K/HTML/img_analysis.html`
`http://rana.Stanford.EDU/software`
`http//www.tigr.org/softlab`

14

Comparative Genomic Hybridisation in Pathology

Marjan M. Weiss, Mario A.J.A. Hermsen, Antoine Snijders, Horst Buerger, Werner Boecker, Ernst J. Kuipers, Paul J. van Diest and **Gerrit A. Meijer**

14.1 Introduction

Cancer is driven by an accumulation of genetic and epigenetic abnormalities, leading to aberrant expression of crucial proteins that ultimately lead to uncontrolled cell growth and behaviour. Genetic abnormalities can occur both at the DNA and the chromosomal level.

Already in 1890, Hansemann observed that mitoses of epithelial tumour cells often were asymmetric. He considered the resulting imbalance between nucleus and cytoplasm to be the cause of malignant growth (Hansemann, 1890). However, it was Boveri who in 1914 hypothesized that cancer was the result of defects in the genetic material, which resides in the nucleus of somatic cells. It took until 1956 before the number of 46 human chromosomes was established (Tjio and Levan, 1956), but the great breakthrough came with the development of chromosome banding techniques in 1969, which made it possible to recognize all chromosomes individually. This development also allowed a characterisation of structural chromosomal rearrangements. The first of these was the Philadelphia chromosome observed in chronic myeloid leukaemia, now identified as the result of a translocation between chromosomes 9 and 22 (Rowley, 1973). Fluorescence *in situ* hybridisation (FISH) was developed in the 1980s, and rapidly became a very powerful technique, both for research and diagnostic purposes (Waters *et al.*, 1998). With this technique a small DNA fragment of known origin (a probe) is fluorescently labelled and hybridised to a metaphase chromosome

Molecular Biology in Cellular Pathology. Edited by John Crocker and Paul G. Murray
 ISBN: 0-470-84475-2

spread or interphase nuclei. The probe binds to homologous sequences within the chromosomes, and this can be visualised by fluorescence microscopy. With the appropriate probes, FISH can be used to confirm a suspected chromosomal aberration. This technique is very useful for investigating tumour series for specific chromosomal loci; however, for screening purposes of the whole genome it is inferior to other techniques.

Solid tumours, which comprise around 95% of all human cancers, commonly exhibit many and complex chromosomal changes, but it is difficult to appreciate the meaning of recurrent and specific chromosomal changes against a background of a multitude of alterations (Jallepalli and Lengauer, 2001; Rowley, 2001). The observation that gross numerical and structural chromosome aberrations occur in high number in late stage tumours (invasive cancers), led some researchers to believe they are merely a side-effect of genetic instability driven by preceding mutations in oncogenes and tumour suppressor genes. This controversy remained for a long time mainly because data on chromosomal aberrations in early, pre-invasive tumours were scarce.

This changed with the introduction of Comparative Genomic Hybridisation (CGH) as a new chromosome analysis technique in 1992 (Kallioniemi *et al.*, 1992). In a single experiment, it provides information on losses or gains of chromosome material throughout the whole tumour genome, requiring only genomic DNA extracted from the tumour. CGH was very difficult to reproduce in less specialised laboratories; only after the publication of an article reviewing the method in great detail did CGH become more widely applied (Kallioniemi *et al.*, 1994a). At present, a considerable number of CGH studies have been performed on many types of solid tumours, using tumour cell lines, fresh or frozen tissues and archival, formaldehyde-fixed and paraffin-embedded tissues. CGH on archival material provided an opportunity to study tumours with clinical follow-up data.

Using CGH, it is now clear that chromosomal abnormalities occur at all stages of tumour development, including early premalignant stages (Bardi *et al.*, 1995; Meijer *et al.*, 1998; Ried *et al.*, 1996). Gains of whole chromosomes or whole chromosome arms are perhaps the most common early chromosomal changes in many tumour types (Ried *et al.*, 1999). Such events may seem rather harmless, and indeed for many years the relevance of low-level gains (as opposed to high-level amplification) had not been recognised. However, it has recently been demonstrated that a single copy gain of a certain gene can have a strong effect on its expression levels (Redon *et al.*, 2001). In general, the number of chromosomal aberrations increases with tumour progression, and non-random chromosome events have been found to correlate with specific tumour types and stages. These results show that changes at the chromosomal level cannot simply be set aside as mere side-effects of tumour progression. In fact, specific chromosomal aberrations correlating to tumour type and stage are now being used as prognostic markers (e.g. in haematological malignancies), and may in the future play a role in decision-making in cancer therapy.

14.2 Technique

CGH is performed as follows: tumour DNA is labelled by nick translation with biotin-16-dUTP, mixed with digoxigenin-11-dUTP-labelled normal (diploid) DNA and unlabelled Cot-1 DNA and then hybridised to normal metaphase preparations. The tumour and normal DNA compete for hybridisation to the chromosomes. After washing and detection with the antibodies avidin-FITC (green fluorescent) and sheep-anti-digoxigenin-TRITC (red fluorescent), the green to red fluorescence ratio along the chromosomal axis is measured by digital image analysis using dedicated software. A gain or amplification in the tumour genome will be visualised by an excess of the green signal (and a fluorescence ratio >1.0), whereas a deletion or loss in the tumour genome is seen by an excess of red signal (and a fluorescence ratio <1.0).

The principle of the technique is shown in Figure 14.1. Labelled tumour DNA competes with differentially labelled normal DNA for hybridising to normal human metaphase chromosomes. Using fluorescence microscopy and digital image processing, the ratio of the two is measured along the chromosomal axes. Digital image processing includes the following steps: (a) background subtraction, (b) segmentation of chromosomes and removal of non-chromosome

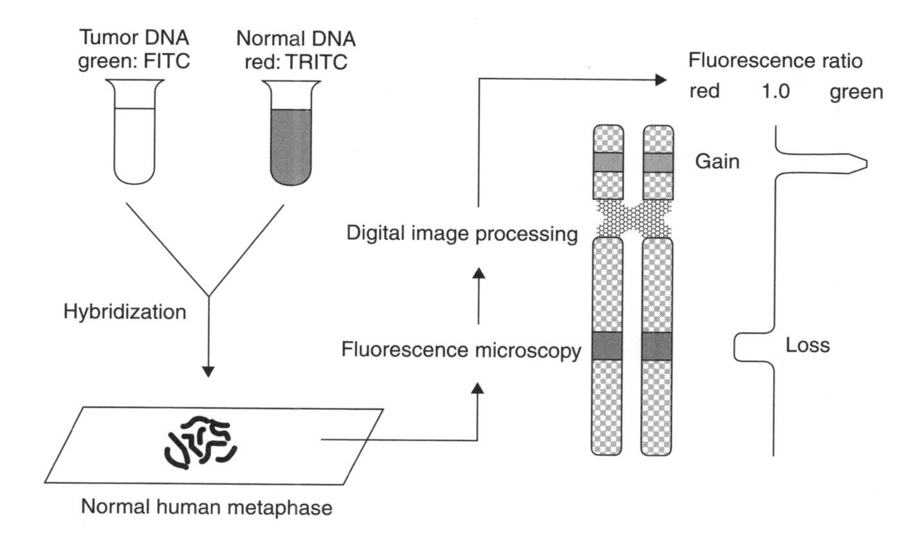

Figure 14.1 Schematic overview of the CGH technique. Tumour and reference DNA are labelled with a green and red fluorochrome, respectively, and hybridised to normal metaphase spreads. Images of the fluorescent signals are captured and the green-to-red signal ratios are digitally quantified for each chromosomal locus along the chromosomal axis. (Repr. from *Human Pathology*, 27, 342–349 (1996) with permission of W.B. Saunders Company)

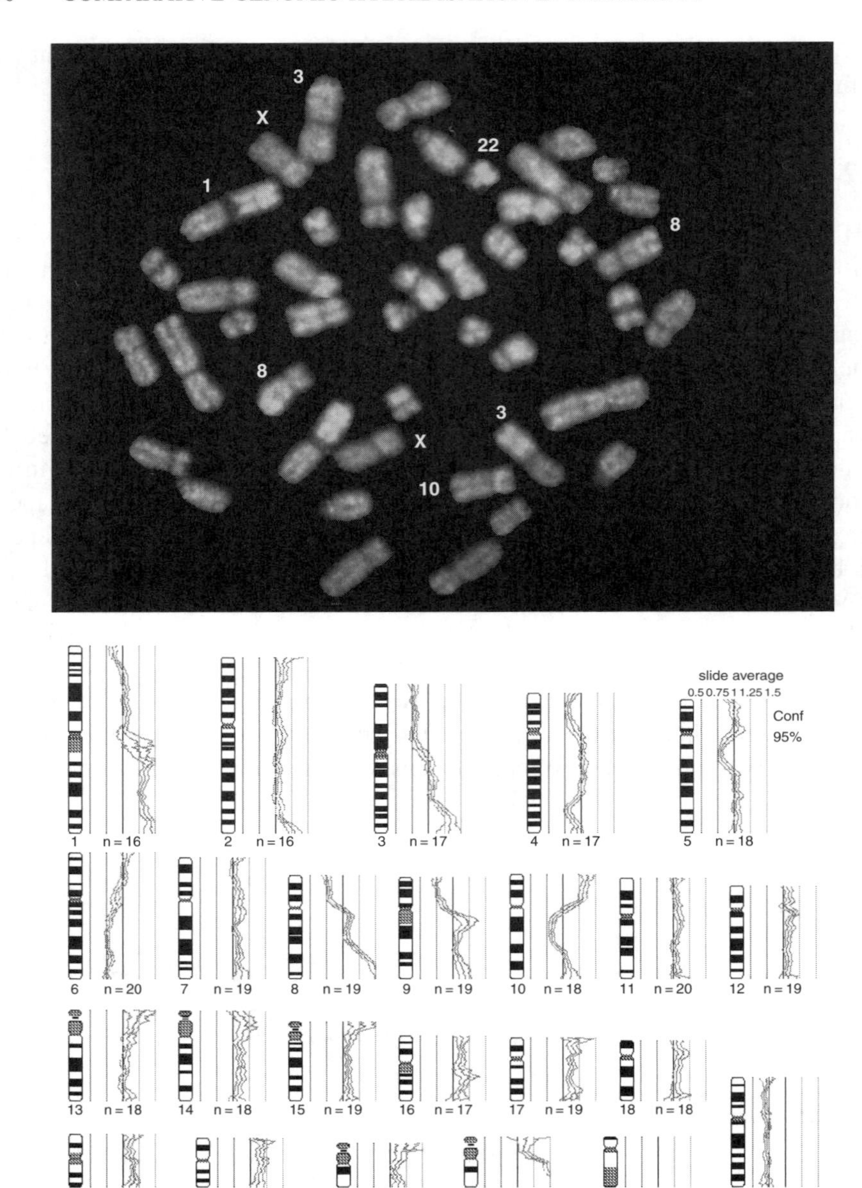

Figure 14.2 (Top) Green and red overlay image of a representative metaphase after hybridisation with tumour DNA from a laryngeal squamous cell carcinoma and normal DNA. Clearly visible amplifications (green) are present at chromosome arms 1q, 3q, 8q, 10p and 22q, and clear losses (red) can be seen at 3p, 4p, 5q, 6q, 8p, 10q and Xp/q. (Bottom) Relative copy number karyotype showing the quantitative analysis of the same tumour. The mean green-to-red fluorescence ratios of the chromosomes of multiple metaphase spreads are plotted in a graph corresponding to the chromosome ideograms, together with the 95% confidence interval: gains at 1p, 1q, 2q, 3q, 7q, 8q, 9q, 10p, 14q, 15q, 16p/q, 17p/q, 19q, 20q, 21q and 22q; losses at 1p, 3p, 4p, 4q, 5q, 6q, 8p, 9p, 10q and Xp/q. Modified from *Genomics Protocols* (Hermsen *et al*., 2001c). A colour version of this figure appears in the colour plate section

objects, (c) normalisation of the FITC/TRITC ratio for the whole metaphase, (d) interactive karyotyping and (e) scaling of chromosomes to a standard length. Deviations from the normal ratio of 1.0 at certain chromosome regions represent amplification or deletion of genetic material in the tumour, and may sometimes already be seen in a green and red overlay image of the hybridised metaphase (Figure 14.2, top). However, digital image processing is necessary for adequate evaluation. The final result is a so-called relative copy number karyotype (see Figure 14.2, bottom) that shows an overview of chromosomal copy number changes in the tumour. The sensitivity of CGH depends on the purity of the tumour sample; admixture with normal cells will reduce the sensitivity of CGH and inherently its resolution. For deletions the limit of detection is 10 Mb, which is about the size of an average chromosome band, but smaller amplifications (down to 250 kb) may be detected when the number of copies is high (Kallioniemi *et al.*, 1994b) (see Figure 14.3). The following steps are required to perform CGH: normal metaphase preparation, DNA labelling, hybridisation and washings, fluorescence microscopy, and capturing and analysing images with dedicated computer software, including karyotyping. A detailed description of the technique is presented elsewhere (Weiss *et al.*, 1999).

When using tissue sections to isolate DNA from tumour cells, admixture of normal cells (stroma or infiltrating lymphocytes) may present a problem. When a sample contains more than 25% normal cells microdissection could become necessary, depending on the ploidy of the tumour (Weiss *et al.*, 1999). This can be done manually or by advanced laser microdissection equipment (Fend and Raffeld, 2000; Zitzelsberger *et al.*, 1998). However, the latter yields only a limited number of cells (hence DNA), possibly necessitating universal DNA amplification techniques (Kuukasjarvi *et al.*, 1997; Lucito *et al.*, 1998). These techniques are time consuming, expensive and the user must perform good

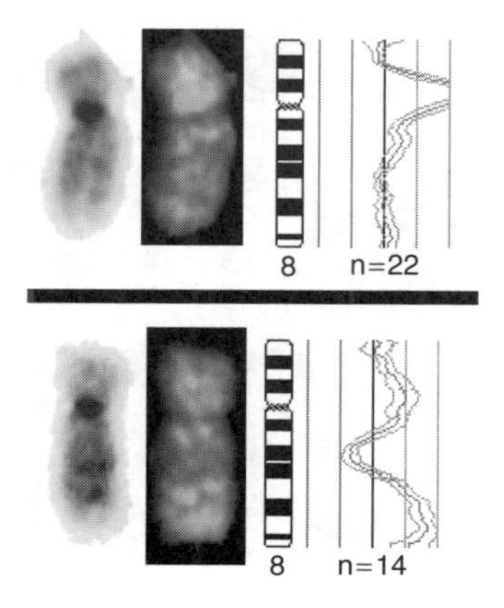

Figure 14.3 (Top) Example of a small high-level amplification in chromosome band 8p11; (bottom) example of a small deletion (approx. 10 Mb) in chromosome band 8q21. Modified from *Genomics Protocols* (Hermsen *et al.*, 2001c). A colour version of this figure appears in the colour plate section

control experiments to ensure the reliability of CGH results. Another approach could be cell sorting (e.g. antibodies attached to magnetic beads or flow cytometric sorting), which may enable selection (and extraction) of tumour cells, or elimination of inflammatory cells from a tissue sample.

14.3 Data analysis

There are two ways to interpret the relative copy number karyotypes. Some researchers use fixed ratio limits, for example 0.85/1.15 or 0.75/1.25, depending on the quality of the hybridisation (Barth *et al.*, 2000). Others prefer to use the 95% confidence interval (CI), which takes into account the quality of the signal. According to the latter definition, deviations from normal are interpreted as loss or gain when the average ratio, together with the 95% CI, is clearly below or above the ratio 1.0. We prefer the CI method. In addition, the green-to-red fluorescent ratios at 1p32-pter, 16p, 19p/q and 22q are often unreliable owing to a high content of repetitive sequences, which would therefore lead to false positive interpretation (Kallioniemi *et al.*, 1994a). In general, these chromosomal regions are excluded from the CGH profile interpretation if it concerns low-level gains or losses.

Analysing Series of Tumours

CGH data of series of tumours are most frequently analysed and presented in terms of frequencies (the common region of overlap) of gains and losses, and

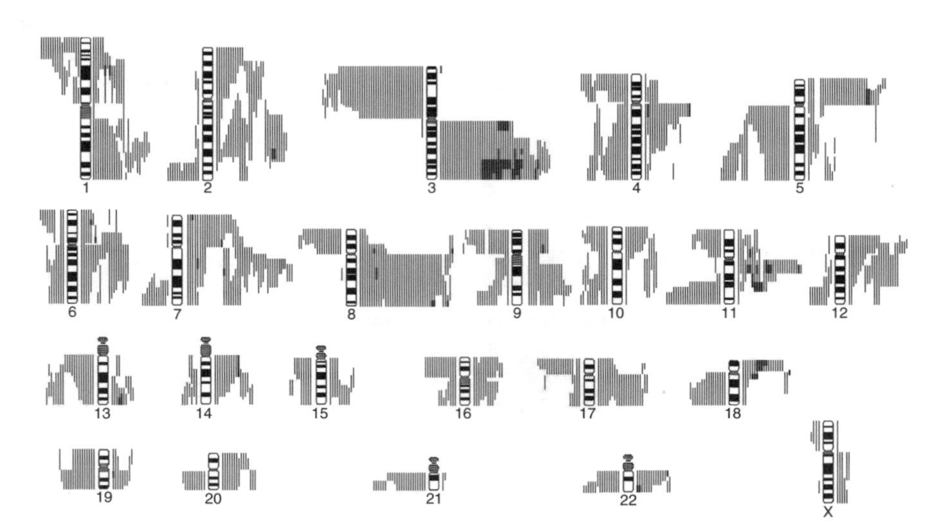

Figure 14.4 Overview of the CGH results of 42 primary [laryngeal ($n = 22$) or pharyngeal ($n = 20$)] tumours. Bars to the right of the ideograms represent gains and bars to the left represent losses. High-level amplifications are shown by bold bars. (Modified from Hermsen *et al.*, 2001a)

presented as shown in Figure 14.4. Especially in smaller series this is the most sensible approach. In this way, however, no information is obtained about possible correlations between different chromosomal changes (which may point to important biological information), nor does it allow the detection of different subsets of tumours with specific patterns of chromosomal aberrations. To this end, two-dimensional hierarchical cluster analysis of the CGH data can be used as microarray data. Software for this purpose, e.g. Cluster and Treeview (`http://rana.lbl.gov/EisenSoftware.htm`), is available in the public domain. For hierarchical cluster analysis, chromosomal gains can be coded as 1, and chromosomal losses as −1. In studies on colorectal adenomas and prostate cancer, this approach turned out to be very informative (Hermsen *et al.*, 2002; Mattfeldt *et al.*, 2001).

14.4 Applications

CGH has been applied for various purposes, ranging from the identification of novel cancer-related genes and unravelling pathogenetic mechanisms of diseases to clinical application for the classification of multiple tumours within one patient.

Identification of Cancer-related Genes

CGH has been helpful in mapping the loci of new cancer-related genes. For example, in the case of Peutz–Jeghers (P–J) syndrome, which is a hereditary disorder with an increased cancer risk characterised by intestinal hamartomatous polyposis and pigmented spots of the lips, buccal mucosa and digits, the tumour suppressor gene involved was first mapped on chromosome 19p using CGH combined with loss of heterozygosity (LOH) analysis (Hemminki *et al.*, 1997). Subsequently, STK11, a serine threonine kinase, could be identified as the gene involved (Jenne *et al.*, 1998).

In prostate cancer, comparison of pre- and post-treatment samples from patients receiving androgen deprivation therapy, CGH showed that amplification of the Xq11-q13 region harbouring the androgen receptor (AR) gene commonly (30%) occurred in tumours that showed recurrence under androgen deprivation therapy, while this was not seen in specimens taken prior to therapy. This suggests that AR amplification emerges during androgen deprivation therapy and is responsible for endocrine treatment failure (Visakorpi *et al.*, 1995).

In malignant fibrous histiocytoma (MFH) CGH revealed various chromosomal aberrations including distinct high-level amplifications at 8p21-pter. Within the 8p amplicon a novel gene designated MASL1 (MFH-amplified sequences with leucine-rich tandem repeats 1) was identified, which was significantly overexpressed in MFH with the 8p amplification. The MASL1 gene has important structural elements for interactions with proteins related to the cell cycle, which suggests that overexpression of MASL1 could be oncogenic in MFH (Sakabe *et al.*, 1999).

Pathogenesis and Tumour Progression

Complex chromosomal aberrations occur in many solid tumours, and the issue still remains whether these aberrations merely reflect 'genetic noise', or whether they relate to specific biological processes relevant to tumour development. To this end, multiple studies have analysed the development of chromosomal changes during tumour progression, and the comprehensive data produced by CGH have largely contributed to our understanding of the complexity of tumour progression. For instance, in a large series of colorectal adenomas and carcinomas, CGH demonstrated that the accumulation of losses in specific chromosomal regions (i.e. 8p21-pter, 15q11-q21, 17p12-13 and 18q12-21, and gains in 8q23-qter, 13q14-31 and 20q13), was strongly associated with adenoma to carcinoma progression, independent of the degree of dysplasia (Hermsen *et al.*, 2002). Furthermore, by analysing the chromosome CGH data by means of hierarchical cluster analysis the presence of three distinct, frequently occurring combinations of genetic aberrations in the adenomas (17p loss and K-ras mutation, 8q and 13q gain, and 18q loss and 20q gain, respectively) could be shown. Carcinomas appeared to have accumulated additional specific chromosomal aberrations, resulting in two distinct patterns of chromosomal instability. One pattern was marked by losses of 11q, 12q, 17p, 17q, 18q and 21q, and a second pattern by gains of 7p+q and 20q, and loss of 18q. Gains of 8q and 13q occurred in both patterns. This provided evidence that chromosomal abnormalities in colorectal adenoma to carcinoma progression occurred in specific combinations of a few abnormalities, rather than as a mere accumulation of events, indicating the existence of multiple independent pathways of colorectal cancer progression.

In breast cancer, it became also evident that the progression of breast cancer from a normal unaltered breast cell towards an invasive carcinoma does not happen in a straightforward manner. It could be shown that at least two different cytogenetic pathways exist, each correlated with a specific morphology (Buerger *et al.*, 1999a, 2000; Vos *et al.*, 1999) and invasive nature (Buerger *et al.*, 1999b, 2001; Roylance *et al.*, 1999). Whereas the morphologically well-differentiated pathway is mainly characterised by the loss of chromosomal material of 16q, the poorly differentiated breast cancer cases show a multitude of chromosomal alterations, including high-level amplifications, such as those involving c-*erb*B-2. Surprisingly, it could be shown that the cytogenetic alteration patterns in ductal carcinoma *in situ* (DCIS) and different subtypes of invasive breast cancer, as well as in the respective lymph node metastasis, did not significantly differ, so that most of the clinical progression steps could not be reproduced at the cytogenetic level (Buerger *et al.*, 2000; Nishizaki *et al.*, 1997). Also, lobular invasive breast cancer and its assumed precursor, lobular carcinoma *in situ*

(LCIS), could be defined as a separate cytogenetic subgroup, again with losses of 16q, including the chromosomal locus of E-Cadherin as the predominant additional cytogenetic feature (Buerger *et al.*, 2000; Gong *et al.*, 2001). Comparative studies of hereditary (BRCA1 and BRCA2 related) breast cancer identified these tumours as an additional tumour subgroup, characterised by a genetic alteration pattern with a significantly higher degree of cytogenetic instability (Tirkkonen *et al.*, 1997). In contrast, no cytogenetic alteration could be defined in the proposed precursor lesions of invasive breast cancer such as ductal hyperplasia (Boecker *et al.*, 2001).

Hereditary vs. Sporadic Cancers

For many hereditary cancer syndromes the causative initiating genetic defect has been resolved (e.g. BRCA1 and BRCA2 in hereditary breast cancer, Rb in retinoblastoma, mismatch repair genes in hereditary non-polyposis colorectal cancer, etc.). Less is known about the additional genetic changes required for the development of these cancers, and to what extent these differ from their sporadic counterparts.

CGH in hereditary ovarian cancer revealed that the majority of the genetic alterations found, are also common in sporadic ovarian cancer. Deletions of 15q11-15, 15q24-25, 8p21-ter, 22q13, 12q24 and gains at 11q22, 13q22 and 17q23-25, however, appeared to be specific to hereditary ovarian cancer. Aberrations at 15q11-15 and 15q24-25, had not been described in familial ovarian cancer. In these regions, important tumour suppressor genes, including the hRAD51 gene, are located. These and other yet unknown suppressor genes may be specifically involved in the pathogenesis of familial ovarian cancer and may explain the distinct clinical presentation and behaviour of familial ovarian cancer (Zweemer *et al.*, 2001).

In retinoblastoma, CGH revealed that the number of chromosomal aberrations shows a bimodal distribution, with a low-level chromosomal instability (CIN) group and a high-level CIN group. In the low-level CIN group the mean age was half the mean age of the high-level CIN group, there were fewer male patients, and there were more hereditary and bilateral cases (Herzog *et al.*, 2001; Van der Waal *et al.*, 2003).

Genotype–Phenotype Correlations

A central concept in molecular pathology is that genotype drives phenotype. As a consequence, the question arises: What is going on in tumours that show a heterogeneous phenotype? For carcinosarcomas (or sarcomatoid carcinomas) that show both an epithelial and a spindle cell histology, a polyclonal and monoclonal origin have been postulated, but the latter has been favoured. To test this, the epithelial and spindle cell components were microdissected from a small series of these mixed tumours from different origins and separately analysed for their respective chromosomal aberrations with CGH (Torenbeek *et al.*, 1999). A high level homology (of up to 91%) in chromosomal aberrations between different components in each tumour was seen, which supports the concept of a monoclonal origin of these tumours.

In line with these investigations is the comparison of the patterns of chromosomal aberrations in different types of cancer arising at the gastro-oesophageal junction, i.e. squamous cell carcinomas, Barrett's carcinomas and adenocarcinomas of the gastric cardia (Weiss *et al.*, 2003a). Typical chromosomal aberrations for the squamous cell carcinoma type were gains at 3q and 11q13, and losses at 3p, 4q, 9p, 11q, and 13q. In contrast, typical copy number changes for both cardia and Barrett's adenocarcinomas were gains at 2q, 7p, 13q, and losses at 17p. High-level amplifications occurred in all three groups, but their frequency in the carcinomas of the cardia was lower than in the other two groups. Therefore, squamous cell carcinomas are characterised by distinct chromosomal aberrations compared with both cardia and Barrett's adenocarcinomas. With respect to Barrett's cancer, chromosomal aberrations reflect an adenocarcinoma phenotype rather than a squamous origin.

Cancer Profiling

Cancer profiling has perhaps been the most widely used application of CGH in pathology, and larger or smaller series of virtually any type of tumour have been studied, and the number of published CGH studies by now has exceeded 17 000. Therefore, a detailed overview would go beyond the scope of this chapter. Instead, a few examples are presented here. In addition, Figure 14.5 (derived and modified from Struski *et al.*, 2002) summarises the chromosomal imbalances detected by CGH in 430 articles (11 984 cases) of human solid and haematological malignancies. When CGH data from different studies are combined, a pattern of non-random genetic aberrations appears. Some of these gains and losses are common to different types of pathologies (e.g. 8q+, 20q+, 13q− etc.), while others are more specific to individual tumour types (e.g. 2p+, 4p−, 10p−, etc.).

Several reports of CGH in breast cancer have been published. In a series of 53 lymph-node-negative breast carcinomas the most frequent chromosomal gains were, in descending order of frequency: 8q, 1q, Xq, 5q, 4q and 3q. Recurring losses were observed at chromosomal arms 19p, 1p, 17p, 22q, 4q and 8p. Of these, gain of chromosome 8q was strongly correlated with high values of mean nuclear area. When comparing only those cases that, according to their cytometric and morphometric features, had either the worst or the best prognosis, gains occurred mainly in the 'poor prognostic features' group, in particular at 8q, 11q13, 17q and 20q. It is hypothesised that these gains could be late, progression-related events and may be associated with aggressive clinical

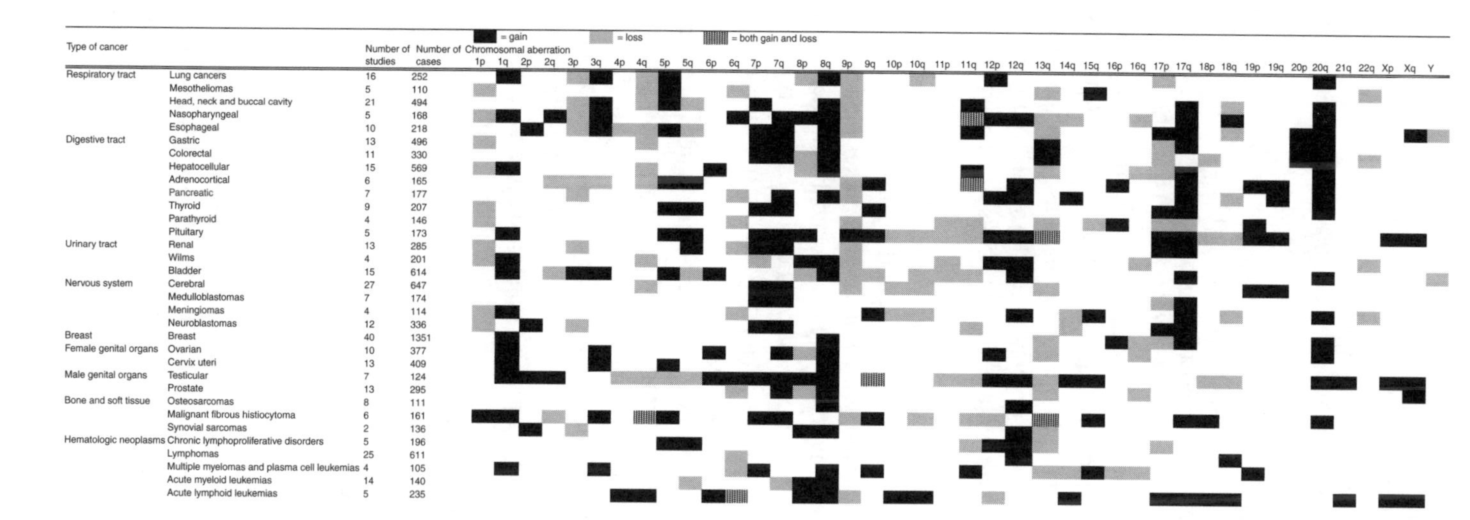

Figure 14.5 Summary of chromosomal imbalances detected by CGH in 430 articles (11 984 cases) of human solid and haematological malignancies. The types of cancer are divided into subgroups (left column), the number of publications and total number of studies cases are indicated. Black squares are gains and grey squares are losses, included when this aberration was present in more than 40% of cases. At least 100 tumours had to be studied in each considered group. (Modified from Struski *et al.*, 2002)

behaviour. Therefore, these four chromosomal regions may be of potential prognostic value (Hermsen *et al.*, 1998).

In gastric cancer the genetic mechanisms underlying carcinogenesis are largely unknown. In a study of 46 gastric cancers (M/F 35/11, age 27–85 yr), considering chromosomal aberrations with frequencies of 20% or higher to be non-random, the mean number of chromosomal events per tumour was 9.7 (range: 0–27), with a mean of 3.2 gains (range: 0–16) and 6.5 losses (range: 0–15). Gains were most frequently found at chromosome 8q and 13q (24% and 26%, respectively). Losses were predominantly found on chromosome arms 2q, 9p, 12q, 14q, 15q, 16p, 16q, 17p, 17q, 19p, 19q and 22q (22%, 30%, 43%, 22%, 33%, 50%, 28%, 50%, 39%, 33%, 39% and 37%, respectively). Common regions of overlap narrowed down to 2q11–14, 8q23, 9p21, 12q24, 13q21–22, 14q24 and 15q11–15. The mean number of gains was higher in tumours with metastases compared with localised tumours (4.1 vs. 1.9, $p = 0.04$). Tumours with a loss at 17p showed a higher number of losses than tumours without a 17p loss (9.5 vs. 4.7 on average, $p < 0.001$). Neither *H. pylori* status (+, $n = 25$; −, $n = 21$) nor *H. pylori* strain were correlated with the total number of events or any specific chromosomal aberrations, nor differences between intestinal ($n = 30$) and diffuse ($n = 15$) cancers or any other clinicopathological variable were found. This makes clear that a complex of chromosomal aberrations is involved in gastric cancer, but their pattern does not depend on *H. pylori* status or strain, nor on the histological type of the tumour. The exact biological meaning of these aberrations in carcinogenesis needs further clarification (van Grieken *et al.*, 2000).

Squamous cell carcinomas show a rather different pattern of aberrations. CGH gains and losses in 42 predominantly late-stage larynx and pharynx squamous cell carcinomas were detected in high frequencies at 1q, 3q, 5p, 7q, 8q, 11q13, 17q and 18p, and losses at 3p, 4p, 5q, 11qter and 18q. Neither the number nor the type of abnormalities or the occurrence of specific chromosome changes were found to be related to DNA ploidy, stage or degree of differentiation of the tumours. Apart from low-level gains, a large number of high-level amplifications were identified. The segments most frequently amplified were 3q24-qter, 11q13, 18p, 18q11.2, 8q23-24 and 11q14-22. Many of these amplified regions had not been reported before. More than half of all loci involved in amplifications harbour genes coding for growth factors and growth factor receptors, suggesting an important role for such genes in squamous cell tumorigenesis and progression of late stage tumours (Hermsen *et al.*, 2001a).

Squamous cancers of the head and neck in patients with Fanconi anaemia (FA) showed multiple chromosomal alterations comparable with those observed in sporadic tumours (Hermsen *et al.*, 2001b). These results suggest that the process leading to early occurrence of oral cancer in FA patients follows a similar pathway as in non-FA cancer patients, which would support a caretaker function for FA genes in protection against oral carcinogenesis. Since FA patients are uniquely hypersensitive to DNA cross-linking agents, while oral cancer in the general population is thought to be largely environmentally induced, these results also suggest that environmental DNA cross-linkers may be causally involved in oral carcinogenesis.

14.5 Clinical applications

A particular problem in pathology practice is provided by patients with multiple tumour localisations. Of course, in many cases the second tumour can be proven – at least for all practical purposes – to be a metastasis of the primary by the clinical context and standard additional techniques such as immunohistochemistry. However, in other cases the situation is less obvious and the decision whether the patient has a metastasis of the primary cancer, or that a second primary cancer has occurred, is difficult to answer. At the same time this question is highly relevant for clinical decision-making: should the patient be regarded to suffer from (high-stage) metastatic disease, or has a low-stage second primary occurred, and should the patient be treated accordingly with an intention to cure? Together with other molecular techniques, CGH can help in reaching a diagnosis, as is illustrated with the following case.

A 49-year-old female presented with a medullary type breast cancer which was resected. Nine years later she presented with a poorly differentiated adenocarcinoma of the left/right ovary. Again one year later she had a metastatic tumour removed from the omentum, which proved to be a poorly differentiated adenocarcinoma, and two years after that she underwent partial duodenectomy because of an obstructing tumour mass, which on microscopic examination also proved to be poorly differentiated adenocarcinoma. The ovarian tumour was regarded clinically as a second primary (FIGO stage), although histologically, discrimination between the tumours was not completely straightforward. Immunopositivity for BRST2 in the breast tumour and the metastases in the omentum and duodenum were an important indication. Genetic analysis by CGH, however, was conclusive: the two metastases showed an almost identical pattern which was very similar to the breast cancer and different from the ovarian cancer, proving that the lesions in the omentum and duodenum were metastases of the breast cancer, and that the ovarian cancer was a second primary (van Diest *et al.*, 2001).

14.6 Screening for chromosomal abnormalities in fetal and neonatal genomes

Besides applications in tumour pathology, CGH analysis has also been used to study chromosomal aberrations in fetal and neonatal genomes (Daniely *et al.*, 1998; Fritz *et al.*, 2001; Kalousek, 1998; Lomax *et al.*, 2000; Tabet *et al.*, 2001). Spontaneous abortions (SAs) in over 50% of cases are associated with chromosomal abnormalities. Conventional cytogenetic analysis of SAs depends on tissue culture and is hampered by a significant tissue culture failure rate and overgrowth of maternally derived cells. CGH can detect numerical and unbalanced structural chromosomal abnormalities associated with SAs while avoiding the technical problems associated with tissue culture. Since CGH is not suited to detect ploidy changes, it can be combined with flow cytometry (FCM). In the large series studied ($n = 301$) by Lomax *et al.* routine cytogenetic and CGH results could be compared in 253 cases. Concordance was found in 92.8% of the cases. Of the 18 discordant cases, 12 samples had chromosomal changes by CGH while cytogenetics showed a 46,XX karyotype indicating maternal contamination of the tissue cultures. These results demonstrate that CGH supplemented with FCM can readily identify chromosomal abnormalities associated with SAs and, by avoiding maternal contamination and tissue culture artefacts, can do so with a lower failure rate and higher accuracy than conventional cytogenetic analysis (Lomax *et al.*, 2000).

14.7 Future perspectives

In the last decade CGH has not only provided genomic copy number profiles for many tumours and syndromes, it has also aided in the identification of numerous genes involved in many of these diseases. However, one of the limitations of CGH is the limited resolution by which these copy number aberrations can be detected. As a consequence, small aberrations frequently were missed and the genomic boundaries of the aberrations were not well defined, limiting the ability to correlate specific genes to a certain aberration. The use of an array comprised of large insert genomic bacterial artificial chromosome (BAC) clones as a substrate for CGH hybridisations (microarray CGH), overcame many of the limitations of classical CGH (Lomax *et al.*, 2000; Pinkel *et al.*, 1998; Snijders *et al.*, 2000). Apart from a higher resolution, array CGH analysis also provides increased sensitivity. In chromosome CGH, small peaks of high-level gains frequently are masked by averaging effects, differences in centromere position, and segmentation. Owing to its 'digital' nature, this problem is solved with array CGH. Figure 14.6 shows an example where chromosome CGH detected a low-level gain of chromosome arm 20q, while array CGH revealed a narrow high-level amplification (Weiss *et al.*, 2003b).

Genome-wide measurement of DNA copy number variation using an array of mapped BAC clones would be of particular interest for diagnostic pathology as well as medical genetics since a single BAC clone provides sufficient signal for the accurate and quantitative detection of low-level copy number aberrations as well as high-level amplifications (Snijders *et al.*, 2001). Moreover, most of the BAC clones are cytogenetically mapped and contain a *sequence tagged site*, which allows linkage of specific BAC clones to the exact location in the genome sequence. This is crucial if diagnostic conclusion or treatment decisions

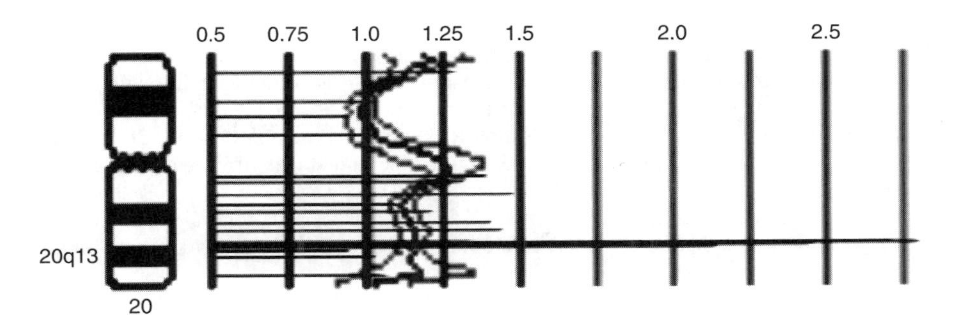

Figure 14.6 Example of 20q amplification: chromosome CGH (line-graph) detected a low-level gain of chromosome arm 20q, while microarray CGH (bar-graph) revealed a narrow high-level amplification

are based upon the presence of absence of a genomic copy number aberration of a certain locus.

Since the human genome is ~3000 Mb, an array of 3000 BAC clones with an average insert size of 100 kb would result in ~1 Mb resolution. A minimum tiling path of the human genome would require ~30 000 BAC clones. Preparation of sufficient BAC DNA for robotic printing is challenging since BACs are single-copy vectors and the yield of BAC DNA from bacteria cultures is relatively low. Moreover, robotic printing of high-density arrays containing whole BAC DNA is difficult. Preparation of sufficient well-represented BAC DNA in the right consistency for robotic printing requires specialised amplification methods such as ligation mediated polymerase chain reaction (PCR) (Snijders *et al.*, 2001).

At the moment initiatives are underway to have a genome wide contig array, covering all chromosomes completely. This would require a microarray of 35 000 spots.

14.8 Acknowledgements

This work was supported by grants NWO-MW 901-02-241, NKB 2002–2618, and a VU University USF grant.

14.9 References

Bardi G, Sukhikh T, Pandis N, Fenger C, Kronborg O and Heim S (1995) Karyotypic characterization of colorectal adenocarcinomas. *Genes, Chromosomes Cancer* **12**(2): 97–109.

Barth TF, Benner A, Bentz M, Dohner H, Moller P and Lichter P (2000) Risk of false positive results in comparative genomic hybridisation [In Process Citation]. *Genes, Chromosomes Cancer* **28**(3): 353–7.

Boecker W, Buerger H, Schmitz K, Ellis IA, van Diest PJ, Sinn HP *et al.* (2001) Ductal epithelial proliferations of the breast: a biological continuum? Comparative genomic hybridisation and high-molecular-weight cytokeratin expression patterns. *J. Pathol.* **195**(4): 415–21.

Boveri T (1914) *Zur Frage der Entstehung maligner Tumoren*. Gustav Fischer, Jena.

Buerger H, Mommers EC, Littmann R, Diallo R, Brinkschmidt C, Poremba C *et al*. (2000a) Correlation of morphologic and cytogenetic parameters of genetic instability with chromosomal alterations in *in situ* carcinomas of the breast. *Am. J. Clin. Pathol.* **114**(6): 854–9.

Buerger H, Mommers EC, Littmann R, Simon R, Diallo R, Poremba C *et al*. (2001) Ductal invasive G2 and G3 carcinomas of the breast are the end stages of at least two different lines of genetic evolution. *J. Pathol.* **194**(2): 165–70.

Buerger H, Otterbach F, Simon R, Poremba C, Diallo R, Decker T *et al*. (1999b) Comparative genomic hybridisation of ductal carcinoma *in situ* of the breast-evidence of multiple genetic pathways. *J. Pathol.* **187**(4): 396–402.

Buerger H, Otterbach F, Simon R, Schafer KL, Poremba C, Diallo R *et al*. (1999a) Different genetic pathways in the evolution of invasive breast cancer are associated with distinct morphological subtypes. *J. Pathol.* **189**(4): 521–6.

Buerger H, Simon R, Schafer KL, Diallo R, Littmann R, Poremba C *et al*. (2000b) Genetic relation of lobular carcinoma *in situ*, ductal carcinoma *in situ*, and associated invasive carcinoma of the breast. *Mol. Pathol.* **53**(3): 118–21.

Daniely M, Aviram-Goldring A, Barkai G and Goldman B (1998) Detection of chromosomal aberration in fetuses arising from recurrent spontaneous abortion by comparative genomic hybridisation. *Hum. Reproduct.* **13**(4): 805–9.

Fend F and Raffeld M (2000) Laser capture microdissection in pathology. *J. Clin. Pathol.* **53**(9): 666–72.

Fritz B, Hallermann C, Olert J, Fuchs B, Bruns M, Aslan M *et al*. (2001) Cytogenetic analyses of culture failures by comparative genomic hybridisation (CGH) – re-evaluation of chromosome aberration rates in early spontaneous abortions. *Eur. J. Hum. Genet.* **9**(7): 539–47.

Gong G, DeVries S, Chew KL, Cha I, Ljung BM and Waldman FM (2001) Genetic changes in paired atypical and usual ductal hyperplasia of the breast by comparative genomic hybridisation. *Clin. Cancer Res.* **7**(8): 2410–4.

Hansemann D (1890) Uber asymmetrische Zellteilung in Epithelkrebsen und deren biologischen Bedeutung. *Virchow's Arch. Path. Anat.* **119**: 299–326.

Hemminki A, Tomlinson I, Markie D, Jarvinen H, Sistonen P, Bjorkqvist AM *et al*. (1997) Localization of a susceptibility locus for Peutz-Jeghers syndrome to 19p using comparative genomic hybridisation and targeted linkage analysis. *Nat. Genet.* **15**(1): 87–90.

Hermsen M, Baak JP, Meijer GA, Weiss JM, Walboomers JW, Snijders PJ *et al*. (1998) Genetic analysis of 53 lymph node-negative breast carcinomas by CGH and relation to clinical, pathological, morphometric, and DNA cytometric prognostic factors. *J. Pathol.* **186**(4): 356–62.

Hermsen M, Guervos MA, Meijer G, Baak J, van Diest P, Marcos CA *et al*. (2001a) New chromosomal regions with high-level amplifications in squamous cell carcinomas of the larynx and pharynx, identified by comparative genomic hybridisation. *J. Pathol.* **194**(2): 177–82.

Hermsen M, Postma C, Baak J, Weiss MM, Rapallo A, Sciutto A *et al*. (2002) Colorectal adenoma to carcinoma progression follows three distinct pathways of chromosomal instability. *Gastroenterology* **123**(4): 109–19.

Hermsen M, Weiss MM, Meijer GA and Baak JP (2001) Detection of chromosomal abnormalities by comparative genomic hybridisation, in: *Genomics Protocols* (ed. M Starkey), pp. 47–55. The Humana Press Inc.

Hermsen M, Xie Y, Rooimans M, Meijer G, Baak J, Plukker J *et al*. (2001b) Cytogenetic characteristics of oral squamous cell carcinomas in Fanconi anemia. *Familial Cancer* **1**: 39–43.

Herzog S, Lohmann DR, Buiting K, Schuler A, Horsthemke B, Rehder H *et al*. (2001) Marked differences in unilateral isolated retinoblastomas from young and older children studied by comparative genomic hybridisation. *Hum. Genet.* **108**(2): 98–104.

Jallepalli PV and Lengauer C (2001) Chromosome segregation and cancer: cutting through the mystery. *Nat. Rev. Cancer* **1**(2): 109–17.

Jenne DE, Reimann H, Nezu J, Friedel W, Loff S, Jeschke R *et al.* (1998) Peutz–Jeghers syndrome is caused by mutations in a novel serine threonine kinase. *Nat. Genet.* **18**(1): 38–43.

Kallioniemi OP, Kallioniemi A, Piper J, Isola J, Waldman FM, Gray JW *et al.* (1994a) Optimizing comparative genomic hybridisation for analysis of DNA sequence copy number changes in solid tumours. [Review] [17 refs]. *Genes, Chromosomes Cancer* **10**(4): 231–43.

Kallioniemi A, Kallioniemi OP, Piper J, Tanner M, Stokke T, Chen L *et al.* (1994b) Detection and mapping of amplified DNA sequences in breast cancer by comparative genomic hybridisation. *Proc. Natl. Acad. Sci. USA* **91**(6): 2156–60.

Kallioniemi A, Kallioniemi OP, Sudar D, Rutovitz D, Gray JW, Waldman F *et al.* (1992) Comparative genomic hybridisation for molecular cytogenetic analysis of solid tumours. *Science* **258**(5083): 818–21.

Kalousek DK (1998) Clinical significance of morphologic and genetic examination of spontaneously aborted embryos. [Review] [44 refs]. *Am. J. Reproduc. Immunol.* **39**(2): 108–19.

Kuukasjarvi T, Tanner M, Pennanen S, Karhu R, Visakorpi T and Isola J (1997) Optimizing DOP-PCR for universal amplification of small DNA samples in comparative genomic hybridisation. *Genes, Chromosomes Cancer* **18**(2): 94–101.

Lomax B, Tang S, Separovic E, Phillips D, Hillard E, Thomson T *et al.* (2000) Comparative genomic hybridisation in combination with flow cytometry improves results of cytogenetic analysis of spontaneous abortions. *Am. J. Hum. Genet.* **66**(5): 1516–21.

Lucito R, Nakimura M, West JA, Han Y, Chin K, Jensen K *et al.* (1998) Genetic analysis using genomic representations. *Proc. Natl. Acad. Sci. USA* **95**(8): 4487–92.

Mattfeldt T, Wolter H, Kemmerling R, Gottfried HW and Kestler HA (2001) Cluster analysis of comparative genomic hybridisation (CGH) data using self-organizing maps: application to prostate carcinomas. *Anal. Cell. Pathol.* **23**(1): 29–37.

Meijer GA, Hermsen MAJA, Baak JPA, Diest van JP, Meuwissen SGM, Belien JAM *et al.* (1998) Progression from colorectal adenoma to carcinoma is associated with non-random chromosomal gains as detected by comparative genomic hybridisation. *J. Clin. Pathol.* **51**: 901–9.

Nishizaki T, Chew K, Chu L, Isola J, Kallioniemi A, Weidner N *et al.* (1997) Genetic alterations in lobular breast cancer by comparative genomic hybridisation. *Int. J. Cancer* **74**(5): 513–7.

Pinkel D, Segraves R, Sudar D, Clark S, Poole I, Kowbel D *et al.* (1998) High resolution analysis of DNA copy number variation using comparative genomic hybridisation to microarrays. *Nat. Genet.* **20**(2): 207–11.

Redon R, Muller D, Caulee K, Wanherdrick K, Abecassis J and du MS (2001) A simple specific pattern of chromosomal aberrations at early stages of head and neck squamous cell carcinomas: PIK3CA but not p63 gene as a likely target of 3q26-qter gains. *Cancer Res.* **61**(10): 4122–9.

Ried T, Heselmeyer-Haddad K, Blegen H, Schrock E and Auer G (1999) Genomic changes defining the genesis, progression, and malignancy potential in solid human tumours: a phenotype/genotype correlation. *Genes, Chromosomes Cancer* **25**(3): 195–204.

Ried T, Knutzen R, Steinbeck R, Blegen H, Schrock E, Heselmeyer K *et al.* (1996) Comparative genomic hybridisation reveals a specific pattern of chromosomal gains and losses during the genesis of colorectal tumours. *Genes, Chromosomes Cancer* **15**(4): 234–45.

Rowley JD (1973) Letter: A new consistent chromosomal abnormality in chronic myelogenous leukaemia identified by quinacrine fluorescence and Giemsa staining. *Nature* **243**(5405): 290–3.

Rowley JD (2001) Chromosome translocations: dangerous liaisons revisited. *Nat. Rev. Cancer* **1**(3): 245–50.

Roylance R, Gorman P, Harris W, Liebmann R, Barnes D, Hanby A *et al.* (1999) Comparative genomic hybridisation of breast tumours stratified by histological grade reveals new insights into the biological progression of breast cancer. *Cancer Res.* **59**(7): 1433–6.

Sakabe T, Shinomiya T, Mori T, Ariyama Y, Fukuda Y, Fujiwara T *et al.* (1999) Identification of a novel gene, MASL1, within an amplicon at 8p23.1 detected in malignant fibrous histiocytomas by comparative genomic hybridisation. *Cancer Res.* **59**(3): 511–5.

Snijders AM, Meijer GA, Brakenhoff RH, van den Brule AJ and van Diest PJ (2000) Microarray techniques in pathology: tool or toy?. *Mol. Pathol.* **53**(6): 289–94.

Snijders AM, Nowak N, Segraves R, Blackwood S, Brown N, Conroy J *et al.* (2001) Assembly of microarrays for genome-wide measurement of DNA copy number. *Nat. Genet.* **29**(3): 263–4.

Struski S, Doco-Fenzy M and Cornillet-Lefebvre P (2002) Compilation of published comparative genomic hybridisation studies. *Cancer Genet. Cytogenet.* **135**(1): 63–90.

Tabet AC, Aboura A, Dauge MC, Audibert F, Coulomb A, Batallan A *et al.* (2001) Cytogenetic analysis of trophoblasts by comparative genomic hybridisation in embryo – fetal development anomalies. *Prenat. Diagn.* **21**(8): 613–8.

Tirkkonen M, Johannsson O, Agnarsson BA, Olsson H, Ingvarsson S, Karhu R *et al.* (1997) Distinct somatic genetic changes associated with tumour progression in carriers of BRCA1 and BRCA2 germ-line mutations. *Cancer Res.* **57**(7): 1222–7.

Tjio JH and Levan A (1956) The chromosome number of man. *Hereditas* **42**(1).

Torenbeek R, Hermsen MAJA, Meijer GA, Baak JPA and Meijer CJLM (1999) Analysis by comparative genomic hybridisation of epithelial and spindle cell components in sarcomatoid carcinoma and carcinosarcoma: histogenetic aspects. *J. Pathol.* **189**(3): 338–43.

Van der Wal J, Hermsen MAJA, Gille H, Schouten-van Meeteren N, Moll A, Imhof S *et al.* (2003) Comparative genomic hybridisation divides retinoblastomas into a high- and a low-level chromosomal instability group. *J. Clin. Pathol.* **56**(1): 26–30.

van Diest PJ, Oudejans JJ, Meijer GA and van den Brule AJ (2001) Moleculaire diagnostiek in de pathologie, in: *Moleculaire Diagnostiek* (eds Van Pelt-Verkuil E, Van Berlo MF, Van Belkum A and Niesters HGM), pp. 171–86. Bohn Stafleu Van Loghum, Houten/Diegem.

van Grieken NC, Weiss MM, Meijer GA, Hermsen MA, Scholte GH, Lindeman J *et al.* (2000) *Helicobacter pylori*-related and non-related gastric cancers do not differ with respect to chromosomal aberrations. *J. Pathol.* **192**(3): 301–6.

Visakorpi T, Hyytinen E, Koivisto P, Tanner M, Keinanen R, Palmberg C *et al.* (1995) *Invivo* amplification of the androgen receptor gene and progression of human prostate cancer. *Nat. Genet.* **9**(4): 401–6.

Vos CB, ter Haar NT, Rosenberg C, Peterse JL, Cleton-Jansen AM, Cornelisse CJ *et al.* (1999) Genetic alterations on chromosome 16 and 17 are important features of ductal carcinoma *in situ* of the breast and are associated with histologic type. *Br. J. Cancer* **81**(8): 1410–8.

Waters JJ, Barlow AL and Gould CP (1998) Demystified...FISH. *J. Clin. Pathol.* **51**: 62–70.

Weiss MM, Hermsen MA, Meijer GA, van Grieken NC, Baak JP, Kuipers EJ *et al.* (1999) Comparative genomic hybridisation. *Mol. Pathol.* **52**(5): 243–51.

Weiss MM, Kuipers EJ, Hermsen MAJA, van Grieken NC, Offerhaus GJ, Baak JPA *et al.* (2003a) Barrett's adenocarcinoma resemble adenocarcinomas of the gastric cardia in terms of chromosomal copy number changes, but relate to squamous cell carcinomas of the distal oesophagus with respect to the presence of high-level amplifications. *Journal of Pathology* **199**: 157–65.

Weiss MM, Snijders AM, Meijer GA, Meuwissen SGM, Baak JPA, Pinkel D *et al.* (2003b) Microarray comparative genomic hybridisation reveals a narrow gain at 20q13 in human gastric carcinomas. *Journal of Pathology*, in press.

Zitzelsberger H, Kulka U, Lehmann L, Walch A, Smida J, Aubele M *et al.* (1998) Genetic heterogeneity in a prostatic carcinoma and associated prostatic intraepithelial neoplasia as demonstrated by combined use of laser-microdissection, degenerate oligonucleotide primed PCR and comparative genomic hybridisation. *Virchows Archiv.* **433**(4): 297–304.

Zweemer RP, Ryan A, Snijders AM, Hermsen MA, Meijer GA, Beller U *et al.* (2001) Comparative genomic hybridisation of microdissected familial ovarian carcinoma: two deleted regions on chromosome 15q not previously identified in sporadic ovarian carcinoma. *Lab. Invest.* **81**(10): 1363–70.

15

DNA Sequencing and the Human Genome Project

Philip Bennett

15.1 Introduction

For many people, the excitement of entering a new millennium was heightened by the anticipation of simultaneously entering a new era in biology and medicine. This new era was of course that marked by the joint press conferences of Prime Minister Tony Blair and President Bill Clinton held on 26 June 2000. Here, they both announced that the human genome had at last been sequenced. Of course, this was greeted by great fanfare, with other heads of states and persons of note the world over looking for ever more colourful metaphors and analogies to describe it; yet for many working in the field of human genetics, the day was in itself no 'big deal'. This is in no way a detraction; indeed, analogies such as 'equivalent to putting man on the moon' are probably well founded. Instead, it is a positive reflection of the way in which the publicly funded Human Genome Project (HGP) has been run, with data being made immediately and freely available to the entire scientific community via the Internet. Put another way, it was not as though millions of bases of nucleotide sequence had suddenly appeared on the public access databases at midnight on 25 June 2000. Furthermore, there was the fact that this date only marked completion of a 'working draft' sequence, i.e. there were still some gaps of missing or poor quality sequence and most of it had yet to be annotated.

At the time and in light of the points mentioned above, for many scientists the most interesting landmarks in the human genome story still lay ahead. The first of these came in February 2001 with the publication of two seminal scientific papers, one in *Nature* (Lander *et al.*, 2001) and one in *Science* (Venter *et al.*,

Molecular Biology in Cellular Pathology. Edited by John Crocker and Paul G. Murray
 ISBN: 0-470-84475-2

2001). These explained how the work had actually been done, along with preliminary annotations of the sequence and the results of initial analyses regarding total gene number, distribution, degrees of sequence conservation between different individuals and across species, etc. At the time of writing, the remaining landmarks, of a complete final sequence and its full annotation, have yet to come. Annotation means ascribing the many features of each gene and the protein(s) it encodes to specific stretches of DNA sequence. Without it, the raw sequence is little more than a meaningless string of As, Cs, Gs and Ts. Although general pictures are emerging fast, accurate and comprehensive annotation of the whole human genome will take some time yet, and is in many ways a more formidable task than its initial sequencing. Nevertheless, it will be worth the wait, as with it we should finally start to see a considerable rate of increase in the HGP's long promised rewards of major advances to health and medicine. The purpose of this chapter is to review the technique of DNA sequencing, starting with the basics, before moving on to cover some specific applications and, finally, its utilisation in the HGP. In the final section the focus will be predominantly on the science of genome sequencing and on the key discoveries to come out of the HGP so far. There will, however, also be a brief discussion of the history and politics of the HGP and how it effectively ended up as a race between the publicly funded HGP consortium and the private biotechnology company, Celera Genomics.

15.2 DNA sequencing: the basics

In 1980 two individuals shared the Nobel prize for chemistry for their independently developed methods of DNA sequencing. One was Walter Gilbert who, along with his colleague Allan Maxam, had described a method using chemical agents to cleave DNA at specific bases. This, combined with electrophoresis and autoradiography, could then be used to deduce the original sequence (Maxam and Gilbert, 1980). The other was Fred Sanger, whose technique utilised dideoxynucleotide analogues (see Figure 15.1) to terminate the extension of a growing DNA chain at specific bases. Again this was followed by

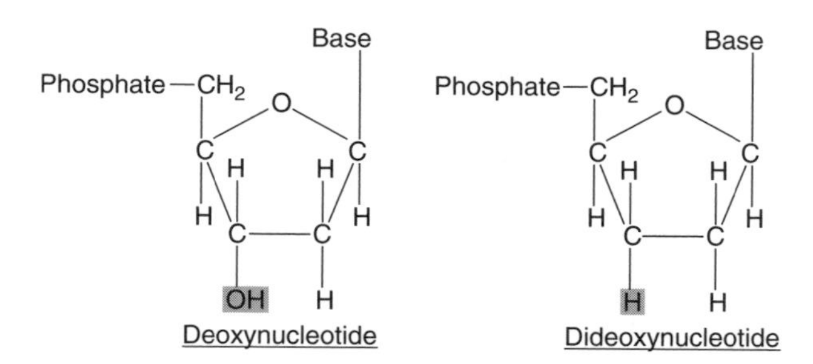

Figure 15.1 Chemical structures of deoxy- and dideoxynucleotides. The replacement of the hydroxyl group by a hydrogen (shown in shading) prevents dideoxynucleotides from undergoing further chain extension during DNA synthesis

electrophoresis and autoradiography in order to determine the original sequence (Sanger and Coulson, 1975). For a number of reasons, mainly to do with the chemicals involved and the complexity of analysis associated with Gilbert's method, Sanger's technique soon established dominance and virtually all modern sequencing, either manual or automated (see below), now uses the Sanger dideoxy chain termination method. Indeed, all of the nucleotide sequence data comprising the UK's contribution to the HGP (chromosomes 1, 6, 9, 10, 11, 13, 20, 22 and X) was produced at the Wellcome Trust funded institute which now bears his name. In light of its now almost universal use, dideoxy sequencing is the only technique that will discussed further in this chapter.

It should be noted at this point that despite being described a number of years earlier, dideoxy sequencing bears considerable similarity to the polymerase chain reaction (PCR), in so much as both involve the enzyme mediated extension of a short oligonucleotide primer annealed to a single stranded DNA template. Furthermore, PCR technology has enabled the easy production of sequencing templates and the ability to sequence smaller amounts of template by way of 'cycle sequencing' protocols (see below). Readers not familiar with the PCR may benefit from reading more about the basics of this technique (see Chapter 9) before proceeding further.

The Sequencing Reaction

Individual DNA molecules consist of a long polymer of nucleotide subunits. Each nucleotide subunit comprises a five carbon sugar (deoxyribose) covalently linked to a nitrogenous base (Adenine 'A', Guanine 'G', cytosine 'C' or Thymine 'T') and a phosphate moiety (see Figure 15.1). Subunits are linked to each other via a phosphodiester bond between the phosphate moiety of one and the free oxygen (shown in red in Figure 15.2) on the deoxyribose of the next. This forms a linear chain, much like beads threaded on a string which, in intact chromosomes, can be up to hundreds of millions of subunits long. *In vivo*, DNA usually exists as a double helix in which two complementary strands are held together by weak hydrogen bonds. When cells divide, their DNA is replicated by enzymes known as DNA polymerases. These polymerases first require a whole battery of other enzymes to manipulate the double helix in order to generate short areas of double-stranded DNA, needed to start or 'prime' the copying process, followed by areas of single-stranded sequence to use as a template for new synthesis. As the DNA polymerase processes along the template, it binds successive free nucleotide triphosphates, linking each to the previous one in the order specified by the template strand. During each iteration a new phosphodiester bond is formed and inorganic pyrophosphate released. This process can be mimicked in the test tube, using heat to separate the double-stranded template and a small synthetic oligonucleotide called a 'primer'. By gradually lowering the temperature, the primer, which is present in excess, can anneal to

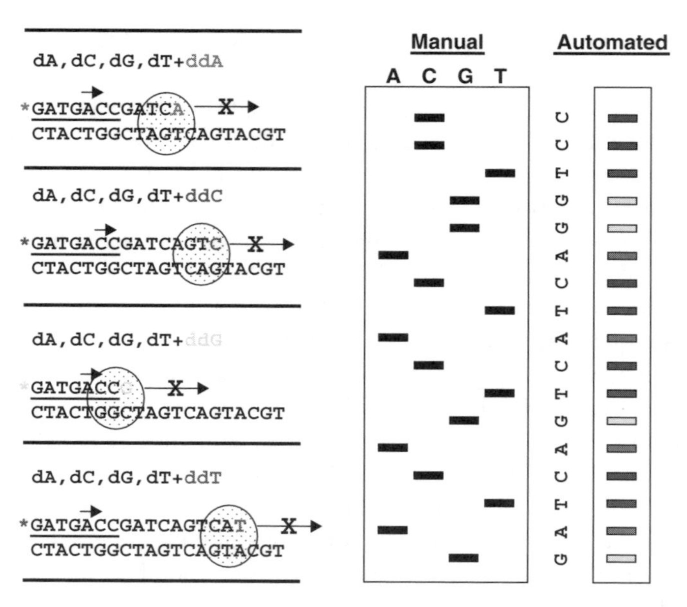

Figure 15.2 Illustration of 'Sanger sequencing' using dideoxynucleotide chain terminators. On the left, DNA polymerase is shown extending a short synthetic primer (underlined). Extension is halted as soon as the first dideoxynucleotide analogue (shown as coloured letters) has been incorporated. There are three basic sequencing formats which differ according to how the extended strands have been labelled for subsequent visualisation. (1) In manual sequencing, identically labelled termination products cannot be distinguished, except by size. Consequently, sequencing reactions (one for each individual dideoxynucleotide terminator) must be performed and electrophoresed separately. This results in a gel image like that shown in the centre of the figure. (2) Using semi-automated 'dye–primer' chemistry, primers are labelled with different fluorophores, as indicated by the coloured asterisks. Individual terminator reactions must still be performed separately, but reaction products may be pooled and electrophoresed in a single gel lane. Termination products are detected in real time, resulting in a virtual gel image, as shown on the right. (3) With semi-automated 'dye–terminator' chemistry, the dideoxynucleotide terminators themselves are labelled with different fluorophores. This allows the sequencing reaction to be performed in a single tube. Subsequent electrophoresis and analysis is performed in a similar way to that used for 'dye–primer' chemistry. A colour version of this figure appears in the colour plate section

its complementary sequence within the template (before the original full length template strands can fully recombine) in order to create the short stretch of double-stranded DNA needed for the DNA polymerase to bind. If a mixture of all four deoxynucleotide triphosphates and purified DNA polymerase is added, then the primer will be extended and a new strand of DNA complementary to the original template will be produced.

Whilst this means of copying DNA *in vitro* was very elegant, the procedure itself obviously gave no information regarding the order in which nucleotides had been incorporated into the new strand. Sanger, however, realised that if he could stop the primer extension at specific bases throughout the copying process, the relative size of the truncated fragments would tell him the order in which those bases had been incorporated. To achieve this he repeated the above procedure four times, all were identical except in each he also added a small amount of a different dideoxynucleotide analogue (ddA, ddC, ddG, or ddT). Thus in the ddA sequencing reaction, DNA polymerase would continue the primer extension process until it came to a 'T' in the template, most of the time it would then incorporate the complementary deoxyadenosine triphosphate (dA) and continue to the next 'T'. Sometimes, though, the polymerase would by chance incorporate a ddA, and the chain extension would terminate (see Figure 15.2). This is because the dideoxyadenosine has no free oxygen on the ribose moiety in order to form the next phosphodiester bond (see Figure 15.1). If enough template molecules were present and the ratio of deoxyadenosine triphosphate to dideoxyadenosine triphosphate was correct, then a population of truncated extension products representing a termination event at every 'T' within the target would result. If the products of all four base-specific sequencing reactions were then loaded into adjacent lanes on a polyacrylamide gel and separated according to size by electrophoresis, then the order and lanes in which successive bands appeared would allow the nucleotide sequence to be determined (see Figure 15.2). Incidentally, as the amount of DNA in each band of such a gel was very small, the truncated fragments were usually made radioactive by either radio-labelling the primer prior to the sequencing reactions or by incorporating radioactive nucleotides within them. The presence of radioisotopes did not affect the chemistry in any way, but it did mean that after electrophoresis, the gel could be exposed to a photographic film for as long as was necessary for an image of the bands to be acquired, a process known as autoradiography.

Problems, Solutions and Developments

The fully manual method of DNA sequencing described above worked well in most cases. However, a number of problems, which are detailed below, meant that tackling templates more than a few thousand bases long could be a considerable undertaking.

1. The process of manually reading gel images and transcribing the sequence onto paper or computer was laborious and prone to error. Also, the large photographic films required careful labelling, and archiving hundreds or even thousands could present both storage and logistical problems. In later years, some of these problems were partially alleviated with the availability of computerised scanners and semi-automated transcribing devices.

2. Even using long polyacrylamide gels, usually it was only possible to read about 300–400 bases of sequence data for each sample. This was because larger fragments at the top had not been run far enough to resolve the single base size difference between them, yet further electrophoresis would have

resulted in smaller fragments at the bottom being lost as they ran off the end. There was an obvious physical limit to the size of gels that could be poured successfully and safely manhandled, yet this problem could be partially overcome by reloading the same set of samples in new lanes with a time delay between. This meant that longer fragments could be resolved on the first set, with the smaller fragments that were inevitably lost visible on the set loaded later. The two sets of transcribed sequences were then overlaid to give a longer read that could have been obtained from either one alone.

3. In addition to only being able to read several hundred bases at a time, it could take anywhere between a few days and even a few weeks after the sequencing gel had been run for the autoradiographs to develop. During this time one did not even know if the process had worked. Even if it had, one could not proceed with further sequencing of a novel template until the new sequence data had been read and transcribed. This was because without it, it was not possible to synthesise the new primer (further along the template) that would be required to read the next several hundred base segments. The only potential solution to this problem was to fragment the target DNA into smaller pieces before cloning. This meant that most of the insert could be read in the first round of sequencing using universal flanking primers within the vector DNA (see the section 'Sequencing Templates and Their Preparation'). This, however, only really moved the problem elsewhere as it then became harder to reconstruct the original full length sequence from the increased number of smaller sub-clones.
4. Polyacrylamide gels are rarely completely flawless or entirely homogeneous. Thus, although loaded in adjacent lanes, the products of the four termination reactions may not run uniformly, making the gels difficult to interpret. Furthermore, if say a small air bubble had disrupted one of the lanes, then the entire sequence would be useless, even if the other three lanes were perfect.
5. Prior to the development of the PCR, no thermostable DNA polymerases were available. Consequently, the sequencing reaction had to be performed at relatively low temperatures (typically 37–45°C). This meant that sometimes the previously separated template strands could begin to re-anneal, either to each other or inwardly upon themselves (forming a secondary structure). This could inhibit the binding of the sequencing primer, and/or the procession of the DNA polymerase. A number of techniques, including genetic systems using bacterial viruses (phage), were developed which allowed single-stranded DNA templates to be produced. Also, additional chemicals, such as dimethylsulphoxide (DMSO), which reduced secondary structure formation, could be added to the sequencing reaction. Nevertheless, certain templates still remained difficult to sequence.
6. The necessary use of radioisotopes in DNA sequencing represented a less than ideal situation. In addition to obvious safety concerns, their limited

half-lives meant that reagents deteriorated quickly and had to be reordered. Later, protocols using non-radioactive silver compounds to stain very small amounts of DNA were utilised by some as an alternative. However, the use of radioisotopes continued in many laboratories until they were able to gain easy and/or affordable access to the fluorescence-based technologies described below.

Two major developments contributed towards advancing the basic dideoxy sequencing technique into one capable of undertaking very large sequencing projects. One of these was the PCR (Saiki *et al.*, 1985); the other was the synthesis of fluorescent oligonucleotide derivatives (Smith *et al.*, 1985). With the development of the PCR, new thermostable DNA polymerases isolated from organisms living in hot springs became available. This meant not only could the sequencing reaction be completed at a higher temperature, thereby reducing re-annealing and problems with secondary structure, but also that the process could be repeated over and over again in cyclic fashion. This so-called 'cycle sequencing' is analogous to the PCR but with a single primer, and resulted in the generation of more signal from less template. Fluorescent labelling of the oligonucleotide primers and hence their extended products from the sequencing reaction, meant that small amounts of DNA could be detected in real time, rather than waiting for the autoradiograph. Furthermore, labelling the primer in each of the four termination reactions with moieties that fluoresced at different wavelengths (colours) enabled the products of the four reactions to be pooled and run in the same lane (see Figure 15.2), eliminating problems of lane to lane variation. All this had the effect of making DNA sequence analysis far more amenable to automation, and very soon, use of the first automated DNA sequencer had been reported (Horvath *et al.*, 1987).

Although modern automated DNA sequence analysis is essentially identical to that described above, there have been constant and considerable improvements in hardware, software and chemistry. The former two are discussed in the next section, and the latter is mentioned below. However, considering all three together, the net effect has been to dramatically increase output. In the late-1980s a laboratory equipped with a state of the art automated sequencer could expect to turn out about 500 bases a day. Today some automated sequencers can produce up to 500 000 bases per day, and of course, large sequencing facilities may have many of these machines, further increasing their capacity to tens of millions of bases per day.

Advances in automated sequencing chemistry have largely been in two areas. The first concerns the thermostable DNA polymerases; over the years more and more have been discovered, each with subtly different properties. Also for some, these properties have been further manipulated by genetic engineering. Consequently, modern enzymes now are very stable, reliable and able to give consistently long read lengths using different chemistries (dye–primer and dye–terminator, see below) on a wide range of templates including many that were previously considered 'difficult'.

The second area of development concerns fluorophore chemistry; successive versions have been released with ever-decreasing overlaps in their emission spectra. Put simply in an example, new 'blue' and 'green' dyes look blue and green, old 'blue' and 'green' dyes looked more bluey green and greenish blue, presenting problems for the software in interpreting the fluorescence data. It also became possible to attach the fluorescent dyes to the dideoxynucleotide analogues themselves (Prober *et al.*, 1987). This meant that sequencing no longer needed to be performed using four separate reactions which were subsequently pooled, so called 'dye–primer' chemistry. It also meant that any template could be sequenced using an ordinary (inexpensive) oligonucleotide primer, without the delay and cost of having to synthesise the sequencing primer in quadruplicate, each one end labelled with a different fluorophore. This so-called 'dye–terminator' chemistry (see Figure 15.2), does have some potential drawbacks in that the dyes can interfere with the polymerase during incorporation. This results in the different base terminators having different incorporation efficiencies, and hence producing bands of different intensities. An example of where this may be problematic is when one is looking for heterozygous mutations. Two bands are present in the same place but one is much brighter than the other. Sometimes this is simply spurious noise; other times, both sequences are indeed present in equal number (a true mutation), but the brightness difference results from their different incorporation efficiencies. Correctly distinguishing the two situations can be very difficult. Consequently, ongoing developments are attempting to combine the benefits of dye–terminator chemistry with the even peak intensities and very long read lengths more commonly seen with dye–primer chemistry.

Equipment

In addition to everyday laboratory apparatus, manual dideoxy sequencing required little more than a fairly long polyacrylamide gel tank. Once run, the two glass plates, between which the thin gel was sandwiched, were carefully separated and the gel normally lifted off onto a sheet of filter paper (so that the glass plates could be promptly reused), before it was dried down and placed in a cassette to expose the photographic film placed on top of it. First generation automated sequencers were essentially exactly the same – long vertical polyacrylamide gel tanks, except that near the bottom of the gel an enclosed laser excitation and fluorescence detection system scanned a narrow window the width of the gel. Fluorescence data were then output to a computer where a virtual gel image could be constructed. In addition to the advantage of this process occurring in real time, the virtual gel image showed much more linear fragment separation. This is because at the point of detection all fragments have run the same distance, large fragments are not still compressed and unresolvable at the top of the gel, nor does it matter that small ones have run off, as the information they contained has already been collected. This facilitated easier automated analysis and much longer read lengths per run. Whilst this was a huge improvement over the fully manual procedure, some of the old drawbacks still remained. First, repeatedly pouring high quality polyacrylamide gels was tedious, laborious and an 'acquired art'. Also, the gels needed an hour or so to set, and in between uses, the large, expensive and fragile plates needed to be cleaned scrupulously and kept free from scratches or other damage. Secondly, after the gel had been run and the virtual gel image collected, a process known as 'tracking' had to be performed. This involved positioning markers for each sample lane at numerous intervals along the length of the gel (see Figure 15.3); without

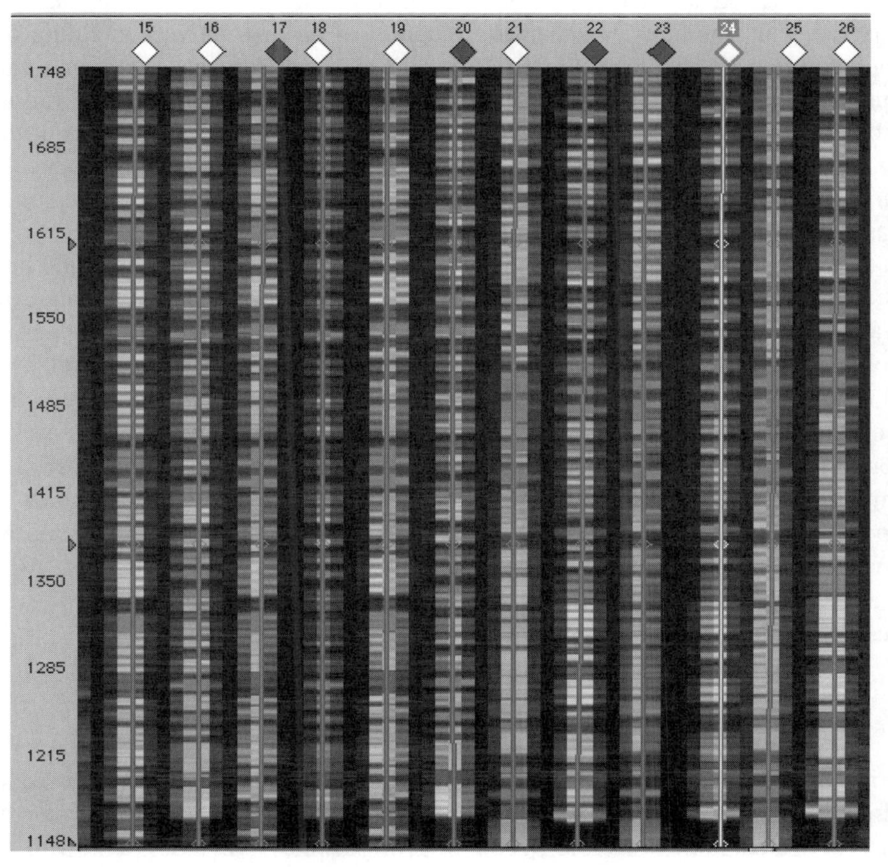

Figure 15.3 Part of a virtual gel image acquired on a 'slab gel' automated DNA sequencer. Gel lane numbers are shown at the top. The Y axis represents the cumulative number of real-time fluorescence excitation/data collection scans across the width of the gel. Note the thin grey lines, with tracking markers, running down the centre of each sample lane (see 'Equipment' section of main text). A colour version of this figure appears in the colour plate section

it the software would not know exactly where to extract the fluorescence data for that sample, and hence be unable to analyse it. Successive algorithms were written in an attempt to automate this task, but at best the results needed manual checking and refinement; at worst, the process had to be completed entirely by hand, dragging and dropping what could be hundreds of marker points with a mouse. The current generation of sequencers have done away with both these problems. Gels have been replaced with thin liquid polymer-filled capillaries; between runs these can be purged of old polymer and refilled in minutes, using automated syringe drivers containing a reservoir of fresh polymer. Once filled, no time is required for 'setting' and the sample can be loaded immediately and automatically. As each specimen runs in a separate capillary, tracking is irrelevant since all the fluorescence data collected from the small optical window towards the end of each capillary must have come from the sample that was injected. For simplicity, a schematic diagram of an instrument containing one such capillary is shown in Figure 15.4. However, instruments containing various numbers of capillaries, from 1 to 96, are currently available to suit varying laboratories depending on their throughput requirements. In combination with robots and other equipment, the larger of these machines can run around the clock, requiring the minimum of human intervention and producing the sort of bases per day output mentioned previously.

It is important to remember that all the advances described above would have been of little use without complementary advances in computing power and software. Ever faster computers have been required for the analysis of the increasing volumes of data. Software algorithms are better at dealing with spectral overlap and gel mobility shifts caused by the use of large fluorophore moieties. This has resulted in much lower base calling error rates, which are essential for large-scale projects where manually checking more than a tiny fraction is impossible. It must also be remembered that accurate base calling is not the end of the line; the downstream storage and manipulation of ever-increasing amounts of sequence data presented many new computational challenges. All of these have had to be overcome to enable the successful completion of large-scale genome-wide sequencing projects.

Sequencing Templates and their Preparation

Prior to the advent of the PCR, all sequencing templates were produced by cloning. The DNA to be sequenced was cut with a restriction enzyme and pieces ligated into a suitable vector (e.g. a plasmid) previously cut with the same restriction enzyme. In a process known as transformation, bacteria exposed to an antibiotic could be induced to take up the modified plasmid which also contained a gene for antibiotic resistance. Colonies derived from individual bacteria could then be grown in culture, producing many millions of copies of one original modified plasmid molecule. These cultures were harvested and

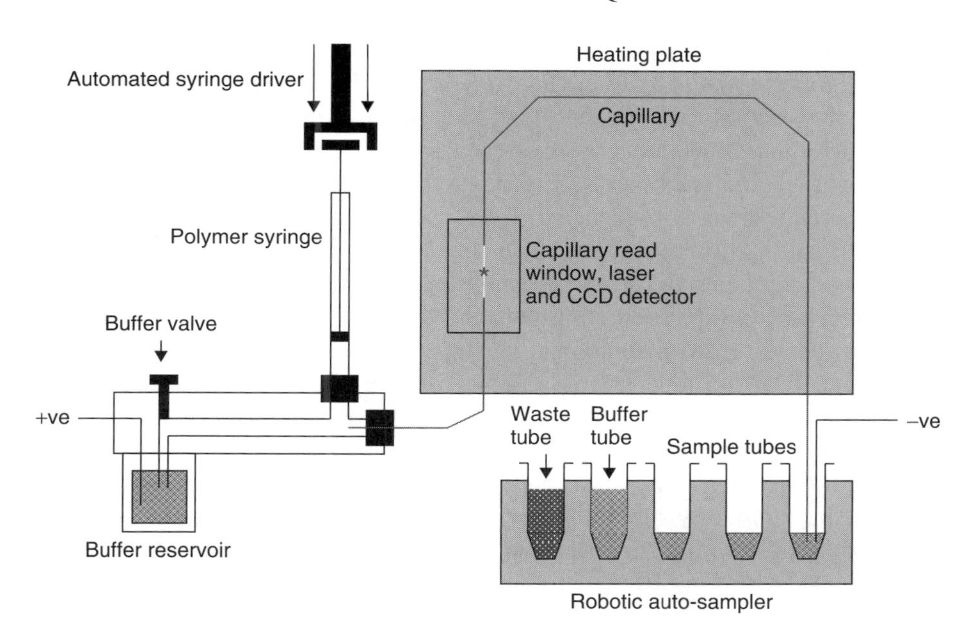

Figure 15.4 Schematic representation of a single capillary automated sequencer. Within a single sample analysis cycle the following events occur. (1) The buffer valve is depressed, preventing polymer from leaking back into the buffer reservoir. (2) The robotic auto-sampler moves such that the capillary/electrode are immersed in the waste tube solution. (3) The automated syringe driver slowly depresses the polymer syringe plunger. This forces fresh polymer into the capillary displacing any old polymer into the waste tube. (4) The robotic auto-sampler moves such that the capillary/electrode are immersed within the first sample. (5) The buffer valve is reopened, and current is applied for a short period. This draws a small aliquot of sample into the capillary (analogous to loading a well on a standard gel). (6) The robotic auto-sampler moves such that the capillary/electrode are immersed within the buffer tube. A current is then applied for 30 min or so, until all the separated DNA fragments have passed the capillary read window. During this time the laser is on, in order to excite the fluorescent dye moieties, and the CCD detector is collecting fluorescence data for subsequent analysis. (7) After data collection is complete, current is discontinued and the capillary/electrode are returned to the waste tube to begin the cycle again with the next sample

the plasmid DNA extracted/purified using standard techniques. This technology was very successful, as even with the relatively large amounts of template DNA required for single round manual sequencing, enough DNA for many reactions could be produced from a single plasmid preparation. Secondly, as the DNA sequence of the vector either side of the inserted fragment was known, a set of universal sequencing primers could be used to sequence several hundred bases into any insert from either end. This feature overcame the problem of initially not knowing any of the template sequence and hence being unable to make the essential first primer.

Even with the advent of the PCR and automated fluorescence-based sequencing, cloning of PCR products prior to sequencing remained popular. Again, this was partly because large amounts of template could be produced easily and cheaply for use in manual sequencing, which was still the only technology available in many laboratories. Furthermore, even though enough template could be generated by the PCR alone, the uncloned fragments were usually quite small and tended to recombine quickly. This made them difficult templates for use with low-temperature manual sequencing. Even in laboratories using early automated sequencers and high-temperature cycle sequencing, dye–primer chemistry was the only one available. Since fluorescently tagged primers were quite expensive, but cloning was cheap, continuing with the use a single universal vector-based primer set could result in significant savings.

Following the introduction of reliable dye–terminator chemistry and with increased access to automated sequencers, most PCR products are now sequenced directly using cycle sequencing. Only a small amount of PCR product is required. It must however be 'cleaned' prior to use as a template. This normally involves the temporary binding of product DNA to a matrix whilst unincorporated deoxynucleotide triphosphates and primers left over from the PCR are washed away. Using a different solution, the cleaned DNA can then be eluted ready for use. Without this step, the leftover nucleotide triphosphates would disturb the finely balanced ratio of deoxy- and dideoxynucleotide triphosphates in the sequencing reaction. Furthermore, the leftover primers would result in competing sequencing reactions starting at several different points, making any meaningful analysis impossible.

15.3 Applications of DNA sequencing

A principal and obvious application of DNA sequencing is the initial characterisation of specific genes, and ultimately whole genomes. The rationale for such *de novo* sequencing is hopefully self-evident, and is discussed with regard to the HGP in the next section. There are, however, many other applications for DNA sequencing which utilise previously characterised targets. Whilst these are too numerous to discuss in full, those described below are intended to illustrate how access to rapid, reliable and affordable DNA sequencing technology can play an important role in disease investigation and diagnosis.

Assay Validation

Many modern diagnostic assays use the PCR to generate what is hopefully a target-specific amplicon. During the validation of such assays, especially those where no further analysis of the PCR product is performed, it is essential to verify that the amplicon has indeed come from the intended target. A common example is pathogen detection in infectious disease diagnosis; PCR primers are designed to amplify specific sequences only from the pathogen in question. The simple presence or absence of a PCR product may be taken as diagnostic for the presence or absence of pathogen within a specimen. During validation, in addition to suitable positive and negative controls, it is essential to repeat this test with a range of samples containing other potential non-specific targets, e.g. other similar pathogens and human genomic DNA. Any PCR products can then be sequenced to ensure their specificity for the intended organism. Without such caution, false positive results may be produced. Even when a test is in routine use, there remains the option to sequence PCR products generated from any atypical specimens as an added safeguard.

Target Identification

Often with cases of suspected infectious disease it can be difficult to decide which specific organisms to test for, as many can produce very similar symptoms. However, with certain bacterial infections, for example, it is possible to use PCR primers that anneal to regions that are highly conserved across all bacterial species, such as the 16S ribosomal RNA gene. Fortunately, these regions flank areas displaying much less conservation, and although the PCR reaction detects the presence of any bacterial species, subsequent sequencing of the PCR product can in most cases provide enough information to unambiguously identify the infecting species. Such information can be essential when deciding on the most appropriate antibiotic to use.

Even when testing for a single type of organism, different strains or subtypes may exist. These may exhibit different degrees of pathogenicity or require different treatment regimes. Although in many cases it may be possible or even desirable to design individual subtype-specific tests, for organisms possessing large numbers of subtypes this may be impracticable. An example is the human papilloma virus (HPV) family, where there are more than 70 known subtypes, yet certain ones are associated with a higher risk of cervical cancer than others. Again it is possible to use a combination of highly conserved and very variable regions of the HPV genome to design a generic PCR/sequencing strategy. This is capable of both detecting and typing HPV infection. For identification by comparison, assays such as the two described above clearly require access to a database of sequences generated from known species or subtypes. They also require access to software capable of rapidly performing many successive pair-wise comparisons, a virtually impossible task to complete manually! Whilst many institutions develop their own specialist databases and software, a common means for many is to perform a 'BLAST' search of 'GENBANK'. This can be done rapidly and for free on the Internet (see useful web sites at the end of this chapter). BLAST stands for Basic Local Alignment Search Tool, and GENBANK is a database which contains the sum total of all publicly available nucleotide sequences submitted to date. This software and the powerful computer servers on which it runs are capable of comparing a submitted sequence to the millions on the database and returning results to the user within seconds. Although by definition 'basic', BLAST searches are usually sufficient in most cases; nevertheless, other more complex (and hence slower!) sequence comparison packages are also freely available.

Mutation Detection

Many small-scale mutations, and in particular single nucleotide polymorphisms (SNPs, pronounced 'snips'), are responsible for a range of human genetic diseases, or can account for the emergence of viral drug resistance. There are many techniques, including sequencing, for screening DNA to identify new mutations. However, most only indicate whether a mutation is present or not, and subsequent sequencing is still required to determine the exact nature of the mutation. This makes sequencing an essential tool when, for example, trying to identify and characterise a novel mutation in BRCA1, the gene responsible for breast cancer in certain families.

Once an important mutation has been characterised, often there is a need to determine routinely whether a sample contains the wild type and/or mutant variants (known as 'mutation calling', or 'SNP typing'). Many methods which are quicker and more cost effective than resequencing now exist. Nevertheless, sequencing remains essential in the validation of assays using such alternative techniques.

15.4 The Human Genome Project

Why?

Why sequence the human genome? Two perfectly reasonable responses to this question would be 'why not?' and 'because we can'. After all, these have been the justification for many other expensive large-scale projects, such as putting a man on the moon. Indeed, the simple quest for greater knowledge is at the heart of human nature, and was probably a great motivator for the individuals whose foresight and determination during the mid-1980s got the HGP off the ground. There were of course a number of somewhat more reasoned arguments. The main one was that the availability of a complete genomic sequence for humans, and a number of other model organisms, could be expected to provide an invaluable scientific resource for advancing basic and biomedical research, and that in the longer term this would have great benefits in understanding and treating disease. It was further argued that carefully coordinating this in a single large-scale project would be the quickest and most efficient way to proceed. In addition to cost and ethical concerns, which are discussed below, the main argument against sequencing all of the genome was that most of it was regarded as 'junk'. A simple response to this could be: 'Well until we have sequenced and analysed it, how do we really know whether it's junk or not?'

That the genome project is now beginning to fulfil its expectations is undoubted. Within days of tracking down potential disease genes to specific genomic regions, individual research groups are now able to download the entire sequence of that area, interrogate it with software to locate what already known or likely additional genes are there and compare these with the genomes of other organisms, immediately highlighting important common or organism-specific genes and giving clues to their function. Finally, they can check polymorphism databases for known variants that may be the source of, or at least very close to the site of, disease susceptibility. In practice, of course, it is not quite this simple. At the end of the day someone has to go back into the laboratory and do some hands-on work to test the results of these computer-generated analyses. Nevertheless, that individual research groups are being saved years of painstaking basic research characterising their particular genomic regions of interest is undisputed.

With hindsight, many other benefits of the HGP are becoming apparent. Not least of these is that without the concerted public effort we could have been in the unfortunate situation of having had large bits of the genome patented, with the rights to use them in diagnosis and treatment in the hands of private companies. Thankfully, finding out whether such patents, on what is regarded by many as the shared property of all humankind, would have held up in court is something that has largely been avoided.

Tackling the Genome

When attempting to sequence the genome of any organism, and especially the three billion bases of the human genome, it is neither possible nor practical to simply start at one end and plod on and on towards the other. As we have seen in the previous sections, any DNA to be sequenced has to be in small enough chunks for it to be amplified either by the PCR or by cloning into a high copy number vector such as a plasmid. The sequencing of these relatively small chunks may be repetitive, laborious and time consuming. It is, however, not that difficult. Instead, the challenge lies with correct reassembly of the original full length sequence from the multitude of little bits. This task is made especially difficult for the human genome because of the large number of repetitive sequences it contains (~50% of the whole genome). In light of these problems, the HGP consortium decided to adopt a so-called 'map based' or 'hierarchical shotgun' approach to genome sequencing (see Figure 15.5). This involved committing large amounts of time and money into generating a high resolution physical map of the genome, long before starting any large-scale sequencing of it. First, purified genomic DNA representing many individual copies of the genome was cut into large overlapping pieces (~100 000–200 000 base pairs long). This was achieved by only allowing partial digestion with restriction enzymes whose recognition sites occurred quite regularly. These large

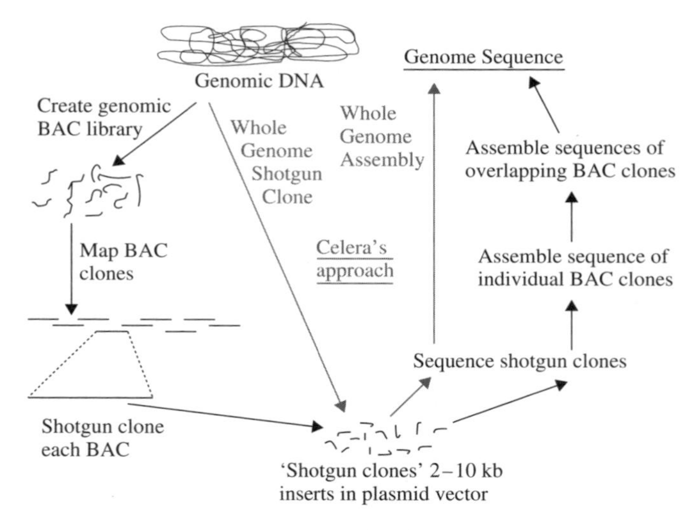

Figure 15.5 The 'map based' or 'hierarchical shotgun' genome sequencing strategy adopted by the HGP consortium. The more direct 'whole genome shotgun' approach adopted by Celera Genomics is shown in red

fragments were then cloned using a special type of vector called a bacterial artificial chromosome (BAC). A big enough collection of these BAC clones (about 20 000) is sufficient to represent the entire genome in overlapping fragments, and is known as a BAC library. Individual BAC clones were then fully digested using additional restriction enzymes to produce a unique pattern or 'fingerprint'. Overlapping clones could be identified by similarities within parts of their fingerprints, and hence ordered with respect to each other. It was also possible to identify which part of which chromosome most BACs had come from by looking for the presence of known genomic landmarks.

Using a higher resolution iteration of the above process, once mapped, each BAC clone was cut into many smaller overlapping fragments (~2000 base pairs long), and these in turn were cloned using a plasmid vector. Each plasmid subclone, also known as a 'shotgun clone' (since it represented a random piece of the original BAC which had been 'blasted' to bits), was then prepared and sequenced from either end. 'Contigs' (pieces of contiguous sequence) could then be constructed using a computer to find the overlaps in data from hundreds of these sequencing runs. As more and more subclones were sequenced, the length of each contig grew until eventually they all merged to represent the complete sequence contained within the original BAC. As more and more BACs were sequenced they also could be overlapped, continuously increasing contig lengths, first to yield the sequence of whole chromosomes and ultimately the whole genome.

Of course in real life things are rarely as simple as they sound, and truly unbroken full length contigs for every chromosome were never really expected.

Owing to the abundance of repetitive sequences, certain regions have proved almost impossible to clone; others, once cloned, have been very hard to sequence. Then there is the 8% or so of the genome which is comprised of heterochromatin. Heterochromatic regions have a compact structure, rich in repeats and very poor in genes, hence their sequencing is both difficult and of low priority. It is for these reasons that the working draft sequence published in June 2000 contained many gaps and ambiguities, representing as it did, between four- and five-fold coverage of about 94% of the genome. The aim of the HGP over the next few years is to produce a gap-free sequence for the entire euchromatic portion of the genome, finished to an accuracy of 99.99%. Although all of the routine sequencing is automated, and assembly is done by computer, many of the final gaps and ambiguities may have to be identified and dealt with manually. Also, even with modern highly sophisticated base calling algorithms, achieving this degree of accuracy will require every single base in the genome to be read an average of nine times.

Celera Genomics, who started their quest to sequence the genome much later than the HGP consortium, used a different strategy, in an attempt to complete the task much more quickly. This was called 'whole-genome shotgunning', and involved cutting the genome into relatively small pieces (2000–10 000 base pairs long) which were cloned directly into plasmids and sequenced at random (see Figure 15.5). Of course this made the subsequent challenge of assembly much more difficult. However, Celera's late start afforded them several advantages. First, they had free access to all the HGP consortium's currently available data, which served to provide them with a reference framework. Secondly, computing power had increased so dramatically since the HGP started, that the previously impossible task of assembling the whole genome entirely *in-silco* had finally become a realistic possibility. Even so, Celera had to specially commission the supercomputers capable of handling over 80 terabytes of data and performing the five hundred million trillion sequence comparisons required for the initial assembly. Although it is debatable whether Celera's 'whole-genome shotgunning' strategy could have worked without access to the HGP data, it was certainly quick, producing a working draft sequence within three years.

One relatively unimportant, but none the less interesting question that often arises with regard to both the HGP's and Celera's genome sequencing strategies is: Whose genome did they sequence? Well in both cases the answer is that no one knows. Both groups selected a small number of samples at random from a larger group of volunteers. With regard to the HGP, all sample identifiers were removed prior to selection, although with Celera it is known that their final five individuals comprised two men and three women from Hispanic, Asian, African–American and Caucasian backgrounds. All this may seem a bit too arbitrary, but it really did not matter. As we shall see below, the thing that has become very clear from the genome project and other work is that at the DNA level we are all remarkably similar.

History of the Human Genome Project

Scientists first began to discuss the possibility of sequencing the human genome in the mid-1980s. Initially, the prospect was not exactly greeted with universal enthusiasm. Some thought the task impossible, others wondered why anyone would want the complete sequence or what use it could be (since it was already known that most of the genome did not encode proteins and hence was regarded as 'junk'). The main concerns, however, focused upon ethical issues and cost. Many scientists worried that so much money would be required for the genome project that there would be none left for any other research interests. These problems were overcome in 1988 with the US congress agreeing to allocate entirely separate budgets to the Department of Energy and National Institutes of Health, with a dedicated fraction devoted to considering the ethical, legal and social issues arising from the undertaking. As such the quest for the genome had officially begun.

By 1990 other centres around the world had joined in and the HGP consortium launched its 15-year plan for producing the genome sequence at an estimated cost of US$3 billion. Over the next five years a great deal of work was performed to generate the necessary high resolution genetic and physical maps. Smaller scale projects such as sequencing the yeast genome were also undertaken as feasibility studies. By 1995 a pilot study to sequence 15% of the genome had begun, the successful completion of which in 1999 marked the move up to full-scale production. This full-scale production was even higher than had originally been intended because in 1998 Craig Venter, who had previously worked at the National Institutes of Health and The Institute for Genomic Research (TIGR), formed Celera Genomics and announced that it would sequence the human genome within three years (considerably ahead of the HGP's anticipated finish date) using a whole genome shotgunning strategy. Despite some doubts over the feasibility of his strategy, Venter's track record in genome sequencing was impressive enough for his claim to be taken very seriously. On top of the obvious disappointment for the HGP at the prospect of being beaten after so much hard work, Celera's intentions raised great concerns regarding the possible patenting of human DNA sequences. Fortunately, the bodies funding the HGP, which in addition to the US congress now also included the UK's Wellcome Trust and Medical Research Council, shared these concerns and very soon agreed to significant budget increases. Although in 1998 the HGP consortium comprised more than 2000 scientists working at 20 institutions in six countries, by early 1999 the urgent drive for increased productivity led to increasing amounts of sequencing taking place in five huge facilities known as the 'G5'. These included the Sanger Centre in England, the Whitehead Institute's Center for Genome Research in Massachusetts, Baylor College of Medicine in Texas, Washington University Genome Sequencing Center in Missouri and the US Department of Energy Joint Genome Institute in California. These increased efforts paid off and the HGP consortium actually completed some 80% of the entire genome sequence in only 15 months, finally releasing their working draft at the same time as Celera. Unlike the HGP, Celera's draft sequence is only available free of charge to registered academic/non-profit users. Furthermore, all 'finished' data resulting from their continuing work towards a completed version is only accessible to paying customers.

Highlights of the Human Genome

Having covered how the genome sequence was produced, what have we learned from it so far? In short, a great deal more than can be discussed here in any detail. Consequently, this section is only intended to cover some of the highlights that have emerged to date. For more detailed information, readers are referred to the special Genome issues of *Nature* and *Science* published in February 2001.

It is now clear that the genome is a far more heterogeneous place than was first expected. Genes as well as some of the repeat sequences that make up so much of our genome are scattered very unevenly, leading to dense clusters and vast seemingly empty expanses. All this is likely to be a reflection of our genome's complex evolutionary history, representing repeated parasitisation and duplication events.

The total number of human genes is far lower than expected, with the consensus currently lying at around 30 000. Of course, most of these have been identified by software which looks for common features present in known genes; thus it is probable that we are missing additional genes that do not adhere to these apparent norms. Nevertheless, the final gene count will be nothing like the 100 000 or more that was originally anticipated. Instead of not having many times more genes than far simpler organisms (about twice that of a worm or fly), is now clear that the genes we do have are more complex. This complexity permits much finer control over gene expression and the use of alternative splicing, meaning that a number of protein products can result from a single gene. Humans and vertebrates in general also seem to have utilised pre-existing protein motifs in more diverse combinations than the invertebrates, again providing for greater complexity.

Analysis of the repetitive sequences comprising about half our genome has provided a unique glimpse into our evolutionary history, via a fossil record dating back hundreds of millions of years. It is clear from this that whilst these repetitions initially arose from the actions of various transposable elements (unstable sequences capable of undergoing replication followed by insertion elsewhere in the genome), their activity has declined dramatically since the divergence of the hominid line. This is in contrast to the mouse genome where many elements are still actively transposing. Why this should be, no one really knows, although it has been suggested that it may reflect the mouse's much shorter generation time.

Another feature of repeat sequences in the human genome is that they seem to be remarkably well tolerated. There are of course some regions which are poor in repeats, suggesting that insertions here may have highly deleterious effects but, on the whole, they seem to be found just about everywhere in one form or another. Whether all repeats can justifiably still be considered as 'junk DNA' as was originally thought, seems increasingly unlikely. This is because just as there are repeat-poor regions, there are other functional regions where repeats seem to be retained preferentially. The most notable example is the clustering of so-called 'Alu' repeats in gene-dense regions, suggesting that these otherwise 'selfish' elements may have a favourable influence on gene expression or evolution. Even repeats and indeed unique sequences (which are probably derived from ancient repeats that have decayed so much as to be unrecognisable) for which there is no evidence of a direct function, may in fact have served a purpose simply by being there. One possibility is that by making ancient genomes much bigger, these were better able to tolerate the rigorous evolutionary events that have ultimately resulted in ourselves and other complex creatures. If this is true, thinking of any of our genome as 'junk' may be a bit harsh.

Although the genome has proved to be very heterogeneous, it is now clear that the entire human population is remarkably homogeneous. On average, comparing a 1000 base-pair stretch of sequence between a group of individuals reveals about one base-pair difference, meaning that we are all 99.9% identical. Of course, this corresponds to there being about three million SNPs across the entire genome, which may seem like an awful lot of scope for inter-individual variation. However, as we have seen, most of the genome is not directly functional; in fact, only about 1.5% actually encodes proteins (see Figure 15.6). Furthermore, like the other heterogeneous features of the genome, SNP density varies considerably. Taking these and other factors into account, probably less than 1% of all SNPs actually affect protein function. Thus, by excluding non-significant variation we can essentially think of ourselves as 99.999% alike. This exceptional homogeneity, which is not seen in other species, is probably a result of population bottlenecks that have occurred during periods in our history, and clearly eliminates any justification for racial or ethnic discrimination at the DNA level.

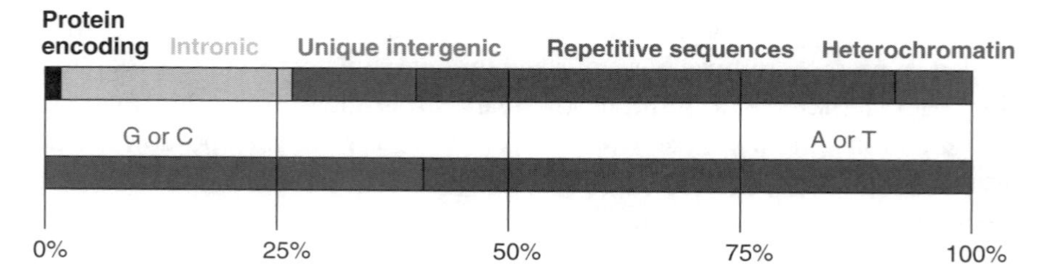

Figure 15.6 Schematic representation of the human genome, showing the relative proportions of different classes of nucleotide sequence components. A colour version of this figure appears in the colour plate section

Unlike racial differences, sex differences have proved far more interesting. Before the evolution of sex differentiation in mammals, the X and Y chromosomes were a matched pair. Now, the Y chromosome is a mere stump containing a smaller number of genes involved in male-specific traits. The genome project has revealed that most of these 'maleness' genes are duplicated many times and that some can vary in their orientation between different men. The former may represent a sort of biological insurance policy to ensure that these genes are not lost completely, and both could have important implications for research in male fertility and virility. Returning to the whole genome, it is now known that the mutation rate in male reproductive cells is twice that of females. This is possibly because the cells that divide to produce sperm do so continuously, with the opportunity to accumulate new mutations each time, whereas a woman's entire complement of eggs is produced in a single round of cell division.

To conclude, access to the genome sequences of a range of species, including our own, provides scientists with an invaluable resource for research. It has already provided some fascinating insights into the very essence of what it means to be human, as well as benefits to our understanding of health and disease. However, discoveries to date are likely to represent only the tip of the iceberg. There will be much more to come before the story of our genome is complete.

15.5 References

Horvath SJ, Firca JR, Hunkapiller T *et al.* (1987) An automated DNA synthesizer employing deoxynucleoside 3′-phosphoramidites. *Methods Enzymol.* **154**: 314–26.

Lander ES, Linton LM, Birren B *et al.* (2001) Initial sequencing and analysis of the human genome. *Nature* **409**(6822): 860–921.

Maxam AM and Gilbert W (1980) Sequencing end-labeled DNA with base-specific chemical cleavages. *Methods Enzymol.* **65**(1): 499–560.

Prober JM, Trainor GL, Dam RJ *et al.* (1987) A system for rapid DNA sequencing with fluorescent chain-terminating dideoxynucleotides. *Science* **238**(4825): 336–41.

Saiki RK, Scharf S, Faloona F *et al.* (1985) Enzymatic amplification of beta-globin genomic sequences and restriction site analysis for diagnosis of sickle cell anemia. *Science* **230**(4732): 1350–4.

Sanger F and Coulson AR (1975) A rapid method for determining sequences in DNA by primed synthesis with DNA polymerase. *J. Mol. Biol.* **94**(3): 441–8.

Smith LM, Fung S, Hunkapiller MW *et al.* (1985) The synthesis of oligonucleotides containing an aliphatic amino group at the 5′ terminus: synthesis of fluorescent DNA primers for use in DNA sequence analysis. *Nucleic Acids Res.* **13**(7): 2399–412.

Venter JC, Adams MD, Myers EW *et al.* (2001) The sequence of the human genome. *Science* **291**(5507): 1304–51.

15.6 Further reading

Dennis C and Gallagher R (eds) (2001) *The Human Genome*. Palgrave, Basingstoke, UK.

15.7 Useful websites

Human Genome Project information web site: www.ornl.gov/hgmis/

National Centre for Molecular Biology Information (NCBI): www.ncbi.nlm.nih.gov/

Access to: GenBank, OMIM, PubMed, BLAST, SNP database, Genome Sequencer Browser, etc.

The Sanger Centre: www.sanger.ac.uk

The Whitehead Institute's Center for Genome Research: www-genome.wi.mit.edu/genome.html

Baylor College of Medicine Human Genome Sequencing Center: www.hgsc.bcm.tmc.edu/

Washington University Genome Sequencing Center: www.genome.wustl.edu/gsc/

US Department of Energy Joint Genome Institute: www.jgi.doe.gov/

Celera's public access web site: http://public.celera.com

The SNP Consortium: http://snp.cshl.org/

16

Monoclonal Antibodies: The Generation and Application of 'Tools of the Trade' Within Biomedical Science

Paul N. Nelson, S. Jane Astley and **Philip Warren**

16.1 Introduction

Monoclonal antibodies serve as medical diagnostic and therapeutic reagents, as well as molecular probes to identify macromolecules and cells. In essence, monoclonal reagents have become 'tools of the trade' in many fields of biomedical science. So what makes a monoclonal antibody (MAb) so special? This is best summarised by the following definition: *a monoclonal antibody is a reagent of exquisite specificity derived from the immortalisation of a single B cell in vitro.* As a consequence, the potential exists to produce an inexhaustible supply of antibody that is monospecific and of reproducible quality and which readily enables the standardisation of both reagent and immunoassay technique.

In brief, MAbs can be produced by fusing ('hybridising') myeloma cells and splenic B cells derived from rodents, e.g. mice or rats, that have been immunised with an antigen (Kohler and Milstein, 1975). Immortalisation with a histocompatible myeloma cell line is necessary since isolated B lymphocytes possess a finite life-span in tissue culture. Alternative methods of generating antibody reagents are also evident, e.g. recombinant phage antibodies, human antibodies, as is the modification of clinically useful antibodies through genetic engineering. Overall, the production of MAbs with desired specificity and reactivity relies on

Molecular Biology in Cellular Pathology. Edited by John Crocker and Paul G. Murray
 ISBN: 0-470-84475-2

a number of key stages that include immunisation, fusion and screening (Nelson *et al.*, 2000). In addition, any novel monoclonal reagent must be thoroughly characterised and assessed 'fit for purpose' within a diagnostic or therapeutic setting. It is noteworthy that prior to developing antibodies in the United Kingdom (under a Home Office Project/Personal Licence) a thorough search must be undertaken to ascertain the validity of developing novel antibodies to a protein or key antigenic determinant. A number of web sites provided in Table 16.1 may prove useful. The specific technical details of generating monoclonal reagents are beyond the scope of this chapter but are available from current methodological publications (Hay *et al.*, 2002; Nelson, 2001). However, relevant background details within the stages of development are provided since these may influence the successful production and application of monoclonal reagents. In addition, this chapter focuses on recombinant reagents and applications. An overview of antibody production is schematically presented in Figure 16.1.

Table 16.1 Web sites of interest for monoclonal and polyclonal antibodies

www.Antibodyresource.Com
Provides sources (online or otherwise) on how to find antibodies or information on antibodies. Free online services are available (e.g. Abcam http://abcam.com/abcam.html)
http://www.mh-hannover.de/aktuelles/projekte/hlda7/hldabase/select.htm
Provides a free service of antibody data for CD antibodies, unclustered antibodies, other defined non-CD antibodies and alphabetical listings of all workshop antibodies relating to the 7th International Workshop on Human Leucocyte Differentiation Antigens
http://www.molbiol.ox.ac.uk/pathology/tig/tac.html
Provides details relating to the Therapeutic Antibody Centre (TAC), Oxford, UK
http://www.path.cam.ac.uk/~mrc7/mikeimages.html
General information on antibodies and reshaping antibodies (Mike Clarke's home page)
www.dako.com
Provides under *immunohistochemistry* information of new primary antibodies
http://serotec.oxi.net/asp/index.html
Provides under 'select category' *histology,* antibodies for immunocytochemistry
http://www.gallartinternet.com/mai/
The monoclonal antibody index is a biotechnology database with updated information on the most important monoclonal antibodies produced for the diagnosis and therapy of human cancer, diseases and infections. Licence fee required
http://www.mrc-cpe.cam.ac.uk/imt-doc/public/INTRO.html
V BASE provides a directory of human variable germline variable region sequences (database of human antibody genes) MRC Centre for Protein Engineering, Cambridge, UK

The table provides a selection of web sites that are available for antibody reagents. Prior to developing antibodies in the UK under a Home Office Project/Personal Licence, a thorough search must be undertaken to ascertain whether antibodies exist for the intended target protein or key antigenic determinants (epitopes). A number of the above web sites may prove useful

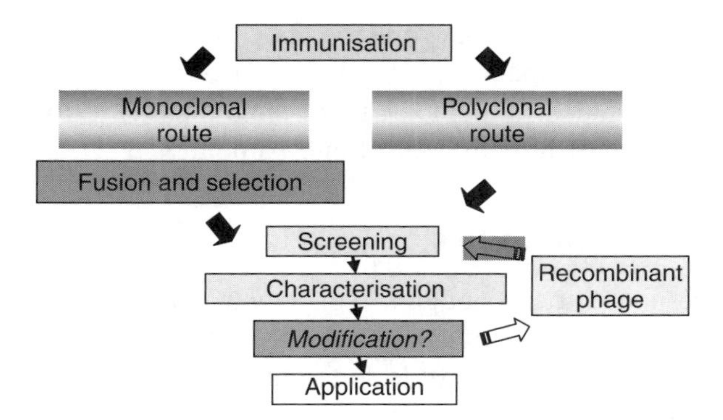

Figure 16.1 Schematic presentation of antibody development. A colour version of this figure appears in the colour plate section

16.2 Antibodies and antigens

In humans, the antibody representing 70% of circulating immunoglobulins is IgG. Electron microscope pictures identified IgG as a 'Y'-shaped structure of about 12 nm in length. Its two tips, denoted as Fab (fragment antigen binding) regions are crucial for interaction with antigen and its stem (denoted as Fc; fragment crystallisable) for biological effector functions. Additional immunoglobulin isotypes that differ in size, shape and biological role, e.g. IgM, IgA, IgD and IgE, constitute the total immunoglobulin pool. Similarly, murine (mouse) immunoglobulins possess a range of isotypes and immunoglobulin subclasses. Of relevance to the production and application of monoclonal reagents is the strength (affinity) of their interaction with antigen and the Fc recognition sites for *in vivo* (and hence therapeutic) biological functions such as antibody dependent complement mediated cytotoxicity (ADCMC) and antibody dependent cellular cytotoxicity (ADCC). For *in vitro* laboratory purposes the Fc region simply provides a docking area for many secondary antibodies that are employed in immunoassay systems.

An antigen provides a generic term relating to any substance, macromolecule, cell or tissue that is largely foreign to a host that may stimulate an immune response. In the 1970s emphasis centred on a limited number of antigenic regions (in the order of 3–5) on selected proteins, e.g. sperm whale myoglobin, hen egg-white lysozyme, although in the 1980s opinion shifted towards antigens possessing a series of overlapping antigenic determinants or epitopes (Lerner, 1982). Advances in the 1990s, especially in techniques such as epitope mapping, identified many major and minor antigenic determinants on proteins. Overall these studies emphasise epitopes as linear (i.e. dependent on the continuous sequence of amino acids) or conformational, composed of discontinuous residues brought into juxtaposition by the tertiary folding of a molecule. Thus antigens should be regarded as three-dimensional shapes with numerous linear or conformational sites to which antibodies might bind. In the development of monoclonal reagents only one epitope may be of paramount importance. These concepts are fundamental to the development and application of MAbs for immunotherapy and in immunoassay systems that serve to detect, monitor and quantify antigens of immunological and pathological interest.

16.3 Polyclonal antibodies

The immune response to an antigen, e.g. tetanus toxoid, is essentially polyclonal. Hence the antiserum reflects a heterogeneous population of antibodies with varying strengths of reactivity to individual target epitopes. For most purposes, animals such as sheep, goats and rabbits are used to generate antibodies for laboratory use. In general, polyclonal reagents are extremely valuable in situations requiring the agglutination of cells, e.g. in haemagglutination and the formation of insoluble antigen–antibody complexes, exemplified by techniques such as nephelometry, radial immunodiffusion and rocket immunoelectrophoresis (Figure 16.2). However, because of their heterogeneity, polyclonal antibodies do not lend themselves to investigations that determine fine structural and antigenic differences between molecules. In addition, polyclonal reagents may exhibit cross-reactivity by virtue of low-affinity antibodies contained within antisera. It is noteworthy that the response to a desired immunogen may represent only a fraction, e.g. <10%, of the antibody population, the remaining antibodies reflecting natural challenges to the host immune system. Of course variability between batches of antisera (even if raised in genetically identical animals) also provides a limitation, as does the finite amount of reagent produced. As a consequence new batches of antiserum need to be compared with old. Despite these limitations, polyclonal reagents are employed in a number

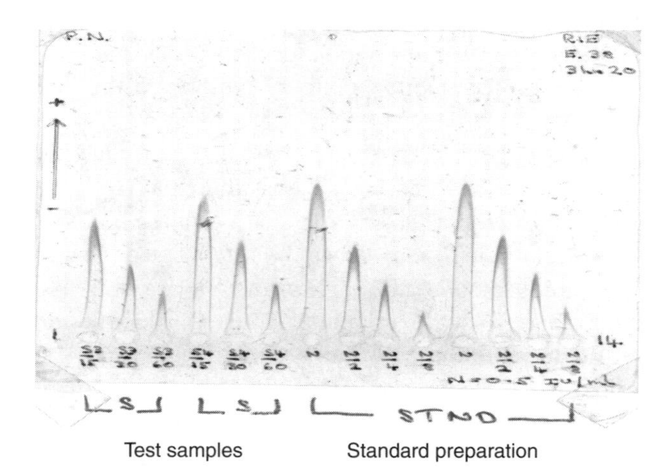

Figure 16.2 Rocket immunoelectrophoresis: determination of levels of growth hormone samples against a standard preparation using a polyclonal antisera. A colour version of this figure appears in the colour plate section

of clinical applications, e.g. blood group and human leucocyte antigen (HLA) typing and detection of micro-organisms. Furthermore, polyclonal antibodies have served therapeutically to neutralise viruses, e.g. rabies, hepatitis A, eliminate bacteria, and provide anti-rhesus D antibodies to postpartum rhesus-negative mothers. In the laboratory setting the availability of polyclonal (e.g. goat or sheep) anti-mouse horseradish peroxidase/alkaline phosphatase conjugates, or fluorescently labelled secondary antibodies, are fundamental to the revealing of primary bound MAbs in immunoassays such as immunocytochemistry (IC), enzyme linked immunosorbent assay (ELISA) and Western blotting.

16.4 Monoclonal antibody development

Immunisation

A contributing factor to the successful production of MAbs is the immunisation strategy. This includes the nature of the immunogen, dose, frequency, site of injection, the inclusion of adjuvant and the strain/species of animal. In any strategy the monitoring of progressive bleeds, e.g. from a mouse tail, is essential to ascertain the effectiveness of a given protocol. For BALB/c mice, native and recombinant proteins (50–100 μg/dose subcutaneous) in adjuvant such as Freunds Complete/incomplete or Titermax Gold are effective. Note that the toxicity of some macromolecules needs to be considered. Cells ($1 \times 10^{6-7}$ in normal saline) administered i.p. (intraperitoneal) provide a satisfactory immune response without the need for adjuvant. In circumstances where antigen is limited, a number of procedures have been adopted. One report (Shakil *et al.*, 2001) highlights the development of MAbs to quail neural crest using intrasplenic immunisation with only 2×10^3 cells. Synthetic peptides which mimic key epitopes also offer an alternative approach. However, the ability to raise monoclonal and polyclonal reagents that bind tenaciously to native/recombinant proteins is possibly overestimated since few authors report failures in generating antibodies to synthetic peptides.

The nuances of synthetic peptides are lack of immunogenicity and the need to covalently link to a carrier protein such as Keyhole limpet haemocyanin (KLH). This step facilitates a 'hapten-carrier response' ensuring antibodies are directed to both peptide and carrier. However, a study developing antibodies to an enzyme tartrate resistant acid phosphate (TRAP) showed that only 1 out of 3 potentially antigenic peptides generated an immune response (Nelson and Chambers, 1995). Thus a hit rate of 1 in 3 might suggest that future strategies ought always to employ a number of peptides from different regions of a protein. As an alternative to carrier-conjugated peptides, multiple antigenic peptides (MAPs) have been adopted in some immunisation protocols. Here a poly-lysine core is used to attach multiple copies of a single peptide that collectively provide a molecular mass capable of inducing an immune response. Overall the configuration of the MAP is important, with a branched chain being more effective for generating immune responses *in vivo* and no carrier protein is involved. Whilst MAPs have proved effective, a recent article (Waldron *et al.*, 2002) suggests that MAPs may indeed generate B-cell responses (ascertained from mouse tail bleeds) but may produce weakly reactive MAbs of IgM (μ) isotype. This situation may arise since MAPs carrying B-cell epitopes alone may behave as T independent antigens. As a result MAbs may be of low affinity and reflect an isotype that is affiliated with a primary immune response. Thus future strategies could employ CD40L or MAPs expressing both B- and T-cell epitopes. Evidently for applications within a pathology laboratory, antibodies raised using synthetic peptides must be able to detect target epitopes on fresh and/or formalin-fixed tissues.

Fusion

The production of the first MAbs relied on the ability of, Sendai virus to act as a fusinogen by encouraging the merging of cell membranes of activated murine B cells and an immortalising carcinoma cell line (Kohler and Milstein, 1975). The hybridomas produced were not homogeneous and the antibody profiles varied in affinity and yield. Similar outcomes have also been achieved using polyethylene glycol (PEG 1500) and further developments have attempted to improve the efficiency of the process. In this respect PEG 1450 is claimed to promote a five-fold increase in antibody yield. Many laboratories continue to use PEG and murine myeloma cell lines, e.g. SP2, NS1 and NSO, as a histocompatible fusion partner. In addition to the fusion itself, the viability of emerging hybridomas is of paramount importance. This phase falls on a selective culture medium called hypoxanthine, aminopterin and thymidine (HAT); the aminopterin eliminating unfused myeloma cells that would otherwise outgrow newly established hybridomas. After a number of weeks stable hybridomas are grown in medium supplemented with hypoxanthine and thymidine (HT medium).

The drive to produce highly selective therapeutic or diagnostic MAbs in high yield has also promoted the development of alternative protocols that have relevance for nuclear transfer technology. One interesting proposal is the use of electrofusion in zero gravity where mixing and sedimentation forces are effectively eliminated. Lower fusion currents, causing less membrane damage, can be selected, although fusion rates are reduced by a factor of 10^6 unless hypo-osmolar reaction conditions are chosen. In earth-bound laboratories, electrofusion is currently under development and shows a potential to overcome the problems of initiating and controlling the fusion of minimal cell numbers with a predicted outcome. When refined it could compete with the alternative approach of humanising murine antibodies, especially if the fusion is preceded by immunological selection and magnetic separation. Investigations using electrofusion enhancers (divalent cations or proteolytic enzymes) have also been performed as has the use of other fusion approaches, e.g. ultrasound and radiofrequency irradiation. In fusing antibody producing human B cells with an immortalising partner, a greater measure of control is required and can be achieved using Simian virus SV40 and limiting dilution. More recently, the advent of a human myeloma cell line that is HAT sensitive (Karpas *et al.*, 2001) will no doubt revolutionise the production of human hybridomas.

Screening

A successful fusion will generate numerous hybridomas: in the order of 100–200+ potential antibody producing cells from a single fusion of mouse B cells plated-out in six 96-well tissue culture plates (Waldron *et al.*, 2002). Important aspects of screening are to identify hybridoma growth (using an inverted microscope) and rapidly determine the reactivity of hybridoma supernatants before cells become fully confluent. Difficulties encountered are that not all hybridomas grow at the same rate and produce different amounts of antibody in culture supernatants (Nelson *et al.*, 2000). The culture medium made available from a 96-well plate for screening is approximately 100 μL, or 1 mL if colonies are transferred to 24-well plates. Clearly, the secret is not to 'expand' a hybridoma until there is sufficient evidence to merit its importance. Of significance is that any novel antibody that is intended for diagnostic purposes should be screened in the immunoassay system for which it is destined, e.g. immunocytochemistry using fresh and/or formalin-fixed tissue/cell preparations.

The process of screening in immunocytochemistry can be accelerated by the use of either multiple sections on single slides, or multiple tissue arrays (TMAs): the production of tissue arrays on slides from multiple tissue blocks (Rimm *et al.*, 2001). These systems, with built-in negative controls, afford rapid assessment by greatly reducing the number of slides and supernatant needed. This robust approach is set to become the new paradigm and provides an invaluable method for improving the efficiency of generating MAbs for cancer research (Wang *et al.*, 2002). Rapid screening can also be achieved by the judicious use of characterised cell lines to manufacture cell micro-arrays (CMAs). Such arrays may include a range of common cancer cell lines, e.g. carcinoma of breast, prostate and colon together with non-transformed ones, to provide a range of positive and negative targets (Hoos and Cordogardon, 2001). The use of TMAs and CMAs provides a considerable cost saving in terms of time and reagents. Naturally this is offset by the cost of commercially available TMA/CMA slides that can range from £80 to £100 depending on the number and type of tissues represented. However, the positive benefit in terms of rapid screening and the avoidance of ethical approval when searching for candidate tissues are enormous.

The criteria for ascertaining the 'positivity' of hybridoma supernatants are broad and vary from technique to technique. The use of HT medium may suffice for background reactivity but should be complemented by positive and negative antibody controls. These should be previously characterised antibody supernatants, i.e. of known reactivity, and preferably of matching isotype. In immunocytochemistry, positive controls may include human leucocyte antigens and cluster of differentiation (CD) markers, whilst a negative control could be a MAb of unrelated reactivity. Overall this approach validates the immunoassay system and enables a grading system of antibody reactivity, i.e. + (weak), ++ (moderate) and + + + (strong), for the comparison and selection of hybridomas. In ELISA systems that readily enable the screening of numerous hybridomas (i.e. 96-well format), antibody tail bleeds may be used as positive controls. Furthermore, the stringency of hybridoma selection can be selected by an arbitrary cut-off point (usually an optical density value) or one that represents a mean plus a couple of standard deviations above a negative reference, e.g. HT medium, or below the positive control. Ultimately the most rigorous of approaches pays dividends in preventing the expansion of unnecessary and poorly reactive hybridomas and the cost of culture medium and cryo-preservatives (not to mention wasted space in liquid nitrogen containers!).

Characterisation

Techniques employed for characterisation often merge with screening systems, although the latter generally are designed as a one-shot method of identifying worthy hybridomas. The successful application of MAbs stems from characterisation of the antibody itself, its potential applications and its target epitope. A given hybridoma is grown in tissue culture medium (ascites production is not readily permitted in the United Kingdom) to provide supernatant, i.e. >30 mL, for testing reactivity and to enable the cryo-preservation of cells. Isotype profiling of hybridoma culture supernatants defines the immunoglobulin class and subclass. For routine laboratory applications gamma murine isotypes, i.e. $\gamma1$, $\gamma2a$, $\gamma2b$ and $\gamma3$, may be preferable to μ isotypes that on repetitive freezing and thawing degrade, and lead to spurious results. For clinical therapy, knowledge of an MAb's isotype is essential for promoting a biological function such as ADCC or CMCC. Isotype profiling also helps to identify double clones (which need re-cloning, e.g. by limiting dilution) and to establish the most appropriate affinity purification method, e.g. Protein-G for γ isotypes.

The cross-reactivity of an antibody can limit applications and must be ascertained by investigating its reactivity profile, i.e. against different antigens/tissues. Ideally, purified MAbs at a fixed concentration (e.g. at 1–30 μg/mL) should be used since this removes a reason for variability often seen when using culture supernatants. Hybridomas, for example with initial immunocytochemistry staining characteristics, can again be assessed using TMAs of the type shown in Table 16.2. Unexpected results may also occur when characterising MAbs against multiple tissue sections/antigens that may be intriguing and offer other avenues of research. This is illustrated by MAbs PNQ312D and PNQ310A

Table 16.2

TMA tissue	Application
Range of tumours – different types	Assess cross reactivity
Range of tumour grade – same tumour type	Assess antigen expression
Range of normal tissues	Assess cross reactivity

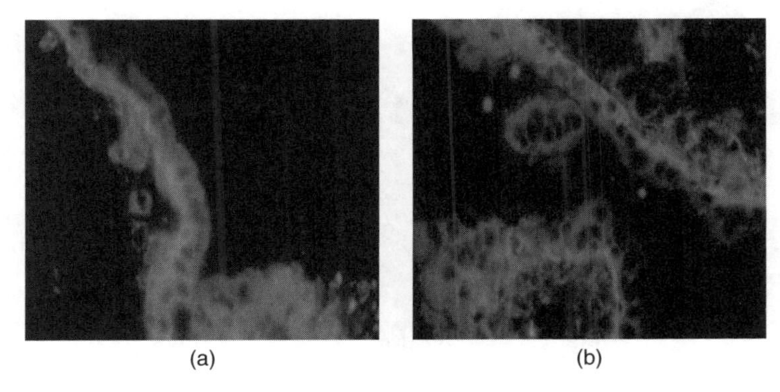

Figure 16.3 (a) MAb PNQ312D. (b) MAb PNQ310A. Both antibodies developed for ELISA purposes that bound weakly to an acid phosphatase. Interestingly, both MAbs exhibit strong reactivity to trophoblastic tissue. A colour version of this figure appears in the colour plate section

that were specifically developed for applications in ELISA (Nelson and Chambers, 1995) but exhibit reactivity in immunocytochemistry to trophoblastic tissue [Figure 16.3(a) and (b). Clearly, the potential for cross-reactivity and an antibody's role in biological systems must be scrutinised for therapeutic purposes. This is exemplified by the MAb PNG211D which identifies human osteoclasts but unfortunately cross-reacts with kidney tubules (Nelson *et al.*, 1991). Furthermore, in a bone resorption assay (employing scanning electron microscopy of bone slices to measure osteoclast pits), no significant inhibition of osteoclast function was observed.

Antibodies must also be characterised for assay restriction: a phenomenon where an MAb may function in one assay system but may prove poor or ineffective in another. Consequently this phenomenon may limit the potential applications of a novel reagent. In essence the knowledge gained from characterisation, including the intensity of reactivity, cross-reactivity and functional ability, are salient to the commercial exploitation of an MAb. More recently, epitope mapping (using a series of overlapping synthetic peptides) of MAbs enables the identification of amino acids that are salient to epitope recognition (Figure 16.4) (Nelson, 2001; Nelson *et al.*, 1997). Furthermore, surface plasmon resonance, e.g. BIACore analysis, enables the affinity of an antibody for its target antigen/epitope to be determined (Hay *et al.*, 2002). Hence MAbs of

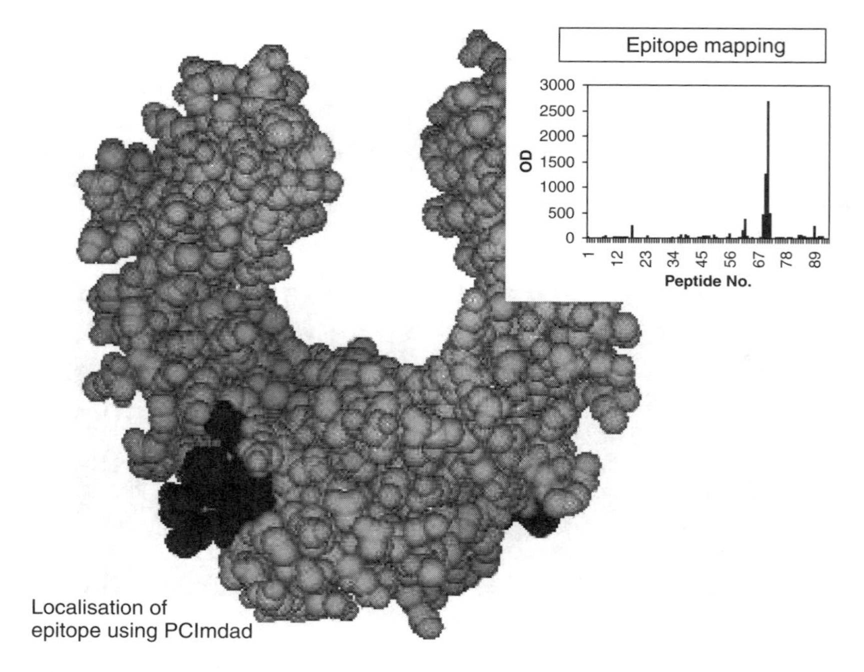

Figure 16.4 Molecular display of an epitope for MAb A57H using PCImdad (Molecular display and design, Molecular Applications Group). Insert shows the epitope mapping profile of the anti-IgG MAb A57H. After Nelson *et al.* (1997). Reproduced by permission of BMJ Publishing Group. A colour version of this figure appears in the colour plate section

high affinity may be selected in preference to low affinity (and possible cross-reactive) MAbs. It is imperative to stress that characterisation is an ongoing process and must go hand-in-hand with routine house-keeping duties. These include the screening of hybridomas for mycoplasma, e.g. by immunofluorescence or polymerase chain reaction, and freezing of multiple vials in different laboratories (and off-site facilities, e.g. the European Cell Culture Collection) in order to maintain healthy stocks of hybridomas.

16.5 Monoclonal antibody variants

Humanised Antibodies

Since MAbs are highly directional reagents, they have potential for *in-vivo* diagnostics and as therapeutic agents, especially when modified to carry reporter or cytotoxic ligands. However, a problem in using murine MAbs for clinical therapy is the inevitable production of human anti-mouse antibodies (HAMAs) that act to reduce antibody effectiveness and patient tolerance to treatment. Consequently 'humanisation' of many rodent MAbs has become the 'holy grail' of many immunologists. Unfortunately the satisfactory production of human hybridomas (except for a few, e.g. monoclonal anti-Rhesus antibodies) has remained elusive with low fusion rates, concerns over infection with viral fusinogens and poor cell stability after Epstein–Barr virus transformation/immortalisation. Clearly, the availability of a human HAT sensitive myeloma cell line will alter this situation (Karpas *et al.*, 2001).

Over a decade elapsed since the production of the first hybridoma and the manufacture of the first acceptable humanised antibody in the mid-1980s. This situation largely reflected technical difficulties and the level of safety testing required. In essence the goal of 'humanisation' has been to alter a murine or rat antibody through molecular engineering so as to include human elements and effectively reduce the immunogenicity to an acceptable level. Over the years a number of approaches have been adopted:

- *First generation/chimeric antibodies:* variable region murine genes are cloned into a mammalian expression vector that contains human heavy and light chain constant region genes. The human constant region component is selected for its intended application. In general this has provided mouse Fab and human Fc regions, respectively, although immunogenicity has still remaining an issue. To minimise anti-allotype responses, an approach has been to produce an immunoglobulin with the most common allotype found within the general population.
- *Second generation/hyperchimeric antibodies:* a further refinement uses the technique of complementarity determining region (CDR) grafting which incorporates the three murine CDRs from both the heavy and light chains into the DNA coding for the framework of the human antibody. This results in a functional antibody with a human content greater than a chimeric antibody which is also less immunogenic.
- *Hyperchimerisation/veneering:* additional refinement comes from matching both the murine and human components of the hyperchimeric antibody to their respective consensus sequences. Thus any mismatch, e.g. of framework residues, is corrected to improve MAb affinity. This is illustrated by therapeutic MAbs to respiratory syncytial virus (RSV19) that has been humanised using this method (Taylor *et al.*, 1991).
- *Humanised bispecific antibodies:* the idea of linking two antibodies that recognise different epitopes to generate a species with a dual specificity is also a possibility. Applications for this hybrid generally relate to a need to simultaneously target the immune response and a malignant tumour. An example of this approach is the combination of the HER2 receptor antibody and an anti-CD3 fragment that is used predominantly in the treatment of unresponsive and aggressive breast cancer (Shalaby and Shepard, 1992).

An illustration of the gradual refining of an antibody is the development of Campath anti-CD52 (Alemtuzumab) that is useful in treating inflammatory and auto-immune disease processes, modifying rejection after transplantation and controlling leukaemia:

1. Campath-1M Rat IgM
2. Campath-1G Rat IgG.2b
3. Campath-1H Human IgG.G1

Recombinant Phage Antibodies

Antibody development has entered new realms through the focus on recombinant phage technology, which has the potential to construct succinct antibody fragments of human origin and hopefully prevent sensitisation (Liang *et al.*, 2001). The technique also provides an alternative answer to the controversial use of animals for developing monoclonal reagents. The concept of phage display is to develop an epitope library that is applicable to antibody production, a discovery encountered during the mid-1980s following research into mimotopes. The idea was exploited and variable light and heavy genes were cloned and utilised as 'construction blocks' for single chain Fv (scFv) or Fab genes. These repertoires were transferred to bacteriophage genomes and finally the gene III protein, expressed on the tip of the phage (four copies per phage), manipulated to display peptide products on the phage surface.

In brief, a recombinant phage antibody (or scFv) is an antigen-binding protein, composed of variable heavy and variable light chains joined by a flexible peptide linker. Their use is advantageous owing to their small size (25 kDa) facilitating assimilation and tissue penetration *in vivo.* The smaller size of antibody fragments has advantages in cancer treatment since they can penetrate tumours effortlessly and will vacate healthy tissue much more promptly. Importantly the selective isolation of transformed phage particles, or phage libraries, can be used to infect the bacterium *Escherichia coli.* As an expression vector, *E. coli* offers ease of manipulation, fast growth and easy fermentation but is limited because of the lack of post-translational modification tasks (Liang *et al.*, 2001). Nevertheless, this vector can be applied to store human antibody libraries that have been synthesised by phages. Recent efforts have also demonstrated the benefits of using the insect expression vector system baculovirus (which enables glycosylation) for reproducing heavy and light genes of Fab or scFv antibody fragments. Such vectors may include light and heavy chain signal sequences, mutant in-frame cloning sites and human Fc regions that facilitate cloning of required genes. Recent studies also demonstrate the success of 'helper phages', also known as *hyperphages* that act to enhance the efficiency of infection.

Recombinant phage technology continues to develop and can be employed to modify or 'rescue' existing hybridomas or simply provide the means to select an appropriate binding specificity from a phage library. In modifying an MAb's binding characteristics, the reliance on an error-prone Taq polymerase (within the polymerase chain reaction) may suffice, although added benefit has come from using site-directed mutagenesis (Coulon *et al.*, 2000). MAbs in general are renowned for their assay-restriction (Nelson *et al.*, 2000) and is a significant area of concern for phage-derived antibodies. It is noteworthy that phage antibodies that are 'selected' via purified antigen-coated plastic perform efficiently in ELISA experiments but not so effectively in other assays, presumably because of epitope conformation. A solution to this problem is 'capture panning' the recombinant antibody using a target antigen that has formerly been bound by a polyclonal or monoclonal antibody. For applications in immunocytochemistry, the success of an scFv antibody is wholly reliant on the affinity for its target epitope. In comparison with intact IgG or $F(ab')_2$ fragments, only one combining site is available and thus a weak affinity scFv reagent will leach-off during washing procedures. Evidently phage display antibodies have proved useful research tools and are epitomised in the characterisation of autoimmune disorders such as type-1 diabetes, Primary Biliary Cirrhosis and SLE.

16.6 Monoclonal antibody applications

Diagnosis

The application of MAbs is ubiquitous in all areas of diagnostic pathology and biomedical research including Immunology, Serology, Microbiology, Clinical Chemistry and Cellular Pathology. Their modalities for use are manifold with formats as diverse as ELISA, chemiluminescence, latex agglutination, flow cytometry, confocal microscopy and the humble slide. A rapid scan of any supplier's catalogue (e.g. DAKO Ltd, Novocastra Laboratories Ltd, Serotec, The Binding Site) will reveal a continuing growth in the number of novel MAbs that have relevance to the improved diagnosis and management of patients with cancer, infection and autoimmune and inflammatory diseases.

In vivo

MAbs may be conjugated to a radioactive tracer that can be detected using conventional imaging techniques. The desired outcome is for an intense image corresponding to active binding at relevant antigenic sites without significant background noise. As described previously, consideration must be given to the humanisation of MAbs and the development of other antibody fragments/moieties. Because the latter are smaller than intact immunoglobulins, they are able to penetrate tissues more readily and are rapidly cleared from the patient.

In vitro

The choice of technique and monoclonal reagent needs to be made with the aim of improving the ability to detect early or residual disease more effectively and to predict the behaviour of transformed or infected cells so as to target treatment. Invariably MAbs are used in demonstrating the expression of abnormal proteins or other cell products during a disease process either within the patient, tissue or body fluid sample(s) taken for laboratory diagnosis. In all areas of diagnostic pathology the broad principle is to use an MAb's targeting ability to localise at a defined site a reporter molecule that can be used to ascertain the presence of bound antibody. Detection is enhanced if all steps of sample preparation are optimised.

Sample preparation

MAbs will react with tissues depending on whether they are fresh or have been fixed as part of specimen processing. As a general rule it is easier to produce antibodies that will react with fresh or frozen tissues especially when evaluating new MAbs. Subsequently frozen sections can be lightly fixed using a range of fixatives, e.g. methanol, acetone or formalin, before testing with an MAb to determine its fixation sensitivity.

Antigen retrieval

Specimens that have been fixed are less likely to bind optimally with antibody because of epitope modification through loss of tertiary protein structure. Many antibodies used at low dilution (or neat) may produce a weak localisation signal but can have their performance improved by employing antigen retrieval techniques. These methods have commonly used proteolytic enzymes, e.g. trypsin, chymotrypsin and streptokinase. More recently the introduction of heat-induced epitope retrieval (HIER) has improved epitope retrieval yields considerably (Shi *et al.*, 2001). The technique is simple and reliable, requiring sections to be heated by microwave, pressure cooker or both, and held briefly at an elevated temperature. The benefits include improved localisation and sensitivity together with a reduction in antibody consumption.

Detection systems

There are two main protocols: direct and indirect. The direct approach involves a single incubation step with MAb, and if the reporter molecule is fluorescent, e.g. fluoroisothiocyanate (FITC), then the preparation can be viewed immediately (after a buffer wash) using fluorescence microscopy. The direct approach is applicable to frozen tissues and also unfixed cell cultures that are often multiply labelled with up to three antibodies/fluorochromes prior to using confocal microscopy. A word of caution is necessary since fluorescent reporter molecules are labile and fade relatively quickly under UV light even if an anti-quenching agent is added to the cover-slipping medium. Thus, consideration needs to be given to some form of image capture or permanent localisation of signal using an anti-FITC antibody. The indirect labelling approach has tended to use polyclonal antibodies that are conjugation to enzymes, e.g. horseradish peroxidase (HRP) or alkaline phosphatase (AP). Another approach is the use of a second layer biotinylated anti-mouse antibody followed by a third layer streptavidin biotin complex (ABC)–HRP conjugate. The final localisation is with a substrate such as diaminobenzidene (DAB) that provides an insoluble reaction product. Multiple labelling is possible using different enzyme/reporter systems and the method has been further simplified by the use of a dual or triple universal anti-animal second layer reagent. Alkaline phosphatase has retained some popularity with cytological preparations although some quality may be lost with the use of aqueous mountants. Automated staining and the development of shorter incubation time reagents and methodologies have maintained the demand for increasing numbers of MAbs.

System controls

Modern methodologies have removed much of the unpredictability from immuno-staining, but strict protocols must be followed to ensure adequate internal quality control. Accredited laboratories are also required to take part in the National External Quality Assurance Scheme (NEQAS). The main errors to guard against are false positives and false negatives. This can be achieved by incorporating known positive controls and also an inappropriate tissue to act as a negative control. The methodology can also be checked by omitting the primary antibody or absorbing the primary antibody before incubation.

Specific Applications

Oncology

The early detection and identification of malignant transformation followed by the selective tailoring of treatment for patient sub-populations is central to the practice and development of Oncology. MAbs can be used in many ways and until recently would predominantly have their main role in detecting elevated serum levels of normal products, e.g. prostate-specific antigen (PSA), or presence of unusual proteins, e.g. carcinoembryonic antigen (CEA). Recent developments are in the application of cell cycle markers and determining the level of expression of oncogenes and tumour suppressor genes during tumour cell growth and development. It is recognised that the behaviour of any primary tumour needs to be predicted as in the case of secondary deposits that are often found in the liver in advanced metastatic disease, e.g. colon cancer. It has been demonstrated that a number of cellular proteins (cytoplasmic or nuclear) have significant predictive value in this area. These can be evaluated by using a panel of MAbs to indicate proliferation Ki67 and apoptosis p53, *bcl, fas* and the retinoblastoma protein Rb. Additional information on apoptosis is afforded by staining with MAb M30 which binds to the product of caspase action on cytokeratin 18. Use of these markers highlights that increased Ki67 expression and decreased M30 product and Rb expression are useful markers for the progression of colorectal cancer (Backus *et al.*, 2002). The technique of flow cytometry has also permitted the rapid and sensitive analysis of cell populations using fluorescent tags on MAbs. Of significance is the ability to demonstrate CD markers on stem cells and other circulating cells of myeloid or lymphoid origin. Recent advances have direct application to the study of lymphoproliferative disorders. Permeabilisation of cells prior to labelling allows the demonstration of Cyclin D1 in mantle cell lymphomas and p53 in chronic lymphocytic leukaemia (CLL). Furthermore, a MAb to CD38 is a new prognostic marker which is recommended as a predictor of reduced survival and an aggressive clinical course (Matutes, 2002).

Infection

Whilst traditionally detected by culture methods, an increasing number of infections are being detected by widening panels of MAbs. This is especially true for virus particles that can only be visualised directly through the use of electron microscopy at high magnifications but whose presence can now be demonstrated through serology/immunology or by direct demonstration in infected tissues. It is also possible to detect non-infectious (unassembled) virus particles associated with some cancers and inflammatory conditions, although MAbs to date have been used in blotting techniques rather than tissue preparations (Sauter *et al.*, 1995). Opportunistic infections are a challenge to diagnosis. MAbs to cytomegalovirus are routinely used to demonstrate infection and possible cause of rejection in immuno-suppressed transplant recipients. MAbs to human herpes virus type 8 (HHV8) also have applications in the diagnosis of Kaposi's sarcoma and angioimmunoblastic lymphadenopathies (Castleman's Disease). There is a significant role for MAbs in the detection of human papilloma virus (HPV) in gynaecological smears or colposcopic biopsies. The selective demonstration of viral subtypes HPV (types 16 and 18) is currently under further development with a view to the eventual automated screening of cytological material (Lin *et al.*, 2001). Emerging areas for MAb development or refinement include reagents for human endogenous retroviruses (HERVs) and acid phosphatases (associated with bone-related disorders) (Bull *et al.*, 2002). HERVs that resemble exogenous retroviruses such as human T-cell leukaemia virus and human immunodeficiency virus, have been implicated in certain autoimmune diseases and cancers (Nelson, 1995; Nelson *et al.*, 2002; Sauter *et al.*, 1995).

16.7 Therapy

Diseases can be treated through MAbs bound to drugs, toxins or radiolabels that are extremely efficient at selectively killing their target (usually tumour) cells with minimal co-lateral tissue damage. Furthermore, MAbs can be employed to interrupt cellular proliferation by binding with growth factor receptors. As an example Herceptin is a humanised recombinant MAb against the growth factor receptor p185/Her2neu that is used for chemotherapy-resistant breast carcinoma. A summary of MAbs and therapeutic applications is provided in Table 16.3 (also reviewed by Vaswani and Hamilton, 1998).

Table 16.3 Monoclonal reagents and therapeutic applications

Antibody	Target	Condition	Action	Side effects
Campath (Alemtuzumab)	**CD52**	Marrow transplant Vasculitis Leukaemia Multiple sclerosis	Immune suppression Anti-inflammatory Immune suppression Anti-inflammatory	'Flu-like' symptoms Diarrhea, arthralgia Cytokine syndrome Grave's Thyroiditis
Herceptin (Trastuzumab)	**HER2**	Breast cancer	Anti-cancer	No significant reaction Use with chemotherapy
Remicade (Infliximab)	**TNFa**	Crohn's disease Rheumatoid	Anti-inflammatory	Substantial benefit Well tolerated
Reopro (Abciximab)	**GPIIa/IIIb**	Acute M.I Stroke	Anti-thrombotic	Clotting complications Used + heparin/warfarin
Rituxan (Rituximab)	**CD20**	Non-Hodgkin's		Well tolerated
Synagis (Palivizumab)	**RSV**	Respiratory	Prophylactic, binds to RSV F protein	Safe for neonates Reacts as placebo
Zenapax (Daclizumab) **Simulect** (Basiliximab)	**CD25**	Transplantation	Il-2 antagonist Anti-rejection	Mild symptoms 'flu', Diarrhoea, arthralgia No cytokine syndrome
Orthoclone (OKT3)	**CD3**	Transplantation	Anti-rejection (T cell)	'Flu-like' symptoms Arthralgia

16.8 Specific applications

Oncology

The regulation of oncogenic viruses affords the potential to influence the progression of malignancies in certain organs, although an alternative approach is the suppression of oncogenic products. The availability of Herceptin (Trastuzumab) for the treatment of a selected subset of breast cancer patients is particularly relevant in this context. The strategy is unique in that screening for levels of the proto-oncogene Her2/neu (encoding a tyrosine kinase trans-membrane receptor) provides an indicator of those patients who would respond positively to antibody therapy. The development of MAbs for cancer (and transplant) treatment will continue to accelerate given the pressure exerted for the routine availability of these reagents. It is estimated that about a quarter of all therapeutic biotechnology drugs under development are MAbs, and that currently about 30 products are in regular use or being developed (Breedvelt, 2000). Overall, MAbs are viewed optimistically as ideal tumour targeting agents for cancer immunotherapy. Huls *et al.* (1999) claim that anti-tumour MAbs derived from phage particles can be of high affinity with active tumour killing in *in-vivo* and *in-vitro* assays. Targets such as epithelial cell adhesion molecules (Ep-Cam), frequently over-expressed in some tumours, are under rigorous scrutiny. Roovers *et al.* (2001) comment on the potential of phage technology as a means of modifying and developing Ep-cam antibodies for colon carcinoma therapy. In addition, epithelial glycoprotein 2 (EGP-2) is a prospective immuno-target for colon cancer and small cell lung carcinoma.

Infection, Immune Suppression and Anti-inflammatory MAbs

Research has focused on the application of recombinant MAbs to human cytomegalovirus (CMV) and also as potential vaccines for diseases such as Rabies as a substitute for horse serum. Phage technology features prominently in investigations encompassing diseases where neutralising antibodies are a significant trait. Nguyen *et al.* (2000) have applied phage display to provide antibodies with high binding affinity of $10^8\ M^{-1}$ to an respiratory syncitial virus (RSV) glycoprotein that shows potential for clinical applications. Significant advances have been made in the construction of MAbs that prolong the function and survival of transplanted human organs. The traditional approach for solid organs, e.g. kidney and liver, has been the use of general immuno-suppressants such as azothioprine, cyclosporin or tacrolimus in conjunction with steroids. Treatment regimes varying from single to triple combined therapy are reasonably successful in managing acute or chronic episodes of rejection, particularly when applied in conjunction with the anti-CD3 MAb OKT3. However, CD3 is a pan-T cell marker and hence the antibody is active against both resting and activated T cells which can result in the increased susceptibility to infection and malignancies. Generalised symptoms also occur that correspond to the effects of a 'cytokine release syndrome' (in particular cytokines IL-2, IL-6, TNFα and IFNγ). This situation has been reduced by the use of antibody fragments [OKT3 $F(ab')_2$] and a second generation antibody, HuM291, that retains its specificity for CD3. More recently developed anti-CD25 MAbs Daclizumab and Basiliximab have been used to block the assembly of a functional IL-2 receptor. The anti-CD52 MAb (Campath) was among the first wave of antibodies to be developed for therapeutic use, initially for the treatment of rheumatoid arthritis (RA) and multiple sclerosis (MS). However, an indication of the unpredictability of systemic effects or complications from antibody therapy is highlighted by an increased incidence of Graves Disease in MS patients. Latterly this anti-CD52 was found to be beneficial in controlling graft-vs.-host disease following bone marrow organ transplantation and for treating chronic B-cell leukaemias (although it remains unlicensed in the United Kingdom).

16.9 Conclusions

Monoclonal antibodies and recombinant variants have revolutionised biomedical science by providing valuable tools for diagnosis, therapy and probes for research purposes. Their numerous applications are perhaps a fitting tribute to the late Cesar Milstein (1927–2002). Clearly, the humanisation of rodent MAbs has been crucial for therapeutic applications and the boundless opportunities of phage technology are now evident. Phage libraries can be employed to generate antibodies against an array of antigens almost 'off the shelf' without the need for animals. The development of novel monoclonal reagents will continue to enable the detection of cells and macromolecules, and to elucidate structure–function relationships. Overall, whatever the method of production, novel monoclonal reagents must be 'fit for purpose' which may depend on a sound knowledge in the stages of development.

16.10 Acknowledgements

The lead author would like to thank collaborators for all aspects of monoclonal antibody research. In particular Dr Mei Choy and Dr Isaac Manyonda for analysis of antibodies to trophoblastic tissue and Professor Frank Hay, Dr Olwyn Westwood, Professor Roy Jefferis and Dr Margaret Goodhall for epitope mapping and Imdad analysis.

16.11 References

Backus HHJ, Van Groeningen CJ, Vos W *et al.* (2002) Differential expression of cell cycle and apoptosis related proteins in colorectal mucosa, primary colon tumours and liver metastases. *J. Clin. Pathol. Mol. Pathol.* **55**: 206–11.

Breedvelt FC (2000) Therapeutic monoclonal antibodies. *Lancet* **355**: 735–40.

Bull H, Murray PG, Thomas D *et al.* (2002) Acid phosphatases demystified. *J. Clin. Pathol. Mol. Pathol.* **55**: 65–72.

Coulon S, Mappus E, Cuilleron CY *et al.* (2000) Improving the specificity of an anti-estradiol antibody by random mutagenesis and phage display. *Disease Markers* **16**: 33–5.

Hay FC, Westwood OMR and Nelson PN (eds) (2002) *Practical Immunology*, 4th edn. Blackwell Science Ltd. Osney Mead, Oxford, UK.

Hoos A and Cordogardon C (2001) Tissue microarray profiling of cancer specimens and cell lines; opportunities and limitations. *Lab. Invest.* **81**: 1331–8.

Huls G, Gestel D, Van der Linden J *et al.* (1999) Tumour cell killing by *in vitro* affinity-matured recombinant human monoclonal antibodies. *Cancer Immunol. Immunother.* **50**: 163–71.

Karpas A, Dremucheva A and Czepulkowski BH (2001) A human myeloma cell line suitable for the generation of human monoclonal antibodies. *Proc. Natl. Acad. Sci. USA* **98**: 1799–804.

Kohler G and Milstein C (1975) Continuous culture of fused cells secreting specific antibody. *Nature* **256**: 495–7.

Lerner RA (1982) Tapping the immunological repertoire to produce antibodies of predetermined specificity. *Nature* **299**: 592–6.

Liang M, Dubel S, Li D *et al.* (2001) Baculovirus expression cassette vectors for rapid production of complete human IgG from phage display selected antibody fragments. *J. Immunol. Meth.* **247**: 119–30.

Lin WM, Asffaq R, Michalopolus EA *et al.* (2001) molecular Papanicolaou tests in the twenty first century: molecular analyses with fluid based Papanicolaou technology. *Am. J. Obstet. Gynecol.* **183**: 39–45.

Matutes E (2002) New additions to antibody panels in the characterisation of chronic lymphoproliferative disorders. *J. Clin. Pathol.* **55**: 180–3.

Nelson PN (1995) Retroviruses in rheumatoid diseases. *Ann. Rheum. Dis.* **55**: 441–2.

Nelson PN (2001) Generating monoclonal antibody probes and techniques for characterizing and localizing reactivity to antigenic determinants, in: *Epitope Mapping: A Practical Approach* (eds Westwood OMR and Hay FC). Oxford University Press, Oxford.

Nelson PN, Carnegie PR, Martin J *et al.* (2002) Human endogenous retroviruses demystified. *J. Clin. Pathol. Mol. Pathol.* (submitted).

Nelson PN and Chambers T (1995) Production and characterisation of putative monoclonal antibodies against tartrate-resistant acid phosphatase. *Hybridoma* **14**: 91–4.

Nelson PN, Pringle JAS and Chambers TJ (1991) Production and characterization of new monoclonal antibodies to human osteoclasts. *Calcif. Tissue Int.* **49**: 317–9.

Nelson PN, Reynolds GM, Waldron EE *et al.* (2000) Monoclonal antibodies demystified. *J. Clin. Pathol. Mol. Pathol.* **53**: 111–17.

Nelson PN, Westwood OM, Jefferis R *et al.* (1997) Characterisation of anti-IgG monoclonal antibody A57H by epitope mapping. *Biochem. Soc. Trans.* **25**: 373.

Nguyen H, Hay J, Mazzulli T *et al.* (2000) Efficient generation of respiratory syncytial virus (RSV)-neutralizing human MoAbs via human peripheral blood lymphocyte (hu-PBL)-SCID mice and scFv phage display libraries. *Clin. Exp. Immunol.* **122**: 85–93.

Rimm DL, Camp RL, Charette LA *et al.* (2001) Tissue microarray: a new technology for the amplification of tissue resources. *Cancer J.* **7**: 42–31.

Roovers RC, van der Linden E, de Bruine AP *et al.* (2001) *in vitro* characterisation of a monovalent and bivalent form of a fully human anti Ep-CAM phage antibody. *Cancer Immunol. Immunother.* **50**: 51–9.

Sauter M, Schommer S, Kremmer E *et al.* (1995) Human retrovirus K10: expression of Gag protein and detection of antibodies in patients with seminomas. *J. Virol.* **69**: 414–21.

Shakil T, Richardson MK, Waldron E *et al.* (2001) Generation and characterization of monoclonal antibodies to the neural crest. *Hybridoma* **20**: 199–203.

Shalaby MR and Shepard HM (1992) Development of humanised bi-specific antibodies reactive with cytotoxic lymphocytes and tumour cells overexpressing HER2 oncogene. *J. Exp. Med.* **179**: 217–25.

Shi S-R, Cote R and Taylor C (2001) Antigen retrieval techniques: current perspectives. *J. Histochem. Cytochem.* **49**: 931–7.

Taylor G, Furze J, Tempest P *et al.* (1991) Humanised monoclonal antibody to respiratory syncitial virus. *Lancet* **337**: 1411–12.

Vaswani SK and Hamilton RG (1998) Review article. Humanized antibodies as potential therapeutic drugs. *Ann. Allergy Asthma Immunol.* **81**: 105–19.

Waldron EE, Murray P, Kolar Z *et al.* (2002) Reactivity and isotype profiling of monoclonal antibodies using multiple antigenic peptides. *Hybridoma and Hybridomics* (*submitted*).

Wang H, Zhang W and Fuller GN (2002) Tissue microarrays: applications in research, diagnosis and education. *Brain Pathol.* **12**: 95–107.

17
Proteomics

Kathryn Lilley, Azam Razzaq and **Michael J. Deery**

17.1 Introduction

There are now numerous organisms whose genome sequences are known, for example Arabidopsis (Arabidopsis Genome Initiative, 2000), Drosophila (Adams *et al.*, 2000), human (International Human Genome Sequencing Consortium, 2001) and rice (Yu *et al.*, 2002). The prediction of open-reading frames from these genomic sequences has enabled the comprehensive identification of many putative protein sequences. These proteins can be arranged into three categories, namely those of known function, those with recognisable motifs and hence a vague idea of function, and those with no sequence similarities to any protein (Gabor Miklos and Malenszka, 2001). Many proteins reside in this latter 'functional vacuum' which could represent as much as 30% of the predicted proteins. Determining protein function is key to understanding cellular mechanism. Studying how protein expression is modulated in response to a given set of circumstances, such as infection, disease, developmental stage, senescence or response to drugs, will facilitate the elucidation of disease pathways and thus provide a mechanism for diagnosis and therapy.

DNA chips (mRNA profiling studies) can contribute to the study of gene expression in response to a particular biological perturbation. However, the extrapolation that changes in transcript level will also result in corresponding changes in protein amount or activity cannot always be made. To understand fully, we need integrated data sets from a variety of protein expression studies, providing information on relative abundances, sub-cellular locations, protein complex formation and the profiling of isoforms generated by either alternate mRNA splicing or post-translational modifications. Proteomics is the word now

Molecular Biology in Cellular Pathology. Edited by John Crocker and Paul G. Murray
 ISBN: 0-470-84475-2

commonly used to describe the discipline associated with the acquisition of these data sets.

17.2 Definitions and applications

Before embarking on an overview of the techniques used in the field of proteomics, it is necessary to define a few terminologies; proteomics, functional genomics, structural genomics and post-genomics are terms that have crept into the scientific vocabulary with alarming stealth and are used freely, and in some cases interchangeably, especially in cases where the use of such 'buzz words' is likely to increase the attendance at departmental seminars.

- *Genome:* A genome represents the entire DNA content in a particular cell, whether or not it is coding, non-coding, or is located either chromosomally or extra-chromosomally.
- *Genomics:* Genomics is the study of the genome, interrogating the complete genome sequence using both DNA and RNA methodologies.
- *Proteome:* The proteome represents the complete set of proteins encoded by the genome.
- *Proteomics:* Proteomics is the study of the proteome and investigates the cellular levels of all the isoforms and post-translational modifications of proteins that are encoded by the genome of that cell under a given set of circumstances.
- *Functional genomics:* This is the study of the functions of genes and their inter-relationships.

Whilst a genome is more or less static, the protein levels in a particular cell can change dramatically as genes get turned on and off during the cell's response to its environment. The proteome originally was defined over seven years ago as 'all the proteins coded by the genome of an organism' (Wasinger *et al.*, 1995). Nowadays the term 'proteomics' is used more widely and implies an effort to link structure to function by whatever means are appropriate. We can expect the definition to change again with time as the field and the investigator's view of it evolves.

17.3 Stages in proteome analysis

Proteomic analysis can ascertain function either by looking for changes in the expression of either all or a subset of proteins, or by identifying binding partners for particular proteins and seeing how their interaction is affected by biological perturbation. Whatever the rationale of the investigation, or the number of proteins involved, the study of the proteome can be broken down into the following stages of analysis.

1. *Separation of proteins.* Before analysing protein expression and abundance levels, proteins first have to be isolated into a 'purified' state. Whilst there are a variety of chromatographic procedures that can achieve this, separation by two-dimensional polyacrylamide gel electrophoresis (2D PAGE) has been the method of choice in the recent past. However, other new methodologies are now emerging, each having its own strengths and weaknesses.
2. *Analysis of comparative expression.* Once separated, it is then necessary to carry out some form of analysis to assess the relative abundance of the proteins present.
3. *Identification of protein species.* Once a set of proteins showing differences in abundance between two or more states has been identified, mass spectrometric analysis is be used to determine their identities.
4. *Confirmatory experiments.*

When a protein has been shown to be important in a given process by the above analysis, it may be necessary to perform further experiments to confirm its implied function or involvement in the process.

Protein Separation and Visualisation

For proteomes that encompass the protein content of a given cell or tissue type, or that of a whole organism, there are two main methods that are first used to resolve the protein mixture, and then to visualise the individual components in such a way that their relative abundances can be quantified. The first method utilises two-dimensional (2D) gel electrophoresis followed by a variety of in-gel staining methods, whilst the second couples liquid chromatographic separation to subsequent ultraviolet (UV) and/or mass spectrometric detection.

2D PAGE

The majority of proteome analysis to date has employed 2D PAGE, a technique that has been around for many years (O'Farrell, 1975) and is the subject of several excellent reviews (Gorg *et al.*, 2000). Using this method, proteins are initially separated by isoelectric focusing (IEF) in the first dimension, according to charge (pI), and then by SDS PAGE in the second dimension, according to size (M_r). This type of separation has the capacity to resolve complex protein mixtures, thus permitting the simultaneous analysis of hundreds or even thousands of proteins at a time (Figure 17.1).

One of the most important considerations to make when visualising, quantifying and then determining the identities of spots from 2D gels is which staining method to use for spot detection. There is a variety of staining methods available which differ in their sensitivity and dynamic range of detection. Coomassie staining (Neuhoff *et al.*, 1988) is compatible for mass spectrometric

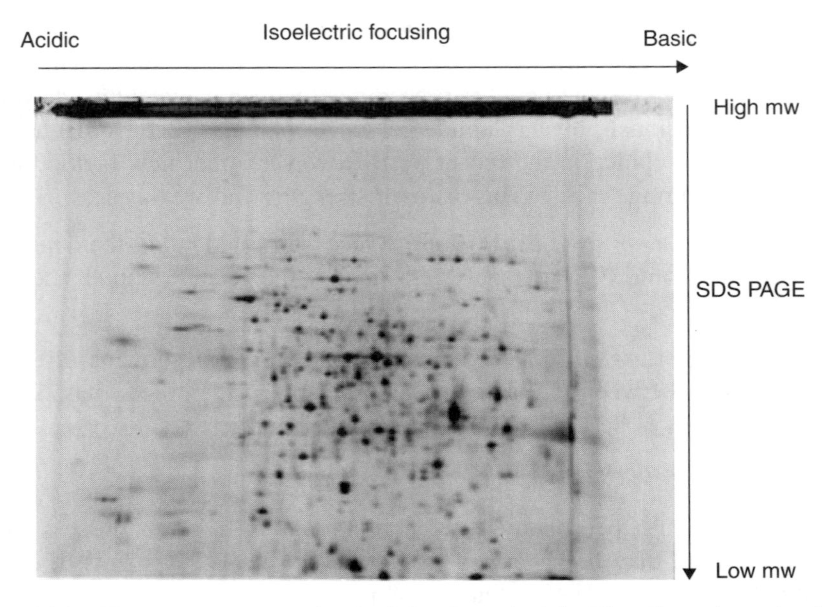

Figure 17.1 The process operates by the following principle: First dimension – isoelectric focusing. Proteins separate on the basis of their charge to a position along the strip where their pI and pH at that position are equivalent and they have no net charge. Second dimension – SDS polyacrylamide gel electrophoresis. Proteins denature in SDS and their migration through the gel matrix is inversely proportional to their mass

identification purposes, but the sensitivity of detection only goes down to the sub-microgram range. Silver staining (Mortz *et al.*, 2001) has been widely used for high sensitivity protein visualisation on 2D gels, detecting sub-nanogram amounts of material. However, the drawbacks of silver staining pertain to its poor dynamic range of staining which renders it unsuitable for quantitative analysis. More recently, the Sypro post-electrophoretic fluorescent stains (Molecular Probes, Oregon, USA) have emerged as alternatives, offering a better dynamic range of detection and ease of use (Malone *et al.*, 2001; Yan *et al.*, 2000). Sypro Ruby has been shown to be more sensitive, as well as compatible for subsequent peptide mass mapping, than silver staining (Lopez *et al.*, 2000).

2D PAGE is not without numerous technical difficulties and inadequacies. Most problems are concerned with the incomplete coverage of the proteome, where proteins with either high molecular weights, or extremes of pI or hydrophobicity, may not be represented on gels. Another problem regards the degree of protein separation achieved by 2D gels. Ideally, every protein in a sample would be resolved as a discrete, detectable spot by 2D PAGE. However, since complex samples can be separated by 2D electrophoresis into as many as 3000 spots, a large proportion of proteins will co-migrate to the same spot position, a property that will subsequently confound quantitation and mass spectrometric identification (Gygi *et al.*, 2000). Additionally, since about 90% of the total protein of a typical cell is made up of only 10% of the 10 000–20 000 different protein species (Gabor Miklos and Malenszka, 2001; Zuo *et al.*, 2001), the less abundant protein species may be obscured by more abundant ones, thus further compounding the proteome coverage problem.

There are strategies designed to overcome the aforementioned limitations of 2D PAGE. For instance, in order to increase the representation of hydrophobic proteins, which are usually membrane or membrane-associated proteins, alternative denaturants or detergents, such as thiourea and ASB14 respectively, can be employed (Molloy *et al.*, 2000; Santoni, *et al.*, 2000). However, although these methods increase the number of hydrophobic proteins represented, protein coverage is still by no means complete.

The loss of low abundance proteins from 2D gels is a problem caused by a combination of limited resolution and sensitivity. These proteins are quite often the most biologically active molecules and if their detection is not masked by a more abundant protein as described above, it may be limited by their scarcity in the starting sample. Zuo *et al.* (2001) estimated that there may need to be at least 1000 copies of a protein within a cell to be detected. Therefore, to increase the representation of low abundance proteins, they may need to be enriched by fractionation of the starting sample. Pre-fractionation of a complex protein sample enables an increase in the amount of sample that can be loaded onto a 2D gel, subsequently increasing the sensitivity of spot detection. Pre-fractionation also enables samples to be simplified into different components which can then each be analysed using separate gels, thus facilitating the ability to achieve complete protein coverage.

One way to fractionate indirectly is to focus greater amounts of sample to a higher resolution on narrow pH range IPG strips such that they may have a greater chance of migrating separately from abundant protein species. One set of researchers has reported displaying 70% of the entire *Escherichia coli* proteome using a combination of six different pH range gels (Tonella *et al.*, 2001). Most fractionation protocols utilise chromatographic, differential extraction or centrifugation principles. Using liquid chromatography, proteins can be separated upon the basis of their hydrophobicity, native size or charge. Additionally, extraction, either chemically using alternative detergents, or manually, by dissection of tissues of interest, can also help to limit the number of proteins in a given sample. Differential centrifugation can be used to isolate or enrich a particular tissue or organelle fraction, this having the added functional advantage of being able to simultaneously isolate groups of proteins that are involved in a common physiological role, for example Arabidopsis mitochondrial proteome (Millar *et al.*, 2001) and Arabidopsis chloroplast (Schubert *et al.*, 2002).

Non-gel-based methods

Although optimisation of 2D PAGE strategies have meant that more of the proteome can be visualised using this method, problems detecting low abundance species, and particularly integral membrane proteins, have prompted several research groups to introduce gel electrophoresis-free approaches. Recently, several researchers have described methods of protein separation and identification that utilise on-line, 2D high performance liquid chromatography (HPLC) separations (Link *et al.*, 1999; Wolters *et al.*, 2001). Wolters *et al.*, who have named the technique multidimensional protein identification technology (MudPIT), use a capillary HPLC column with a strong cation exchange matrix at the front end of the column which is followed by reverse phase packing at the back end. Their approach involves the tryptic digestion of soluble and insoluble protein fractions of the entire yeast proteome, followed by the application of total tryptic peptides from the two fractions onto the strong cation exchange matrix at the top of the column. A salt step gradient is then used to displace a fraction of the peptides onto the reverse phase packing. Displaced peptides are then eluted into the mass spectrometer using a solvent gradient. This procedure is then repeated in steps, each time using an increasing amount of salt to release further peptides from the cation exchange to the reverse phase packing. Each peptide eluted is introduced into a mass spectrometer capable of generating fragmentation data, which in turn are used for automated searches against protein databases and identification. Mass spectrometric techniques are dealt with in more detail below.

The application of this approach to the total yeast proteome is impressive, enabling the identification of approximately 1500 proteins (Washburn *et al.*, 2001), in comparison with 2D PAGE based methods which have only identified around 400 spots (corresponding to about 300 gene products). One advantage of MudPIT is that there are many more protein classes represented in its proteome than are produced from the electrophoresis-derived proteome. It therefore has the potential to provide more information about proteins that do not behave well on 2D PAGE, and also to generate quantitative data when coupled with stable isotope labeling (see below).

Quantitation

Qualitative proteomics enables the investigator to determine whether or not a particular protein shows an increase or decrease in expression. As this provides no measure of the extent of this expression change, this approach is therefore unsuitable for clustered data analysis which ultimately presents an insight into functionality. On the other hand, quantitative proteomics does allow co-expression patterns to be studied, and proteins showing similar expression trends can then be assigned into the same functional groups. An example of this was demonstrated by Grunenfelder *et al.* (2001) who discovered that expression changes during the cell cycle of *Caulobacter* can be grouped into 23 distinct pattern clusters. Proteins within each cluster have been shown, in many cases, to have a similar function. To generate quantitative data, the authors used radioactive labelling of proteins in conjunction with phosphorimaging of 2D gels. One drawback of this method is that the labelling is not as straightforward as that used in alternative methods. One alternative, a non-gel-based method for quantitation, involves mass spectrometry utilising differential isotope coded affinity labeling (see below).

In the case of quantitative analysis using 2D PAGE, this has involved running the two test samples to be compared on separate 2D gels, and then visualising the spots with silver stain. Image analysis software has been used to estimate spot volumes in each gel, which are then expressed as a ratio that provides a measure of any change in expression. However, quantitation using this type of analysis is crude as this technique has a number of inherent drawbacks. First, as no two gels run identically, there are problems with irreproducibility. Additionally, the quantitative process is complicated by the fact that corresponding spots between gels have to be matched prior to quantitation. However, spot matching can be performed by warping algorithms which are built into most gel analysis software. Silver staining, as stated earlier, has a very poor dynamic range which reaches saturation in the low nanogram range, thus rendering it unsuitable for accurate quantitation. These aforementioned factors all add variability into the system that makes this method unsuitable for the accurate quantitation of differences between two test samples.

Difference gel electrophoresis (DIGE), first described some time ago, has the potential to overcome many of the issues described above (Unlu *et al.*, 1997). The technique relies on pre-electrophoretic labelling of samples with one of three spectrally distinct fluorescent dyes, cyanine-2 (Cy2), cyanine-3 (Cy3) or cyanine-5 (Cy5). The samples are all run in one gel and then viewed individually by scanning the gel at different wavelengths, thus circumventing problems with spot matching between gels. Image analysis programs can then be used to generate volume ratios for each spot, which essentially describe the intensity of a particular spot in each test sample, and thus enable expression differences to be identified and quantified (Figure 17.2).

Fluorescent labels bind to lysine residues and labelling is carried out at stoichiometries such that only a small proportion of the protein is labelled. These labelled proteins are compatible with in-gel digestion and mass spectrometric analysis. This method is more sensitive than staining with silver or any Sypro dye, with a detection limit of somewhere in the region of 100–200 pg protein and a dynamic range of labelling of over 5 orders of magnitude (Kernec *et al.*, 2001; Tonge *et al.*, 2001).

Identification

Since the early 1990s, mass spectrometry has evolved into an extremely powerful technique for the identification of proteins from 2D gels. The application of two ionisation techniques to the analysis of peptides and proteins were partly

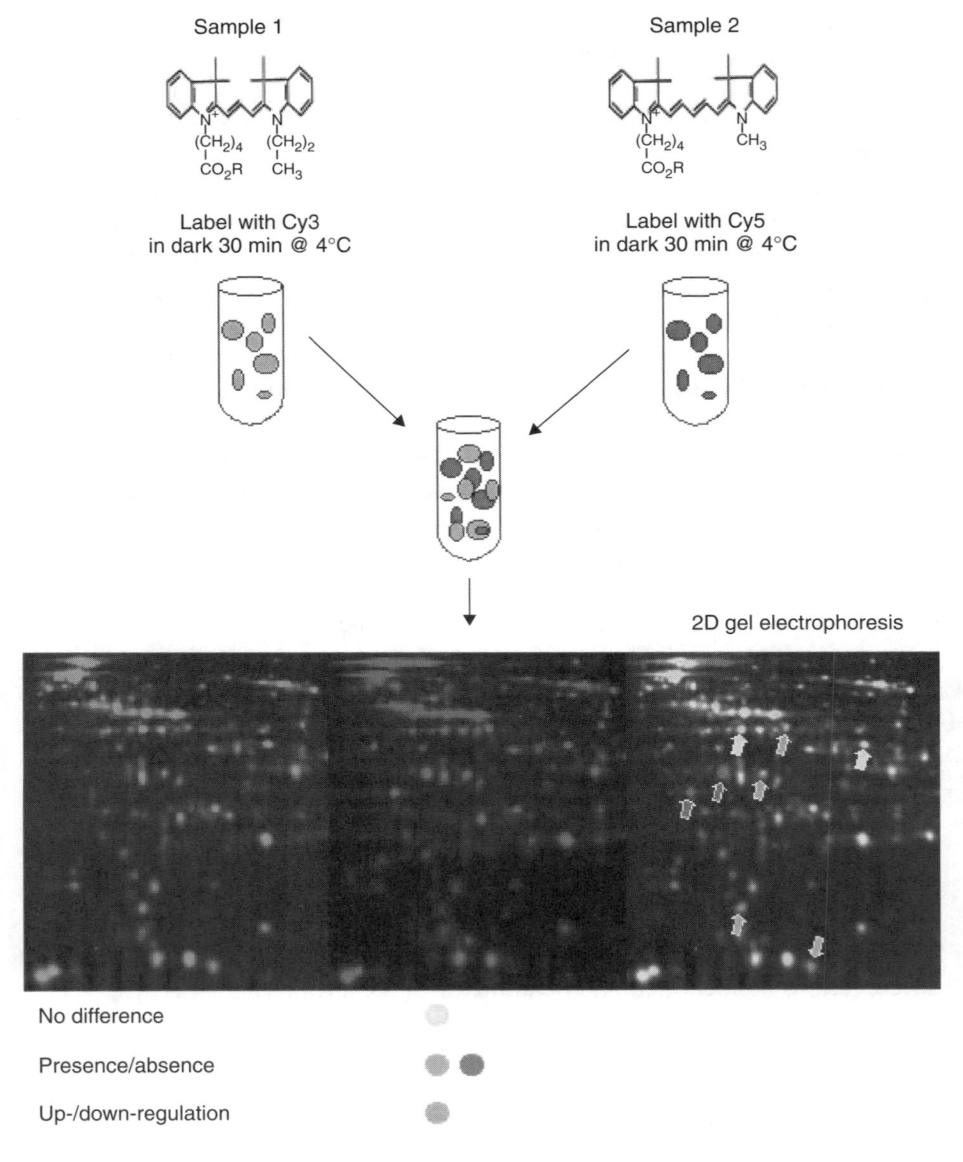

Figure 17.2 Schema for labelling of two samples whose protein profiles are to be compared by the 2D DIGE technique. The images are from a 2D analytical gel (pH 4–7) which was loaded with 50 μg of a total protein extract from a wild-type strain of *Erwinia carotovorea* labelled with Cy3 and 50 μg of total protein extract from a mutant strain of *Erwinia carotovorea* labelled with Cy5. The images were acquired using a 2920–2D Master Imager (Amersham Biosciences). The images were exported as 16-bit TIF images for analysis. To produce this figure, 8-bit TIF versions of the images were imported into ADOBE Photoshop version 5 and false coloured, green for Cy3, red for Cy5 and the two images overlaid. A colour version of this figure appears in the colour plate section

responsible for the surge of interest in biological mass spectrometry. Matrix-assisted laser desorption/ionisation (MALDI) and electrospray ionisation (ESI) were developed by Karas and Hillenkamp (1988) and Fenn *et al.* (1989), respectively. The combination of either of the above mass spectrometric techniques with the separation of proteins by 2D-PAGE is now an established method for proteome analysis. In both cases, identification takes place at the peptide level, not the entire protein. It is therefore necessary to convert proteins in excised gel pieces into peptides which can be extracted for analysis. The preparation of peptides ideally is performed using a fully automated digestion robot. This primarily reduces preparation time, but also prevents contamination by keratins, of which there are many sources ranging from hair, skin, dust and clothing. Excised spots must be destained, depending on the visualisation method used, reduced and alkylated to prevent inter-peptide disulphide bridge formation which could complicate analysis, and finally digested into relatively short peptides using a robust protease such as trypsin. Peptides generated are then extracted in an appropriate solvent compatible with the mass spectrometric technique to be used.

MALDI–TOF mass spectrometry (MALDI)

Peptides can be analysed by matrix assisted laser desorption ionisation (MALDI) to produce peptide mass fingerprints which are then matched against protein databases in order to identify the corresponding proteins (Henzel *et al.*, 1993; Mann *et al.*, 1993; Pappin *et al.*, 1993; Yates *et al.*, 1993). In addition to peptide identification, MALDI may also be used to identify which peptides in a tryptic digest have undergone post-translational modifications such as phosphorylation and glycosylation, which are mediated by kinases (Liao *et al.*, 1994; Zhang *et al.*, 1998) and glycosyl transferases (Colangelo and Orlando, 1999; Mechref and Novotny, 1998).

MALDI may be divided into three separate stages: ionisation, mass separation and detection. One of the main advantages of using MALDI is that very little internal energy is imparted into the ions during ionisation, thus resulting in minimal fragmentation and therefore the formation of intact adduct ions. These are of the type $(M + H)^+$, where M = the biological molecule and H = H^+ (or a proton). MALDI is therefore classified as a 'soft' ionisation technique.

Ions are generated through the use of a laser (usually nitrogen) which is fired at a sample plate containing a dried mixture of matrix and sub-picomole quantities of sample. The matrix, constituted by small organic molecules such as α-cyano-4-hydroxycinnamic acid, absorbs radiation from the laser (337 nm for nitrogen lasers) which results in the excitation of the matrix molecules. A dense plume containing matrix and analyte molecules is then produced, and the analyte molecules interact with hydrogen atoms from the matrix to form predominantly singly charged $(M + H)^+$ ions. These ions are then separated in the mass analyser, with lower mass ions travelling faster than high mass ions at constant energy, until they are detected by a microchannel plate detector (MCP). All proteomic experiments are performed using time-lag-focusing and reflectron mode, both of which act as energy focusing devices that greatly improve the resolution of the ions. Although the mass range of a time of flight (TOF) analyser is theoretically infinite, it does have an upper mass range of around 750 kDa in linear mode and 100 kDa in reflection mode.

A typical MALDI mass spectrum of a typical protein digest mixture obtained from a 2D gel spot is shown in Figure 17.3. The acquisition and processing of MALDI data is automated and includes background subtraction, smoothing, centroiding and the generation of text files. The masses of the centroided peaks (minus contaminant peaks such as keratins) are then entered into a search engine. This program compares the observed, calculated masses against the theoretical calculated masses of all the possible tryptic digest fragments from all proteins in non-redundant protein databases. The main parameters required for the peptide fingerprint search are the database to be searched (e.g. NCBI), type of enzyme used to digest the protein (e.g. trypsin), modifications of the peptides (e.g. carbamidomethylation of cysteine residues) and mass tolerance (usually within 100 ppm for MALDI data). An output file is then generated which displays hits, the scores of which are calculated by means of a probability-based scoring algorithm.

Although MALDI has proved to be an impressive and rapid method for protein identification, it is limited by the fact that a protein sequence searched against may not be represented in the database. Also, MALDI does not cope well with samples that constitute a mixture of proteins and therefore result in heterogeneous digestion products. An alternative identification method involves

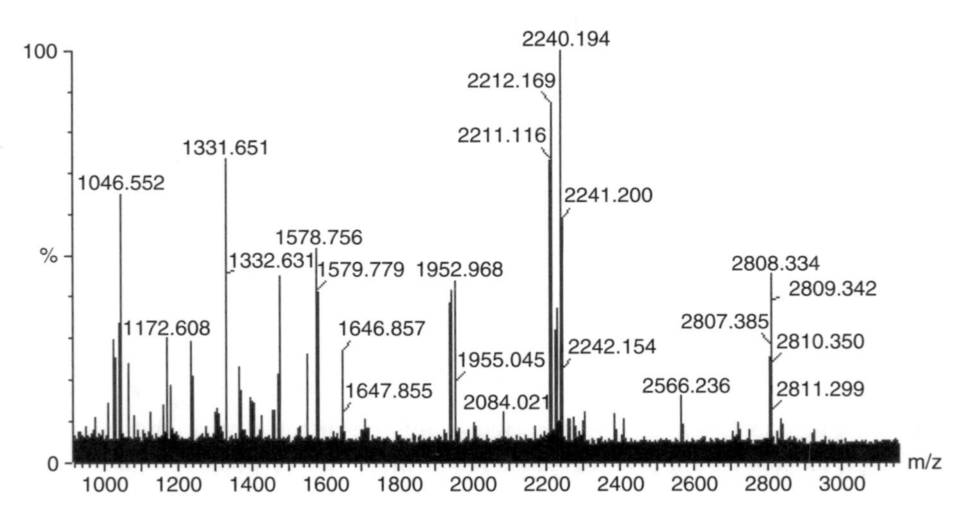

Figure 17.3 MALDI mass spectrum of the digest mixture of peptides obtained from a *Drosophila* 2D gel spot

peptide sequencing using nanospray/LC–MS/MS. This has the advantage over MALDI in that it is more sensitive, fully automated, and can also make positive identifications from mixtures of peptides. However, the combination of mass mapping and sequencing has allowed protein identification to be both rapid (MALDI–TOF) and specific (LC–MS/MS).

Nanospray ionisation mass spectrometry (nano-ES)

An alternative mass spectrometric-based method for protein identification is nanoelectrospray (nanoES). This utilises very fine glass capillaries with gold-coated tips in order to produce a spray (Wilm and Mann, 1996) which subsequently enables extremely low levels (<50 fmoles of total protein) of protein digest mixtures to be analysed at flow rates of approximately 20–50 $nL\,min^{-1}$. For nanoES sample preparation, samples are dissolved in a solvent mixture (usually 50 : 50 acetonitrile : water, or 50 : 50 methanol : water containing 0.1% formic acid) and are then injected into the gold-coated capillary. Usually, samples contain salts and buffers from the digestion process which must be removed prior to electrospray ionisation (ESI) analysis to prevent contaminant peaks and poor peptide signals. Once loaded, the capillary is then held at a potential of $\sim$1–1.5 kV which results in the formation of a very fine spray of solvent droplets containing preformed ions. The different types of ions formed are $(M + nH)^{n+}$, where M is the peptide molecule, nH is the number of protons attached to the molecule, and $n+$ is the net charge of the peptide ion (usually 2–4). Multiply-charged gas-phase ions are then formed as a result of desorption processes that occur due to evaporation of the solvent droplets.

ESI differs from MALDI in that a series of ions are formed that have different numbers of charges attached (i.e. $(M + nH)^{n+}$). The resulting mass spectra obtained therefore display several peaks that correspond to ions which have the same peptide or protein sequence, but different numbers of attached charges. As mass spectrometry is concerned with the measurement of mass-to-charge (m/z) ratios, rather then mass alone, ESI–MS allows the analysis of relatively high molecular weight samples whilst using analysers with only modest m/z ranges. In general, the types of mass analysers used in conjunction with ESI are quadrupoles, quadrupole ion traps, quadrupole time-of-flight (Q-TOF) hybrid instruments and time-of-flight. The advantage of these instruments over MALDI is that the fragmentation of the ions can be very carefully controlled using tandem mass spectrometry (MS/MS), thus leading to precise sequence information. In addition, these types of instruments also have the ability to fragment very low abundance ions.

The fragmentation of ions results from collision-induced dissociation (CID) processes in which low-energy collisions occur within a collision cell between isolated ions of a specific m/z and an inert gas, usually argon. The fragment ions are then separated and detected to give a fragment ion spectrum (MS/MS

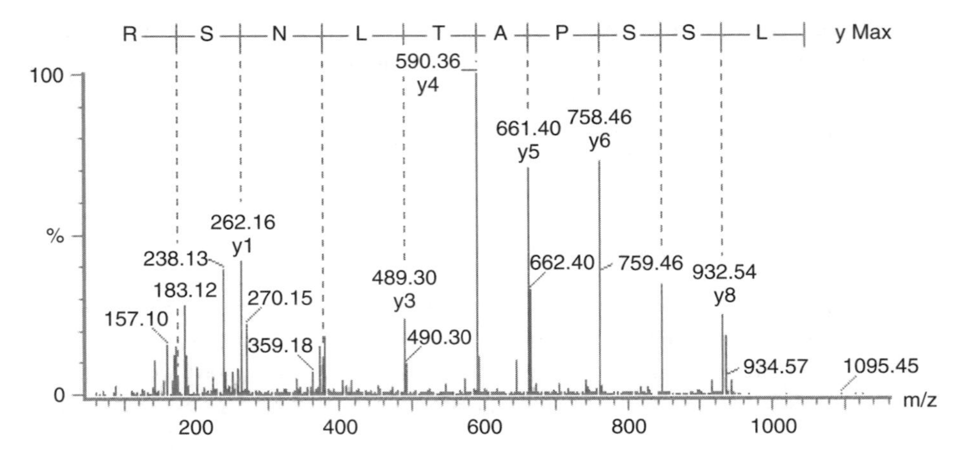

Figure 17.4 ESI–MS/MS spectrum of a doubly charged ion (*m/z* 523.29) of a trypsin autolysis product from porcine trypsin. Subtraction of the masses of adjacent fragment ion peaks (*p*-type) corresponds to the masses of the amino acids in the peptide chain. Hence, the complete sequence of the peptide is LSSPATLNSR

spectrum; Figure 17.4). The manually interpreted sequences and m/z values of the intact peptides, or uninterpreted data in the form of a text file (containing the m/z values and relative abundances of each peak in the MS/MS spectra) are then submitted to a search engine and hits are obtained based on the sequence information along with the masses of the peptides. Usually, the scores of the identified proteins are much higher than the hits obtained from MALDI data because the information submitted is much more specific; i.e. m/z values and sequence information for ESI–MS, as opposed to only m/z values for MALDI.

In addition to nanoES, nano-liquid chromatography mass spectrometry (nanoLC–MS/MS) can also be used for sequencing. In this technique, peptides are separated by reverse phase in a narrow capillary column (i.d. of 50–180 μM) that is packed with C_{18} material. The advantage of this system over nanospray is that the analysis is fully automated. Up to 96 samples can be loaded and injected onto the system, mass spectra and MS/MS spectra acquired and processed, and the resulting data searched against databases without any manual intervention. As with MALDI–TOF, ESI–MS/MS is a powerful tool for the determination of post-translational modifications. The latter technique can be considered more powerful as, in addition to the type of modification, the exact site of the modification can be determined.

Comparative analysis of the proteome using stable isotope labelling

The section 'Quantitation' above dealt with quantitative proteomics using differential labelling in conjunction with 2D PAGE. However, as discussed earlier, a sizeable proportion of proteins cannot be detected using this technology largely because of solubility and abundance issues. Although the MudPIT approach provides a mechanism to interrogate proteins recalcitrant to gel electrophoresis-based methods, it does not yield information about the relative quantities of the identified proteins. One technique which results in the measurement of relative abundances is mass coded abundance tagging (MCAT), where whole samples are digested to peptides, and the terminal lysine residues of peptides are converted to homoarginine using O-methylisourea. The relative abundance of proteins between two samples can be measured if peptides from one sample only are labelled and are then combined with unmodified peptides prepared from the comparative sample. Peptide masses, determined by MS, which differ by 42 atomic mass units (amus) correspond to the modified and unmodified forms of a peptide present in both samples. Relative peak intensities then give information about the relative abundance of the parent protein in the original samples (Cagney and Emili, 2002).

A more robust methodology than MCAT to look at quantitative differences in protein abundance between sets of samples employs the use of stable isotope labelling. Here, a protein sample is prepared such that it contains a particular isotopic form of an amino acid or added label, with a comparative sample similarly prepared, but with an alternative isotope. The samples can then be combined and digested to peptides. The peptides are then separated and sequenced, and the relative distribution of the isotopes assessed by nanoLC–MS/MS. One approach is to use the incorporation of isotopic variants of amino acids into protein samples, although this approach is only suitable in the case of cultured material (Pratt *et al.*, 2002). A more universally applicable method is that of isotope coded affinity tagging (ICAT; Gygi *et al.*, 1999). This involves the use of two probes, each comprising a biotin tag, a linker region and a cysteine reactive iodoacetamide handle. Although the two probes employed have identical chemical properties, they differ by a mass of 8 Da in their linker region, which can therefore be either a light version (8 hydrogen atoms) or a heavy version (8 deuterium atoms).

When comparative analysis of two samples is made, one sample is labelled with the light probe whilst the other is labelled with the heavy probe. The two extracts are then pooled and digested with trypsin. The digest is loaded onto a monomeric avidin affinity column which retains the labelled, cysteine-containing peptides. These are subsequently eluted from the avidin by formic acid. The eluate, which consists of peptide pairs (heavy and light), is then applied to nanoLC–MS/MS. At the first MS stage, each peptide appears as a mass pair separated by 8 Da (provided there is a single cysteine per peptide). Since the starting amount of each peptide pair should be equivalent, as assessed by the mass signal intensity of each, a ratio of these two intensities will enable the identification of peptides that show expression differences. Subsequent fragmentation and MS/MS peptide identification leads to the name of the corresponding protein.

Since cysteine content differs widely from one protein to another, some proteins will therefore be accurately assigned by several peptides, whilst the analysis of others may be slightly ambiguous, as identification relies upon a single peptide. In addition, since one out of seven proteins does not contain a cysteine residue any analysis will therefore not represent the whole proteome.

Protein Function and the Proteome

The analysis of proteome components by 2D gel electrophoresis and multi-dimensional chromatography, as described above, provides information about which proteins are expressed in a given cell or tissue type under a given set of circumstances. Although many of the identified proteins may already have a previously characterised function, there are examples where no function is known. All that can therefore be concluded about these proteins is that they are implicated in the function of the cell in the situation studied.

There are two other types of approaches designed to study the interactions of specific proteins within a proteome. The first involves the analysis of multiprotein complexes which are the functional units that are responsible for the majority of cellular processes. Identifying the protein components of these complexes may help us understand how these multiprotein units function within the proteome. A second approach for a functional study utilises protein arrays. Here, function is assigned by the identification of binding partners from substrates that are applied to ordered arrays of immobilised proteins, peptides, or other capture agents.

Analysis of multiprotein complexes

The study of pairwise protein interactions on a proteome-wide scale has been characterised in many ways. For example, Uetz *et al.* (2000) and Ito *et al.* (2001) have both adapted the yeast 'two-hybrid' assay into a high-throughput method for mapping binary protein interactions on a large scale. Investigating protein interactions that occur within a functional complex, of which the spliceosome, the proteasome and the nuclear pore complex are well-documented examples, is not so straightforward. First, the complex must be purified away from other protein components in a native from. The complex can then be separated by 1D or 2D PAGE, or by a variety of chromatographic techniques, and the individual members subsequently identified by the high throughput mass spectrometric methods described above.

The isolation of the complexes in the first instance is key to this process. This isolation process must be stringent enough to remove non-specifically bound proteins, but not so harsh as to strip off *bona fide* components. The most widely used method to isolate whole complexes is by immunoprecipitation. Here an antibody raised against a known component of a given complex is mixed with cell extract.

Application of this mixture to Protein A sepharose beads subsequently binds the antibody and, consequently, the known component and its associated partners, thus enabling their isolation from other proteins in the cell extract. Although this is a powerful technique, it has its drawbacks. These include the expense of generating the specific antibodies, the inability to isolate low copy number complexes, and also the complications caused by false positives and negatives.

Recently, two groups have published slightly different but effective approaches for the identification of yeast complexes that comprise three or more components (Ho *et al.*, 2002; Gavin, 2002). These studies purify multi-protein complexes away from cell extracts by using a tagged version of one of the protein components of the complex as 'bait' (Figure 17.5). Molecular cloning is used to attach tags to many different bait proteins. These constructs are then introduced into yeast cells where they are expressed and form physiological protein complexes with their respective binding partners. After isolating these complexes by their tags, a combination of 1D PAGE and standard mass-spectrometry methods were used to identify the individual protein components. Ho *et al.* (2002) constructed an initial set of 725 yeast bait proteins which were tagged with a FLAG epitope (FLAG tag Gateway™). A single immunoaffinity purification step using immobilised anti-FLAG subsequently enabled the identification of 3617 interactions involving 1578 different proteins.

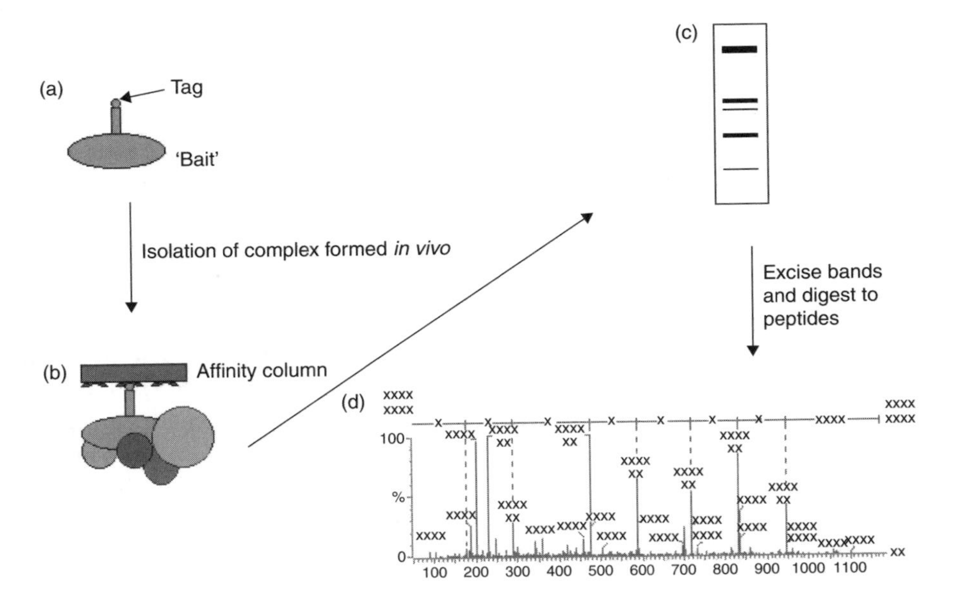

Figure 17.5 In the 'co-precipitation/mass spectrometry' approach used by Gavin *et al.* (2000) and Ho *et al.* (2002) an 'affinity tag' is first attached to the protein (the 'bait') (a). (b) Bait proteins are systematically precipitated, along with any associated proteins, on an 'affinity column'. (c) Purified protein complexes are resolved by 1D SDS–PAGE. (d) Proteins are excised from the gel, digested with the trypsin, and analysed by mass spectrometry. A colour version of this figure appears in the colour plate section

Gavin *et al.* (2002) based their proteome-wide scale analysis on an approach initially described by Rigaut *et al.* (1999). These workers used tandem affinity purification (TAP) tagging, where bait is coupled to two different tags in series to allow binding to either IgG or calmodulin. The advantage of a dual tag approach is that two purification steps are used for the isolation of complexes, thus enhancing the stringency and sensitivity of the system. Using this method, the authors identified 1440 distinct proteins within 232 multiprotein complexes in yeast. As 91% of these complexes contain at least one protein of previously unknown function, this study provides a wealth of new information on 231 previously uncharacterised yeast proteins, and on a further 113 proteins to which the authors ascribe a previously unknown cellular role. Furthermore, Gavin *et al.* found that most of the identified complexes share common protein components. This allows the assembly of a higher-order network of interactions that can be used to coordinate cellular functions.

Although clearly powerful, this kind of approach does have its drawbacks. A significant number of false-positive interactions, as well as a failure to identify many known associations, were problems encountered by both groups. Gavin *et al.* estimate that 30% of the interactions they detect may be spurious, as inferred from duplicate analyses of 13 purified complexes.

Protein arrays

An emerging technology to assign function to proteins involves the use of protein arrays. This technology, which has many similarities with high throughput DNA chip methodology, involves the printing of ordered arrays of thousands of ligand molecules onto glass or silicon-based surfaces. Subsequent functional analysis using these arrays involves exposing the protein chips to a particular test sample, which might be a protein sample if protein–protein interactions are under investigation, or might contain various other potential substrates for binding such as DNA, metabolites, etc.

Protein can be arrayed in several formats. Arrayed molecules can be whole or fragments of recombinant proteins, peptide molecules, either monoclonal, polyclonal, or specific domains of antibodies, and also fragments of a protein expressed by phage display. The production of stable protein arrays is not as straightforward as that of DNA arrays, however. One major problem is that the printing of small volumes of protein ligands onto a support can result in protein denaturation, thus compromising the biological activity of these proteins, and consequently, array function. Another hurdle is that ligand molecules ideally should be attached to the support in roughly the same orientation and also in a hydrophilic environment that will facilitate binding reactions.

There are therefore many different methods for immobilising protein ligands to primarily glass or silicon surfaces, such that a stable functional protein array is generated. Some arrays are created by the non-covalent adsorption of proteins to nitrocellulose-coated glass supports, which involves van der Waal's, hydrophobic and hydrogen bonding interactions. The advantage of this method is that proteins require no form of modification for immobilisation. However, the random attachment, and therefore orientation, means that this method is more prone to steric occlusion effects, as well as to protein denaturation and inactivation.

An alternative method involves the covalent attachment of proteins, such as the treatment of glass surfaces with poly-lysine aldehyde to immobilise proteins via their primary amines (MacBeath and Schreiber, 2000). Other examples include the immobilisation of biotinylated proteins onto streptavidin-coated surfaces or his-tagged proteins onto nickel-chelating surfaces (Zhu *et al.*, 2001). The advantage of these types of attachment is that as well as the protein attaching in roughly the same orientation, the amount of non-specific protein binding upon application of the test sample is also reduced. The most common type of proteins employed in protein arrays are antibodies. With the advent of phage display antibody libraries, it is now relatively easy to obtain sufficient amounts of thousands of pure proteins with varying specificities (de Wildt and Mundy, 2000). The alternative would be to express and purify large quantities of hundreds of different recombinant proteins for immobilisation.

There are a number of ways in which proteins binding to particular positions on the array can be identified. For example, in order to detect the binding of the array to a particular protein of interest, antibodies specific to this protein that are conjugated to either an enzymatic or fluorescent label can be used. Another approach identifies the binding to the array of proteins that have a particular activity. These can subsequently be detected using enzymatically or fluorescently conjugated substrates. An elegant example of this latter application comes from Zhu *et al.* (2001) who cloned and over-expressed 5800 yeast proteins that were then spotted onto glass slides. This group then screened the slides for calmodulin and phospholipid binding and found many new examples of proteins exhibiting these functionalities.

Another more general method employs the use of mass spectrometric techniques to look at proteins or other substrates bound at each position on the array. One application of this technique is known as SELDI (surface enhanced laser desorption ionisation) (Isaaq *et al.*, 2002). Here proteins or substrates interacting with material bound at a certain position undergo laser desorption to form gaseous ions whose masses can then be analysed by the same mechanism as MALDI–TOF MS. Methods such as surface plasmon resonance (SPR) that employ the label-free detection of proteins are also now being used. Here, proteins are printed onto gold-coated glass slides which are then exposed to either the proteins of interest, cell lysates or drug candidates, and kinetic measurements of binding in real time can be obtained.

Protein arrays are starting to be widely used in the analysis of protein–protein and protein–drug interactions, as well as for expression profiling studies of disease-related proteins.

17.4 Future directions

The most common method to date of separating and identifying complex protein mixtures has been to employ 2D PAGE and mass spectrometry, respectively. This methodology is particularly useful for establishing which proteins in any particular cell are up- or down-regulated under particular physiological conditions which, for example, allow the identification of potential target proteins in diseased cells. It is anticipated that these methods will continue to be used routinely for the characterisation of protein expression levels and subsequent identification and that the throughput of samples will continue to increase. In addition, many researchers and instrument manufacturers are seeking ways to automate the identification and characterisation of post-translational modifications of gel-excised proteins, which has until now been a difficult and laborious challenge, particularly for low abundant proteins. It is apparent, however, that only a modest proportion of proteins from the proteome are visualised on 2D gels because of problems associated with protein precipitation during the first dimension. As a result of this, it is likely that an increasing number of labs will invest in technology associated with multidimensional liquid chromatography that has been shown to be an effective method for the identification of both acidic and basic proteins. These techniques may be used in conjunction with labelling methods such as ICAT to enable high-resolution peptide separations, identification and relative quantitation. Furthermore, as the number of protein sequences that are found in databases increases, it is envisaged that there will be a marked shift from routine identification to the analysis of protein function of proteins within complexes. Encouraging results obtained from affinity purification/2D electrophoresis/mass spectrometry and protein array technologies have shown that the rapid analysis of protein–protein interactions is now feasible and potentially will have a large impact on biological science and drug discovery.

17.5 References

Adams MD, Celniker SE, Holt RA *et al.* (2000) The genome sequence of *Drosophila melanogaster*. *Science* **287**(5461): 2185–95.

Arabidopsis Genome Initiative (2000) Analysis of the genome sequence of the flowering plant *Arabidopsis thaliana*. *Nature* **408**: 796–815.

Cagney G and Emili A (2002) *De novo* peptide sequencing and quantitative profiling of complex protein mixtures using mass-coded abundance tagging. *Nat. Biotechnol.* **20**(2): 163–70.

Colangelo J and Orlando R (1999) On-target exoglycosidase digestions/MALDI-MS for determining the primary structures of carbohydrate chains. *Anal. Chem.* **71**(7): 1479–82.

de Wildt RMT and Mundy CR (2000) Antibody arrays for the high-throughput screening of antibody–antigen interactions. *Nat. Biotechnol.* **18**: 989–94.

Fenn JB, Mann M, Meng CK *et al.* (1989) Electrospray ionization for mass-spectrometry of large biomolecules. *Science* **246**(4926): 64–71.

Gabor Miklos GL and Malenszka R (2001) Protein functions and biological contexts. *Proteomics* **1**: 169–78.

Gavin AC, Bosche M, Krause R *et al.* (2002) Functional organisation of the yeast proteome by systematic analysis of protein complexes. *Nature* **415**: 141–7.

Gorg A, Obermaier C, Boguth G *et al.* (2000) The current state of two-dimensional electrophoresis with immobilized pH gradients. *Electrophoresis* **21**: 1037–53.

Grunenfelder B, Rummel G, Vohradsky J *et al.* (2001) Proteomics analysis of the bacterial cell cycle. *Proc. Natl Acad. Sci.* **98**(8): 4681–6.

Gygi SP, Corthals GL, Zhang Y *et al.* (2000) Evaluation of two-dimensional gel electrophoresis based proteome analysis technology. *Proc. Natl. Acad. Sci.* **97**(17): 9390–5.

Gygi SP, Rist B, Gerber SA *et al.* (1999) Quantitative analysis of complex protein mixtures using isotope-coded affinity tags. *Nat. Biotechnol.* **17**: 994–9.

Henzel WJ, Billeci TM, Stults JT *et al.* (1993) Identifying proteins from 2-dimensional gels by molecular mass searching of peptide-fragments in protein-sequence databases. *Proc. Natl Acad. Sci.* **90**: 5011–15.

Ho Y, Gruhler A, Heilbut A *et al.* (2002) Systematic identification of protein complexes in *Saccharomyces cerevisiae* by mass spectrometry. *Nature* **415**(6868): 180–3.

International Human Genome Sequencing Consortium (2001) Initial sequencing and analysis of the human genome. *Nature* **409**: 860–921.

Isaaq HJ, Veenstra TD, Conrads TP *et al.* (2002) The SELDI-T of MS approach to proteomics: protein profiling and biomarker identification. *Biochem. Biophys. Res. Commun.* **292**(3): 587–92.

Ito T, Chiba T, Ozawa R *et al.* (2001) A comprehensive two-hybrid analysis to explore the yeast protein interactome. *Proc. Natl Acad. Sci.* **98**(8): 4569–74.

Karas M and Hillenkamp F (1988) Laser desorption ionization of proteins with molecular masses exceeding 10 000 daltons. *Analy. Chem.* **60**: 2299–301.

Kernec F, Unlu M, Ozawa R *et al.* (2001) Changes in mitochondrial proteome from mouse hearts deficient in creatine kinase. *Physiol. Genomics* **6**: 117–28.

Liao PC, Leykam J, Andrews PC *et al.* (1994) An approach to locate phosphorylation sites in a phosphoprotein – mass mapping by combining specific enzymatic degradation with matrix-assisted laser-desorption ionization mass-spectrometry. *Anal. Biochem.* **219**(1): 9–20.

Link AJ, Eng J, Schlieltz DM *et al.* (1999) Direct analysis of protein complexes using mass spectrometry. *Nat. Biotechnol.* **17**: 677–82.

Lopez MF, Berggren K, Chernokalskaya E *et al.* (2000) A comparison of silver stain and SYPRO ruby protein gel stain with respect to protein detection in two-dimensional gels and identification by peptide mass profiling. *Electrophoresis* **21**: 3673–83.

MacBeath G and Schreiber SL (2000) Printing proteins as microarrays for high-throughput function determination. *Science* **289**(5485): 1760–3.

Mann M, Hojrup P and Roepstorff P (1993) Use of mass-spectrometric molecular-weight information to identify proteins in sequence databases. *Biol. Mass Spectrom.* **22**(6): 338–45.

Malone JP, Radabaugh MR, Leimgruber RM *et al.* (2001) Practical aspects of fluorescent staining for proteomics applications. *Electrophoresis* **22**: 919–32.

Mechref Y and Novotny MV (1998) Mass spectrometric mapping and sequencing of N-linked oligosaccharides derived from submicrogram amounts of glycoproteins. *Anal. Chem.* **70**(3): 455–63.

Millar AH, Sweetlove LJ, Giege P *et al.* (2001) Analysis of the Arabidopsis mitochondrial proteome. *Plant Physiol.* **127**(4): 1711–27.

Molloy MP, Herbert BR, Slade MB *et al.* (2000) Proteomics analysis of *Escherichia coli* outer membrane. *Eur. J. Biochem.* **267**: 2871–81.

Mortz E, Krogh TN, Vorum H *et al.* (2001) Improved silver staining protocols for high sensitivity protein identification using matrix-assisted laser desorption/ionization – time of flight analysis. *Proteomics* **1**(11): 1359–63.

Neuhoff V, Arold N, Taube D *et al.* (1988) Improved staining of proteins in polyacrylamide gels including isoelectric-focusing gels with clear background at nanogram sensitivity using Coomassie brilliant blue G-250 and R-250. *Electrophoresis* **9**(6): 255–62.

O'Farrell PH (1975) High resolution two-dimensional electrophoresis of proteins. *J. Biol. Chem.* **250**: 4007–21.
Pappin DJC, Hojrup P and Bleasby AJ (1993) Rapid identification of proteins by peptide-mass fingerprinting. *Curr. Biol.* **3**(6): 327–32.
Pratt JM, Robertson DHL, Gaskell SJ *et al.* (2002) Stable isotope labelling *in vivo* as an aid to protein identification in peptide mass fingerprinting. *Proteomics* **2**(2): 157–63.
Rigaut G, Shevchenko A, Berthold R *et al.* (1999) A generic protein purification method for protein complex characterization and proteome exploration. *Nat. Biotechnol.* **17**: 1030–2.
Santoni V, Kieffer S, Desclaux D *et al.* (2000) Membrane proteomics: use of additive main effects with multiplicative interaction model to classify plasma membrane proteins according to their solubility and electrophoretic properties. *Electrophoresis* **21**: 3329–44.
Schubert M, Petersson UA, Haas BJ *et al.* (2002) Proteome map of the chloroplast lumen of *Arabidopsis thaliana*. *J. Biol. Chem.* **277**(10): 8354–65.
Tonella L, Hoogland C, Binz PA *et al.* (2001) New perspectives in the *Escherichia coli* proteome investigation. *Proteomics* **1**: 409–23.
Tonge R, Shaw J, Middleton B *et al.* (2001) Validation and development of fluorescence two-dimensional differential gel electrophoresis proteomics technology. *Proteomics* **1**: 377–96.
Uetz P, Giot L, Cagney G *et al.* (2000) A comprehensive analysis of protein–protein interactions in *Saccharomyces cerevisiae*. *Nature* **403**(6770): 623–7.
Unlu M, Morgan ME and Minden JS (1997) Difference gel electrophoresis: a single gel method for detecting changes in protein extracts. *Electrophoresis* **18**: 2071–7.
Washburn MP, Wolters DA and Yates JR (2001) Large-scale analysis of the yeast proteome by multidimensional protein identification technology. *Nat. Biotechnol.* **19**(3): 242–7.
Wasinger VC, Cordwell SJ, Cerpapoljak A *et al.* (1995) Progress with gene–product mapping of the *mollicutes – Mycoplasma genitalium*. *Electrophoresis* **16**(7): 1090–4.
Wilm M and Mann M (1996) Analytical properties of the nanoelectrospray ion source. *Analyt. Chem.* **68**(1): 1–8.
Wolters DA, Washburn MP and Yates JR (2001) An automated multidimensional protein identification technology for shot-gun proteomics. *Analyt. Chem.* **73**(23): 5683–90.
Yan JX, Harry RA and Spibey C (2000) Postelectrophoretic staining of proteins separated by two-dimensional gel electrophoresis using SYPRO dyes. *Electrophoresis* **21**: 3657–65.
Yates JR, Speicher D and Griffin PR (1993) Peptide mass maps – a highly informative approach to protein identification. *Anal. Biochem.* **214**(2): 397–408.
Yu J, Hu SN and Wang J (2002) A draft sequence of the rice genome (*Oryza sativa* L. ssp indica). *Science* **296**(5565): 79–92.
Zhang XL, Herring CJ and Romano PR (1998) Identification of phosphorylation sites in proteins separated by polyacrylamide gel electrophoresis. *Anal. Chem.* **70**(10): 2050–9.
Zhu H, Bilgin M and Bangham R (2001) Global analysis of protein activities using proteome chips. *Science* **293**(5537): 2101–5.
Zuo Z, Echan L and Hembach P (2001) Towards global analysis of mammalian proteomes using sample prefractionation prior to narrow pH range two-dimensional gels and using one-dimensional gels for insoluble large proteins. *Electrophoresis* **22**: 1603–15.

Index